THE ROUTLEDGE HANDBOOK OF EVOLUTION AND PHILOSOPHY

W0113117

In recent years, the relation between contemporary academic philosophy and evolutionary theory has become ever more active, multifaceted, and productive. The connection is a bustling two-way street. In one direction, philosophers of biology make significant contributions to theoretical discussions about the nature of evolution (such as "What is a species?"; "What is reproductive fitness?"; "Does selection operate primarily on genes?"; and "What is an evolutionary function?"). In the other direction, a broader group of philosophers appeal to Darwinian selection in an attempt to illuminate traditional philosophical puzzles (such as "How could a brain-state have representational content?"; "Are moral judgments justified?"; "Why do we enjoy fiction?"; and "Are humans invariably selfish?"). In grappling with these questions, this interdisciplinary collection includes cutting-edge examples from both directions of traffic. The thirty contributions, written exclusively for this volume, are divided into six sections: The Nature of Selection; Evolution and Information; Human Nature; Evolution and Mind; Evolution and Ethics; and Evolution, Aesthetics, and Art. Many of the contributing philosophers and psychologists are international leaders in their fields.

Richard Joyce is Professor of Philosophy at Victoria University of Wellington, New Zealand. He is author of *The Myth of Morality* (2001), *The Evolution of Morality* (2006), and *Essays in Moral Skepticism* (2016), as well as many articles on metaethics and moral psychology. He has co-edited *A World Without Values* (2010) and *Cooperation and its Evolution* (2013).

ROUTLEDGE HANDBOOKS IN PHILOSOPHY

Routledge Handbooks in Philosophy are state-of-the-art surveys of emerging, newly refreshed, and important fields in philosophy, providing accessible yet thorough assessments of key problems, themes, thinkers, and recent developments in research.

All chapters for each volume are specially commissioned, and written by leading scholars in the field. Carefully edited and organized, *Routledge Handbooks in Philosophy* provide indispensable reference tools for students and researchers seeking a comprehensive overview of new and exciting topics in philosophy. They are also valuable teaching resources as accompaniments to textbooks, anthologies, and research-orientated publications.

A full list of titles in this series is available at: https://www.routledge.com/Routledge-Handbooks-in-Philosophy/book-series/RHP

Recently published:

The Routledge Handbook of Embodied Cognition
Edited by Lawrence Shapiro

The Routledge Handbook of Philosophy of Well-Being
Edited by Guy Fletcher

The Routledge Handbook of Philosophy of Imagination
Edited by Amy Kind

The Routledge Handbook of the Stoic Tradition
Edited by John Sellars

The Routledge Handbook of Philosophy of Information
Edited by Luciano Floridi

The Routledge Handbook of the Philosophy of Biodiversity
Edited by Justin Garson, Anya Plutynski, and Sahotra Sarkar

The Routledge Handbook of Philosophy of the Social Mind
Edited by Julian Kiverstein

The Routledge Handbook of Philosophy of Empathy
Edited by Heidi Maibom

The Routledge Handbook of Epistemic Contextualism
Edited by Jonathan Jenkins Ichikawa

The Routledge Handbook of Epistemic Injustice
Edited by Ian James Kidd, José Medina and Gaile Pohlhaus

The Routledge Handbook of Philosophy of Pain
Edited by Jennifer Corns

The Routledge Handbook of Brentano and the Brentano School
Edited by Uriah Kriegel

The Routledge Handbook of Metaethics
Edited by Tristram McPherson and David Plunkett

The Routledge Handbook of Philosophy of Memory
Edited by Sven Bernecker and Kourken Michaelian

The Routledge Handbook of Evolution and Philosophy
Edited by Richard Joyce

The Routledge Handbook of Mechanisms and Mechanical Philosophy
Edited by Stuart Glennan and Phyllis Illari

THE ROUTLEDGE HANDBOOK OF EVOLUTION AND PHILOSOPHY

Edited by
Richard Joyce

Routledge
Taylor & Francis Group

NEW YORK AND LONDON

First published 2018
by Routledge
52 Vanderbilt Avenue, New York, NY 10017

and by Routledge
2 Park Square, Milton Park, Abingdon, Oxon OX14 4RN

First issued in paperback 2020

Routledge is an imprint of the Taylor & Francis Group, an informa business

Library of Congress Cataloging-in-Publication Data
Names: Joyce, Richard, 1966–, editor.
Title: The Routledge handbook of evolution and philosophy / edited by Richard Joyce.
Description: 1 [edition]. | New York : Routledge, 2017. | Series: Routledge
      handbooks in philosophy | Includes bibliographical references and index.
Identifiers: LCCN 2016059095 | ISBN 9781138789555 (hardback)
Subjects: LCSH: Evolution.
Classification: LCC B818 .R83 2017 | DDC 149—dc23
LC record available at https://lccn.loc.gov/2016059095

ISBN 13: 978-0-367-57307-2 (pbk)
ISBN 13: 978-1-138-78955-5 (hbk)

Typeset in Minion Pro
by Apex CoVantage, LLC

Contents

Figures

Tables

Contributors

Louise Barrett is Professor of Psychology and Canada Research Chair in Cognition, Evolution and Behaviour at the University of Lethbridge. Her research interests consider the ways in which sociality has selected for particular kinds of psychological and behavioral traits in both human and non-human primates. She is author of *Beyond the Brain: How Body and Environment Shape Animal and Human Minds* (2011).

Patrick Bateson is Professor (Emeritus) of Ethology at Cambridge University, a Fellow at King's College, Cambridge, and a Fellow of the Royal Society. He has written many books and articles on ethology, animal welfare, behavioral development, and evolution. He is co-author with Paul Martin of *Design for a Life: How Behavior and Personality Develop* (1999) and *Play, Playfulness, Creativity and Innovation* (2013).

Jonathan Birch is Assistant Professor of Philosophy, Logic and Scientific Method at the London School of Economics and Political Science where he specializes in the philosophy of the biological and behavioral sciences. His work mainly concerns the evolution and psychology of social behavior, though he has also written on innateness, teleology, self-locating belief, and the "major evolutionary transitions."

Brian Boyd is a Distinguished Professor in the Department of English at the University of Auckland, New Zealand. He is known primarily as an expert on the life and works of author Vladimir Nabokov and on literature and evolution. He is author of *On the Origin of Stories: Evolution, Cognition, and Fiction* (2009) and *Why Lyrics Last: Evolution, Cognition, and Shakespeare's Sonnets* (2012).

Michael Bradie is Professor (Emeritus) in the Department of Philosophy at Bowling Green State University. He has published numerous articles on the philosophy of science and epistemology. He is author of *The Secret Chain: Evolution and Ethics* (1995).

Ellen Clarke is Lecturer in Philosophy at the University of Leeds. Her work explores the metaphysics and epistemology of biological science, especially the ontology of the living world. In particular, she has written about levels of selection and the problem of defining biological individuals.

Christine Clavien is Senior Lecturer at the Institute for Ethics, History, and the Humanities, University of Geneva. She works on the conceptual and methodological difficulties related to the interdisciplinary treatment of issues such as cooperation, altruism, and normative behavior. She is author of *Je t'aide . . . moi non plus: Biologique, comportemental ou psychologique: l'altruisme dans tous ses états* (2010).

Nicola Clayton is Professor of Comparative Cognition in the Department of Psychology at the University of Cambridge, a Fellow of Clare College, and a Fellow of the Royal Society. Her expertise lies in the contemporary study of comparative cognition, integrating a knowledge of both biology and psychology to introduce new ways of thinking about the evolution and development of intelligence in non-verbal animals and pre-verbal children.

Stephen Davies is a Distinguished Professor of Philosophy at the University of Auckland. His research is primarily about the philosophy of the arts, including literary interpretation, the definition of art, the ontological variety of artworks, expressiveness in music, the nature of performance, and the appreciative understanding of art. His authored books include *Definitions of Art* (1991), *Philosophical Perspectives on Art* (2007), and *The Artful Species: Aesthetics, Art, and Evolution* (2012).

Helen De Cruz is Senior Lecturer in Philosophy at Oxford Brookes University. Her main specialization is philosophy of cognitive science, and she has also published in philosophy of religion, epistemology, and general philosophy of science. She is co-author of *A Natural History of Natural Theology: The Cognitive Science of Theology and Philosophy of Religion* (2015).

Johan De Smedt is Lecturer in Philosophy at Oxford Brookes University. His research interests include philosophy of cognitive science, philosophy of religion, and aesthetics. He is co-author of *A Natural History of Natural Theology: The Cognitive Science of Theology and Philosophy of Religion* (2015).

Stephen M. Downes is Professor of Philosophy at the University of Utah. His research is in philosophy of biology with a main focus on biological approaches to explaining human behavior. He is co-editor of *Arguing About Human Nature* (2013).

Chloë FitzGerald is a Postdoctoral Fellow at the Institute for Ethics, History, and the Humanities, University of Geneva. Her research interests lie at the intersection of moral psychology, ethical theory, and bioethics, with a focus on automaticity, dual process theories of cognition, implicit bias and stereotypes, emotions, and moral intuitions.

Ben Fraser is a Postdoctoral Research Fellow in the School of Philosophy at the Australian National University. His research aims to link empirical claims about the evolution of moral cognition to metaethical claims about the epistemological status of moral judgments.

Nir Fresco is a Kreitman Postdoctoral Fellow in the Philosophy Department, Ben Gurion University of the Negev. His main interest is to understand the role that information theory (broadly construed) and information processing play in cognitive science and cognition, respectively. He is the author of *Physical Computation and Cognitive Science* (2014).

Simona Ginsburg is Professor (retired) in the Department of Natural Sciences, the Open University of Israel. Her interests include neurobiology and philosophy of biology. She has studied the function of ionic channels in the presynaptic nerve terminal and the kinetic and stochastic properties of channels in general; in the past decade her work has focused on early nervous systems and the evolution of experiencing.

Nathalie Gontier is a Postdoctoral Researcher in the Center of Philosophy of Science, University of Lisbon, where she is Director of the Applied Evolutionary Epistemology Lab. Her research interests include philosophy of evolutionary biology with a focus on the extended synthesis (reticulate evolution, macroevolution, and hierarchy theory), applied evolutionary epistemology, anthropology of science, and the origins and evolution of language. She is editor-in-chief of the Springer book series *Interdisciplinary Evolution Research*.

Paul Griffiths is Professor of Philosophy at the University of Sydney. He is a philosopher of science with a focus on biology and psychology, and he is a Fellow of the American Association for the Advancement of Science and the Australian Academy of the Humanities. He is co-author of *Genetics and Philosophy: An Introduction* (2013), co-author of *Sex and Death: An Introduction to Philosophy of Biology* (1999), and author of *What Emotions Really Are* (1998).

Valerie Hardcastle is Professor of Philosophy and Psychology at the University of Cincinnati. She studies the nature and structure of interdisciplinary theories in the cognitive sciences and has focused primarily on developing a philosophical framework for understanding conscious phenomena responsive to neuroscientific, psychiatric, and psychological data. She is author of *The Myth of Pain* (1999) and *Constructing the Self* (2008).

Eva Jablonka is Professor at the Cohn Institute for the History of Philosophy of Science and Ideas, Tel Aviv University, and a member of the Sagol School for Neuroscience. Her main interest is the understanding of evolution, especially evolution that is driven by non-genetic hereditary variations, and the evolution of nervous systems and consciousness. She is co-author of *Epigenetic Inheritance and Evolution: The Lamarckian Dimension* (1995) and *Evolution in Four Dimensions: Genetic, Epigenetic, Behavioral, and Symbolic Variation in the History of Life* (2005).

Daniel Kelly is Associate Professor in the Philosophy Department at Purdue University. His research focuses on issues in the philosophy of mind, cognitive science, and empirical moral psychology. He is author of *Yuck! The Nature and Moral Significance of Disgust* (2011).

Anton Killin is a Postdoctoral Fellow in the School of Philosophy, and the Centre of Excellence for the Dynamics of Language, at the Australian National University. He has written extensively on the evolution of music.

Justine Kingsbury is Senior Lecturer in Philosophy at the University of Waikato, New Zealand. Her areas of research include philosophy of mind, philosophy of biology, aesthetics, applied epistemology, and metaphilosophy. She is co-editor of *Millikan and Her Critics* (2013) and is currently working on a monograph entitled *Locating Millikan*.

Maria Kronfeldner is Associate Professor in Philosophy at the Central European University. She works in the philosophy of the life and social sciences, and is author of *Darwinian Creativity and Memetics* (2011). Her article "Darwinian 'blind' hypothesis formation revisited" won the Karl Popper Essay Prize of the British Society for the Philosophy of Science.

Joseph LaPorte is Professor of Philosophy at Hope College. He has published many articles in the philosophy of language, metaphysics, and the philosophy of science, and is the author of *Natural Kinds and Conceptual Change* (2004) and *Rigid Designation and Theoretical Identities* (2012).

Edward Legg is a Research Associate in the Department of Psychology, University of Cambridge. He is a member of the Comparative Cognition Lab, and his research focus is on theory of mind and social cognition in corvids and humans.

Tim Lewens is Professor of Philosophy of Science in the Department of History and Philosophy of Science, University of Cambridge. He has written extensively on evolution, and his authored books include *Cultural Evolution: Conceptual Challenges* (2015), *Darwin* (2007), and *Organisms and Artifacts: Design in Nature and Elsewhere* (2004).

Elisabeth A. Lloyd holds the Arnold and Maxine Tanis Chair of History and Philosophy of Science at Indiana University. Her research interests include philosophy of biology and climate science, models, and pragmatism. She is author of *The Structure and Confirmation of Evolutionary Theory* (1988), *The Case of the Female Orgasm: Bias in the Science of Evolution* (2005), and *Science, Politics and Evolution* (2008).

Edouard Machery is Distinguished Professor in the Department of History and Philosophy of Science at the University of Pittsburgh, and the Director of the Center for Philosophy of Science at the University of Pittsburgh. He works primarily on the philosophical issues raised by psychology and cognitive neuroscience, including concepts, social cognition and moral psychology, methodological issues, and statistical inference. He is the author of *Doing Without Concepts* (2009) and *Philosophy Within its Proper Bounds* (2017).

Darcia Narvaez is Professor of Psychology at the University of Notre Dame. Her research focuses on ethical development and moral education, and she is the author of *Embodied*

Morality: Protectionism, Engagement and Imagination (2016) and *Neurobiology and the Development of Human Morality: Evolution, Culture and Wisdom* (2014).

Karen Neander is Professor of Philosophy and Linguistics in the Department of Philosophy at Duke University. She has written extensively on philosophy of mind, philosophy of biology, and the conceptual foundations of cognitive science and neuroscience. She is author of *A Mark of the Mental: In Defense of Informational Teleosemantics* (2017).

Ljerka Ostojić is a Research Associate in the Department of Psychology, University of Cambridge. She is a member of the Comparative Cognition Lab. Her research focuses on the different mechanisms by which humans and non-human animals interact with other social agents.

Thomas Polger is Professor of Philosophy at the University of Cincinnati. His research is located at the intersection of philosophy of mind, philosophy of science, and metaphysics. He is interested in the relationships between explanations, models, and entities of different sciences—especially those of the cognitive and brain sciences. He is author of *Natural Minds* (2003) and co-author of *The Multiple Realization Book* (2016).

Ulrich Stegmann is Senior Lecturer in Philosophy at the University of Aberdeen. His research is on causation in biology, mechanistic explanation, and the nature of purportedly informational or representational phenomena. He is editor of *Animal Communication Theory: Information and Influence* (2013) and author of a number of articles on philosophy of biology.

Kim Sterelny is Professor of Philosophy at the Australian National University. His main research interests are philosophy of biology, philosophy of psychology, and philosophy of mind. His authored books include *The Evolved Apprentice* (2012) and *Thought in a Hostile World: The Evolution of Human Cognition* (2003)—the latter of which received the 2004 Lakatos Award for distinguished contribution to the philosophy of science.

Karola Stotz is Senior Lecturer in the Department of Philosophy, Macquarie University. She is currently working on a Templeton World Charity Foundation project entitled "Causal Foundations of Biological Information," which aims to ground the idea of biological information in contemporary philosophical work on the nature of causation. She is co-author of *Genetics and Philosophy: An Introduction* (2013).

Kari Theurer is Assistant Professor of Philosophy at Trinity College, Connecticut. Her research is at the intersection of philosophy of science and philosophy of mind, where she focuses primarily on issues surrounding explanation and reduction in the mind and brain sciences.

John Wilkins is an Honorary Research Fellow at the University of Melbourne, and a Research Scholar at the Ronin Institute, headquartered in Montclair, New Jersey. He is author of *Species: A History of the Idea* (2009) and *Defining Species: A Sourcebook from*

Antiquity to Today (2009), and co-author of *The Nature of Classification: Kinds and Relationships in the Natural Sciences* (2013).

Catherine Wilson is the Anniversary Professor of Philosophy at the University of York. Her research is focused on the relationship between historical and contemporary developments in the empirical sciences, including physics and the behavioral and life sciences. She is author of *Moral Animals: Ideals and Constraints in Moral Theory* (2004), *Epicureanism at the Origins of Modernity* (2008), and *Metaethics from a First-Person Standpoint* (2016), along with a number of papers on moral philosophy from a naturalistic standpoint.

Preface

It is commonplace to note that the separation of philosophy and science is of relatively recent vintage. When a young Charles Darwin set sail on the *Beagle* in 1831, the word "scientist" had not yet entered the English language[1]—his nickname on the voyage was "Philos." But by the time he wrote *The Descent of Man* forty years later, the disconnection was clear: it was apparent to everyone that Darwin was engaged in a very different sort of enterprise than his Victorian contemporaries Herbert Spencer, John Stuart Mill, or Henry Sidgwick. The divorce between philosophy and science—as well as further divisions into separate sciences and separate fields of philosophy—was one of practical necessity as much as the result of intellectual revolution: by the mid-19th century there was simply far too much for any person to know. (The polymath Thomas Young (1773–1829) has been described—with only moderate exaggeration—as "the last man who knew everything.")

While a division of the scholarly labor is necessary and immensely profitable, the intellectual isolationism to which it can give rise is neither. The century following Darwin's death may be characterized (again, with only moderate exaggeration) as a time when academic philosophers decided that empirical findings were largely irrelevant to their central concerns, and when scientists looked upon those concerns with suspicion or dismissiveness. But the romantic image of philosophy as "pure"—as engaged in some exclusively a priori analytic enterprise—turned out to lead nowhere very worthwhile. And much scientific energy was wasted due to a lack of conceptual clarity or the absence of an integrative vision—a clarity and a vision that philosophy is often well suited to supply. Pockets of suspicion and isolationism remain to this day, of course, but it is also fair to claim that the last years of the 20th century saw a much more collaborative and communicative relation emerge between science and philosophy. Nowhere has this been more apparent than in the relation between philosophy and biology.

The connection between contemporary academic philosophy and Darwin's theory of evolution is a two-way street. In one direction we can find philosophers of biology making significant contributions to theoretical discussions about the nature of evolution, addressing such empirically oriented questions such as: "What is a species?"; "What is reproductive fitness?"; "Does selection operate primarily on genes?"; and "What is an evolutionary function?". In the other direction we can find philosophers appealing to Darwinian selection (or some aspect thereof) in an attempt to illuminate traditional and perennial philosophical puzzles, such as: "How could a brain-state have representational content?"; "Are moral judgments justified?"; "Why do we enjoy fiction?"; and "Are humans invariably selfish?".

This collection of thirty chapters includes examples from both directions of traffic. Many of the chapters, especially early in the collection, are by philosophers of biology contributing directly to discussions that are central to evolutionary biology. Many other chapters consist of philosophers turning to evolutionary considerations in an endeavor to make headway on questions that traditionally have been largely considered independent of biology. And a good many chapters are by authors who would not identify as "philosophers" at all, but whose work is nevertheless of great relevance to certain philosophical discussions. It is a testament to how far we have come from the dark days of isolationism that for certain topics it has ceased to matter whether one is reading a paper by a philosopher or a psychologist or an economist or a theoretical biologist. Their papers sit happily side by side, often addressing the same problems, and with complementary voices. (I was tempted to say that often one cannot tell the difference among the voices, but decided that was going too far!)

An introduction to a collection of articles will often contain a general overview of its topic in order to situate the reader in the relevant dialectic: sketching the main theories, identifying key principles, mentioning the strengths and weaknesses of central arguments, and so on. No such introduction seems possible in the present case, for the relation between evolution and philosophy is just far too diverse and miscellaneous to permit it. (In this respect this collection is considerably less concentrated than certain others in this Routledge series, such as the *Handbook of Embodied Cognition* or the *Handbook of German Idealism*.) Rather, I will confine myself to saying a few words on the structure of this collection and then leave the contributing authors to speak for themselves.

The collection is divided into six sections, each with five chapters. The first chapter of each section is designed to be something of an overview of the section's topic. This does not mean that the author of the overview introduces the other chapters specifically, but rather that the first chapter of each section takes a broader perspective and strikes a more introductory tone, in the hope of providing the subsequent more focused chapters with a context. In choosing the six section topics for this collection, I did not aim at a comprehensive coverage of the possible relations between philosophy and evolution, but rather selected some focal areas where rich and groundbreaking work is still very much underway. I preferred to leave certain topics entirely untouched rather than try to include everything and spread the coverage too thin. So there is little discussion of evolution and culture, for example, and no attention paid to the theme of evolution and religion.

Despite these self-enforced restrictions, some of the section topics remain so wide-ranging that focus and coverage had to be traded off against one another in my editorial decisions. The section "Evolution and Information," for example, contains topics that are rarely considered side-by-side: one chapter concerns how genes carry information, another is about the evolution of signaling and the emergence of human language, while others assess analyses of natural information in general. The section "Human Nature" contains chapters focused on the case of humans, of course, but the obvious preliminary question of what it even is for a kind to have a "nature" led me to include chapters on the much more general topic of whether there really are such things as biological species.

Sometimes editors of collections must begin their efforts by searching around to produce a sufficient quorum of contributors; sometimes, by contrast, they are faced with an abundance of riches which they must somehow trim and mold into a coherent shape. In planning this collection my challenge was entirely of the latter sort. I suppose I might have tried harder to impose a tighter structure on the raw material, but I was also keen to display how the relationship between evolution and philosophy—which has always been a complicated one—is nowadays evermore active, multifaceted, and productive.

Richard Joyce

NOTE

1. It was coined by William Whewell in 1834. See S. Ross, "*Scientist*: The story of a word," *Annals of Science* 18 (1962): 65–85.

I

The Nature of Selection

1

The Nature of Selection: An Overview

Tim Lewens

"THE SINGLE BEST IDEA THAT ANYONE HAS EVER HAD"

Near the beginning of his book *Darwin's Dangerous Idea*, Daniel Dennett tells us that "If I were to give an award for the single best idea anyone has ever had, I'd give it to Darwin, ahead of Newton and Einstein and everyone else" (1995: 21). Of course, the idea Dennett had in mind was natural selection. But what, exactly, is that idea? Should we even think of it as a "single" idea? I suggest we should think of natural selection as a family of related modes of explanation, which have changed gradually over the years as the theories in which they are embedded have been reformulated in order to address different problems. In other words, our understanding of natural selection itself has been subject to a process of "descent with modification."

This chapter uses some of these transformations in our thinking to reflect on conceptual puzzles about what natural selection is, and how it works. In particular, I focus on a series of contentious questions. Does natural selection entail "gradualism"? In other words, is Jerry Fodor right when he asserts that "Darwinism can work only if . . . there is some organic parameter the small, incremental variation of which produces correspondingly small, incremental variations of fitness" (2001: 89)? Is sexual selection a different process to natural selection, or just a type of natural selection? In what sense does natural selection involve a "struggle for existence"? Can natural selection work with any form of inheritance, or must inheritance be "particulate"? How does our verdict on these questions affect the prospects of efforts to apply natural selection to cultural change, rather than to organic change? And what, finally, does all of this tell us about how natural selection explains the phenomena that were of central interest to Darwin—namely the emergence and perfection of structures and habits that adapt organisms so well to their conditions of life?

DARWIN'S QUESTION

In the introduction to the *Origin of Species*, Darwin (1859) pointed out how much strong evidence there is in favor of "transformism." This is the view, espoused by others before him such as the then-anonymous author of the 1844 work *Vestiges of the Natural History of Creation*, that the species we see around us are modified descendants of earlier ancestors they hold in common. Transformism can be supported, for example, by pointing to facts about the anatomical similarities of distinct species, their distribution around the globe, and so forth. But this sort of transformism faces a significant problem. How on earth, Darwin asks, have "the innumerable species inhabiting this world . . . been modified, so as to acquire that perfection of structure and coadaptation which most justly excites our admiration" (1859: 3)? By itself the hypothesis of common ancestry contains nothing that might explain, to use just one of Darwin's examples, "the structure . . . of the woodpecker, with its feet, tail, beak, and tongue so admirably adapted to catch insects under the bark of trees" (1859: 3).

Darwin considers a mystical response on behalf of the transformist: "The author of the 'Vestiges of Creation' would, I presume, say that, after a certain unknown number of generations, some bird had given birth to a woodpecker," and that it had been produced "perfect as we know them" (1859: 3–4). Needless to say, Darwin immediately responds that "this assumption seems to me to be no explanation, for it leaves the case of the coadaptations of organic beings to each other and to their physical conditions of life, untouched and unexplained" (1859: 4). Darwin observes, in other words, that transformism is incomplete unless it offers some explanation for the emergence of organic structures that are brilliantly adapted to each other, and to the life of the organisms that bear them. Darwin designs natural selection in such a way that it can serve as an answer to what we might call *Darwin's question*:

> How have all those exquisite adaptations of one part of the organic organisation to another part, and to the conditions of life, and of one distinct being to another being, been perfected?   (1859: 60)

In this chapter I will argue that by focusing on the pragmatic origins of natural selection as a response to the question of adaptation—and by focusing on exactly how Darwin himself understands that question—we can understand why Darwin tends to describe natural selection as he does, and why his descriptions often depart from more recent accounts of what natural selection is. In some cases these differences are superficial, and in other cases they are profound.

DARWIN'S ANSWER

Darwin does not usually define natural selection in any short, pithy way, nor does he give a set of conditions that are necessary and sufficient for natural selection to act. Instead, he tends to give far longer descriptions that illustrate, in a schematic form, how complex adaptations can come to exist. The organic world, he says, is characterized by struggle. In his *Notebooks* he writes of "the dreadful but quiet war of organic beings"

(Barrett et al. 1987: E114), a competition so intense that "a grain of sand turns the balance" between life and death (E115c). This struggle has profound consequences, as he later explains in the *Origin*:

> Owing to this struggle for life, any variation, however slight and from whatever cause preceding, if it be in any degree profitable to an individual of any species, in its infinitely complex relations to other organic beings and to external nature, will tend to the preservation of that individual, and will generally be inherited by its offspring. The offspring, also, will thus have a better chance of surviving, for, of the many individuals of any species which are periodically born, but a small number can survive. I have called this principle, by which each slight variation, if useful, is preserved, by the term of Natural Selection, in order to mark its relation to man's power of selection.   (1859: 61)

For Darwin, "this principle" names a wide set of processes whereby valuable variants are generated, maintained, and refined in a population of organisms.

In contrast to this, modern treatments of evolution are often at pains to give far more compact definitions of natural selection. We might be told, for example, that natural selection occurs whenever organisms—or indeed entities of any kind—vary in their "fitness"—roughly speaking, when they vary in their abilities to leave offspring—and whenever these abilities are passed from parents to their babies (e.g., Lewontin 1970). This very general account allows us to ask whether selection might act at several different levels of natural organization—perhaps at the level of the group, or the species, perhaps at the level of the cell or the gene—and it also allows us to ask whether selection might act on entities outside the organic realm—computer viruses, tools, scientific theories. Call this the "inherited variation in fitness" definition.

We can appreciate one limitation of this definition by imagining a population that obeys all these conditions for the action of natural selection, and which also has very few members. Maybe it is divided into slow and fast runners, babies grow to run at the same speed as their parents, and running speed assists in catching prey. Consistent with this, it might also turn out, let us suppose, that the fastest running predators in this population all happen to die young from infections. These infections are just as likely to affect slow and fast individuals: the fast ones just happen to be unlucky. The result is that the slower individuals dominate. Here, modern biologists will say that "drift" is at work, in addition to selection.

So one drawback of our equation of natural selection with "inherited variation in fitness" is that, taken by itself, it does not help us to distinguish natural selection from drift. Modern theorists often move on to define selection in a way that allows us to ask which evolutionary "forces" are at work on a population, and which allows us to give a quantified description of how strong those forces are (Sober 1984). In this mode, we need to find a way of understanding what "selection" is that distinguishes it sharply from other "forces" including drift, mutation, and migration. A standard way of doing this is to propose that natural selection is a force that tends to make the fitter variant in a population increase in frequency, and whose strength depends on the fitness differences between the variants in the population. Drift, on the other hand, is then understood as a force whose strength is in an inverse relationship with population size. In small populations it can overwhelm selection. The broad issue of whether evolution should be understood in terms of interacting "forces" has been the subject of lively debate in recent years, with defences from Sober (1984), Stephens (2004), Reisman & Forber (2005), and Sober & Shapiro (2007), and dissent from Walsh et al. (2002), Matthen & Ariew (2002), and Lewens (2010a), among several others.

In contrast to these modern theorists, Darwin did not approach evolution in a way that demanded a quantified decomposition of different evolutionary "forces," hence he was not driven to define evolutionary processes in a way that would permit sharp differentiation between selection, drift, mutation, and migration. His strong conceptual linkage between natural selection and the explanation of adaptation meant that he sometimes omitted to distinguish between what we would now think of as mutation, on the one hand, and selection, on the other. Instead, because a constant supply of novel variation is essential if complex adaptations are to be produced at all, he often understood the introduction of variation itself as part of the overall process of selection.

More generally, Darwin thought that a diverse variety of circumstances would tend to augment, or undermine, the production of complex adaptations, and he tended to think of these as factors "favorable" or "unfavorable" to the action of selection. The sorts of factors he mentioned include traumatic environmental shifts that can (he thought) act on reproductive organs to stimulate the production of a wide range of "profitable variations" (1859: 82); increases in population size that increase the chances of beneficial variations arising merely because the population is larger; and the geographical isolation of populations, which can allow new varieties to become established and improved in an environment that is comparatively shielded from competitive immigrants (1859: 101–109).

We should not exaggerate how significant these differences are. Darwin understood, even if he did not approach the topic in a mathematically disciplined way, that factors such as the size of a population and the rate at which variation appears within it can affect the production of complex adaptations. Similarly, even though they might isolate selection as just one evolutionary force among many, more recent theorists have often argued that the question of whether a population is able to produce complex adaptations will depend on many other factors, in addition to the question of whether the population is affected by selection in their own rigorously defined sense. For example, Sewall Wright (1932) is well known for his suggestion that drift can in fact facilitate the production of complex adaptation, roughly speaking because of the way it frees an evolving lineage from the demands of immediate gradual improvement, allowing it to colonize unexplored, and potentially profitable, areas of design space. (For a skeptical assessment of Wright's ideas see, among others, Coyne et al. 1997.)

To take another example, Richard Lewontin (1978) has argued that the production of complex adaptations will be favored if the developmental organization of individual organisms is "quasi-independent." Suppose an organism's developmental processes are so tightly enmeshed and integrated that mutations affecting, for example, the structure of the eye end up having further knock-on effects on the heart, the ears, the brain, the kidneys, and so forth. And suppose the same is true for all traits: a mutation that alters one ends up altering all the others. Lewontin's idea is that under these circumstances, even when a mutation arises which improves the functioning of the eye, the chances are that its overall effects on the fitness of the organism will be negative, because it will most likely damage the functioning of many of those other systems the mutation affects. The end result is that iterated sequences of adaptive improvement will be vanishingly unlikely to arise. Hence Lewontin's notion that developmental processes themselves need to be fairly isolated from each other if complex fitness-enhancing organs like eyes are to be built over time. On this view, natural selection is an important element of the explanation for the emergence of complex adaptations, but it is not the full explanation for how these structures come to be.

To summarize the results of this section, we can say that Darwin and more modern theorists disagree in their definition of natural selection. Darwin tends to favor a conception of selection that is explanatorily expansive, in that it encompasses many processes that contribute to adaptation. The price paid is that selection, as he understands it, is resistant to quantification and comparison with alternative evolutionary "forces." Modern theorists make the opposite choices, defining natural selection in a way that is more narrowly focused on just one aspect of the evolutionary process, but more amenable to quantification because of that. Nonetheless, all agree on the more general and pragmatic point that if we want to understand the production of complex adaptations, we cannot focus solely on the processes that cause the fittest variants to dominate in a population.

GRADUALISM

By considering the explanatory task that Darwin intends natural selection to discharge, we can also understand why he describes natural selection in a way that makes it an essentially gradual process. Darwin tells us, for example, that, "As natural selection acts solely by accumulating slight, successive, favourable variations, it can produce no great or sudden modification; it can only act by very short and slow steps" (1859: 471). This might seem like another contrast with modern understandings of natural selection. If we think of natural selection as a process that acts to favor fitter variants in a population, then selection can be at work regardless of whether the fitter variants have sprung forth as fully formed functional macromutations, or whether they are instead tiny modifications of what has gone before. Indeed, if we think—as modern population geneticists tend to—of selection as a force whose strength increases in proportion with fitness differences in the population (see Sober & Shapiro 2007), then selection will be more powerful in a population of eyeless organisms when a fully formed eye comes into existence all in one go, than in a population where a novel variant is only a slight improvement on the eyeless variants that characterize the bulk of the population.

Of course, the reason that Darwin insists that selection must work on successive, slight variations is that it is only in this way that his whole explanatory schema can do the work it needs to. If he could regularly appeal to large "saltations," or "macromutations," then his own theory would offer no better explanation of how adaptations come to be—how, that is, the woodpecker comes to be so good at catching insects—than the mystical version of transformism that he sneers at in the *Origin*'s introduction (see Ariew 2003).

This also helps to explain the otherwise puzzling, and seemingly varied, ways in which Darwin discusses the relationship between natural selection and "use-inheritance." Darwin thought it was an empirically established fact that habits acquired during the life of an adult organism would, if practiced enough, develop as instincts in the adult's offspring. In the *Expression of Emotions*, Darwin (1872) seems to suggest that use inheritance is an alternative explanation—and a better explanation—than natural selection for many of our forms of emotional expression. Indeed, natural selection hardly features at all in Darwin's explanation for the nature of our facial expressions and their associated physiological dispositions (Lewens 2007). However, in the *Origin*, Darwin suggests that natural selection and use and disuse might often be complementary, hazarding that the diminished and poorly functioning eyes of moles probably owe their reduced state "to gradual reduction from disuse, but aided perhaps by natural selection" (1859: 137). These differing relationships asserted between use-inheritance and natural selection can be understood once we see that if practiced habit in the life of an adult is able to account for the origination of a complex functional trait, then it is use-inheritance, rather than natural selection, which explains adaptation. Here, habit passed on to offspring effectively constitutes a saltationist mode of explanation, albeit one that Darwin thinks is well supported by evidence, and hence a legitimate way of accounting for emotional reactions. If disuse works in a gradual way to reduce the eye of a creature that lives in darkness, even though this reduction does nothing to assist the organism in the struggle for life, then again use-inheritance explains a phenomenon that natural selection does not address. (It is striking, and surprising, that Darwin thinks "it is difficult to imagine that eyes, though useless, could be injurious to animals living in darkness," but it is because of this that he attributes their loss "wholly to disuse," and not at all to selection [1859: 137].) But use-inheritance can also be the mechanism whereby Darwin's "successive, slight, favourable variations" can be introduced into a population: here, use-inheritance works as a form of inheritance underpinning the natural selection of complex functional traits. Darwin believes that natural selection aids the effect of disuse in cases where the reduction of a structure—perhaps because a reduced structure will be less likely to attract damaging infections—is of benefit in the struggle for existence (1859: 137).

SEXUAL SELECTION

Our pragmatic focus on "Darwin's Question" also helps us understand why Darwin is so often at pains to distinguish natural selection from sexual selection. For modern theorists, this distinction can often seem unimportant: what matters is whether organisms vary in their inherited abilities to have offspring, regardless of whether these abilities are conferred by the organisms' resilience to disease, their relative superiority compared with other members of their population in evading predators or competing over food, or their relative attractiveness to the opposite sex. These modern theorists will not deny that many traits are best explained as adaptations to the capricious preferences of potential mates—in that sense, they think of sexual selection as an important process—but they also tend to think, since natural selection is always about differences in fitness, that "sexual selection" names just one of the many ways in which selection pressures might act on populations.

For Darwin, on the other hand, natural selection is formulated in order to explain how structures come into existence that are of benefit to individual organisms. He tells us that "Natural selection works solely by and for the good of each being" (1859: 489), and that it acts on variations that are "useful to each being's welfare" (1859: 127). It is no surprise, then, that when Darwin turns to consider a process that "depends, not on a struggle for existence, but on a struggle between the males for possession of the females" (1859: 88), and which can result in traits that are highly injurious to an animal's prospects for survival, but contribute to perceived attractiveness—think of the over-sized and encumbering antlers of the Irish Elk—he does not think of that process as natural selection at all.

This is what explains Darwin's insistence that if the structures that enable an oceanic crustacean to hold on to its mate in a buffeting sea should turn out to be "absolutely required . . . in order to propagate their kind," then they should be attributed to natural selection. If, on the other hand, these structures simply give a relative advantage, compared with other males, with respect to the speed or ease with which mates can be secured, then "sexual selection must have come into action, for the males have acquired their present structure, not from being better fitted to survive in the struggle for existence, but from having gained an advantage over other males" (1877: 244). Again, we see a difference between Darwin's conception of natural selection and the modern view. For Darwin, natural selection concerns that which is "absolutely required," sexual selection that which gives "a relative advantage." For the modern theorist, natural selection is always about differences in fitness, hence it is always about relative advantage.

FITNESS

I have suggested that, for Darwin, natural selection is understood as a process that favors variation that aids individual organisms in the struggle for existence. Sexual selection, on the other hand, works to promote those traits that assist in the struggle for mates. As we will now see, modern theorists think of natural selection in a more generalized manner that encompasses not only Darwin's natural and sexual selection, but also further processes that Darwin would not have recognized as selection at all. That is because of the intimate link in modern theories between natural selection and fitness, and because of the greatly expanded manner in which fitness is now understood.

This point is easiest to appreciate if we look at a curious case initially explored by the population geneticist John Gillespie (1974), and discussed by Elliott Sober (2001), Denis Walsh (2010), and myself (Lewens 2010a), among several others. Sober (2001) explains the case in simplified terms like this. Imagine that reproduction is asexual, and that offspring resemble parents perfectly. Now suppose a finite population contains individuals with one of two reproductive strategies. Type A individuals always have two offspring. Type B individuals either have one offspring or three, with equal probability. So both types have exactly the same expected number of offspring, namely two. But in spite of this, type A—which has narrower variance in offspring number—will reliably increase its frequency in the population.

To see why this will happen, imagine that in generation one there are just two As and two Bs in our population. In the second generation there will be four As. What about the Bs? Each one has a 50 percent chance of having just one offspring, and a 50 percent chance of having three. So there is a 25 percent chance there will be just two Bs, a 50 percent chance there will be four, and a 25 percent chance there will be 6. The point that is crucial to this example is that the number of Bs also makes a different to the overall population size, hence to the frequency of A. The overall population size (As plus Bs) can be six, eight, or ten. So the expected frequency of A is calculated by a weighted average of

4/6, 4/8, and 4/10. A calculation will reveal that the expected frequency of A in generation two is 0.52. In other words, we should expect A to increase in frequency.

Gillespie (1974) thinks of this as a case where there is selection for lower variance. That is justified by the fact that lower variance reliably increases its frequency in the population, hence lower variance is thought of as fitter than higher variance. And if lower variance is fitter, it must be being favored by selection. But it is a considerable stretch to argue that type A's reproductive strategy offers some kind of advantage in the struggle for existence, or the struggle for mates. Using the A strategy does not confer any sort of advantage to individual "welfare," and Darwin would surely not have thought of the A strategy as being favored by natural selection at all.

This one example illustrates how, in modern population genetic treatments of evolution, natural selection is understood in a way that is far more general than it was for Darwin. In modern treatments, natural selection and fitness differences are synonymous: if a trait has higher fitness than another, that is equivalent to saying that the former is favored by selection over the latter. Moreover, population geneticists tend to understand fitness itself in an expansive way, such that a trait has higher fitness than an alternative under a very broad variety of circumstances where it can be reliably expected to increase its frequency over that alternative (see, e.g., Frank & Slatkin 1990). We can now see that this expansion in how natural selection is understood means that natural selection no longer has such a tight conceptual linkage with adaptation—in the specific sense of structures that are of benefit to individual survival and individual welfare—that prompted Darwin's theorizing in the first place. As evolutionary theory turns its attention to understanding the factors that explain the reliable increase in frequency in traits in a population, rather than the structures that improve an individual organism's welfare, the notion of natural selection is reformulated and made more general.

STRUGGLE

We have seen repeatedly in this chapter how Darwin tends to think of natural selection as promoting those variations that assist in the struggle for existence. Darwin puts great stress on the importance of extreme struggle when he explains the workings of natural selection. Some modern textbook treatments begin by quoting Darwin on selection and struggle, before moving on to give their own definitions of natural selection in terms of "inherited variation in fitness," in a way that seems to suggest the "inherited variation in fitness" condition is merely a short summary of Darwin's own explanation for the workings of selection (e.g., Ridley 2003: 72–74). But they fail to notice that a population might show inherited variation in fitness in spite of the fact that there is no struggle for existence going on at all.

This is a point that has been noted by many commentators, from Fisher (1930) to Lewontin (1970) to Sober (1984). When Darwin talks about the struggle for existence, he stresses that it inevitably results from a situation in which "more individuals are born than can possibly survive" (1859: 63). He leans heavily on Malthus's earlier argument (1798), which aimed to establish that populations will always grow so as to outstrip the food that is available to them. But a population that is so blessed with food and other resources that all of the individuals born into it end up surviving might still undergo selection in the modern "inherited variation in fitness" sense, just so long as the individuals in the population differ with respect to how many babies they have. Is this a profound difference between Darwin's theorizing and modern theorizing, or just a superficial one?

The difference seems to me to be quite profound. Darwin is careful to remind us that he uses the notion of struggle "in a large and metaphorical sense" (1859: 62). But he does not say this in order to shrug off the ideas that resources are scarce, and that organisms of all kinds consequently find themselves in situations where all but the best adapted will perish. Instead, he simply means to remind his readers that he is not asserting that organisms literally do battle with each other, hence plants can "struggle" for water just as dogs can struggle over a bone. Darwin's picture of natural selection is "the doctrine of Malthus applied with manifold force to the whole animal and vegetable kingdoms" (1859: 63). Why is this Malthusian doctrine so important to him?

Part of the answer to this question lies in a surprising place. Modern commentators sometimes suggest that Darwin's idea of natural selection has persisted into modern biology more or less unchanged, but they qualify this with the assertion that we now know far more about the mechanisms of inheritance than Darwin ever suspected. Poor Darwin was hampered by his ignorance of genetics. Dennett's summary of the conceptual history of evolutionary theory is a good example of this approach:

> In all his brilliant musings, Darwin never hit upon the central concept, without which the theory of evolution is hopeless: the concept of a gene. Darwin had no proper unit of heredity, and so his account of the process of natural selection was plagued with entirely reasonable doubts about whether it would work. Darwin supposed that offspring would always exhibit a sort of blend or average of their parents' features. Wouldn't such "blending inheritance" always simply average out all differences, turning everything into a uniform gray?   (Dennett 1995: 22)

Darwin did indeed note that blending might undermine the action of selection. He acknowledged this worry in the first edition of the *Origin*, but he thought that exceptionally strong struggle would counteract the erosion that blending would otherwise occasion: "The process [of selection] will often be greatly retarded by free intercrossing. Many will exclaim that these several causes are amply sufficient wholly to stop the action of natural selection. I do not believe so" (1859: 108).

As usual, Darwin makes his case by using an analogy between what talented breeders can do on the farm, and what nature can achieve in the wild. He did indeed suppose that the traits of offspring were often just an average of their parents' traits. But he supposed this to be just as true of baby lambs in a barn as it would be of baby wolves in a forest. Even so, animal breeders had been able to improve wool and meat in sheep. This suggested to Darwin that it must be possible for improvements to occur even with blending: the important thing was that breeders made sure that the best animals only mated with each other. If struggle in the wild was exceptionally intense, then nature could achieve something similar to the demanding breeder. Only the best specimens would then be able to mate—hence they would only mate with each other—because intense competitive struggle would ensure that all the other inferior specimens perished. That is why, when Darwin reminds us in his 1868 book *The Variation of Plants and Animals under Domestication* of his argument in the *Origin*, the strength of competition is so boldly underlined:

> It was there shown that all organic beings, without exception, tend to increase at so high a ratio, that no district, no station, not even the whole surface of the land or the whole ocean, would hold the progeny of a single pair after a certain number of generations. The inevitable result is an ever-recurrent Struggle for Existence . . . [The] severe and often-recurrent struggle for existence will determine that those variations, however slight, which are favourable shall be preserved or selected, and those which are unfavourable shall be destroyed.   (1868: 5–6)

Our interpretation of Darwin's own understanding of natural selection needs to make sense of his repetitive insistence on the severity of struggle. We can do so by noting that severe struggle ensures not merely differential rates of survival and reproduction, but also that all but the very best adapted forms die. This allows Darwin to argue for the exceptionally discerning eye of nature, which determines who will be allowed to mate, and who will not. That, in turn, allows him to make a case for thinking that natural selection will be no less efficacious—indeed, it will be far more efficacious—than famed animal and plant breeders, in spite of the fact that (as he thought) all labored against a backdrop of blending inheritance.

PARTICLES

Dennett's comments about Darwin's failure to understand the mechanism of inheritance are not especially unusual. Peter Godfrey-Smith, too, has recently written that "One of the weaker points in Darwin's work was his understanding of reproduction and inheritance" (2014: 9). Dennett, remember, suggests that Darwin was confused by his insistence that inheritance would have a "blending" character. Similarly, Godfrey-Smith informs us that R. A. Fisher "argued in 1930 that inheritance *had* to operate in a 'particulate' manner, with discrete and stable genes, in order for sustained Darwinian evolution to be possible" (2014: 97).

It would be easy to infer from all of this that there has been an important transition from Darwin's image of selection as a process that works against a background of blending inheritance, to a more modern understanding that it would be impossible for selection to work in such a way, and that selection instead requires particulate inheritance. Occasionally modern commentators even suggest that Darwin himself may have dimly appreciated the importance of particulate inheritance for the efficacy of selection: the thought goes that nearly ten years after the *Origin* was first published, Darwin formulated a theory of inheritance based on the transmission of particles from parents to offspring, and that he did so in order to save natural selection from the problems posed by blending (e.g., Charlesworth & Charlesworth 2009).

The historical picture sketched in the preceding paragraph is misleading. We have already seen that Darwin was not much troubled by blending. He thought that because the struggle for existence was exceptionally powerful, blending would not overwhelm the tendency of selection to discern and retain beneficial variations. Moreover, the theory of inheritance that Darwin published in 1868 was mentioned only once in the fifth and sixth editions of the *Origin* (Peckham 1959). Surely if Darwin had formulated that theory in order to deal with problems posed by blending, he would have mentioned it incessantly in those later editions. It seems that Darwin himself thought that his hypothesis of inheritance—which did indeed explain parent–offspring resemblance in terms of the transmission of particles—was wholly irrelevant to the case he wanted to make for the efficacy of natural selection.

Darwin called his theory of inheritance the hypothesis of "pangenesis." He thought that all the cells in the body: "throw off minute granules which are dispersed throughout the whole system . . . They are collected from all parts of the system to constitute the sexual elements, and their development in the next generation forms a new being" (Darwin 1868). Much later, R.A. Fisher would write that it was "universally admitted" that "Darwin accepted the fusion or blending theory of inheritance" (1930: 1). This may seem like an odd pronouncement, for we have just seen that Darwin explicitly thought that offspring acquired a set of particles from their parents, passed on at conception. He thought these particles matured during the growth of the new organism in such a way as to explain trans-generational resemblance. But Darwin also thought that the traits of offspring would often—although not always—be an average of their parents' traits.

Darwin seems, then, to have held a particulate theory regarding the mechanism of inheritance, while simultaneously holding a blending theory about the relationship between the characteristics of parents and offspring. Understood like this, blending and particulate views of inheritance are obviously compatible: they address phenomena at different levels. Fisher famously wrote of them in a way that suggests one must choose between the two. He argued that "one of the main difficulties felt by Darwin is resolved by the particulate theory" (1930: 12). But Darwin himself espoused a particulate theory. So what is going on here?

Fisher did not make a mistake, or somehow overlook Darwin's theory of pangenesis (Lewens 2015a). When Fisher described "the great contrast between the blending and the particulate theories of inheritance" (1930: 4), his exploration of the consequences of the two systems addressed *phenomenal* patterns of inheritance. Fisher entertained no theory of the *mechanism* that might underlie a blending system. He showed us how quickly variance will disappear in a system whereby offspring trait values are always intermediate between the values of the parents, regardless of what the underlying processes that explain such a pattern might look like. He then contrasted this purely phenomenal model with a different phenomenal model, whereby offspring trait values follow what he called "the modern scheme of Mendelian or factorial inheritance" (1930: 7). Evidently a population characterized by the simple blending model will behave differently to a population that follows a Mendelian scheme. In a simple blending model, for example, the offspring of a given pair of parents are always identical, and reversion to the character of a grandparent is impossible. In the Mendelian case, the traits of siblings can differ, and characters can disappear for a generation before reappearing again. And so, Fisher argued that "the mechanism of particulate inheritance" results in "no inherent tendency for the variability to diminish" (1930: 9).

It is important to recognize that Fisher did not argue that selection could not *possibly* work with this blending pattern. Instead, he pointed to the demanding conditions that would need to be in place for selection to work with blending:

The important consequence of the blending is that, if not safeguarded by intense marital correlation, the heritable variance is approximately halved in every generation . . . If variability persists, as Darwin rightly inferred, causes of new variability must continually be at work. Almost every individual of each generation must be a mutant.   (1930: 5)

In other words, if inheritance follows a blending pattern, then for selection to be efficacious it must also be the case that like organisms mate with like ("intense marital correlation"), or that new variations are constantly arising, or both. We have already seen that Darwin thought a condition very much like the first was satisfied. He thought that the struggle for existence was frequently so intense that only the very best adapted individuals would survive, hence they would end up mating with each other. He also thought something close to the second condition was satisfied, too. He thought that "sports"—that is, rare variations, of large magnitude—were of little significance for evolutionary change when compared with what he called "individual differences." These were the "many slight differences" which Darwin thought regularly appeared in populations (Vorzimmer 1963). In short, Fisher argued that improbably strong conditions would need to be in place for natural selection to work with blending inheritance. Darwin was untroubled by blending because he felt these strong conditions were satisfied (Lewens 2010b).

CULTURAL SELECTION

We have just seen some significant differences between Darwin's detailed conception of selection and more modern conceptions that build on Fisher's population genetics. But our discussion of the relationship between selection and inheritance is instructive for other reasons. We have seen that Fisher allowed that it was possible—just highly unlikely in practice—that selection could work with a "blending" pattern of inheritance. This concession is important when we think about applying evolutionary thinking in non-standard domains. Theorists of cultural evolution, for example, often propose that techniques, ideas, and so forth evolve by natural selection (see Richerson & Boyd 2005 and Mesoudi 2011 for overviews, and Lewens 2015b for philosophical evaluation). There are many different ways of making pots, one generation's pottery techniques often resemble the techniques of earlier generations, and techniques are adopted or rejected according to how well they match the demands of users and manufacturers. So something like our "inherited variation in fitness" conditions are satisfied. Even so, one might worry that techniques are not transmitted in a manner analogous to the genetic transmission of organic traits: one person's pot-making technique might be an inferred mish-mash of the techniques of many teachers—a blend, that is—with the result that techniques are not at all "particulate" in their mode of transmission. If cultural transmission is nothing much like genetic transmission, we might then wonder whether it makes sense to think that culture evolves by a process of natural selection.

We now see that the mere fact that techniques may follow a "blending" pattern of inheritance does not suffice to show that techniques cannot evolve by natural selection. The evolutionary theorists of culture Richerson & Boyd (2005: 88–90) point out, for example, that even though biologists follow Fisher in thinking that organic mutation rates are too low to sustain organic evolution with blending, we might yet wonder whether cultural entities such as techniques enjoy much higher rates of mutation. They also suggest that features of how we influence each other socially—such as our apparent tendency to conform with the majority views that we encounter, and our practices of actively policing social norms—might help to maintain group-level variation that selection can act on, even when single individuals rarely copy each other in ways that parallel faithful genetic inheritance. Natural selection might work in the cultural domain, but in a manner that is quite different than its operation on genetic variation.

Fisher did not show that natural selection requires that inheritance be particulate, or gene-like: instead, he showed that the character of inheritance affects what else needs to be the case for natural selection to be effective. What is more, Fisher's appreciation of the significance of inheritance required him to develop a mathematically sophisticated way of figuring out how patterns of parent–offspring resemblance would be reflected in the makeup of populations over time. It did not depend on any detailed understanding of the molecular basis of genetic transmission. The historical development of evolutionary theory consequently offers encouragement to those who want to develop an illuminating evolutionary theory of cultural change even in the absence of a detailed understanding of the precise processes by which cultural transmission works.

ORIGINS

Darwin introduced the concept of natural selection in order to explain the phenomenon of adaptation. It is essential for this task that natural selection does not merely explain why beneficial adaptive traits, once they arise as variants, become widely distributed in a population. Darwin also took it that natural selection could explain why these adaptive traits come to exist in the first place. In Peter Godfrey-Smith's (2009, 2014) language, natural selection is supposed to answer demands for *origin* explanations, as well as demands for *distribution* explanations. If natural selection cannot discharge this first role, it is unclear how natural selection is supposed to be an improvement either over the mystical version of transformism that Darwin rejects at the beginning of the *Origin*, or over the "special creationist" view that says a beneficent creator has fashioned adaptations.

We have seen that modern accounts of natural selection tend to characterize it as a force that can increase trait frequencies. How, then, does natural selection also explain the origination of beneficial adaptations? The answer must be that by increasing trait frequencies, selection makes the emergence of further adaptations more likely. Several writers, including Neander (1995), myself (Lewens 2004), and Godfrey-Smith (2014) have sketched how this can happen.

First, suppose that genomic bases X, Y, and Z produce eyes of increasing functionality. Second, suppose—and this is not at all trivial—that it is more likely that Z will be produced by mutation from Y, than it is that Z will be produced by mutation from X. Now, imagine that Y appears by chance mutation in a population of organisms that all have X. Natural selection will favor Y, and it will tend to increase the number of organisms with Y in the population. Selection has now made it more likely that Z will appear, by increasing the number of organisms with Y. Selection has explained the origination of a more functional eye.

The problem with this story is that it presupposes that selection acts in one specific way. In general, modern theorists tend to think that selection increases the *frequency* of one trait over another. But if the chances of Z appearing are to be increased by selection, then it seems necessary that selection specifically increases the *absolute* number of organisms with Y. Selection need not act in this way, even though it sometimes might do, because Y might increase its frequency over X even when the absolute numbers of both are declining (Lewens 2004, 2015a; Godfrey-Smith 2009, 2014).

Godfrey-Smith has recently suggested that it is under conditions of scarce resources that selection ends up increasing absolute numbers of the favored variant. This suggestion seems to give a further boost to Darwin's own insistence on Malthusian struggle as a key component of any explanation for the origination of complex adaptations. Godfrey-Smith writes that, "the fact of scarce resources—when it is a fact—ties relative reproductive success and absolute reproductive success together" (2014: 42). Unfortunately this linkage can break, for if resources are in exceptionally short supply, the result might be that all of the different types in a population end up decreasing in absolute numbers, with some decreasing less quickly than others, and increasing their frequency as a result. When this is the case, we will find that selection ends up lowering the absolute numbers of the favored variant, precisely because resources are so scarce.

A better way to think about the relationship between selection and the origin of adaptive traits requires us to distinguish two ways of understanding what it means to say that selection increases the chances of adaptation (Lewens 2004, 2015a). Let us return to our example of eyes. Suppose, again, that resources are exceptionally scarce. Because of this, the absolute numbers of Y variants are decreasing, but less quickly than the absolute numbers of X variants. This means that in each generation the chances of a Z variant appearing in the population get lower. In that sense, selection makes adaptation less likely. Even so, Y is increasing in frequency: in that sense, it is favored by selection. Crucially, the population would have been even less likely to have produced Z variants if X, rather than Y, had been increasing in frequency. So even when resources are scarce, and even when the chances of a Z variant are constantly diminishing, we can say that selection explains the origination of adaptation, in the sense that the chances of Z arising are higher if Y increases its frequency than they would have been if X had increased its frequency.

This brings us back to Darwin. Nature, he said, works in a similar manner to a skilled breeder. But nature is far more discerning in its choices, and far more devoted in performance of its duties, than any human:

> It may be said that natural selection is daily and hourly scrutinizing, throughout the world, every variation, even the slightest; rejecting that which is bad, preserving and adding up all that is good; silently and insensibly working, wherever and whenever opportunity offers, at the improvement of each organic being in relation to its organic and inorganic conditions of life.   (1859: 84)

We have now seen one way of making sense of this. A population is more likely to produce novel beneficial variations when that population is primarily composed of fitter, rather than less fit, variants. Natural selection helps to ensure that populations are continually transformed in ways that make them apt to produce valuable new traits. In that sense, natural selection does indeed preserve and add up what is good, in a way that gives an answer to Darwin's question. Natural selection can, it turns out, explain "that perfection of structure which most justly excites our admiration" (1859: 3).

ACKNOWLEDGMENTS

I am grateful to Richard Joyce and Jonathan Birch for very helpful comments on an earlier draft of this chapter. The research leading to these results has received funding from the European Research Council under the European Union's Seventh Framework Programme (FP7/2007–2013)/ERC Grant agreement no 284123.

REFERENCES

Ariew, A. 2003. "Natural selection doesn't work that way: Jerry Fodor vs. evolutionary psychology on gradualism and saltationism." *Mind and Language* 18: 478–483.

Barrett, P., Gautrey, P., Herbert, S., Kohn, D., & Smith, S. 1987. *Charles Darwin's Notebooks, 1836–1844* (Cambridge University Press).

Charlesworth, B. & Charlesworth, D. 2009. "Darwin and genetics." *Genetics* 183: 757–766.

Coyne, J., Barton, N., & Turelli, M. 1997. "A critique of Sewall Wright's shifting balance theory of evolution." *Evolution* 51: 643–671.

Darwin, C. [1859] 1964. *On the Origin of Species* (Harvard University Press).

Darwin, C. [1868] 1998. *The Variation of Animals and Plants Under Domestication, Vols. 1 & 2* (Johns Hopkins University Press).

Darwin, C. [1872] 1998. *The Expression of Emotions in Man and Animals* (Oxford University Press).

Darwin, C. [1877] 2004. *The Descent of Man* (Penguin).

Dennett, D. 1995. *Darwin's Dangerous Idea* (Allen Lane).

Fisher, R. 1930. *The Genetical Theory of Natural Selection* (Clarendon Press).

Fodor, J. 2001. *The Mind Doesn't Work That Way* (MIT Press).

Frank, S. & Slatkin, M. 1990. "Evolution in a variable environment." *American Naturalist* 136: 244–260.

Gillespie, J. 1974. "Natural selection for within-generation variance in offspring number." *Genetics* 76: 601–606.

Godfrey-Smith, P. 2009. *Darwinian Populations and Natural Selection* (Oxford University Press).

Godfrey-Smith, P. 2014. *Philosophy of Biology* (Princeton University Press).

Lewens, T. 2004. *Organisms and Artifacts: Design in Nature and Elsewhere* (MIT Press).

Lewens, T. 2007. *Darwin* (Routledge).

Lewens, T. 2010a. "The natures of selection." *British Journal for the Philosophy of Science* 61: 313–333.

Lewens, T. 2010b. "Natural selection then and now." *Biological Reviews* 85: 829–835.

Lewens, T. 2015a. "The nature of philosophy and the philosophy of nature." *Biology and Philosophy* 30: 587–596.

Lewens, T. 2015b. *Cultural Evolution: Conceptual Challenges* (Oxford University Press).

Lewontin, R. 1970. "The units of selection." *Annual Review of Ecology and Systematics* 1: 1–18.

Lewontin, R. 1978. "Adaptation." *Scientific American* 239: 159–169.

Malthus, T. [1798] 1986. *An Essay on the Principle of Population* (Penguin).

Matthen, M. & Ariew, A. 2002. "Two ways of thinking about fitness and natural selection." *Journal of Philosophy* 99: 55–83.

Mesoudi, A. 2011. *Cultural Evolution* (University of Chicago Press).

Neander, K. 1995. "Pruning the tree of life." *British Journal for the Philosophy of Science* 46: 59–80.

Peckham, M. 1959. *The Origin of Species by Charles Darwin: A Variorum Text* (University of Pennsylvania Press).

Reisman, K. & Forber, P. 2005. "Manipulation and the causes of evolution." *Philosophy of Science* 72: 1113–1123.

Richerson, P. & Boyd, R. 2005. *Not by Genes Alone: How Culture Transformed Human Evolution* (University of Chicago Press).

Ridley, M. 2003. *Evolution* (Blackwell).

Sober, E. 1984. *The Nature of Selection* (Chicago University Press).

Sober, E. 2001. "The two faces of fitness." In R. Singh, C. Krimbas, D. Paul, & J. Beatty (eds.), *Thinking About Evolution* (Cambridge University Press) 309–321.

Sober, E. & Shapiro, L. 2007. "Epiphenomenalism: The dos and don'ts." In P. Machamer & G. Wolters (eds.), *Studies on Causality: Historical and Contemporary* (University of Pittsburgh Press) 235–264.

Stephens, C. 2004. "Selection, drift, and the 'forces' of evolution." *Philosophy of Science* 71: 550–570.

Vorzimmer, P. 1963. "Charles Darwin and blending inheritance." *Isis* 54: 371–390.

Walsh, D. 2010. "Not a sure thing: Fitness, probability, and causation." *Philosophy of Science* 77: 147–171.

Walsh, D., Lewens, T., & Ariew, A. 2002. "The trials of life" *Philosophy of Science* 69: 429–446.

Wright, S. 1932. "The roles of mutation, inbreeding, crossbreeding and selection in evolution." *Proceedings of the 6th International Congress of Genetics* 1: 356–366.

2

Multilevel Selection and Units of Selection Up and Down the Biological Hierarchy

Elisabeth A. Lloyd

INTRODUCTION

There has been considerable debate in both biology and philosophy about which entities undergo natural selection and what it is that fits them for that role. For nearly forty years, some participants in the "units of selection" debates have argued that more than one issue is at stake. Richard Dawkins, for example, introduced "replicator" and "vehicle" to stand for different roles in the evolutionary process (1982). David Hull (1980) broke Dawkins's category of "replicator" up into "replicator" and "interactor."

In an essay in *Keywords in Evolutionary Biology* (Lloyd 1992), and later in the *Stanford Encyclopedia of Philosophy* (2012), I delineated two further roles which I called the "beneficiary" and the "manifestor of adaptation." In sections 2–6 of this chapter, four quite distinct questions will be isolated that have, in fact, been asked in the context of considering what is a unit of selection. The following four sections return to the sites of several very confusing debates about "the" unit of selection, which are analyzed utilizing the taxonomy of my distinct questions.

FOUR BASIC QUESTIONS

The term *replicator*, originally introduced by Dawkins, is used to refer to any entity of which copies are made. Dawkins also introduced the term *vehicle*, which he defines as "any relatively discrete entity . . . which houses replicators, and which can be regarded as a machine programmed to preserve and propagate the replicators that ride inside it" (1982: 295). According to Dawkins, most replicators' phenotypic effects are represented in vehicles, which are themselves the proximate targets of natural selection.

Hull observed that Dawkins's theory has replicators interacting with their environments in two distinct ways: they produce copies of themselves, and they influence their own survival and survival of their copies through production of secondary products that ultimately have phenotypic expression. Hull suggested the term *interactor* for the latter entities. An *interactor* denotes that entity which interacts, as a cohesive whole, directly with its environment in such a way that replication is differential—in other words, an entity on which selection acts directly (Hull 1980: 318; see Brandon 1982: 317–318). The terms *replicator* and *interactor* will be used in Hull's sense in the rest of this chapter.

THE INTERACTOR QUESTION

The interactor question is: what units are being actively selected in a process of natural selection? As such, this question is involved in the oldest forms of the units of selection debates (Darwin [1859] 1964; Haldane 1932; Wright 1945). In his classic review, Lewontin's purpose was "to contrast the levels of selection, especially as regards their efficiency as causers of evolutionary change" (1970: 7).

Questions about interactors focus on the description of the selection process itself, that is, on the interaction between entity, trait, and environment, and on how this interaction produces evolution; they do not focus on the outcome of this process (Wade 1977). Here, the interactor is possibly at any level of biological organization, including a group, a kingroup, an organism, a gamete, a chromosome, or a gene. Some portion of the expected fitness of the interactor is directly correlated with the value of the trait in question. The expected fitness of the interactor is commonly expressed in terms of genotypic fitness parameters, that is, in terms of the fitness of combinations of replicators, that is, replicator success. Several statistical methods are available for expressing such a correlation between interactor trait and (genotypic or genic) fitness, including partial regression, variances, and covariances.[1]

Note that the "interactor question" does not involve attributing adaptations or benefits to the interactors or their replicators. Interaction at a particular level involves only the presence of a trait at that level with a special relation to genic or genotypic expected success that is not reducible to interactions at a lower level. The most common error made in interpreting many of the interactor-based approaches is that the presence of an interactor at a level is taken to imply that the interactor is also a manifestor of an adaptation at that level (see "Group Selection" below).

THE REPLICATOR QUESTION

The focus of discussions about replicators concerns just which organic entities actually meet the definition of replicator. Hull refined and restricted Dawkins's meaning of "replicator," which he defined as "an entity that passes on its structure directly in replication" (Hull 1980: 318).

Hull's definition of replicator corresponds more closely than Dawkins's to a longstanding debate in genetics about how large or small a fragment of a genome ought to count as a replicating unit—something that is copied, and which can be treated separately in evolutionary theory (Lewontin 1970). This debate revolves critically around the issue of linkage disequilibrium and led Lewontin to advocate the use of parameters referring to the entire genome in genetical models (1974; Brandon 1982; Slatkin 1972; Wimsatt 1980). That is, with much linkage disequilibrium, individual genes cannot be considered as replicators because they do not behave as separate units during reproduction.

This is not to suggest that the replicator question has been solved; it is simply no longer a significant part of the units of selection debates.[2] Philosophers working on the units problems often tacitly adopted Dawkins's suggestion that the replicator, whatever it turned out to be, be called the "gene" (see section "Genic Selection: The Originators"). Recent work by Paul Griffiths and Karola Stotz on the concept of the "gene" has turned up many multiple meanings; nevertheless, they remain outside the rubric of the units of selection debates (2013).

In addition, James R. Griesemer has rejected the role of replicator as misconceived. He proposes in its place the role of "reproducer," which focuses on the material transference of genetic and other matter from generation to generation, and incorporates development into heredity and the evolutionary process (Griesemer 2000a, b). Thinking in terms of reproducers allows for both epigenetic and genetic inheritance to be dealt with within the same framework, and can be central to work on the units of evolutionary transition (Griesemer 2000c). Peter Godfrey-Smith has also since introduced a notion of a reproducer that relaxes the material overlap requirement and focuses on an understanding of "who came from whom, and roughly where one begins and another ends" (2009: 86).

THE BENEFICIARY QUESTION

Who benefits from a process of evolution by selection? There are two predominant interpretations of this question: "Who benefits ultimately in the long term, from the evolution by selection process?" and "Who gets the benefit of possessing adaptations as a result of a selection process?".

Take the first of these, the issue of the ultimate beneficiary. There are two ways of characterizing the long-term survivors and beneficiaries of the selection process. One might say that the species or lineages are the ultimate beneficiaries of the evolutionary process. Alternatively, one might say that the surviving alleles are the relevant long-term beneficiaries. I have not located any authors holding the first view (perhaps Hull 1980), but, for Dawkins, the latter interpretation is the *primary fact* about evolution. To arrive at this conclusion, Dawkins adds the requirement of agency to the notion of beneficiary; it must function as the initiator of a causal pathway. The replicator is causally responsible for all of the various effects that arise further down the biochemical or phenotypic pathway, irrespective of which entities might reap the long-term rewards (1982).

The evolution by selection process may also be said to "benefit" a particular level of entity under selection, through producing adaptations at that level (Williams 1966; Maynard Smith 1976). On this approach, the level of entity actively selected (the interactor) benefits from evolution by selection at that level through its acquisition of adaptations. This sense of "beneficiary" that concerns adaptations will be treated as a separate issue, discussed in the next section.

THE MANIFESTOR OF ADAPTATION QUESTION

At what level do adaptations occur? Or, as Elliott Sober puts this question, "When a population evolves by natural selection, what, if anything, is the entity that does the adapting?" (1984: 204).

The presence of adaptations at a given level of entity is sometimes taken to be a requirement for something to be a unit of selection (e.g., Brandon 1985; Burian 1983; Maynard Smith 1976; Mitchell 1987; Vrba 1984). The *manifestor of adaptation* demonstrates an adaptation at a specified level.

Much confusion is the result of a very important but neglected duality in the meaning of *adaptation* (in spite of useful discussions in Brandon 1985, Burian 1983, Krimbas 1984, Sober 1984). Sometimes *adaptation* is taken to signify any trait at all that is a direct result of a selection process at that level: the *selection-product* view (e.g., Sober 1984; Brandon 1985, 1990; Arnold & Fristrup 1982). Sometimes, on the other hand, *adaptation* is reserved for traits that are "good for" their owners, that is, those that provide a "better fit" with the environment, and that intuitively satisfy some notion of "good engineering" (e.g., Ariew & Lewontin 2004; Gould & Lewontin 1979; Hull 1980; Lewontin 1978; Williams 1966). These two meanings of adaptation, the *selection-product* and *engineering* definitions respectively, are distinct, and in some cases, incompatible (Lloyd 2008).

The peppered moth case is an example of a *selection-product* adaptation, where there is no new evolution of a newly "engineered" trait by the selection process, because the adaptive dark colored moth is simply a variant already present in the population, and presents no newly evolved trait. The camera eye, on the other hand, is an *engineering* adaptation, providing a "better fit" with the environment and responding to particular "challenges" from the environment. The *manifestor of adaptation* refers only to *engineering* adaptations, and not to *selection-products*.

In sum, when asking whether a given level of entity possesses adaptations (rather than asking about which unit is undergoing selection, i.e., the interactor), it is necessary to state not only the level of selection in question but also which notion of adaptation—either *selection-product* or *engineering*—is being used. It is crucial that the *interactor question* does not involve attributing engineering adaptations to the interactors, as does the *manifestor-of-adaptation question*.

Summary

Four distinct questions have been described that appear under the rubric of "the units of selection" problem: What is the interactor? What is the replicator? What is the beneficiary? What entity manifests engineering adaptations resulting from evolution by selection? (Lloyd 1992, 2001, 2012). Commenting on this analysis, John Maynard Smith wrote in *Evolution*: "[Lloyd 2001] argues, correctly I believe, that much of the confusion has arisen because the same terms have been used with different meanings by different authors . . . [but] I fear that the confusions she mentions will not easily be ended" (Maynard Smith 2001: 1497). In the following sections, this taxonomy of questions is used to sort out some of the most influential positions in five debates: group selection, species selection, genic selection, genic pluralism, and holobiont selection and evolution.

GROUP SELECTION

George Williams's famous near-deathblow to group panselectionism was, oddly enough, about benefit. He was interested in cases in which there was selection among groups where the groups as a whole *benefited from* organism-level traits (including behaviors) that seemed disadvantageous to the organism (similarly for Maynard Smith 1964). Williams argued that the presence of a benefit to the group was not sufficient to establish the presence of group selection, nor is it enough to show that they are adaptations; in order to be a group adaptation, under Williams's view, the trait must be an *engineering* adaptation that evolved by group selection. Williams argued that group benefits do not, in general, exist *because* they benefit the group; that is, they do not have the appropriate causal selective history (see Brandon 1985: 81; Sober 1984: 262 ff.; Sober & Wilson 1998).

Implicit in Williams's discussion is the assumption that being a unit of selection at the group level requires two things: (1) having the group as an interactor, and (2) having a group-level engineering-type adaptation. That is, Williams combines two different questions, the interactor question and the manifestor-of-adaptation question, and calls this combined set *the* unit of selection question, just as his prime target, Vero Wynne-Edwards, did. This combined requirement of engineering group-level adaptation in addition to the existence of an interactor at the group level has been a very popular version of the necessary conditions for being a unit of selection within the group selection debates (see Hull 1980: 325; Maynard Smith 1976: 278).

In contrast to the preceding authors, Sewell Wright distinguished group selection for "group advantage" from group selection per se (1980: 840; see Wright 1931). In essence, he claimed that the combination of the interactor question with the question of what entity had adaptations had created a great deal of confusion in the units of selection debates (Lewontin 1978; Gould & Lewontin 1979). Wright takes Maynard Smith, Williams, and Dawkins to task for mistakenly thinking that because they have successfully criticized group selection for group advantage, they can conclude that "natural selection is practically wholly genic." This is a fair criticism of Maynard Smith, Williams, and Dawkins; these authors failed to distinguish between two questions—the interactor question and the manifestor-of-adaptation question. Wright's interdemic group selection model involves groups only as interactors, not as manifestors of group-level adaptations. Modelers following Wright's interest in structured populations have created a set of group selection models (Uyenoyama & Feldman 1980; Wade 1985, 2016). (See Goodnight & Stevens (1997) for an excellent summary of the significant empirical and theoretical discoveries enabled by Wright-style group selection models.)

For a period spanning two decades, however, Maynard Smith, Williams, and Dawkins did not acknowledge that the position they attacked, namely Wynne-Edwards's, is significantly different from other available approaches to group selection, such as Wright's, Wade's, D.S. Wilson's, Uyenoyama and Feldman's, or Lewontin's. Ultimately, however, both Williams and Maynard Smith recognized the significance of the distinction between the interactor question and the manifestor-of-an-adaptation question (Williams 1985: 7–8; Maynard Smith 1987: 123).

Debates about the relationship of kin selection and inclusive fitness theory to group selection have populated the evolution literature since the 1960s, but one of the most significant contributions, by W.D. Hamilton (1975), in which he said that kin selection was a form of group selection, was widely ignored by his followers, and still is, by Dawkins (2012). More recently, Martin Nowak co-authored a controversial paper on the evolution of eusociality that appealed to group selection (Nowak et al. 2010), and D.S. Wilson and E.O. Wilson published several commentaries affirming the role of group selection in evolution (2007; E.O. Wilson 2005; E.O. Wilson & Holldobler 2005; see Sober & Wilson 1998).

Wilson and Sober's book, *Unto Others* (1998), brought the attention of the larger philosophical community to group selection models, and explained them in an accessible fashion. Godfrey-Smith provides a considerably different view of the necessary conditions for group selection, one that rejects many of the currently accepted cases of the phenomenon. For him, a "Darwinian population" must exist at the level of selection being described, which requires the presence of both an interactor and a reproducer at that level, thus putting together what others have pulled apart (2009: 39, 112).

Taking the evolution of altruism, commonly attributed to group selection, it should not be considered as such, on Godfrey-Smith's account, because of the lack of a true reproducer at the group level (2009: 119). He argues that his system presents an advantage over other available analyses of units of selection because it can account for previously neglected examples such as epigenetic inheritance systems. But Griesemer's theory of reproducers accounts for epigenesis without doing violence to the theoretical results from hierarchical selection theory (Griesemer 2000a, b, c).

SPECIES SELECTION

Ambiguities about the definition of a unit of selection have also snarled the debate about selection processes at the species level. The combining of the interactor question and the manifestor-of-adaptation question (in the engineering sense) led to the rejection of research aimed at considering the role of species as interactors, *simpliciter*, in evolution. Once it is understood that species-level interactors may or may not possess design-type adaptations, it becomes possible to distinguish two research questions: "Do species function as interactors, playing an active and significant role in evolution by selection?" and "Does the evolution of species-level interactors produce species-level engineering adaptations and, if so, how often?".

For much of the history of the species selection debate, asking whether species could be "units of selection" meant asking whether they fulfilled *both* the interactor and the manifestor-of-adaptation roles. For example, Elisabeth Vrba used Maynard Smith's treatment of the evolution of altruism as a touchstone in her definition of species selection (1984). Vrba argued that the spread of altruism at the group level should not be considered a case of group selection because "there is no group adaptation involved; altruism is not emergent at the group level" (1984: 319). Vrba's view was that evolution by selection is not happening at a given level unless there is both an interactor and a benefit or engineering adaptation at that level (1983: 388). Niles Eldredge also merges selection and adaptation, rejecting certain cases as higher-level *selection processes* because "frequencies of the properties of lower-level individuals which are part of a high-level individual simply do not make convincing higher-level adaptations," as does Stephen Jay Gould (Eldredge 1985: 133–134, 196; Vrba & Eldredge 1984; Vrba & Gould 1986).

But consider the lineage-wide trait of variability. Treating species as interactors has a long tradition (Dobzhansky 1937; Thoday 1953; Lewontin 1958). Lineages able to endure environmental stresses would be selected for, and the trait of variability itself would be selected for; it would spread in the population of populations. In other words, lineages were treated as interactors. The theorists at the time were explicitly concerned with the effect on species-level characters such as speciation rates, lineage-level survival, and extinction rates of species. I argue that this sort of case represents a perfectly good form of species selection, even though Vrba and Eldredge would balk at the thought that variability would then be considered as an aggregate species-level property (Lloyd 1988; Lloyd & Gould 1993). Paleontologists David Jablonski and Gene Hunt (2006) used the Lloyd and Gould approach to species selection in their research on fossil gastropods, and the approach has also appeared in the leading text on speciation, by Jerry Coyne and H. Allen Orr (Coyne & Orr 2004). Vrba eventually recognized the advantages of keeping the interactor question separate from a requirement for an engineering-type adaptation (Vrba 1989), acknowledging variability as a lineage-level trait.

GENIC SELECTION: THE ORIGINATORS

One may understandably think that Dawkins is interested in the replicator question because he claims that the unit of selection ought to be the replicator. This would be a mistake. Dawkins does not specify how large a chunk of the genome he will allow as a replicator. He argues that if linkage disequilibrium is very strong, then the "effective replicator will be a very large chunk of DNA" (Dawkins 1982: 89). We can conclude from this that Dawkins is not interested in the replicator question at all; his framework can accommodate any of its possible answers. Rather, Dawkins is interested primarily in a specific ontological issue about benefit. His is a special version of the beneficiary question, and his answer to that question dictates his answers to the other three questions flying under the rubric of the "units of selection."

Briefly, Dawkins argues that because replicators are the only entities that "survive" the evolutionary process, they must be the beneficiaries. What happens in the process of evolution by natural selection happens *for their sake*, for their benefit. Hence, interactors interact for the replicators' benefit, and adaptations belong to the replicators. Replicators are the only entities with real agency as initiators of causal chains that lead to the phenotypes; hence, they accrue the credit and are the real units of selection. He argues that people who focus on interactors are laboring under a misunderstanding of evolutionary theory.

Dawkins believes that interactors, which he calls "vehicles," are not relevant to *the* units of selection problem. Organisms or groups as "vehicles" may be seen as the unit of function in the selection process, but they should not, he argues, be seen as the units of selection because the characteristics they acquire are not passed on (1982: 99). Here, he is following Williams's line that genotypes are destroyed in every generation by meiosis and recombination in sexually reproducing species; they are only temporary (Williams 1966: 109). Hence, replicators ("genes") are the only units that survive in the selection process. The replicator is the real unit because it is an "indivisible fragment," it is "potentially immortal" (Williams 1966: 23–24; Dawkins 1982: 97).

The second key aspect of Dawkins's views on interactors is that he seems to want to do away with them entirely. Dawkins is aware that the vehicle concept is "fundamental to the predominant orthodox approach to natural selection" (1982: 116). When Dawkins argues against "the vehicle concept," he is only arguing against the desirability of seeing the individual organism as the one and only possible vehicle: "Theoretical dangers attend the assumption that adaptations are for the good of . . . the individual organism" (1982: 91). His target is explicitly those who hold what he calls the "Central Theorem," which says that *individual organisms should be seen as maximizing their own inclusive fitness* (1982: 5, 55). Dawkins's arguments are indeed damaging to the Central theorem, but they are ineffective against other approaches that define units of selection as interactors.

Consider, for example, Dawkins's argument that groups should not be considered units of selection:

> To the extent that active germ-line replicators benefit from the survival of the group of individuals in which they sit, over and above the [effects of individual traits and altruism], we may expect to see adaptations for the preservation of the group. But all these adaptations will exist, fundamentally, through differential replicator survival. The basic beneficiary of any adaptation is the active germ-line replicator.   (1982: 85)

Notice that Dawkins begins by admitting that groups can function as interactors, and even that group selection may effectively produce group-level adaptations. The argument that groups should not be considered real units of selection amounts to the claim that the groups are not the ultimate beneficiaries. To counteract the intuition that the groups do, of course, benefit, in some sense, from the adaptations, Dawkins uses the terms "fundamentally" and "basic," thus signaling what he considers the most important level. Thus, the replicator is the unit of selection, because it is the beneficiary, and the real owner of all adaptations that exist.

Saying all this does not, however, address the fact that other researchers investigating group selection are asking the interactor question and sometimes also the manifestor-of-adaptation question. Dawkins gives no additional reason to reject these other questions as legitimate; he simply reasserts the superiority of his own preferred unit of selection. In sum, Dawkins has identified three criteria as necessary for something to be a unit of selection: it must be a replicator; it must be the most basic beneficiary of the selection process; and it is automatically the ultimate manifestor of adaptation through being the beneficiary. In the next section, we will consider some work in which genic selectionism is defended through a pluralist approach to modeling.

GENIC SELECTION: THE PLURALISTS

Dawkins had particular problems with his treatment of the interactor. While he admitted that the "vehicle" was necessary for the selection process, he did not want to accord it any weight in the units of selection debate because it was not the beneficiary. Starting with Ken Waters's work in 1986, though, there emerged a new angle available to genic selectionists, pursued by Sterelny and Kitcher (1988), Waters (1991), Kitcher and colleagues (1990).

Kim Sterelny and Philip Kitcher proposed, with the new "genic pluralism" in 1988, that there are two "images" of natural selection, one in which selection accounts are given in terms of a hierarchy of entities and their traits' environments, the other of which is given in terms of *genes* essentially acting as interactors, having properties that affect their abilities to leave copies of themselves (Kitcher et al. 1990, Sterelny 1996a, b; Waters 1991). The big payoff follows from the claim that hierarchical models or selection processes can be reformulated in terms of the genic level: "Once the possibility of many, equally adequate, representations of evolutionary processes has been recognized, philosophers and biologists can turn their attention to more serious projects than that of quibbling about the real unit of selection" (Kitcher et al. 1990: 161; Waters 1991, 2005). Moreover, these descriptions are claimed to be on equal ontological footing.[3]

The pluralists seem to be arguing against the utility of the notion of the interactor in studying the selection process. Echoing Dawkins, their idea is that the whole causal story can be told at the level of genes, and that no higher-level entities need be considered in order to have an accurate and complete explanation of the selection process (Kitcher et al. 1990). Thus, the functional claim of the pluralists is that anything that a hierarchical selection model can do, a genic selection model can do just as well. Let us recall what the interactor question in the units of selection debate amounts to: What levels of entities interact with their environments through their traits in such a way that it makes a difference to replicator success?

Take the classic account of the efficacy of group selection, which even Williams acknowledged was hierarchical selection. Lewontin and Dunn (1960; Lewontin 1962), in investigating the house mouse, found first that there was segregation distortion, a genic-level trait, in that over 80 percent of the sperm from mice heterozygous for the t-allele also carried the t-allele, whereas the expected rate would be 50 percent. They also found that male homozygotes (those with two t-alleles) tended to be sterile, a selective trait occurring at the organismic level. Finally, they also found a substantial effect of group extinction based on the fact that female mice would often find themselves in groups in which all males were sterile, and the group itself would therefore become extinct. This, then, is how a genuine and empirically robust three-level hierarchical model was developed.

Waters tells how to re-construct a genic model of the causes responsible for the frequency of the t-allele. He writes that we must first distinguish "genetic environments that are contained within female mice that are trapped in small populations with only sterile males from genetic environments that are not contained within such females. In effect, the interactions at the group level would be built in as a part of one kind of genetic environment" (1991: 563). In order to determine the invariant fitness parameter of a specific allele, let's call it "A," we would need to know what kind of environment it is in at the allelic level; for example, whether it is paired with a t-allele. Then we would need to know a further layer of the environment of A, such as what the sex is of the "organismic environment" it is in. If it is in a t-allele arrangement, and it is also in a male environment, the allelic fitness of A would be changed. Finally, we need to know the type of subpopulation or deme the A allele is in. Is it in a small deme with many t-alleles? Then it is more likely to become extinct.

Waters writes, thus, "What appears as a multiple level selection process (e.g., selection of the t-allele) to those who draw the conceptual divide [between environments] at the traditional level, appears to genic selectionists of Williams's style as several selection processes being carried out at the same level within different genetic environments" (1991: 571). The "same level" here means the "genic level," while the "genetic" environments include the nested environments, from the other allele at the locus, to whether the genotype is present in a male or female mouse, to the size and composition of the deme the mouse is in. This completes the sketch of the genic pluralist position. We now turn to its reception.

Within philosophy, the genic pluralist view has been widely disseminated, taught, and critiqued. Pragmatic critiques of genic pluralism primarily note that in any given selective scenario the genic perspective provides no information that is not also available from the hierarchical point of view (Glymour 1999; Shanahan 1997). This state of affairs is often taken as a sufficient reason to prefer whichever perspective is most useful at the time. The major critiques are based on the causal structure of selective episodes. While genic pluralism gets the "genetic book-keeping" (i.e., the input/output relations) correct, it does not accurately reflect the causal processes that bring about the result in question. This line of argument was first broached by William Wimsatt (1980, 1981; see Sober & Wilson 1998; Shanahan 1997; Stanford 2001; Glennan 2002). The weakness of this line of criticism is its inability to isolate a notion of cause that is both plausibly true of hierarchical but not genic level models. My own critique denies that genic selectionists have any distinct and coherent genic level causes at all (Lloyd 2005).

Genic pluralism presents alleles as independent causal entities, claiming that the availability of such models makes hierarchical selection models—and the ensuing debates about how to identify interactors in selection processes—moot. However, in each case of the causal allelic models, these models are directly and completely derived from precisely the hierarchical models the authors reject. Moreover, causal claims made on behalf of alleles are utterly dependent on hierarchically identified and established interactors *as causes*, thus undermining their claims that the units of selection (interactor) debates are mere "quibbles" and are irrelevant to the representation of selection processes. Like myself, Sahotra Sarkar has argued that, contrary to the claims of pluralists, cases of frequency-dependence, such as in heterosis and in game-theoretic models of selection, necessitate selection at higher than genic levels because the relevant properties of the entities at the genic level are only definable relative to higher levels of organization. Thus, they cannot be properly described as properties of alleles nor are they "even definable at the allelic level" (Sarkar 2008: 219).

Thus, despite the pluralists' repeated claims, we can see *from their own calculations and examples* that theirs are *derivative* models, and thus that their "genic level causes" are derivative from and dependent on higher-level causes. Their genic level models depend for their empirical, causal, and explanatory adequacy on entire mathematical structures taken from the hierarchical models and refashioned. Furthermore, Sterelny, Kitcher, and Waters all use the same methods for isolating relevant genic-level environments as others do for the traditional isolating of interactors (Waters 1991: 563; Lloyd 1988). Perversely, given where they have gotten all their information from, the genic pluralists propose doing away with interactors altogether. Are we to think that renaming changes the meta-physics of the situation?

The issue concerning renaming model structures is especially confusing in the genic pluralists' presentations, because they repeatedly rely on an assumption or intuition that, given an allelic state space, we are dealing with allelic causes. This last assumption is easily traced back to Williams's and Dawkins's views that alleles are the ultimate beneficiaries of any long-term selection process; thus, the genic pluralist argument rests substantially on a view regarding the superior importance of the beneficiary question, which has been clearly delineated from the interactor question, above.

There is a further complication with respect to the nature of the genic selection mod-els put forward by genic pluralists. These models function under the presupposition that they are at least mathematically equivalent to hierarchical models. This claim has largely depended on the work of Dugatkin and Reeve in establishing this equivalence (Dugatkin & Reeve 1994; Sterelny 1996b; Sober & Wilson 1998; Sterelny & Griffiths 1999; Kerr & Godfrey-Smith 2002; Waters 2005). However, recent work has indicated that this equivalence between genic and hierarchical models does not in fact hold (Lloyd et al. 2008).

HOLOBIONTS AS UNITS OF SELECTION

It has recently become clear that each "individual" human being is actually a commu-nity of organisms co-evolved together for mutual benefit. Our microbiota (the collec-tion of bacteria, viruses, and fungi living in our gut, mouth, and skin) is necessary for our survival and development, and our species is also often necessary in turn for the survival of the microbiota species. Bacterial symbionts help induce and sustain the immune system, T-cells, and B-cells (Lee & Mazmanian 2010; Round et al. 2010), as well as providing essential vitamins to the host human being. Gut bacteria are neces-sary for cognitive development (Sampson & Mazmanian 2015), for the development of blood vessels in the gut, lipid metabolism, detoxification of dangerous bacteria, viruses, and fungi, and regulation of colonic pH and intestinal permeability (Nicholson et al. 2012). Scott Gilbert, Eugene Rosenberg, and Ilana Zilber-Rosenberg appeal to the fact that the holobiont—the combination of the host mammal and its microbiota—functions as a unique biological entity anatomically, metabolically, immunologically and during development (Gilbert 2011). Similarly, Gilbert and colleagues emphasize that the holobiont is an "integrated community of species, [which] becomes a unit of nat-ural selection" (Gilbert et al. 2012: 334). This is, in essence, to claim that the holobiont can function as an interactor, since it has features that bind it together as an interac-tive whole, in such a way that it can interact in a natural selection process. So what ties the different species together to produce an interactor? According to John Dupré, who has pioneered philosophical thought about holobionts, it is the community's "functioning whole," that characterizes it as an evolutionary interactor, "objects between which natural selection selects" (Dupré 2012: 160; see Zilber-Rosenberg & Rosenberg 2008). Gilbert and colleagues describe this community as a "team" of consortia undergoing selection (Gilbert et al. forthcoming; see also Gilbert 2011). This puts a nice emphasis on their *common fate*, and also contrasts well with regular group selection (Wade 2007). Dupré and O'Malley's use of "collaborators" or "polygenomic consortia" has the advantage of encompassing both competition and cooperation (2013: 314).

Holobionts are also reproducers, where the host usually reproduces vertically and the microbiota reproduces either or both vertically and horizontally, a situation that has provoked some discussion among philosophers (Wade 2007; Godfrey-Smith 2009; Drown et al. 2013; Griesemer 2014; Booth 2014; Drown & Wade 2014; Theis et al. 2016). Holobionts' microbiota can reproduce outside the context of the original host organism, and Godfrey-Smith has declared that such holobionts are not "Darwinian populations" under his definition (2009), and therefore not units of selection (see Booth 2014). Contrast this approach with Griesemer's "reproducer" (2000a, 2014), which includes retroviruses excluded under Godfrey-Smith's account. Godfrey-Smith's exclusion rests on the merging of the interactor with the reproducer requirements, and, as such, will not hold sway over those who do not buy such a confounding. This is yet another case wherein distinguishing the interactor question from the replicator/reproducer question can do a world of good in sorting out evolutionary outcomes, as Dupré and Maureen O'Malley argue (2013; see Gilbert et al. 2012; Lloyd forthcoming).

Finally, holobionts can also be manifestors of adaptations, as in the case of the evolution of placental mammals, in the acquisition by horizontal gene transfer from a retrovirus of a crucial gene coding for the protein *syncytin* (Dupressoir et al. 2012). What syncytin does is allow fetuses to fuse to their mother's placenta, a role crucial to the evolution of placental mammals. Moreover, it seems that several retrovirally derived enhancers played critical roles in the formation of a key cell in the uterine wall, also crucial for maintaining pregnancy, enhancing the holobionts' role as manifestors of adaptation (Wagner et al. 2014).

CONCLUSION

It makes no sense to treat different answers as competitors if they are answering different questions. We have reviewed a framework of four questions with which the debates appearing under the rubric of "units of selection" can be classified and clarified. Dawkins, Hull, and Brandon separated the classic question about the level of selection or interaction (the interactor question) from the issue of how large a chunk of the genome functions as a replicating unit (the replicator question). The interactor question should also be separated from the question of which entity should be seen as acquiring adaptations as a result of the selection process (the manifestor of adaptation question). In addition, there is a crucial ambiguity in the meaning of "adaptation" that is routinely ignored in these debates: adaptation as a selection product and adaptation as an engineering design. Finally, we can distinguish the issue of the entity that ultimately benefits from the selection process (the beneficiary question) from the other three questions.

This set of distinctions has been used to analyze leading points of view about the units of selection and to clarify precisely the question or combination of questions with which each of the protagonists is concerned. There are many points in the debates in which misunderstandings may be avoided by a precise characterization of which of the units of selection questions, or which combination of them, is being addressed.

NOTES

1 Methods of relating interactor traits with genotypic or genic fitness include Arnold & Fristrup 1982; Damuth & Heisler 1988; Heisler & Damuth 1987; Lloyd 1988; Sober & Wilson 1998; Wade 1985, as well as Hamilton 1975; Lande & Arnold 1983; Li 1967; Price 1972; Uyenoyama & Feldman 1980; Wade & McCauley 1980; Wilson & Colwell 1981; Wimsatt 1981; see Lloyd 2012. See discussion of some of the model differences in Lloyd 1988 and Wade 2016.

2 Susan Oyama, Paul Griffiths, and Russell Gray have been leading thinkers in formulating a radical alternative to the interactor/replicator dichotomy known as Developmental Systems Theory (Oyama 1985; Griffiths & Gray 1997; Oyama et al. 2001). Here the evolving unit is understood to be the developing system as a whole, privileging neither the replicator nor the interactor.

3 Significantly, the authors help themselves to all the higher-level information. See Lloyd (1988, 2005), and Lloyd et al. (2008) for a discussion of structured population models and the consequences of parameter changes.

REFERENCES

Ariew, A. & Lewontin, R. 2004. "The confusions of fitness." *British Journal for the Philosophy of Science* 55: 347–363.

Arnold, A.J. & Fristrup, K. 1982. "The theory of evolution by natural selection: A hierarchical expansion." *Paleobiology* 8: 113–129.

Booth, A. 2014. "Symbiosis, selection, and individuality." *Biology and Philosophy* 29: 657–673.

Brandon, R. 1982. "The levels of selection." *Proceedings of the Philosophy of Science Association* 1: 315–323.

Brandon, R. 1985. "Adaptation explanations: Are adaptations for the good of replicators or interactors?" In D. Depew & B. Weber (eds.), *Evolution at a Crossroads: The New Biology and the New Philosophy of Science* (MIT Press) 81–96.

Brandon, R. 1990. *Adaptation and Environment* (Princeton University Press).

Burian, R. 1983. "Adaptation." In M. Grene (ed.), *Dimensions of Darwinism* (Cambridge University Press) 287–314.

Coyne, J. & Orr, H. 2004. *Speciation* (Sinauer).

Damuth, J. & Heisler, I. 1988. "Alternative formulations of multilevel selection." *Biology and Philosophy* 3: 407–430.

Darwin, C. [1859] 1964. *On the Origin of Species* (Harvard University Press).

Dawkins, R. 1982. *The Extended Phenotype* (Oxford University Press).

Dawkins, R. 2012. "The descent of Edward Wilson." *Prospect Magazine.* www.prospectmagazine.co.uk/magazine/edward-wilson-social-conquest-earth-evolutionary-errors-origin-species.

Dobzhansky, T. 1937. *Genetics and the Origin of Species* (Columbia University Press).

Drown, D. & Wade, M. 2014. "Runaway coevolution: Adaptation to heritable and nonhertiable environments." *Evolution* 68: 3039–3046.

Drown, D., Zee, P., Brandvain, Y., & Wade, M. 2013. "Evolution of transmission in obligate symbionts." *Evolutionary Ecology Research* 15: 43–59.

Dugatkin, L. & Reeve, H. 1994. "Behavioral ecology and levels of selection: Dissolving the group selection controversy." *Advances in the Study of Behavior* 23: 101–133.

Dupré, J. 2012. "Post genomic Darwinism." In J. Dupré (ed.), *Processes of Life: Essays in the Philosophy of Biology* (Oxford University Press) 143–160.

Dupré, J. & O'Malley, M. 2013. "Varieties of living things: Life at the intersection of lineage and metabolism." In S. Normandin & C. Wolfe (eds.), *Vitalism and the Scientific Image in Post-Enlightenment Life Science, 1800–2010* (Springer) 311–344.

Dupressoir, A., Lavialle, C., & Heidmann, T. 2012. "From ancestral infectious retroviruses to bona fide cellular genes: Role of the captured syncytins in placentation." *Placenta* 33: 663–671.

Eldredge, N. 1985. *Unfinished Synthesis: Biological Hierarchies and Modern Evolutionary Thought* (Oxford University Press).

Gilbert, S. 2011. "Symbionts as genetic sources of hereditable variation." In S. Gissis & E. Jablonka (eds.), *Transformations of Lamarckism: From Subtle Fluids to Molecular Biology* (MIT Press) 283–293.

Gilbert, S., Sapp, J., & Tauber, A. 2012. "A symbiotic view of life: We have never been individuals." *Quarterly Review of Biology* 87: 325–341.

Gilbert, S., Rosenberg, E., & Zilber-Rosenberg, I. forthcoming. "Holobiont with their hologenomes as levels of selection in evolution." In S. Gissis, E. Lamm, & A. Shavit (eds.), *Landscapes of Collectivity in the Life Sciences* (MIT Press).

Glennan, S. 2002. "Contextual unanimity and the units of selection problem." *Philosophy of Science* 69: 118–137.

Glymour, B. 1999. "Population level causation and a unified theory of natural selection." *Biology and Philosophy* 14: 521–536.

Godfrey-Smith, P. 2009. *Darwinian Populations and Natural Selection* (Oxford University Press).

Goodnight, C. & Stevens, L. 1997. "Experimental studies of group selection: What do they tell us about group selection in nature?" *American Naturalist* 150: S59–S79.

Gould, S.J. & Lewontin, R. 1979. "The spandrels of San Marco and the Panglossian paradigm: A critique of the adaptationist programme." *Proceedings of the Royal Society of London, Series B* 205: 581–598.

Griesemer, J. 2000a. "Development, culture, and the units of inheritance." *Philosophy of Science* 67: S348–S368.

Griesemer, J. 2000b. "Reproduction and the reduction of genetics." In P. Beurton, R. Falk, & H.-J. Rheinberger (eds.), *The Concept of the Gene in Development and Evolution: Historical and Epistemological Perspectives* (Cambridge University Press) 240–285.

Griesemer, J. 2000c. "The units of evolutionary transition." *Selection* 1: 240–285.

Griesemer, J. 2014. "Reproduction and the scaffolded development of hybrids." In L. Caporael, J. Griesemer, & W. Wimsatt (eds.), *Developing Scaffolds in Evolution. Culture, and Cognition* (MIT Press) 23–55.

Griffiths, P. & Gray, R. 1997. "Replicator II: Judgement day." *Biology and Philosophy* 12: 471–492.

Griffiths, P. & Stotz, K. 2013. *Genetics and Philosophy: An Introduction* (Cambridge University Press).

Haldane, J.B.S. 1932. *The Causes of Evolution* (Longmans, Green).

Hamilton, W.D. 1975. "Innate social aptitudes in man: An approach from evolutionary genetics." In R. Fox (ed.), *Biosocial Anthropology* (Wiley) 133–155.

Heisler, I. & Damuth, J. 1987. "A method for analyzing selection in hierarchically structured populations." *American Naturalist* 130: 582–602.

Hull, D. 1980. "Individuality and selection." *Annual Review of Ecology and Systematics* 11: 311–332.

Jablonski, D. & Hunt, G. 2006. "Larval ecology, geographic range, and species survivorship in Cretaceous mollusks: Organismic versus species-level explanations." *American Naturalist* 168: 556–564.

Kerr, B. & Godfrey-Smith, P. 2002. "Individualist and multi-level perspectives on selection in structured populations." *Biology and Philosophy* 17: 477–517.

Kitcher, P., Sterelny, K., & Waters, C. 1990. "The illusory riches of Sober's monism." *Journal of Philosophy* 87: 158–161.

Krimbas, C. 1984. "On adaptation, Neo-Darwinian, tautology, and population fitness." *Evolutionary Biology* 17: 1–57.

Lande, R. & Arnold, S. 1983. "The measurement of selection on correlated characters." *Evolution* 37: 1210–1227.

Lee, Y. & Mazmanian, S. 2010. "Has the microbiota played a critical role in the evolution of the adaptive immune system?" *Science* 330: 1768–1773.

Lewontin, R. 1958. "A general method for investigating the equilibrium of gene frequency in a population." *Genetics* 43: 421–433.

Lewontin, R. 1962. "Interdeme selection controlling a polymorphism in the house mouse." *American Naturalist* 96: 65–78.

Lewontin, R. 1970. "The units of selection." *Annual Review of Ecology and Systematics* 1: 1–18.

Lewontin, R. 1974. *The Genetic Basis of Evolutionary Change* (Columbia University Press).

Lewontin, R. 1978. "Adaptation." *Scientific American* 239: 156–169.

Lewontin, R. & Dunn, L. 1960. "The evolutionary dynamics of a polymorphism in the house mouse." *Genetics* 45: 705–722.

Li, C. 1967. "Fundamental theorem of natural selection." *Nature* 214: 505–506.

Lloyd, E.A. 1988. *The Structure and Confirmation of Evolutionary Theory* (Greenwood).

Lloyd, E.A. 1992. "Unit of selection." In E.F. Keller & E.A. Lloyd (eds.), *Keywords in Evolutionary Biology* (Harvard University Press) 334–340.

Lloyd, E.A. 2001. "Units and levels of selection: An anatomy of the units of selection debates." In R. Singh, C. Krimbas, D. Paul, & J. Beatty (eds.), *Thinking About Evolution: Vol. 2 – Historical, Philosophical, and Political Perspectives* (Cambridge University Press) 267–291.

Lloyd, E.A. 2005. "Why the gene will not return." *Philosophy of Science*, 72: 287–310.

Lloyd, E.A. 2008. "An open letter to Elliott Sober and David Sloan Wilson, regarding their book, *Unto Others: The Evolution and Psychology of Unselfish Behavior.*" In her *Science, Politics, and Evolution* (Cambridge University Press) 95–105.

Lloyd, E.A. 2012. "Units and levels of selection." In E. Zalta (ed.), *Stanford Encyclopedia of Philosophy.* www.plato.stanford.edu/entries/selection-units/

Lloyd, E.A. forthcoming. "Holobionts as units of selection." In S. Gissis, E. Lamm, & A. Shavit (eds.), *Landscapes of Collectivity in the Life Sciences* (MIT Press).

Lloyd, E.A. & Gould, S.J. 1993. "Species selection on variability." *Proceedings of the National Academy of Sciences USA* 90: 595–599.

Lloyd, E.A., Lewontin, R., & Feldman, M. 2008. "The generational cycle of state spaces and adequate genetical representation." *Philosophy of Science* 75: 140–156.

Maynard Smith, J. 1964. "Group selection and kin selection: A rejoinder." *Nature* 201: 1145–1147.

Maynard Smith, J. 1976. "Group selection." *Quarterly Review of Biology* 51: 277–283.

Maynard Smith, J. 1987. "How to model evolution." In J. Dupré (ed.), *The Latest on the Best: Essays on Evolution and Optimality* (MIT Press) 119–131.

Maynard Smith, J. 2001. "Reconciling Marx and Darwin." *Evolution* 55: 1496–1498.

Mitchell, S. 1987. "Competing units of selection? A case of symbiosis." *Philosophy of Science* 54: 351–367.

Nicholson, J., Holmes, J., Kinross, R., Burcelin, G., Gibson, W., Jia, S., & Pettersson S. 2012. "Host-gut microbiota metabolic interactions." *Science* 336: 1262–1267.

Nowak, M., Tarnita, C., & Wilson, E.O. 2010. "The evolution of eusociality." *Nature* 466: 1057–1062.

Oyama, S. 1985. *The Ontogeny of Information* (Cambridge University Press).

Oyama, S., Griffiths P., & Gray, R. 2001. *Cycles of Contingency: Developmental Systems and Evolution* (MIT Press).

Price, G. 1972. "Extension of covariance selection mathematics." *Annals of Human Genetics* 35: 485–490.

Round, J., O'Connell, R., & Mazmanian, S. 2010. "Coordination of tolerogenic immune responses by the commensal microbiota." *Journal of Autoimmunity* 34: J220–225.

Sampson, T. & Mazmanian, S. 2015. "Control of brain development, function, and behavior by the microbiome." *Cell Host Microbe* 17: 565–576.

Sarkar, S. 2008. "A note on frequency dependence and the levels/units of selection." *Biology and Philosophy* 23: 217–228.

Shanahan, T. 1997. "Pluralism, antirealism, and the units of selection." *Acta Biotheoretica* 45: 117–126.

Slatkin, M. 1972. "On treating the chromosome as the unit of selection." *Genetics* 72: 157–168.

Sober, E. 1984. *The Nature of Selection* (MIT Press).

Sober, E. & Wilson, D.S. 1998. *Unto Others: The Evolution and Psychology of Unselfish Behavior* (Harvard University Press).

Stanford, P. 2001. "The units of selection and the causal structure of the world." *Erkenntnis* 54: 215–233.

Sterelny, K. 1996a. "Explanatory pluralism in evolutionary biology." *Biology and Philosophy* 11: 193–214.

Sterelny, K. 1996b. "The return of the group." *Philosophy of Science* 63: 562–584.

Sterelny, K. & Griffiths, P. 1999. *Sex and Death: An Introduction to the Philosophy of Biology* (University of Chicago Press).

Sterelny, K. & P. Kitcher 1988. "The return of the gene." *Journal of Philosophy* 85: 339–61.

Theis, K., et al. 2016. "Getting the hologenome concept right: An ecoevolutionary framework for hosts and their microbiomes." *mSystems* 1: E00028-16. doi:10.1128/mSystems.00028-16.

Thoday, J. 1953. "Components of fitness." *Symposia of the Society for Experimental Biology* 7: 96–113.

Uyenoyama, M. & Feldman, M. 1980. "Evolution of altruism under group selection in large and small populations in fluctuating environments." *Theoretical Population Biology* 17: 380–414.

Vrba, E. 1983. "Macroevolutionary trends: New perspectives on the roles of adaptation and incidental effect." *Science* 221: 387–389.

Vrba, E. 1984. "What is species selection?" *Systematic Zoology* 33: 318–328.

Vrba, E. 1989. "Levels of selection and sorting with special reference to the species level." *Oxford Surveys in Evolutionary Biology* 6: 111–168.

Vrba, E. & Eldredge, N. 1984. "Individuals, hierarchies and processes: Towards a more complete evolutionary theory." *Paleobiology* 10: 146–171.

Vrba, E. & Gould, S. J. 1986. "The hierarchical expansion of sorting and selection: Sorting and selection cannot be equated." *Paleobiology* 12: 217–228.

Wade, M. 1977. "An experimental study of group selection." *Evolution* 31: 134–153.

Wade, M. 1985. "Soft selection, hard selection, kin selection, and group selection." *American Naturalist* 125: 61–73.

Wade, M. 2007. "The co-evolutionary genetics of ecological communities." *Nature Reviews Genetics* 8: 185–195.

Wade, M. 2016. *Adaptation in Metapopulations: The Role of Interactions in Evolution* (University of Chicago Press).

Wade, M. & McCauley, D. 1980. "Group selection: The phenotypic and genotypic differentiation of small populations." *Evolution* 34: 799–812.

Wagner, G., Kin, K., Muglia, L., & Pavlicev, M. 2014 "Evolution of mammalian pregnancy and the origin of the decidual stromal cell." *International Journal of Developmental Biology* 58: 117–126.

Waters, C. 1991. "Tempered realism about the force of selection." *Philosophy of Science* 58: 553–573.

Waters, C. 2005. "Why genic and multilevel selection theories are here to stay." *Philosophy of Science* 72: 311–333.

Williams, G. 1966. *Adaptation and Natural Selection* (Princeton University Press).

Williams, G. 1985. "A defense of reductionism in evolutionary biology." *Oxford Surveys in Evolutionary Biology* 2: 1–27.

Wilson, D. S. & Colwell, R. 1981. "Evolution of sex ratio in structured demes." *Evolution* 35: 882–897.

Wilson, D. S. & Wilson, E. O. 2007. "Rethinking the theoretical foundation of sociobiology." *Quarterly Review of Biology* 82: 327–348.

Wilson, E. O. 2005. "Kin selection as the key to altruism: Its rise and fall." *Social Research* 72: 159–166.

Wilson, E. O. & Holldobler, B. 2005. "Eusociality: Origin and consequences." *Proceedings of the National Academy of Sciences USA* 102: 13367–13371.

Wimsatt, W. 1980. "Reductionist research strategies and their biases in the units of selection controversy." In T. Nickles (ed.), *Scientific Discovery: Case Studies* (Reidel) 213–259.

Wimsatt, W. 1981. "Units of selection and the structure of the multi-level genome." *Proceedings of the Philosophy of Science Association* 2: 122–183.

Wright, S. 1931. "Evolution in Mendelian populations." *Genetics* 10: 97–159.

Wright, S. 1945. "Tempo and mode in evolution: A critical review." *Ecology* 26: 415–419.

Wright, S. 1980. "Genic and organismic selection." *Evolution* 34: 825–843.

Zilber-Rosenberg, I. & Rosenberg, E. 2008. "Role of microorganisms in the evolution of animals and plants: The hologenome theory of evolution." *FEMS Microbiology Review* 32: 723–735.

3

Adaptation, Multilevel Selection, and Organismality: A Clash of Perspectives

Ellen Clarke

INTRODUCTION

What is an adaptation? Although it is easy to suggest examples of adaptation—a bird's wing, an eye, a startle reflex—to define one is much more challenging. The status of adaptation always belongs to traits: structures or behaviors, parts of a phenotype. But what sort of trait? An adaptation is a trait that has a particular value or purpose. It is good at something, or good *for* something. Nowadays natural selection is implicated somehow in expanding upon these sentences. Natural selection explains adaptations, in that natural selection brings adaptations into existence. But little more can be said without entering into controversy.

Philosophical disputes about adaptation have tended to focus on just how important adaptation is. Adaptation*ism* is accused of exaggerating the prevalence of adaptations amongst the traits of organisms, and thus of exaggerating the role of natural selection in explaining the appearance of the living world. The anti-adaptationist Stephen Jay Gould famously complained about the telling of what he called "just-so stories," which characterize traits as perfectly fulfilling some concocted function (Gould 1978). Gould assigned most traits a much lowlier status: as imperfect; as lucky accidents; or as barely functioning salvage-jobs. Instead of elevating natural selection as an all-powerful designing force, anti-adaptationists see the living world as testament to a history of drift, of developmental constraints, and of brute luck. So the traditional problem of adaptation concerns, very broadly speaking, the extent to which the traits of organisms are shaped by natural selection, rather than by other, non-selective, causes.

I will be looking at an orthogonal dimension of the problem of adaptation: whether adapted traits must be considered always as possessed *by organisms*. Our default view used to be that adaptations are present at a fixed hierarchical level—the level whose units we refer to as "dog," "person," or "tree." In fact, the traditional picture assumed a cozy proximity between adaptations, organisms, and selection—they all occurred at the same level. However, beginning in the 1970s this view was revolutionized by the realization that these units have been constructed, in a process of "major evolutionary transition," out of smaller building blocks (Margulis 1970; Maynard Smith & Szathmáry 1995). Our best current science tells us that most of the living things we observe around us are serially compounded entities: nested aggregations of lower-level units which have been pushed through a major transition by natural selection. So even if it were the case that selection presently acts only at the level of dogs, it is impossible that it has always acted at just this one level, because this level has not always been around (Okasha 2001). The reality of major transitions therefore forces us to accept that selection can act at different levels of a compositional hierarchy. To put the same point another way, group selection is real, because "the organisms of today were the groups of past ages" (Wilson 2015: 29).

The transitions revolution has transformed the levels of selection debate. My question is: How does acceptance of multilevel selection oblige us to revise our ideas about adaptation? In order to extend the concept of adaptation to a hierarchical setting, we must relax the assumption that adaptations occur only at one fixed hierarchical level and determine whether they can be borne by the units at higher and lower compositional levels. I call this an additional dimension to the problem of adaptation because as well as settling the question whether selection is behind the emergence of some trait, as opposed to something like constraint, we are forced additionally to determine the hierarchical level at which selection is acting.

I focus on two pairs of authors, both of whom have, in light of major transitions, reconfigured the traditional relationship between bearing adaptations, being an organism, and being acted on by natural selection. But the pairs have arrived at differing configurations, and their differences have recently led them into explicit conflict with one another. Arbitrating this dispute is valuable because it illuminates the different ways in which the three core concepts of evolutionary biology can be situated with respect to one another. It is also interesting because the two pairs represent the state of the art of opposite poles in a long-standing dichotomy concerning hierarchical evolution and the levels of selection.

Andy Gardner and Alan Grafen are biologists in what I will call a "reductionist tradition," whose members have included Bill Hamilton, George Williams, Richard Dawkins, and Stuart West. The tradition includes "kin selection" or "inclusive fitness" theory (Hamilton 1964), the "Selfish Gene" tradition (Williams 1966; Dawkins 1982), and the formal darwinism project (Grafen 2007, 2014). All of these are reductionistic in so far as they reserve a privileged role for lower-level units, especially genes, although in recent versions the focus of agency is shifted up to the level of the common-sense organism (our "dog" and "tree"), rather than to genes.[1]

Opposing the reductionists are what I will call the "holists"—biologists and philosophers who defend the reality or significance of higher levels of selection. Sometimes called "group selectionists," sometimes "multilevel selectionists," this tradition can be traced through Stephen Jay Gould, Richard Lewontin, Rick Michod, and Samir Okasha, as well as my focal authors, Elliott Sober and David Sloan Wilson.

The two sides have tended to favor different ways of carving up natural selection. Reductionists carve selection up into the correlation between an individual's traits and its fitness, on the one hand, and a correlation between the individual's traits and the fitness of other individuals with which it share genes, on the other hand. This partition allows the reductionists to "capture how individuals can influence the transmission of their genes to future generations by influencing their own reproductive success or that of related individuals" (West & Gardner 2013: R577).

My "holists," by contrast, carve selection up into a competition between individuals within groups, and a competition between groups.[2] I call them "holists" because they prefer not to give priority to lower-level units such as genes or organisms. The holistic model is level-neutral, in that the members of the groups can be genes, or cells, or unicellular organisms, or multicellular organisms, and even colonies of multicellular organisms.[3] Like most dichotomies, this one is messy, and there has been much diversity and change within each side. Gardner and Grafen, and Sober and Wilson, express some of the most recent sophistications of each view.

The transitions revolution has transformed the levels of selection debate but, perhaps surprisingly, has not made it go away (Okasha 2001). In recent years there has been convergence on the view that reductionistic models (in other words, kin selection or inclusive fitness models) are mathematically equivalent to holistic models (group selection or multilevel models). For example, Gardner writes, "It is now increasingly understood that separation of natural selection into within-group and between-group components (the 'group-selection partition') is a perfectly appropriate alternative to the separation of natural selection into direct and indirect components (the 'kin-selection partition')" (2013: 105). Despite this, the two sides remain in opposition. The attacks launched by each side are interesting in their own right, but the analysis I offer also provides some reason to doubt the extent to which formal equivalence between two models can be understood as trivializing their differences.

TWO VIEWS ABOUT MULTILEVEL ADAPTATION

Let's begin with consensus. Both sides agree that adaptation stands in a very special relationship with selection, in that only selection can explain or give rise to adaptation. Adaptation *entails* that there has been selection, in philosophical speak. Both sides connect the notion of adaptation to that of purpose or function, and understand that function as being fixed by natural selection in some way.

Both sides allow that selection can occur at multiple levels, and assume that we should treat the special relation between selection and adaptation as level-specific, so that adaptation *at a level* implies selection *at that same level* (Sober & Wilson 2011: 463; Gardner 2013: 109).[4]

Both sides are content, furthermore, to understand Price's covariance approach as answering questions about the hierarchical level at which selection is acting (Price 1970, 1972; Okasha 2006). On this approach, selection acts at the individual/lower level only if there is covariance between the traits and fitnesses of particles within groups. Selection acts at the group/higher level only if there is covariance between the traits and fitness of whole groups.[5] The two levels of selection, in other words, correspond to different terms of Price's multilevel equation (Price 1972). This agreement is significant, in that it amounts to the reductionists agreeing to utilize the holists' manner of carving up selection into within-group and between-group terms. Of course, such agreement seems perfectly reasonable, on the assumption that the different methods of carving are mathematically equivalent anyway.

Distance emerges between the two sides in respect of exactly *how much* selection needs to occur at the focal level in order for traits at that level to qualify as adaptations. Gardner and Grafen provide an account that uses a formal optimization program to derive a claim about when we can expect adaptations to appear at a particular hierarchical level. They conclude that adaptations will appear at a level only when selection at the level below is completely suppressed. In practice, this occurs in two scenarios: if the parts of the unit at the focal level are genetic clones of one another, or if the unit at the focal level has evolved policing mechanisms which function to eliminate competition amongst its parts (Gardner & Grafen 2009: 660). Gardner codifies this supplement to Williams's Principle as "Maynard Smith's Principle: Adaptation of an entity at any level of biological organization requires the absence of selection within entities at that level" (Gardner 2013: 109).

Sober and Wilson also make the existence of adaptation at a level conditional on facts about how much selection takes place at the level below, but they say, "We think it is more useful to use 'group adaptation' to label traits that evolved because group selection dominated the selection process" (Sober & Wilson 2011: 465). In other words, they argue that more than 50 percent, or half, of the total selection force (the sum of both Price partitions) should act at the relevant level before we call the traits at that level "adaptations."

What is behind these different decisions? Everyone conceives of adaptations in teleological terms, as serving a purpose or function. Gardner and Grafen can be understood as motivated primarily to identify what this purpose is—to find the "maximand." On their account, all adaptations serve the same purpose, which is the maximization of inclusive fitness. In other words, an adaptation is a trait that is *good for* increasing the representation of the bearer's genes in future populations.

We might call this an "ultimate" function, in addition to the proximate functions carried out by particular traits. For example, a polar bear's white fur serves the function of retaining body heat, but it has only evolved to serve this function in so far as the retention of body heat has helped polar bear white fur genes to increase in frequency. Focusing on this ultimate function allows Gardner and Grafen to capture two aspects of adaptation: the fit between organisms and their environments, and the fit of the parts of an organism to one another. "Individual organisms appear contrived as if towards some purpose; a quality that is evident only because all the adaptations wielded by the individual appear contrived for the same purpose" (Gardner & Grafen 2009: 665). Gardner and Grafen run these two together, assuming that the organism can be well fitted to its environment only if its parts are well fitted to one another, in the same way that a pocket watch achieves its function of reliable time keeping only if each of its cogs and springs works in harmony with the others (Paley 1802; Gardner 2009, 2013).

It is in order to preserve this commonality of purpose that Gardner and Grafen argue that a trait is an adaptation only when its bearer is an exclusive level of selection, with no possibility of selection taking place at a lower level. A group trait might be rightly called a group adaptation, they think, but only under very special circumstances—in "scenarios where groups comprise genetically identical individuals, or where within-group competition is repressed" (Gardner & Grafen 2009). In fact, Gardner recognizes that real-life organisms will not meet this ideal and weakens the view to accommodate "negligible selection." What is crucial, he argues, is that there is not *significant* conflict between the parts of the unit, because only in the absence of conflicts can the unit evolve the harmonious common-purpose that they take to be characteristic of true adaptation (Gardner 2013: 111).

While Gardner and Grafen identify a single maximand which any selected population is always moving toward, Sober and Wilson's vision has conflict at its heart. They picture adaptations as able to serve *different* purposes, at different hierarchical levels. In choosing their favored 50 percent threshold, Sober and Wilson are motivated by cases in which there is oppositional selection acting at two hierarchical levels simultaneously. Altruism, on their definition, is a trait which is favored by between-group selection, but disfavored by within-group selection (Sober & Wilson 1999: 26; Sober & Wilson 2011: 463). They conceptualize altruism as pulled in different directions by two opposing forces. Altruism can spread, therefore, only if the between-group force is stronger than the within-group force.[6]

In summary, both sets of authors assume a one-way dependence between bearing adaptations at a level and being selected at that level. But they make different decisions about how much selection needs to act at a particular level before we can apply the special label "adaptation" to traits at that level. We will delve deeper into the conceptual distance between the two pictures by examining two conflicts between their authors.

TWO DISPUTES

Punch One: Gardner and Grafen Stand Accused of Contradiction

In their 2011 paper, Sober and Wilson argue that Gardner and Grafen hold a logically inconsistent set of commitments:

1. Williams's Principle: Adaptation at a level implies selection at that same level.
2. Maynard Smith's Principle: Adaptation at a level implies no selection at the lower level.
3. Reductionism: Adaptation always occurs at the level of the individual organism.
4. Groups can bear adaptations.

Are Gardner and Grafen guilty? It certainly seems that they need something like proposition 3 in order to maintain their commitment to inclusive fitness theory. They want to say that all adaptations are adaptations at the individual level—because they maximize the inclusive fitness of individuals—even while some rare phenomena, such as the waggle dance language of a honeybee colony (Gardner, pers. comm.), or the policing behaviors of the same (Gardner & Grafen 2009: 666), might properly be considered group-level adaptations *as well.*

However, the logical contradiction depends on the meaning of "selection at the lower level" or "selection of the individual" staying fixed. Sober and Wilson always use "lower-level selection" and "individual selection" to refer to within-group selection, as opposed to between-group selection.[7] When Gardner states Maynard Smith's Principle in 2013, he is careful to use this same meaning: "Adaptation of an entity at any level of biological organization requires the absence of selection within entities at that level" (Gardner 2013: 109).

However, there is a second way in which "individual selection" can be used: to refer to the global selection acting on individuals, ignoring population structure (Sober

2011: 223); this has been called "broad individualism".[8] This assumes a single-level measure of selection, which collapses the two components of the Price partition into a single term. Muddles often arise as a consequence of different authors using these terms differently.[9] It may be that Gardner and Grafen sometimes use this second meaning, although they certainly do not flag it as such. When Gardner and Grafen claim that adaptation always obtains at the level of the individual, perhaps what they mean is that individuals' traits are always produced by *global* individual selection. If so, then we seem to have two rather different senses of adaptation in play. Adaptation[multilevel] is a trait that is produced by selection *at a level*—one component of the Price partition. But Adaptation[global] is a trait that is produced by *global* individual selection.

If we assume that propositions 1 and 2 involve only Adaptation[multilevel], and we reinterpret Proposition 3 as saying that all Adaptations[multilevel] are simultaneously Adaptations[global], then the purported contradiction is removed. But the interpretive work here is rather radical. Why do Gardner and Grafen not say this?

It is the movement between these two notions that allows Gardner and Grafen to hold on to the reductionist claim that lower levels have some sort of evolutionary priority, but without having to deny group selection. What would Sober and Wilson make of my reinterpreted Proposition 3? They will agree that broad individualism correctly predicts the outcome of multilevel selection processes. A trait will increase in the global population only if on average its bearers are fitter than the bearers of the competing trait. But nobody will deny this—it is tautologous. Broad individualism, they would say, leaves out the interesting details. For example, Wilson defined "weak altruism" as the consequence of a trait which benefits the bearer as well as other members of the bearer's group (Wilson 1980). To call weak altruism an individual adaptation would miss the point that altruism enhances the absolute fitness of only those individuals that find themselves in groups of the right sort—groups with a high enough frequency of altruists. In groups with a low frequency of altruists, altruism is costly to absolute fitness. If we call their altruism an "individually adaptive trait," then we obscure all the important information about the mechanism in virtue of which altruists can prosper (Sober and Wilson call this the "averaging fallacy" (1999: 33)). So Sober and Wilson have no interest in Adaptation[global]—to them the outcome of global selection is mere book keeping, and tells us nothing about the relevant causal processes.

So does acceptance of multilevel selection oblige us to accept multilevel adaptation? While it appears that both pairs of authors say yes, in fact only Sober and Wilson present an account in which there is truly adaptation coexisting at multiple hierarchical levels. For in Gardner and Grafen's picture, there is Adaptation[multilevel] which is level-neutral, in that it can emerge at any level of biological hierarchy, but it can only ever exist at one level *at a time*. Adaptation[global], with which Adaptation[multilevel] does coexist, is a different notion.

Punch Two: Sober and Wilson Stand Accused of Paradox

Gardner and Grafen, for their part, accuse Sober and Wilson's position of paradox, because they give a particular kind of mechanism the status of both cause and consequence of adaptation (Gardner & Grafen 2009: 667). A "policing mechanism," both sides agree, is a mechanism possessed by some living unit which acts to police or suppress selective conflict within that unit. Germ-soma separation, bottleneck life cycles, and worker policing behaviors in social insects, are all examples of policing mechanisms, which determine the extent to which some unit is able to undergo within-unit selection (Clarke 2014). Policing mechanisms are implicated by both sides in the evolution of higher-level adaptations, because they control the all-important quantity of lower-level selection. But what does this imply about the ordering of events during a major transition?

In 1999 and 2011, Sober and Wilson say that some policing mechanisms are adaptations: "the mechanisms that currently limit within-individual selection are not a happy coincidence but are themselves adaptations that evolved by natural selection" (1999: 97). They give the example of a weakly altruistic punishing behavior, which benefits the group by discouraging cheaters, but imposes a cost on the individual. The cost to the individual entails that the trait cannot evolve by within-group selection, as a matter of definition. It can evolve only by between-group selection, they say, and must therefore, by Williams's Principle, be a group adaptation: "Superorganisms [they use Gardner and Grafen's term here, to mean 'adapted group'] are a possible product of the group selection process, not a precondition for the process" (2011: 466).

Gardner and Grafen claim that Sober and Wilson have things the wrong way round. Policing mechanisms cannot be understood as group adaptations, contra Sober and Wilson, because until they are in place there can be no higher-level adaptations: "The superorganism [i.e., the adapted group] comes into existence only after these mechanisms are already established" (Gardner & Grafen 2009: 667). But, of course, it is only on Gardner and Grafen's own definition of a group adaptation that policing is a necessary precondition.

Sober and Wilson require only majority group selection, and this they assume to be possible in the absence of policing mechanisms, in which case there is no problem in their supposing that adaptations such as policing mechanisms evolve in the absence of prior policing mechanisms. Furthermore, Sober and Wilson's picture of the temporal priority between policing mechanisms and adaptations contains an extra element. While Gardner and Grafen define higher-level organisms in terms of adaptations, Sober and Wilson assume that higher-level adaptation can occur in the absence of a higher-level organism. A *group* can bear adaptations, if there is sufficient group selection. And policing mechanisms are among the adaptations that a group can bear. In order to qualify as a (super)organism, on the other hand, the group must meet further, independent, criteria. The possession of policing mechanisms is necessary—but not sufficient. An adapted, policed group qualifies as an organism, according to Sober and Wilson, only if it meets the extra criterion of functional organization (Wilson & Sober 1989: 350).

As a consequence, our two pairs of authors imagine a different temporal ordering of key events in an evolutionary transition. Sober and Wilson assume that you start with groups of individuals.[10] They imagine some selection acting on the groups. If selection acting between the groups *outweighs* selection acting within the groups, then you can start to see the appearance of group-level adaptations. Amongst these might be policing mechanisms. And if the groups additionally begin to exhibit functional organization, then we can start to call them "organisms."

Let us assume Gardner and Grafen also start with groups of individuals and some selection acting on groups. They assume that the next available step is the appearance of policing mechanisms. After these are in place we can assume that selection acts exclusively between the groups, rather than within them, and so the groups qualify as superorganisms. Finally, as a last step, we can expect that selection acting on the superorganisms will begin to create higher-level adaptations.

These are different pictures, but neither is obviously internally inconsistent. On Sober and Wilson's picture, policing mechanisms are higher-level adaptations, but on Gardner and Grafen's they are not. Sober and Wilson's verdict is conditional on their assuming that a group can bear adaptations even if there is some selective conflict within the group. But this is a conceptual choice, and in no way commits them to logical error, as Gardner and Grafen accused. As long as one makes the right adjustments elsewhere in the overall puzzle, both of these views are viable.

Sober and Wilson say that the appearance of policing mechanisms has to be a "happy coincidence" for Gardner and Grafen, because, unlike them, they cannot appeal to higher-level selection to explain the evolution of policing (Sober & Wilson 1999: 97). Yet one option would be for Gardner and Grafen to appeal to within-group kin selection to explain policing. For example, kin selection theory expects worker insects to prefer egg-laying by the queen to egg-laying by their fellow workers, to whose offspring they would be less closely related. Kin selection acting *within* the group can therefore explain the evolution of egg-eating amongst worker insects (Foster & Ratnieks 2001; Bourke 2011[11]). This could give rise to what Birch called a "crane of coercion," in which within-colony conflict generates policing mechanisms that then suppress within-colony conflict (Birch 2012). Sober and Wilson, on the other hand, are committed to the possibility of between-group selection being dominant over within-group selection, even in the absence of policing mechanisms, so that higher-level adaptations (such as policing mechanisms) can occur.

Furthermore, Sober and Wilson have introduced a dislocation between organisms and adaptations, such that there can be adapted groups that are not organisms. This is drastic. Our intuitive sense that being adapted coincides with being an organism is so strong that it has led some authors to *define* organisms as bearers of adaptations (Gould & Lloyd 1999). And it is not clear that the dislocation is necessary, or what is achieved by it. The concept of the organism ends up as something of a spare part in Sober and Wilson's picture. They hold that some collectives come to be functionally organized, in addition to being adapted, but we are left with no explanation for why. They say that "When a group of organisms is functionally organized, its members coordinate their activities for a common purpose, just like the organs of an organism and the parts of a can opener" (Wilson 2015: 9). We will need some account from Sober and Wilson of what sort of common purpose they have in mind. We can assume they do not mean inclusive fitness maximization. An alternative would be to lean on Wright's selected effects account of function, on which purpose emerges as a consequence of a selection process (Wright 1973). The problem is this would imply that functional organization emerges as a consequence of a common selection pressure acting on all the parts of the organism. But in that case it looks like "functional organization" would end up playing the same role for Sober and Wilson as does "adaptation" for Gardner and Grafen.

We have seen that while neither side commits the simple logical mistakes they accuse each other of, each of them is committed to some further differentiating commitments, summarized in Table 3.1. The punches mostly missed their targets, although we saw that Gardner and Grafen duck one blow only by switching, unannounced, between two different notions of "adaptation" in order to reconcile their reductionism with multilevel selection. The price of Sober and Wilson's escape from paradox, on the other hand, is that they have severed the concept "adaptation" from that of "organism."

THE BIG PICTURE

We can now see that, in spite of claims for the formal equivalence of their views about how to model selection at multiple hierarchical levels, the focal pairs of authors hold a number of divergent commitments (detailed in Table 3.1). At this point it is fruitful to step back from the details of the criticisms made by each party against the other in order to examine the reasons each party arrived at the picture they defend.

Sober and Wilson's central motivation is to maintain space for conflict between adaptations at different levels: "There is a robust alternative to portraying natural selection as a process that always leads toward the optimization of individual fitness. Natural selection can also be understood as the net effect of opposing forces" (Wilson 2011).[12] They want to emphasize that oppositional levels of selection will often lead to evolutionary trade-offs, in which the optimal outcome for the individual differs from the optimal outcome for the group. It is paramount in their overall schema that our concept of "adaptation" should reflect the reality of such conflict.

For Gardner and Grafen, by contrast, all the details are in the service of maintaining the focus on selection as a maximizing or optimizing process. They maintain that if we allow talk of adaptations as being *bad* for individuals, then we lose the ability to account for the teleology, the purpose, apparent in the design of living things. So while Sober and Wilson hold that the adaptations at different levels can have *different* functions, and that this will sometimes bring them into conflict with one another, Gardner and Grafen insist that adaptations at different levels can only ever be harmonious with one another, in the service of one universal biological function.

In some ways these divergent motivations are symptoms of the traditional adaptationism disputes. On one side we have the adaptationists, focused on optimization, on the other we have those with a commitment to evolutionary outcomes that are intermediate, imperfect and conflicted. For Sober, the connection of adaptation to selection is an important semantic commitment—to call something an adaptation is to make a claim about its history, *rather than* about its current adaptive value (Sober 1984: 208). This distinction is required to allow room for traits to serve a particular function

Table 3.1 Comparison of Sober and Wilson's and Gardner and Grafen's commitments

Sober and Wilson's commitments	Gardner and Grafen's commitments
Adaptation at a level implies selection at that same level ("Williams's Principle")	Adaptation[multilevel] at a level implies selection at that same level ("Williams's Principle") Adaptation[global] implies global individual selection
Adaptation at a level implies less than 50 percent selection at the lower level	Adaptation[multilevel] at a level implies zero selection at the lower level ("Maynard Smith's Principle")
Adaptation sometimes occurs at higher levels in opposition to lower levels (Conflict assumption)	Adaptation[global] always occurs at the level of the individual (by which they mean the common sense "organism") (Reductionism)
Policing mechanisms are higher-level adaptations	Policing mechanisms are not higher-level adaptations
Non-organisms can bear adaptations	Adaptation[multilevel] at a level implies (super)organismality at that level.

without ever having been selected for that role—for Gould's spandrels, in other words (Gould & Lewontin 1979).

Yet, there is more going on than a simple resurgence of those old arguments about adaptationism. It would not be correct to accuse Gardner and Grafen of supposing that organismal traits are generally optimal, for example. Gardner and Grafen say we should expect natural selection to push systems *toward* optimality, not that we should expect systems to be optimal. Furthermore, both sides make use of optimality reasoning—in which we assume that some of the nearby trait space consists of less fit phenotypes. Sober and Wilson utilize this sort of reasoning in their counterfactual method, which asks whether the individual would be fitter if a selection component were eliminated to suppress a trade-off (Sober & Wilson 1999: 103–118). In fact, because Sober and Wilson set a lower bar for higher-level adaptations, their picture admits many *more* adaptations than does Gardner and Grafen's.

How much, if anything, turns on this confrontation of perspectives? We can shift between the perspectives without any loss in our ability to predict evolutionary change. It is a short step to disparaging the whole matter as involving only "rival metaphors for the very same evolutionary logic and [being] thus empirically empty" (Reeve & Keller 1999: 4). But we should be aware that even if the two perspectives are formally equivalent to one another, there are also clear differences between them. What sort of differences are they? How substantive are they? One effect, of switching from Sober and Wilson's description to Gardner and Grafen's, is that some adaptations will have to be redescribed as obtaining at a different hierarchical compositional level.[13] There is a sort of quantitative disagreement also, in that the holist's world will contain many more adaptations than the reductionist's, as a consequence of Sober and Wilson's lower threshold. We might, further, think there is some sort of empirical disagreement to be gotten out of Gardner and Grafen's assumption that complex design is possible only in the context of unified selection pressures, but given the inchoate nature of "design," as well as the caveats about how much lower-level selection qualifies as "negligible," we will not be able to construct tests to decide these matters.

There is always a danger of slipping from a local to a more global pluralism once different viewpoints are shown to be equally valid. It is important to distinguish changes in evolutionary *process* from changes in mere *perspective* (Sober & Wilson 1999: 57). It is important, furthermore, to appreciate that there are limits to perspectival pluralism. When considering a population comprising groups within each of which there is conflict, the inclusive fitness approach and the multilevel selection approach constitute different perspectives— different ways of thinking about the same thing. As a matter of *fact*, rather than perspective, this first system is different from one in which there is no conflict within groups. We are not at liberty to choose whether to treat the groups as if they are subject to conflict or not. But once we turn our attention to the latter sort of system—one in which groups are exclusive, unconflicted, levels of selection—a new choice of perspectives becomes available. When selection acts only at the group level, then we can choose to frameshift a standard single-level evolutionary model up to the group level. We can choose to treat the groups as if they are organisms.[14] This second switch of perspective, like the first, preserves any predictions about what will happen (evolutionarily speaking) in the system.

Perhaps, then, we can understand Gardner and Grafen as having provided formal justification for the latter sort of perspective-switch. When policing mechanisms are in place, what is good for the higher-level group is good for all of the lower-level units in the group too. So no matter which of these two levels we look at, we will make the same prediction about what will be favored. And this is a way to explain what is special about paradigm organisms. In effect we perform exactly this trick when we treat multicellular organisms as individuals. We do not bother counting up all the lower-level units—the cells. In so far as organisms such as pigs have policing mechanisms, it is possible for us to just ignore most of the millions of cells in the pig's body and count it once, without introducing any distortion into our picture of pig evolutionary dynamics. And so with the units even lower down—genes. We do not have to count all the copies of a particular allele—we just assume there is one in each cell. And this same trick works whenever some level is an exclusive level of selection: if *groups* of pigs were to evolve sufficient policing mechanisms to make them exclusive levels of selection, we could dispense with tracking the fitness of separate pigs and just plug the values for the whole group into a single-level model.

So we might understand Gardner and Grafen as having accomplished the explanatory feat of explaining what was in the background of Williams's thoughts: that certain sorts of organisms really are special, and it really is a neat trick to describe selection as acting on them. Nonetheless, we must not forget that not all living things are exclusive levels of selection. When selection acts at multiple levels, we cannot apply a single-level approach. We face a choice, instead, to use an inclusive fitness view[15] taken down to the highest level at which there is no lower-level selection, or to use a multilevel model.

CONCLUSIONS

Our traditional ideas about evolutionary adaptation have to be revised in light of the facts about multilevel selection and the major transitions. But the details of that revision depend on our favored perspective on multilevel selection. I reviewed the positions adopted by proponents of two rival perspectives. I hope to have shed some light on a rather opaque disagreement between these authors in Gardner and Grafen's 2009 paper and Sober and Wilson's 2011 reply. But in addition, the exercise has demonstrated that there is a complex set of relations between three concepts—*adaptation*, *organism*, and *selection*—that are pivotal to modern evolutionary biology. These constitute moveable pieces in the overall puzzle, so that each pair of authors achieves a picture that is internally consistent, but in order to do so they commit to various further claims that create distance between the two camps. Thus I also have shown that, despite its being widely accepted that the different perspectives are formally equivalent, there are concrete differences in the way they conceptualize the furniture of the biological realm.

Note that I have been arbitrating a rather limited class of disputes between the two sides here. I have not considered problems about how to distinguish between higher-level covariance that is caused by lower-level selection and that caused by higher-level selection (Okasha & Paternotte 2012). I have not considered contextual approaches to measuring group selection (Goodnight 2015), or neighbor-modulated approaches to social evolution (Frank 1998) which constitute third and fourth rival camps. There may be other problems that are fatal to the views assessed here.

There is a *decision* to be made regarding whether or not adaptation is conceptualized such that it can serve different purposes at different levels. There is no fact of the matter that can settle the choice for us. But once the choice is made, there are additional onto-logical details which we are obliged to accept, so that significant conceptual divergence emerges from the two formally equivalent perspectives.

ACKNOWLEDGMENTS

I am very grateful to Cedric Paternotte, Jonathan Birch, and the audience at "Individuals across the Sciences 2012" for helpful feedback.

NOTES

1. Sober and Wilson name this camp "individual-level functionalists" (Sober & Wilson 1999: 10).
2. Note that a third approach exists which has also been called "group selection," but which uses contextual analysis, sometimes known as multiple regression, to carve selection up into different components (Goodnight 2015). I do not discuss this approach here.
3. For example in the case of Argentine ant supercolonies (Giraud *et al.* 2002).
4. Both sides refer to this as "Williams's Principle" after the early adaptationist George Williams.
5. Where, importantly, the group values may simply be averages of the member values. This is important because it means that neither Gardner and Grafen nor Sober and Wilson require groups to be understood as undergoing differential extinction (Maynard Smith 1976) or giving rise to offspring groups (Godfrey-Smith 2009), in order to manifest adaptations.
6. Sober and Wilson's definition includes "weak altruism," where the altruists pay a cost to their relative fitness, not their absolute fitness. The opposing camp usually insists that only the "strong" form is real altruism, and that weak altruism is better understood as evolving thanks to mutual benefits (West *et al.* 2007). But it underlines the porosity of the dichotomy to point out that an argument similar to Sober and Wilson's was made by Richard Dawkins. He wrote that a genetically heterogeneous plant could be expected to evolve plant-level improvements only if "the inter-plant selection pressure [is sufficiently] strong to outweigh selection among cells within plants" (Dawkins 1982: 262).
7. It refers to one of the two partitions in the multilevel Price equation, in other words (Price 1972).
8. "Broad individualism" describes a view which models selection as acting globally on individuals (Sterelny & Griffiths 1999: 167).
9. Indeed, there is even a third use, which distinguishes higher- from lower-level selection according to contextual analysis, but I do not discuss that meaning here.
10. Defined in terms of fitness-affecting interactions between the members (Sober & Wilson 1999: 92).
11. With thanks to Jonathan Birch for this point. Note that, in any case, worker egg-eating could be selected purely for its direct nutritional benefits.
12. Note that while it has been the subject of recent criticism, this view of evolution as involving conflict between different forces enjoys continued support beyond this particular debate (Velasco & Hitchcock 2014).
13. For example, worker egg-eating changes from being a colony-level adaptation to an adaptation of certain individuals within the colony.
14. In other words, in the second case, both the inclusive fitness view and the multilevel view can be compressed—the statistical partitions collapsed—so that selection is described as the simple covariance between the traits and the fitnesses of units, *where the units may be either individuals or groups.*
15. It is worth stating that although Gardner and Grafen might call the inclusive fitness approach a single-level view, because it redescribes everything from a single level, I do not consider it to be a single-level view, because the indirect component of inclusive fitness separates effects on fitness which are concealed by the single-level Price equation—they are higher-level effects.

REFERENCES

Birch, J. 2012. "Collective action in the fraternal transitions." *Biology and Philosophy* 27: 363–380.

Bourke, A. 2011. *Principles of Social Evolution* (Oxford University Press).

Clarke, E. 2014. "Origins of evolutionary transitions." *Journal of Biosciences* 39: 303–317.

Dawkins, R. 1982. *The Extended Phenotype* (Oxford University Press).

Foster, K. & Ratnieks, F. 2001. "Convergent evolution of worker policing by egg eating in the honeybee and common wasp." *Proceedings of the Royal Society, Series B: Biological Sciences* 268: 169–174.

Frank, S. 1998. *Foundations of Social Evolution* (Princeton University Press).

Gardner, A. 2009. "Adaptation as organism design." *Biology Letters* 5: 861–864.

Gardner, A. 2013. "Adaptation of individuals and groups." In F. Bouchard & P. Huneman (eds.), *From Groups to Individuals* (MIT Press) 99–116.

Gardner, A. & Grafen, A. 2009. "Capturing the superorganism: A formal theory of group adaptation." *Journal of Evolutionary Biology* 22: 659–671.

Giraud, T., Pedersen, J., & Keller, L. 2002. "Evolution of supercolonies: The Argentine ants of southern Europe." *Proceedings of the National Academy of Sciences USA* 99: 6075–6079.

Godfrey-Smith, P. 2009. *Darwinian Populations and Natural Selection* (Oxford University Press).

Goodnight, C. 2015. "Multilevel selection theory and evidence: A critique of Gardner." *Journal of Evolutionary Biology* 28: 1734–1746.

Gould, S. J. 1978. "Sociobiology: The art of storytelling." *New Scientist* 80: 530–533.

Gould, S. J. & Lewontin, R. 1979. "The spandrels of San Marco and the Panglossian paradigm: A critique of the adaptationist program." *Proceedings of the Royal Society of London B: Biological Sciences* 205: 581–598.

Gould, S. J. & Lloyd, E. 1999. "Individuality and adaptation across levels of selection: How shall we name and generalize the unit of Darwinism?" *Proceedings of the National Academy of Sciences USA* 96: 11904–11909.

Grafen, A. 2007. "The formal Darwinism project: A mid-term report." *Journal of Evolutionary Biology* 20: 1243–1254.

Grafen, A. 2014. "The formal darwinism project in outline." *Biology and Philosophy* 29: 155–174.

Hamilton, W. D. 1964. "The genetical evolution of social behaviour I." *Journal of Theoretical Biology* 7: 1–16.

Margulis, L. 1970. *Origin of Eukaryotic Cells* (Yale University Press).

Maynard Smith, J. 1976. "Group selection." *Quarterly Review of Biology* 51: 277–283.

Maynard Smith, J. & Szathmáry, E. 1995. *The Major Transitions in Evolution* (Freeman).

Okasha, S. 2001. "Why won't the group selection controversy go away?" *British Journal for the Philosophy of Science* 52: 25–50.

Okasha, S. 2006. *Evolution and the Levels of Selection* (Oxford University Press).

Okasha, S. & Paternotte, C. 2012. "Group adaptation, formal Darwinism and contextual analysis." *Journal of Evolutionary Biology* 25: 1127–1139.

Paley, W. 1802. *Natural Theology* (Wilks & Taylor).

Price, G. 1970. "Selection and covariance." *Nature* 227: 520–521.

Price, G. 1972. "Extension of covariance selection mathematics." *Annals of Human Genetics* 35: 485–490.

Reeve, K. & Keller, L. 1999. "Levels of selection: Burying the units-of-selection debate and unearthing the crucial new issues." In L. Keller (ed.), *Levels of Selection in Evolution* (Princeton University Press) 3–14.

Sober, E. 1984. *The Nature of Selection: Evolutionary Theory in Philosophical Focus* (MIT Press).

Sober, E. 2011. "Realism, conventionalism, and causal decomposition in units of selection: Reflections on Samir Okasha's *Evolution and the Levels of Selection*." *Philosophy and Phenomenological Research* 82: 221–231.

Sober, E. & Wilson, D. S. 1999. *Unto Others: The Evolution and Psychology of Unselfish Behavior* (Harvard University Press).

Sober, E. & Wilson, D. S. 2011. "Adaptation and natural selection revisited." *Journal of Evolutionary Biology* 24: 462–468.

Sterelny, K. & Griffiths, P. 1999. *Sex and Death: An Introduction to the Philosophy of Biology* (University of Chicago Press).

Velasco, J. & Hitchcock, C. 2014. "Evolutionary and Newtonian forces." *Ergo* 1: 39–77.

West, S. & Gardner, A. 2013. "Adaptation and inclusive fitness." *Current Biology* 23: R577–R584.

West, S., Griffin, A., & Gardner, A. 2007. "Social semantics: Altruism, cooperation, mutualism, strong reciprocity and group selection." *Journal of Evolutionary Biology* 20: 415–432.

Williams, G. 1966. *Adaptation and Natural Selection* (Princeton University Press).

Wilson, D.S. 1980. *The Natural Selection of Populations and Communities* (Benjamin/Cummings).

Wilson, D.S. 2011. "Homage to George Williams and the last gasp of individualism IV: The last gasp." January 17. http://scienceblogs.com/evolution/2011/01/17/homage-to-george-williams-and-2/

Wilson, D.S. 2015. *Does Altruism Exist? Culture, Genes and the Welfare of Others* (Yale University Press).

Wilson, D.S. & Sober, E. 1989. "Reviving the superorganism." *Journal of Theoretical Biology* 136: 337–356.

Wright, L. 1973. "Functions." *Philosophical Review* 82: 139–168.

4

Fitness Maximization

Jonathan Birch

Adaptationist approaches in evolutionary ecology often take it for granted that natural selection maximizes fitness. Consider, for example, the following quotations from standard textbooks:

> The majority of analyses of life history evolution considered in this book are predicated on two assumptions: (1) natural selection maximizes some measure of fitness, and (2) there exist trade-offs that limit the set of possible [character] combinations.
> (Roff 1992: 393)

> The second assumption critical to behavioral ecology is that the behavior studied is adaptive, that is, that natural selection maximizes fitness within the constraints that may be acting on the animal.
> (Dodson et al. 1998: 204)

> Individuals should be designed by natural selection to maximize their fitness. This idea can be used as a basis to formulate optimality models.
> (Davies et al. 2012: 81)

Yet there is a long history of skepticism about this idea in population genetics. As A.W.F. Edwards puts it:

> [A] naive description of evolution [by natural selection] as a process that tends to increase fitness is misleading in general, and hill-climbing metaphors are too crude to encompass the complexities of Mendelian segregation and other biological phenomena.
> (Edwards 2007: 341)

Is there any way to reconcile the adaptationist's image of natural selection as an engine of optimality with the more complex image of its dynamics we get from population genetics? This has long been an important strand in the controversy surrounding adaptationism.[1] Yet debate here has been hampered by a tendency to conflate various different ways of thinking about maximization and what it entails. In this chapter I distinguish, at a deliberately coarse grain of analysis, four varieties of maximization principle.[2] I then discuss the logical relations between these varieties, arguing that, although they may seem similar at face value, none entails any of the others. I then turn briefly to the status of each variety, arguing that, while each type of maximization principle faces serious problems, the problems are subtly different for each type.

In the last section, I reflect on what is at stake in this debate. Defenders of fitness maximization are often motivated by a desire to defend adaptationist, optimality-based approaches in evolutionary ecology of the sort described in the quotations at the start of this chapter. I argue, however, that the value of optimality-based approaches as tools for hypothesis generation does not depend on the existence of a universal maximization principle describing the action of natural selection. The need for such a principle arises only for those who hold a more epistemically ambitious view about what these approaches can achieve.

FOUR VARIETIES OF MAXIMIZATION

Any maximization principle, to be worthy of the name, must spell out what is meant by a fitness maximum, and must assign a special status to such a point in the dynamics of evolution by natural selection. This, however, leaves many options open regarding the nature of the maximum and its significance in the dynamics. We should not be surprised, then, to find many quite different fitness maximization principles in evolutionary biology.

I suggest that two distinctions lead to a useful taxonomy of such principles. First, we should distinguish between maximization principles that concern *what happens at equilibrium* and those that concern *the direction of change*. Second, we should distinguish between maximization principles that concern the *population mean fitness* and those that concern the *behavioral strategies of individual organisms*.

As a preliminary, I want to introduce Sewall Wright's (1932) adaptive landscape metaphor, to which Edwards alludes in the above quotation. This controversial metaphor looms large in debates about fitness maximization. Wright imagined the mean fitness of a population moving through a multidimensional gene frequency space.[3] Flattening this space to three dimensions for ease of visualization, he pictured a landscape characterized by "adaptive peaks" representing mean fitness maxima, and he pictured evolution by natural selection as a "hill-climbing" process that drives a population toward the nearest maximum. In this vision of evolution, natural selection sometimes drives populations to the highest peak (the global maximum) but it may also cause populations to become marooned on local maxima, separated from the global maximum by fitness valleys.

The adaptive landscape metaphor combines two seductive ideas about the dynamics of evolution by natural selection: an idea about equilibrium and an idea about change. First, it pictures the stationary points of evolution by natural selection as points at which mean fitness is maximized, such that any change in the frequency of any allele will decrease mean fitness. Second, it pictures a population out of population-genetic equilibrium as moving reliably upward, in the direction of greater mean fitness.

These two claims are conceptually distinct. To help us keep these ideas separate, let us denote them with the labels "MAX-A" and "MAX-B":

> **MAX-A (*Mean fitness, equilibrium*):** A population undergoing evolution by natural selection is at a stable population-genetic equilibrium if and only if its mean fitness is maximized, such that any change in allele frequencies will reduce mean fitness.

> **MAX-B (*Mean fitness, change*):** If a population is not in population-genetic equilibrium, then natural selection will reliably change allele frequencies in a way that leads to greater mean fitness, even if other factors prevent the population from reaching a maximum.

In both MAX-A and MAX-B, the variable that is maximized is the population mean, averaged over genotypes or over individuals, of some fitness measure. In this sense, MAX-A and MAX-B are population-centered: they focus on the properties and dynamics of populations, making no explicit reference to the properties of individuals in those populations. But this is not the only way to think about fitness maximization. Behavioral ecologists commonly start with the assumption that an individual organism will behave as if attempting to maximize its own individual fitness or (in the case of social behavior) its inclusive fitness. They then ask: which strategy, from the range of feasible options, would it be rational for the organism to adopt, given its apparent goal?

We can say (following Alan Grafen) that behavioral ecologists who think in this way are employing an "individual as maximizing agent" analogy (Grafen 1984, 1999). Agential thinking of this sort is widespread in many areas of evolutionary ecology, including inclusive fitness theory, life history theory, and evolutionary game theory (e.g., Maynard Smith 1982; Parker & Smith 1990; Davies et al. 2012). The analogy does not involve any literal attribution of rational agency to non-human organisms. Instead, the thought is that organisms, regardless of their degree of cognitive sophistication, can be modeled *as if* they were rational agents attempting to maximize their individual fitness (or inclusive fitness), because natural selection tends to lead to equilibria at which organisms adopt strategies that maximize their individual fitness (or inclusive fitness) within the set of feasible options. This leads to a third conception of fitness maximization:

> **MAX-C (*Individual fitness, equilibrium*):** A population undergoing evolution by natural selection is at a stable population-genetic equilibrium if and only if all organisms adopt the phenotype that maximizes their individual fitness (or inclusive fitness) within the set of biologically feasible phenotypic options.

This notion of maximization clearly bears some resemblance to MAX-A, in that it posits a close relationship between population-genetic equilibria and fitness maxima, but it differs in that it defines these maxima not in terms of the mean fitness of the population, but rather in terms of optimal strategy choice, within the set of biologically feasible options, on the part of individual organisms (the reference to "biologically feasible options" makes it clear that we are talking here about optimization *subject to constraints*, not unconstrained maximization). Despite the superficial similarities, this way of thinking about maximization has little to do with Wright's adaptive landscape metaphor. It is much closer to the notion of maximization which appears in economics, in which humans are typically modeled as rational agents maximizing utility subject to constraints.

MAX-C, like MAX-A, is a claim about what happens at equilibrium. However, the equilibrium/change distinction cross-cuts the mean fitness/individual fitness distinction. This leads to our fourth variety, an individual-level analogue of MAX-B concerning the direction of change:

> **MAX-D (*Individual fitness, change*)**: If a population is not in population-genetic equilibrium, then natural selection will reliably drive it in the direction of a point at which all organisms adopt the phenotype that maximizes their individual fitness (or inclusive fitness) within the set of biologically feasible phenotypic options, even if other factors prevent the population from reaching this point.

RELATIONS BETWEEN THE VARIETIES

We now have four varieties of fitness maximization on the table (Table 4.1). I claim that none of them entails any of others. I will defend this claim piecemeal, looking first at the rows in Table 4.1 and then at the columns. I assume that if there is no entailment along the rows or the columns, then there is no serious prospect of entailment across the diagonals.

The first non-entailment I want to consider concerns the first row. The key points here can be expressed in terms of the adaptive landscape metaphor. In principle, it might be that adaptive peaks are always stationary points and yet selection might be ineffectual at driving populations up slopes toward them. Conversely, selection might drive populations reliably upward whenever they are out of equilibrium, and yet the population might stably stop at least some of the time at points that are not peaks. Hence MAX-A does not entail MAX-B, or vice versa.

The broader point here is that claims about what happens at equilibrium do not entail claims about the direction of out-of-equilibrium change, or vice versa. This carries over to the second row. In principle, it might be that a stable stationary point in the dynamics of evolution by natural selection occurs if and only if all organisms in the population have optimal phenotypes, and yet natural selection is ineffectual at driving populations

Table 4.1 Four varieties of fitness-maximization

	Equilibrium	Change
Mean fitness	**MAX-A**	**MAX-B**
Individual fitness	**MAX-C**	**MAX-D**

toward such optima. Conversely, selection might reliably drive populations toward such optima, only to reach a stable stationary point part way there. Hence MAX-C does not entail MAX-D, nor vice versa.

The columns are a little more subtle. MAX-A does not entail MAX-C, or vice versa, because there can be mean fitness maxima (in allele frequency space) at which suboptimal phenotypes are present in the population. MAX-A says that these points constitute stable population-genetic equilibria, whereas MAX-C says they do not. Consider, for example, the polymorphic equilibrium in the standard model of heterozygote advantage, illustrated by the famous case of sickle-cell anaemia and malarial resistance. In regions with a high incidence of malaria, an allele that causes sickle-cell anaemia in the homozygote (i.e., the genotype with two copies of the allele) is nonetheless present at a low frequency at equilibrium because it causes malarial resistance in the heterozygote (i.e., the genotype with one copy). In the standard model of this situation, the equilibrium is a mean fitness maximum—any change in allele frequencies lowers the mean fitness— but it is not a point at which every organism has an optimal phenotype within the range of feasible options (Hedrick 2011).

This suggests that the relationship between MAX-A and MAX-C, far from being one of logical entailment, is actually one of logical incompatibility: they imply contradictory claims about the status of mean fitness maxima at which suboptimal phenotypes are present. However, MAX-A and MAX-C can be made compatible if interpreted as claims about different evolutionary timescales. MAX-A-type maximization principles have usually been studied and discussed in the context of models of short-term "microevolution," such as the heterozygote advantage model discussed above. Yet when applying the "individual as maximizing agent" analogy, evolutionary ecologists often have a longer timescale in mind: the timescale of what Hans Metz (2011) and Peter Godfrey-Smith (2012) have called "mesoevolution." The idea here is that we should think of the attainment of phenotypic optimality as occurring over a timescale long enough for populations to escape short-term equilibria, such as the sickle-cell equilibrium, at which suboptimal phenotypes may be present. This move is central to Peter Hammerstein's (1996) "streetcar theory," which I consider below. For now, I simply want to note that MAX-A, read as a claim about the equilibria of short-term microevolution, is logically independent of MAX-C, read as a claim about the "mesoevolutionary" long run.

The broader point here is that there is a logical gap between claims about short-term changes in gene frequency and claims about longer-term phenotypic evolution (cf. Wilkins & Godfrey-Smith 2009). This carries over to the second column. Read as claims about the direction of short-term change, MAX-B and MAX-D seem to disagree about what will happen in cases in which a population stands to increase its mean fitness by reducing the frequency of an optimal phenotype. We see this in the sickle-cell model, in which an initially high frequency of malarial resistance is reduced by selection, owing to the adverse fitness consequences of the same gene in the homozygote. Mean fitness increases, but there is no convergence on universal malarial resistance.

As with MAX-A and MAX-C, however, thinking about timescales can help remove this apparent tension. We can read MAX-D as the claim that *over the long term* the dynamics of a population evolving by natural selection will converge on a point at which the population realizes an optimal phenotypic profile. This claim about long-term convergence is logically independent of MAX-B, read as a claim about the short-term direction of change. It is compatible with selection reliably driving a population in the direction of greater mean fitness in the short term, even if this sometimes means driving it away from phenotypic optimality, provided the population converges on phenotypic optimality in the long run. It is also compatible with the direction of short-term change in mean fitness being highly variable and context-dependent.

STATUS OF THE VARIETIES: MAX-A AND MAX-B

All four varieties of fitness maximization are controversial, but for different reasons. Let us start with MAX-A and MAX-B. While these may look innocuous to biologists trained to think of evolution in terms of adaptive landscapes, they are contentious in population genetics (Ewens 2004; Edwards 2007). MAX-A is challenged by models in which evolution stops at a point that, on any reasonable measure of fitness, is not a mean fitness maximum, even though natural selection is the only evolutionary process at work. Meanwhile, MAX-B is challenged by models in which, on any reasonable measure of fitness, natural selection drives the mean fitness of a population downwards over time.

Models of both sorts have a long history in population genetics. In one-locus models that satisfy various other assumptions (random mating, frequency-independent fitness, selection on viability differences only), the mean fitness does reliably increase and stable equilibria do correspond to mean fitness maxima (Scheuer & Mandel 1959; Mulholland & Smith 1959; Edwards 2000). But relax any of the assumptions of these models and the result is no longer valid. A standard citation in this context is P.A.P. Moran's 1964 article, wherein he constructed a two-locus model in which mean fitness decreases over time, and in which population-genetic equilibrium occurs far from any "adaptive peak." Moran took this result to debunk the very idea of an "adaptive topography." Warren Ewens (1968) and Samuel Karlin (1975) reinforced Moran's conclusions with further results along similar lines. The overall message of this work is that both MAX-A and MAX-B are extremely dubious in the multilocus case (see also Hammerstein 1996; Eshel et al. 1998; Ewens 2004).

Intuitively, the source of the trouble in multilocus models is that Mendelian segregation, recombination, and epistasis complicate the transmission of fitness between parents and offspring. Offspring, while resembling their parents on the whole, inherit a combination of genes that is not a simple replica of either parent. Consequently, a gene that promotes the fitness of a parent can, on finding itself in a new genomic context, detract from the fitness of the offspring by whom it is inherited, with adverse consequences for the population mean fitness. Unfortunately, natural selection only "sees" whether current bearers of an allele are fitter, on average, than non-bearers; it does not "see" what the mean population fitness will be after the vagaries of Mendelian inheritance have taken their course.

In the models referenced above, the fitness of a genotype is assumed to be independent of population gene frequencies. Matters are even worse for mean fitness maximization when we introduce frequency-dependent genotypic fitness. Here, the intuitive problem is that frequency-dependence makes it possible for an allele to be selected even when an increase in its frequency would, via knock-on effects on genotypic fitness values in the next generation, detract from the mean fitness of the population. The moral of over fifty years of work in this area is that, when genotypic fitness depends on gene frequency, the mean fitness does not reliably increase and is rarely maximized at equilibrium. Indeed, in an early treatment of frequency-dependence, Jerome Sacks (1967) showed that frequency-dependent selection can lead to a stable equilibrium that is also a fitness minimum. This point has been underlined by recent work in the field of adaptive dynamics, which suggests an important role for fitness minimization in evolution. The idea is that mean fitness minima act as "evolutionary branching points" at which a population fragments, causing different subpopulations to pursue divergent evolutionary trajectories (Geritz et al. 1998; Doebeli & Dieckmann 2000; Doebeli 2011).

To be clear, the problem these models pose for MAX-A is not simply that the population stops at a local maximum rather than finding its way to the global maximum. The problem is that the population stops at a point that is *not a maximum at all*, whether local or global. If we insist on employing the "adaptive landscape" metaphor in such cases, we should say that the stopping point lies on a "slope" or in a "valley" rather than on a "peak." Likewise, note that the problem these models pose for MAX-B is not simply that the "uphill push" of natural selection is counteracted by other causes of gene frequency change. The problem is that, even when there is no cause of gene frequency change other than natural selection, the mean fitness still decreases.

FISHER'S FUNDAMENTAL THEOREM

From Wright onwards, defenders of MAX-B have often cited R.A. Fisher's fundamental theorem of natural selection (Fisher 1930, 1941) in support of their claims, even though Fisher himself never regarded the theorem as a maximization principle (Edwards 1994). The theorem states that the rate of change in the mean fitness in a population "ascribable to a change in gene frequency" is equal to the additive genetic variance in fitness. Although there has long been uncertainty over its mathematical validity, later reconstructions show clearly that it is a correct result, given a particular interpretation of what Fisher meant by the rate of change "ascribable to a change in gene frequency" (Price 1972; Ewens 1989; Lessard 1997). Since variance cannot be negative, the theorem seems at first glance to imply that the rate of change in mean fitness cannot be negative either, apparently contradicting the results Moran and others have obtained in specific models.

A lot depends, however, on what is packed into Fisher's rather obscure concept of a rate of change "ascribable to a change in gene frequency." In informal terms, the quantity that Fisher proved can never be negative is a quantity that captures what the total rate of change in mean fitness *would* be, *if* we could hold the average effects of alleles on fitness at their current values as natural selection changes their frequencies. The trouble is that, except in cases of perfectly additive genetics (no dominance, epistasis or linkage), the average effects of alleles depend on genotype frequencies, and therefore on allele frequencies, and therefore on the action of natural selection. So as natural selection changes allele frequencies, it changes the average effects of alleles, creating a gap between the total rate of change in mean fitness and the "partial" rate of change with which Fisher's fundamental theorem is concerned.

There is in fact no theoretical guarantee that the total rate of change in mean fitness will be non-negative. To use a potentially misleading metaphor, the picture we get from the fundamental theorem, when we interpret it correctly, is of natural selection pushing the population "uphill" with one hand while it reshapes the landscape with the other. The total action of natural selection may leave the population higher, lower or at the same level, depending on the details. Of course, as Moran (1964) pointed out, this arguably casts doubt on the utility of the adaptive landscape metaphor.[4]

STATUS OF THE VARIETIES: MAX-C AND MAX-D

MAX-C-type maximization principles, which switch the focus from the population mean fitness to individual phenotypes and their fitness consequences, have two main cards up their sleeve to help them deal with the traditional problem cases for MAX-A and MAX-B. First, in cases of strategic interaction, an equilibrium that is not a mean fitness maximum can still be reconciled with MAX-C, as long as it is a Nash equilibrium. For, at a Nash equilibrium, organisms are best-response maximizers: they adopt the phenotype (or a phenotype, in cases of weak Nash equilibrium) that is fitness-optimal conditional on the phenotypes of their social partners. This is true even if the Nash equilibrium is a mean fitness minimum.

Second, polymorphic equilibria in which one of the phenotypes present is clearly sub-optimal, such as the sickle-cell equilibrium, can be reconciled with MAX-C provided MAX-C is understood as a claim about the stable equilibria of long-term phenotypic evolution, not the stable equilibria of short-term gene frequency change. The key here is to adopt a particularly demanding conception of stability when defining a stable equilibrium of long-term phenotypic evolution, so that sickle-cell-type polymorphic equilibria do not qualify as stable. Crucially, the sickle-cell equilibrium is vulnerable to invasion by a mutant that produces malarial resistance in the heterozygote without producing sickle-cell anaemia in the homozygote. So if we define stability in terms of resistance to invasion in the long run, this equilibrium may not be stable after all.

Peter Hammerstein's (1996) "streetcar theory" has been particularly influential in this context (see also Eshel & Feldman 1984, 2001; Liberman 1988; Hammerstein & Selten 1994; Eshel et al. 1998; Hammerstein 2012). On Hammerstein's picture, "an evolving population resembles a streetcar in the sense that it may reach several temporary stops that depend strongly on genetic detail before it reaches a final stop which has higher stability properties and is mainly determined by selective forces at the phenotypic level" (Hammerstein 1996: 512). The "final stop," he argues, will be a Nash equilibrium. Hence we arrive at a tenable version of MAX-C, provided we interpret "stable" equilibria as only those which correspond to Hammerstein's "final stops" achieved in the evolutionary long run, as opposed to the "stops along the way" described by standard microevolutionary theory.[5]

Hammerstein's argument, however, does not establish (or attempt to establish) MAX-D: it characterizes a special sort of long-term stable equilibrium and shows that it corresponds to a fitness maximum in a certain sense, but it does not give us a reason to think that a population evolving by natural selection will reliably converge toward such a point. This is ultimately an empirical matter, because it depends on the rate of mutation and the rate at which the selective environment changes. As Ilan Eshel and Marcus Feldman note, arguments of this general sort predict optimal outcomes in the long run only if "the regime of selection acting on the trait under study remains invariant during the slow process of transitions between genetic [i.e. short-term] equilibria" (Eshel & Feldman 2001: 186). By contrast, "for shorter-lived processes of conflict (e.g., in a newly colonized niche) we expect the population to be close to a short-term stable equilibrium, but not to one that is long-term stable" (Eshel & Feldman 2001: 186).

In light of this, it also seems clear that the streetcar theory, although it does provide support for a specific, long-term version of MAX-C, does not support the idea that natural selection has any tendency to maximize fitness in the absence of other causes of gene frequency change. On the contrary, the argument concedes that natural selection often will not be able to do so unless another cause of gene frequency change, that is, mutation, is powerful enough to circumvent genetic barriers to optimality (cf. Sober 1987). So, to the extent that the streetcar theory supports a version of adaptationism, it is a version that recognizes the importance of both mutation and selection in determining evolutionary outcomes.

FORMAL DARWINISM

This is where Alan Grafen's ongoing "Formal Darwinism" project enters the scene (Grafen 2002, 2006, 2007, 2014). Grafen aims to show that, even in models in which we assume the absence of mutation,[6] there are strong formal links between population genetic equilibrium and phenotypic optimality, where the optimal phenotype is defined as that which maximizes inclusive fitness within a set (X) of specified alternative options.

The assuming away of mutation in Grafen's models marks one important difference with Hammerstein's project. The other notable difference is that Grafen's formal links concern the direction of short-term change as well as the nature of long-term equilibrium. In broad terms, what Grafen has shown is that, across a wide range of (mutation-free) models, a population is at a point at which there is no "scope for selection" (roughly, no expected change in any gene frequency) and no "potential for positive selection" (roughly, no phenotype in X that is selected-for or that would be selected-for if present) if and only if all organisms have the optimal phenotype in X. He also proves links (which I will not discuss here) concerning changes in gene frequency in populations in which some or all individuals are suboptimal. Grafen (2014: 166) glosses these results as showing that "there is a very general expectation of something close to fitness maximization, which will convert into fitness maximization unless there are particular kinds of circumstances."[7]

I have criticized the Formal Darwinism project on other occasions, and I cannot do justice to this complex topic here (Birch 2014, 2016). I will, however, explain briefly why I think that some of Grafen's informal glosses, such as that in the above quotation, overstate the implications of his formal results for fitness maximization.

We should first ask: which varieties of maximization are at stake? The maximization in which Grafen is interested is the maximization of individual fitness by individual phenotypes: it does not directly involve population means. In effect, he claims to have shown that versions of MAX-C and MAX-D *would* be true in a world without mutation (and in which various other idealizations he makes in his models, such as the absence of meiotic drive and gametic selection, also obtain). Here I will focus on MAX-C.[8]

A natural reaction to this claim is to ask: how could MAX-C possibly be true in a world without mutation? Assuming away mutation seems to make things worse, not better, for fitness maximization. For in a world without mutation, there is no way to get around the constraints imposed by genetic architecture. A population can get permanently stuck at a sickle-cell type polymorphic equilibrium at which suboptimal phenotypes are present. Yet Grafen proves that all of his formal links between gene frequency change and optimal strategy choice still hold in such a context. This is surprising at face value, and it leaves two possibilities: either these cases are not really incompatible with MAX-C after all, despite the apparent presence of suboptimal phenotypes, or else Grafen's formal links do not really imply a version of MAX-C after all, even though his informal gloss suggests they do.

It takes a bit of untangling to see what is going on here (Grafen 2014; Okasha & Paternotte 2014; Birch 2016). The key is to see that Grafen's links do not explicitly refer to population-genetic equilibrium: instead, they characterize an equilibrium as a point at which there is no "scope for selection" and no "potential for positive selection." It turns out that the sickle-cell equilibrium does not qualify as an equilibrium in this sense, because there is a phenotype—malarial resistance—that is being selected-for. By characterizing evolutionary equilibrium in partly phenotypic terms, Grafen is able to disqualify equilibria in which gene frequencies are stably constant but suboptimal phenotypes are present.

However, this unorthodox way of thinking about equilibrium has some odd consequences. For example, an initial population composed of 100 percent heterozygotes, all with the optimal malarial resistance phenotype, qualifies as an equilibrium in the sense that matters for Grafen's links. It qualifies because it has no expected change in gene frequencies in the initial time step and no phenotype that is or would be selected-for, even though selection will inevitably start altering gene frequencies as soon as homozygotes appear (Grafen 2014).

As Grafen himself notes, the way in which the links hold in cases of heterozygote advantage "seem[s] to contain an element of evasion, and call[s] into question the meaning and value of the links themselves" (Grafen 2014: 165). The question is what this means for the relationship between the links and our MAX-C. Here is one way to go: MAX-C is clearly false in sickle-cell type models without mutation, but Grafen's links are true; so, despite appearing at face value to do so, Grafen's links do not imply MAX-C. This is the response I advocated on an earlier occasion (Birch 2016).

However, there is, I think, another way of reading this: a way more sympathetic to Grafen's aims. This is to say that Grafen, like Hammerstein, has found a way of constructing a non-standard equilibrium concept so that equilibria at which suboptimal phenotypes are present do not qualify as equilibria. The novelty of Grafen's approach is to appeal to phenotypic considerations in constructing the equilibrium concept, where Hammerstein appeals to assumptions about the rate of mutation and the long-run malleability of genetic architectures. If we formulate MAX-C using Grafen's non-standard equilibrium concept, then it comes out true (see "MAX-C**" in Birch 2016). What remains up for debate is whether evolution by natural selection has any reliable tendency to arrive at equilibria, thus construed.

LIVING WITHOUT MAXIMIZATION

For both Hammerstein and Grafen, the project of pursuing fitness maximization principles, in the face of widespread skepticism from population geneticists, is justified by the need to provide a theoretical foundation for adaptationist, optimality-based approaches in behavioral ecology (and evolutionary ecology more generally). The same need is clearly felt by those behavioral ecologists who have pounced on Grafen's links (too hastily, in my view) as providing "an extremely solid theoretical grounding" for the field (West & Burton-Chellew 2013: 1043).

What drives this need? Why are behavioral ecologists simply unable to accept the message from population genetics that the dynamics of natural selection are messy and complicated, and revise their models accordingly? The problem with this suggestion is that the vast majority of work in behavioral ecology relies on what Grafen (1984) has termed "the phenotypic gambit": the bet that the evolution of complex phenotypes can be understood in ignorance of the complex genetic architectures that underlie them. Approaches as diverse as inclusive fitness theory, life history theory, multilevel selection theory and evolutionary game theory all have this much in common.

The precise nature of the gambit varies depending on the details of the approach; for example, the inclusive fitness approach aims to understand the evolution of a trait by looking at its fitness effects on an organism and its social partners, the patterns of genetic relatedness between social partners, and the trait's heritability. Maximization-based techniques are often employed, but need not be. In virtually all cases, however, researchers make a fundamental bet that they can explain evolutionary outcomes without detailed knowledge of the genotype-phenotype map. The rationale for this bet is a practical one. We may be living in a "post-genomic" age, but, for the vast majority of traits in the vast majority of species, we still lack the sort of data concerning the genetic architectures underlying complex behavior that ecologists would need in order to do without the phenotypic gambit. This is what drives the desire to show that the long-term equilibria of the evolutionary process are governed by, in Hammerstein's words, "selective forces at the phenotypic level," which can be understood in the absence of detailed knowledge of genetics.

There is, I think, a real danger that this holy grail of foundational work in behavioral ecology will prove mythical. The early models of Moran and others should already be enough to convince us that there can be no purely theoretical guarantee that evolutionary equilibria will be fitness maxima. This inevitably depends on the ability of mutation to alter genetic arrangements. There may be nothing further to say here except that sometimes this happens and sometimes it does not. In some cases, the genetic architecture underlying a trait will preclude its optimization; sometimes it will be favorable. Sometimes an unfavorable architecture will be made more favorable by a change in the genetics; in other cases it may persist longer than the selective environment. It all depends on the details.

I suggest, however, that we can make peace with the phenotypic gambit without having to deny the dependence of real-world evolutionary outcomes on genetic detail. The key is to recognize that, while the gambit really is a gambit—an opening bet—and not a "solid theoretical grounding" for which we have compelling independent evidence, it is not always problematic to rest a scientific research program on a bet of this sort. This depends on the epistemic ambitions of the program. If optimality modeling aims to yield, by itself, knowledge of the evolutionary processes that have shaped phenotypic traits, then its reliance on a bet is indeed a problem. For this suggests that, even when the hypotheses it generates are true, they are only luckily true (i.e., true because the assumptions of the phenotypic gambit happened to be true in this case), and this undermines the idea that they constitute knowledge.

However, if we see the goal of optimality-based approaches as primarily one of hypothesis generation, the reliance on a bet is unproblematic.[9] After all, it is a bet that has led consistently to the generation of serious and credible evolutionary hypotheses—hypotheses that are plausible given everything we currently know. This is not a trivial achievement. The phenotypic gambit, in all its forms, represents a very well-designed heuristic for this purpose. On the one hand, it permits modelers to idealize away potential complications about which they are unavoidably ignorant, while, on the other hand, it demands sensitivity to the knowable empirical facts about fitness effects, population structure, heritabilities, coefficients of relatedness and so on.

The upshot is that whether optimality modelers should be worried about the absence of a theoretical justification for fitness maximization depends on the function they intend their models to serve. The lack of such a justification challenges more epistemically ambitious claims about their function, but it does not undermine their value as sources of credible empirical hypotheses: hypotheses that should not be regarded as knowledge until the underlying genetic architecture of the trait in question—and its compatibility or otherwise with the hypothesis—is known.[10]

I suspect many evolutionary ecologists would want to resist this epistemically modest conception of the function of optimality modeling. But I think we can embrace it while still recognizing the scientific value of this kind of work. Serious and credible evolutionary hypotheses are hard to find, and we should not be dismissive of methodological approaches that have consistently generated them.

ACKNOWLEDGMENTS

I thank Anthony Edwards, Warren Ewens, Alan Grafen, Rufus Johnstone, Tim Lewens, Samir Okasha, Steven Orzack, Cedric Paternotte, and John Welch for very helpful discussions and/or email exchanges on these issues. I thank Wiley-Blackwell for permitting the re-use of some material from Birch, J. (2016) "Natural selection and the maximization of fitness," *Biological Reviews*, © 2016 Cambridge Philosophical Society. The present chapter is intended as a companion piece to this longer article, in which Fisher's fundamental theorem and Formal Darwinism are discussed in greater mathematical detail. This work was supported by a Philip Leverhulme Prize from the Leverhulme Trust.

NOTES

1. For excellent introductions to these wider debates, see Lewens 2007, 2009; Godfrey-Smith & Wilkins 2008; Orzack & Forber 2012.
2. The basic taxonomy here is set out in greater detail in Birch 2016.
3. Wright originally envisaged "genotypes [. . .] packed, side by side [. . .] in such a way that each is surrounded by genotypes that differ by only one gene replacement" (Wright 1988: 116). On such a landscape, populations would be represented by clouds of genotypes. But the version of the metaphor that now features in standard textbooks represents a population as a single point moving through a space defined by population gene frequencies (Ridley 2004; Futuyma 2013). See Pigliucci & Kaplan 2006 and Kaplan 2008 for discussion of the different versions of the metaphor.
4. See Price 1972; Ewens 1989; Frank & Slatkin 1992; Frank 1997; Edwards 1994; Ewens 2004; Plutynski 2006; Okasha 2008; Ewens 2011; Edwards 2014; Grafen 2015; Ewens & Lessard 2015; Birch 2016 for further detail on, and discussion of, these complex issues.
5. This argument does not, however, give us a tenable version of MAX-A, since a Nash equilibrium need not be a mean fitness maximum.

6. The key assumptions of Grafen's framework are that there is "no mutation, no gametic selection, fair meiosis and that all the loci contributing to the p-score have the same mode of inheritance" (Grafen 2002: 82). I previously described the absence of mutation as a "limitation" of the framework (Birch 2016), but I now suspect that this not the right way to think about it. A more charitable reading is that Grafen intentionally assumes away mutation in the hope of proving links between selection and optimality that do not rely on assumptions about mutation, as Hammerstein's (1996) results do.

7. Others cite Grafen in support of stronger claims. See, for example, West & Burton-Chellew (2013: 1043): "The success of the behavioral ecology approach is built on an extremely solid theoretical grounding (Davies et al. 2012). Darwin ([1859] 1964) argued that traits that increase fitness will accumulate in populations, leading to organisms that behave as if they are trying to maximize their fitness. Our modern most general genetical interpretation of this is that organisms should behave as if they are trying to maximize their inclusive fitness (Hamilton 1964; Grafen 2006)."

8. Similar considerations complicate the relationship between Grafen's links and MAX-D, though I will not discuss this issue here.

9. Anna Alexandrova (2008) argues that we should understand models in experimental economics as tools for hypothesis generation and constructs a detailed account of how this works. Roughly, the idea is that models provide "open formulae" for causal hypotheses: they generate schemas for causal hypotheses that do not assert anything until we add either a quantifier or a singular instance. I suspect this sort of account would fit many optimality models in evolutionary ecology quite well, but I do not pursue this in detail here.

10. In a similar vein, Angela Potochnik (2009) distinguishes strong and weak uses of optimality models, where the "strong use" involves the claim that selection was the only important influence on the evolution of the trait, and the "weak use" involves the weaker claim that the model accurately represents the role played by selection in the evolution of the trait. However, even Potochnik's "weak use" strikes me as epistemically ambitious, since it relies on the idea that optimality models "accurately represent the selection dynamics involved in producing the target evolutionary outcome" (Potochnik 2009: 187). My proposal is more akin to the "even weaker" use that Potochnik attributes to Seger & Stubblefield 1996.

REFERENCES

Alexandrova, A. 2008. "Making models count." *Philosophy of Science* 75: 383–404.

Birch, J. 2014. "Has Grafen formalized Darwin?" *Biology and Philosophy* 29: 175–180.

Birch, J. 2016. "Natural selection and the maximization of fitness." *Biological Reviews* 91: 712–727.

Darwin, C. [1859] 1964. *On the Origin of Species* (Harvard University Press).

Davies, N., Krebs, J., & West, S. 2012. *An Introduction to Behavioural Ecology* (Wiley-Blackwell).

Dodson, S., Allen, T., Carpenter, S., Ives, A., Jeanne, R., Kitchell, J., & Langston, N. 1998. *Ecology* (Oxford University Press).

Doebeli, M. 2011. *Adaptive Diversification* (Princeton University Press).

Doebeli, M. & Dieckmann, U. 2000. "Evolutionary branching and sympatric speciation caused by different types of ecological interactions." *American Naturalist* 156: S77–S101.

Edwards, A. 1994. "The fundamental theorem of natural selection." *Biological Reviews* 69: 443–474.

Edwards, A. 2000. *Foundations of Mathematical Genetics* (Cambridge University Press).

Edwards, A. 2007. "Maximisation principles in evolutionary biology." In M. Matthen & C. Stephens (eds.), *Handbook of the Philosophy of Science: Philosophy of Biology* (Elsevier) 335–347.

Edwards, A. 2014. "R. A. Fisher's gene-centred view of evolution and the fundamental theorem of natural selection." *Biological Reviews* 81: 135–147.

Eshel, I. & Feldman, M. 1984. "Initial increase of new mutants and some continuity properties of ESS in two locus systems." *American Naturalist* 124: 631–640.

Eshel, I. & Feldman, M. 2001. "Optimality and evolutionary stability under short- and long-term selection." In S. Orzack & E. Sober (eds.), *Adaptationism and Optimality* (Cambridge University Press) 161–190.

Eshel, I., Feldman, M., & Bergman, A. 1998. "Long-term evolution, short-term evolution and population genetic theory." *Journal of Theoretical Biology* 191: 391–396.

Ewens, W. 1968. "A genetic model having complex linkage behaviour." *Theoretical and Applied Genetics* 38: 140–143.

Ewens, W. 1989. "An interpretation and proof of the fundamental theorem of natural selection." *Theoretical Population Biology* 36: 167–180.

Ewens, W. 2004. *Mathematical Population Genetics* (Springer).

Ewens, W. 2011. "What is the gene trying to do?" *British Journal for the Philosophy of Science* 62: 155–176.

Ewens, W. & Lessard, S. 2015. "On the interpretation and relevance of the fundamental theorem of natural selection." *Theoretical Population Biology* 104: 59–67.

Fisher, R. A. 1930. *The Genetical Theory of Natural Selection* (Clarendon Press).

Fisher, R. A. 1941. "Average excess and average effect of a gene substitution." *Annals of Human Genetics* 11: 53–63.

Frank, S. 1997. "The Price equation, Fisher's fundamental theorem, kin selection, and causal analysis." *Evolution* 51: 1712–1729.

Frank, S. & Slatkin, M. 1992. "Fisher's fundamental theorem of natural selection." *Trends in Ecology and Evolution* 7: 92–95.

Futuyma, D. 2013. *Evolution* (Sinauer).

Geritz, S., Kisde, É., Meszéna, G., & Metz, J. 1998. "Evolutionarily singular strategies and the adaptive growth and branching of the evolutionary tree." *Evolutionary Ecology* 12: 35–57.

Godfrey-Smith, P. 2012. "Darwinism and cultural change." *Philosophical Transactions of the Royal Society of London B: Biological Sciences* 367: 2160–2170.

Godfrey-Smith, P. & Wilkins, J. 2008. "Adaptationism." In S. Sarkar & A. Plutynski (eds.), *A Companion to the Philosophy of Biology* (Blackwell) 186–201.

Grafen, A. 1984. "Natural selection, kin selection and group selection." In J. Krebs & N. Davies (eds.), *Behavioral Ecology* (Blackwell) 62–84.

Grafen, A. 1999. "Formal Darwinism, the individual as maximizing agent analogy, and bet-hedging." *Proceedings of the Royal Society, Series B: Biological Sciences* 266: 799–803.

Grafen, A. 2002. "A first formal link between the Price equation and an optimization program." *Journal of Theoretical Biology* 217: 75–91.

Grafen, A. 2006. "Optimization of inclusive fitness." *Journal of Theoretical Biology* 238: 541–563.

Grafen, A. 2007. "The formal darwinism project: a mid-term report." *Journal of Evolutionary Biology* 20: 1243–1254.

Grafen, A. 2014. "The formal Darwinism project in outline." *Biology and Philosophy* 29: 155–174.

Grafen, A. 2015. "Biological fitness and the fundamental theorem of natural selection." *American Naturalist* 186: 1–14.

Hamilton, W. D. 1964. "The genetical evolution of social behaviour." *Journal of Theoretical Biology* 7: 1–52.

Hammerstein, P. 1996. "Darwinian adaptation, population genetics and the streetcar theory of evolution." *Journal of Mathematical Biology* 34: 511–532.

Hammerstein, P. 2012. "Towards a Darwinian theory of decision making: Games and the biological roots of behavior." In S. Okasha & K. Binmore (eds.), *Evolution and Rationality: Decisions, Co-operation and Strategic Behavior* (Cambridge University Press) 7–22.

Hammerstein, P. & Selten, R. 1994. "Game theory and evolutionary biology." In R. Aumann & S. Hart (eds.), *Handbook of Game Theory with Economic Applications*, vol. 2 (Elsevier) 929–993.

Hedrick, P. 2011. *Genetics of Populations* (Jones and Bartlett).

Kaplan, J. (ed.). 2008. *The Adaptive Landscape: Metaphors and Models. Biology and Philosophy* [special issue] 23.

Karlin, S. 1975. "General two locus selection models: Some objectives, rules and interpretations." *Theoretical Population Biology* 7: 364–398.

Lessard, S. 1997. "Fisher's fundamental theorem of natural selection revisited." *Theoretical Population Biology* 52: 119–136.

Lewens, T. 2007. "Adaptation." In D. Hull & M. Ruse (eds.), *The Cambridge Companion to the Philosophy of Biology* (Cambridge University Press) 1–21.

Lewens, T. 2009. "Seven types of adaptationism." *Biology and Philosophy* 24: 161–182.

Liberman, U. 1988. "External stability and ESS: Criteria for initial increase of new mutant allele." *Journal of Mathematical Biology* 26: 477–485.

Maynard Smith, J. 1982. *Evolution and the Theory of Games* (Cambridge University Press).

Metz, J. 2011. "Thoughts on the geometry of meso-evolution: Collecting mathematical elements for a postmodern synthesis." In F. Chalub & J. Rodrigues (eds.), *The Mathematics of Darwin's Legacy* (Birkhauser) 193–232.

Moran, P. 1964. "On the non-existence of adaptive topographies." *Annals of Human Genetics* 27: 383–393.

Mulholland, H. & Smith, C. 1959. "An inequality arising in genetical theory." *American Mathematical Monthly* 66: 673–683.

Okasha, S. 2008. "Fisher's 'fundamental theorem' of natural selection: A philosophical analysis." *British Journal for the Philosophy of Science* 59: 319–351.

Okasha, S. & Paternotte, C. 2014. "Adaptation, fitness and the selection-optimality links." *Biology and Philosophy* 29: 225–232.

Orzack, S. & Forber, P. 2012. "Adaptationism." In E. Zalta (ed.), *Stanford Encyclopedia of Philosophy*. http://plato.stanford.edu/entries/adaptationism.

Parker, G. & Smith, J. 1990. "Optimality theory in evolutionary biology." *Nature* 348: 27–33.

Pigliucci, M. & Kaplan, J. 2006. *Making Sense of Evolution: The Conceptual Foundations of Evolutionary Biology* (University of Chicago Press).

Plutynski, A. 2006. "What was Fisher's fundamental theorem of natural selection and what was it for?" *Studies in History and Philosophy of Biological and Biomedical Sciences* 37: 59–82.

Potochnik, A. 2009. "Optimality modeling in a suboptimal world." *Biology and Philosophy* 24: 183–197.

Price, G. 1972. "Fisher's fundamental theorem made clear." *Annals of Human Genetics* 36: 129–140.

Ridley, M. 2004. *Evolution* (Wiley-Blackwell).

Roff, D. 1992. *The Evolution of Life Histories: Theory and Analysis* (Chapman and Hall).

Sacks, J. 1967. "A stable equilibrium with minimum average fitness." *Genetics* 56: 705–708.

Scheuer, P. & Mandel, S. 1959. "An inequality in population genetics." *Heredity* 31: 519–524.

Seger, J. & Stubblefield, J. 1996. "Optimization and adaptation." In M. Rose & G. Lauder (eds.), *Adaptation* (Academic Press) 93–123.

Sober, E. 1987. "What is adaptationism?" In J. Dupré (ed.), *The Latest on the Best: Essays on Evolution and Optimality* (MIT Press) 105–118.

West, S. & Burton-Chellew, M. 2013. "Human behavioral ecology." *Behavioral Ecology* 24: 1043–1045.

Wilkins, J. & Godfrey-Smith, P. 2009. "Adaptationism and the adaptive landscape." *Biology and Philosophy* 24: 199–214.

Wright, S. 1932. "The roles of mutation, inbreeding, crossbreeding and selection in evolution." *Proceedings of the Sixth International Congress of Genetics* 1: 356–366.

Wright, S. 1988. "Surfaces of selective value revisited." *American Naturalist* 131: 115–123.

5

Does Biology Need Teleology?

Karen Neander

To ask the function of short-term memory one might ask, "What is short-term memory for?"[1] Or, to ascribe a function to eyelashes one might say, "Eyelashes divert airflow to protect the eye."[2] If a function of x is to z, it is *for z-ing* or is there *to z*. This manner of speaking has a teleological flavor, but do biologists really use a teleological notion of function in contemporary biology, and, if so, what (if any) scientific purpose is it serving?

TELEOLOGY

The word "teleology" invokes ideas of intentional design or purpose, and mainstream contemporary biologists do not believe that organic systems result from intentional design or purpose, setting aside special cases of domestic breeding and genetic engineering and the like. But a more general characterization of a teleological explanation is that it is *forward-looking*. It purports to explain the means by the ends.

In contrast, ordinary causal explanations cite preceding causes, or maybe phenomena occurring at the same time as what is explained. In an ordinary causal explanation that describes a sequence of events at roughly the same level of analysis, Sally's throwing a rock at the window explains the window's breaking only if she threw the rock before the window broke; if she threw it later, her rock throwing was not the cause. In explaining how mechanisms perform a process (e.g., photosynthesis), we describe contributions by components (e.g., chlorophyl) that contribute to the process. These contributions might be made during the process being explained. But, again, if we start describing what occurs after the process has run to completion (e.g., the later release of chemical energy to fuel the plant), we are no longer explaining how the process (in this case, photosynthesis) occurred.

Yet teleological explanations refer to a result of what is supposed to be explained. If we are told that Lauren jumped into the water to save the child, that the cat prowls to catch a bird, that the lever on the sewing machine is there to raise and lower the needle, or that eye lashes are for reducing air flow across the eye, the means is seemingly explained by the ends. The item explained (such as the jumping, prowling, machine lever, or set of eyelashes) is seemingly explained by what it brings about or might bring about.

At a glance this could seem to invoke backward causation, but a second look soon reveals that at least some of these explanations do not invoke backward causation. If Lauren tells us that she jumped into the pool to save the child from drowning, her explanation is correct if it points to a precedent of the jump: her *intention* to save the child. The explanation of the cat's prowling can be given a similar treatment, if the cat has intentional mental states. Functional explanations of artifacts are similar. How can the lever's raising and lowering the needle explain the lever's presence on the machine if it can raise and lower the needle only once it is already on the machine? The answer is, plausibly, that when we ascribe functions to artifacts we ascribe intentions to those who design or use them.[3] For example, we imply that someone added the lever to let the user raise and lower the needle.

So, in *purposive-teleological* explanations, such as purposive explanations of behavior or functional explanations of artifacts, the looking forward to ends served by the means is a way to look back to past intentions on the part of those supplying the means. Purposive-teleological explanations have Form 1 on their surface, but dig deeper and Form 2 is revealed.

Form 1: End at time t_3 explains means at earlier time t_2.
Form 2: Event at t_1 (involving an intentional attitude to end at t_3) explains means at t_2.

Teleological explanations in contemporary biology are prima facie more problematic. A creationist might think that plants and animals are God's artifacts and seek to explain their traits in terms of God's intentions, but mainstream post-creationist neo-Darwinian biologists do not. Once special cases (domestic breeding and genetic engineering and the like) are set aside, the traits of organisms are not thought to depend on agent intentions. Biologists still speak of the "species' designs," the "purpose" of naturally occurring traits, the "reason" why traits were selected by natural selection, and so on; however, they know perfectly well that natural selection is a blind mechanical force. Their talk of "design" in this context is not intended to refer to intentional design. Such talk is metaphorical, or has by now become dead metaphor (i.e., it has acquired a new literal meaning).

How, then, can teleological functional explanations be respectable in contemporary biology? One answer is that they cannot. For instance, Morton Beckner (1959: 112) declares: "Only the most Paleozoic reactionary would claim that 'plants have chlorophyll' is explained by 'plants perform photosynthesis." And, along similar lines, Robert Cummins (1975) maintains that such explanations are a hangover from a creationist past or a result of a basic misunderstanding of evolutionary theory. These strong claims precede more recent philosophical analyses in support of the opposing view, which is that Darwin did not eliminate teleology from biology, but instead provided a naturalistic interpretation of it. A now popular view (though still to some extent controversial) is that there is a respectable naturalistic *teleonomic* notion of function in use. A teleonomic function is like a teleological function of an artifact, except it is (as a matter of stipulation) mind independent in the way naturalism requires.

TELEONOMY AND THE ETIOLOGICAL THEORY

One way to develop this idea is to argue that functional explanations of biological traits have a similar although non-identical structure to that of functional explanations for artifacts (Neander 1991). Consider the kangaroo's pouch, seemingly explained by its function to carry and protect joeys. Obviously, a pouch can carry and protect joeys only once a kangaroo has already inherited and developed the pouch. So, if we explain the pouch in terms of its functions, we explain the means by the ends, and this is the forward-looking aspect of this explanation. But there is also a backward-looking aspect if the function ascription implies that the pouch is an adaptation for carrying and protecting joeys.

That something is an adapt*ation* is a historical fact about it, whereas whether it is adap-*tive* depends on its fit with its current environment. Something is an adaptation for z-ing only if it was selected for z-ing in the past. Something is adaptive only if it presently con-tributes to fitness. Since natural selection operates over types, which increase or decrease in proportional representation in a population, x is an adaptation for z-ing only if items of x's type were selected for z-ing. For this to be the case, at least these three conditions must be met: (i) traits of x's type did z, (ii) their z-ing was on average adaptive for the individuals with the x type of trait in the relevant population, and (iii) in consequence, there was selection of the mechanisms responsible for the inheritance and development of xs. (More needs to be said, for instance to accommodate changes in the direction of selection over time—think here of such cases as the emu's vestigial wings and the pen-guin's flippers, which are no longer adaptations for flight, despite past selection for flight in the lineage of these forelimbs.)

This understanding of teleonomy in biology is supported by *etiological theories* of function (e.g., Millikan 1989; Neander 1991). The details vary with different versions, but these theories generally tell us that a/the function of a naturally occurring biological trait depends on its history of selection: (roughly) an item's function is to do z if items of the type were selected to do z. Given this type of theory, function ascriptions of the form "the function of x is to z" can (in part) explain xs because "the function of x is to z" *entails* that xs were selected for z-ing. In the case of teleological functions of artifacts, intentional selection is involved (and it can apply to token artifacts as well as to types, depending on what was intentionally selected). But in the case of the teleonomic func-tions of naturally occurring biological traits, a non-intentional (and in that sense) *natural* process of selection is involved.

Evolution by natural selection involves the random generation of heritable variations that differ in fitness and in their rates of replication due to such differences. (See Lewens, Chapter 1 this volume, for discussion of how ideas regarding natural selection have devel-oped since Darwin.) Phylogenetic natural selection involves generations of individuals, but analogous ontogenetic processes occur within the lifespan of a single organism (e.g., antibody selection and perhaps some of the processes involved in learning) and these can also ground teleonomic functions, on this view.[4]

Reasonable concerns about how the process of evolution by natural selection is being understood may still arise, but they should now take a subtler form. For instance, one might worry whether the purported teleonomic explanations are overly adaptationist. But they need not be, because the friends of the etiological theory of functions can read-ily agree that not every trait has a function, and that those with functions are only in part explained by selection for them. They can readily agree that natural selection operates within the constraints of a changing environment on variants that happen to randomly arise and are not eliminated by drift, and within restrictions imposed by hard to change developmental pathways and architectural and physical requirements. But many good explanations are partial, and an explanation's being partial is no good reason to consider it illegitimate as opposed to incomplete.

A second worry is sometimes expressed concerning what exactly natural selection explains. Can it contribute to creating complex adaptations or does it affect only their distribution in a population? Does it really help answer Paley's question about how wondrous adaptations arise, or not? Clearly, a once off "sieving" of pre-existing traits can affect only their proportional representation in a population. But cumulative selection does more. By selecting an adaptive trait, selection increases the chances that further random alterations to it will arise (since, given selection, the relevant mechanisms of inheritance and development are replicated more frequently than they would otherwise have been. Then, subsequent rounds of selection can select the beneficial alterations and eliminate the deleterious ones (Neander 1995). In any event, even if natural selection explained only the distribution of traits in a population, as some claim (Sober 1984: ch. 5), function ascriptions could explain the presence of traits of a type by explaining their preservation in a population over time.

Cummins (2002) raises a third kind of worry when he argues that, for example, avian wings cannot have the function to enable flight in virtue of being selected for enabling flight because they were *not* selected for enabling flight. He rightly points out that natural selection requires alternatives from which to select. And, more tendentiously, he claims that if wings were selected for flight, then there must have been other forelimbs in the same dinosaur population at the same time that did not enable flight, against which wings that enabled flight were selected. Flight-enabling avian wings evolved gradually (probably from forelimbs adapted for gliding from tree to tree). Cummins claims that there was never selection for flight as opposed to no flight. There was only ever selection for incremental improvements, such as more energy-efficient flight, faster flight, or more maneuverable flight.

There are two things to keep in mind here. One is that even traits "gone to fixation" require maintenance selection to weed out deleterious mutations, or else capacities will tend to deteriorate. In the case of wings, deleterious mutations that prevent flight can still occur. The other thing to keep in mind is that the fineness of the lens with which we view a selection history will need to match the fineness of the description of the relevant function. To speak of avian wings having the function of flight is an extremely coarse-grained way of speaking, warranting a sweeping view across many lineages and vast spans of time (a hundred million years or so). From that perspective, selection for more energy-efficient flight, faster flight, or more maneuverable flight, is selection for flight. The coarse-grained function ascription is also consistent with more fine-grained ascriptions of function to particular features of wings in specific lineages during specific periods of time.

The etiological theory of functions descends from an early account offered by Larry Wright (1973, 1976) in which teleological-purposive explanations as well as teleonomic explanations in contemporary biology appeal to "consequence etiologies." A consequence etiology is a history of an item in which one or more of its consequences plays a role. In a consequence etiology, the item is where it is or in the form that it is in owing to one or more of its effects. The details of Wright's account of functions proved to be problematic (see e.g., Boorse 1976), but his core idea of a consequence etiology is in effect retained in later versions.

ALTERNATIVE ACCOUNTS OF THE TELEONOMIC NOTION OF FUNCTION

Proponents of other theories of function also claim that their theories can account for or allow for the teleological flavor of function ascriptions in contemporary biology. For instance, some argue that the functions of traits are those of their effects that make them presently adaptive. On this view, functions are (roughly) their present species-typical contributions to the survival and/or reproduction of the individuals who possess them, or their typical contributions within a more restricted reference-class, such as a sex and/ or age group in a species (Boorse 1977, 2002).

If a theory of this type were true, "x has the function to z" would not entail "xs were selected for z-ing." Nevertheless, knowing a trait's function could still cast some light on its history, given a background understanding of evolutionary theory and the past environment. If a polar bear's fur has the function to keep it warm because it keeps polar bears warm and this is adaptive on average now, learning the function of the fur could suggest that it might have been adaptive in the past too, and might have been selected for that reason. What is presently adaptive might not have been adaptive in the past and what was adaptive in the past might not be adaptive now, since environments change, but the current functions of traits can, on this type of ahistorical-statistical theory, provide clues to relevant selection histories.

A related suggestion is that the function of a trait is what it does that makes it apt for selection in the future in a creature's "natural habitat" (a notion not well elucidated). John Bigelow and Robert Pargetter (1978) contend that this propensity theory of functions best captures the "forward looking" nature of functional explanations. Note, though, that it does not capture the way in which teleological explanations look back by looking forward.

One last type of theory that is of interest in this context draws to some extent on other parallels between teleonomic functional explanations in biology and teleological explanations. In this case the focus is on goal-directed processes or behaviors and cybernetic or homeostatic systems. Homeostasis is the property of a system such that it regulates its inner states to preserve stability in the face of perturbations on some dimension, and cybernetics is the study of feedback mechanisms that can be used to do so. For example, a thermostat turns a house into a homeostatic system when it monitors the temperature and turns the heating or cooling on or off to keep the temperature within a set range. Similarly, there are many somatic homeostatic systems that contribute to maintaining stable states (e.g., a stable body temperature or a stable level of glucose in the blood, despite fluctuations in the surrounding temperature or supply of glucose). Goal-directed processes or behaviors are not always directed at maintaining a stable state, but they tend to involve feedback mechanisms that produce resilience (persistence and plasticity) in the pursuit of a goal. For example, one creature chasing another shows resilience when it moves around obstacles to continue the chase. The performance of some functions (involved, for instance, in maturation or reproduction) will disrupt as opposed to pre-serve a stable state in an organism. Thus some versions of this type of theory consider survival and reproduction or the maintenance of a recurring life cycle to be the apical goal to which the parts of a living system have the function to contribute, and hence this type of theory might be blended with the ahistorical-statistical theory.[5]

In any event, the teleological flavor of function ascriptions might again be viewed as due to an implied or suggested consequence-etiology. For, on this view, when a trait performs its function it tends to contribute to its own preservation by maintaining the individual organism whose trait it is, or it tends to contribute to the inheritance and development of traits of the type by perpetuating the relevant life cycle. For instance, when a token heart performs its function, it tends to help maintain the individual to whom the heart belongs, and so tends to help preserve itself and/or (depending on how the details of the theory are spelt out) it tends to perpetuate the life cycle of individuals of the same kind and so furthers the production of hearts of that type.

My view has long been that an etiological theory is the best theory of the relevant notion of function in biology, but it is not the only theory that can lay claim to explaining the notion's teleological or teleonomic flavor. Which is the best overall theory? This is not a question to try to settle here, since it calls for a more detailed and lengthy comparison of the main theories that can be provided here (but see Garson 2016). In what follows, this chapter instead outlines some aspects of the theoretical role that this notion of function might be serving in contemporary biology. Even if the relevant notion of function is sci-entifically respectable, it remains up in the air whether it plays a significant scientific role. I believe it does but this remains controversial.

THE FUNCTION/DYSFUNCTION AND
FUNCTION/NON-FUNCTION DISTINCTION

At this point it helps to acknowledge two distinctions to which a teleonomic notion of function is sensitive: the function/dysfunction and function/non-function distinctions.

There can in this sense of "function" be malfunction. A statement of the form "x has the function to z" is consistent with "x lacks the capacity to z (owing to dysfunction on x's part)."[6] Even without malfunction, a token trait might not perform its function because the opportunity never arises or the environment is uncooperative, but a trait that malfunctions will lack the capacity to perform one or more of its normal or proper functions (or lack the disposition to perform one or more of them normally or properly) even when the opportunity arises and the world cooperates. A pancreas, for example, can have the function to produce insulin even if it lacks any capacity to produce insulin. Nor is there any apparent conceptual incoherence in the idea that functional impairment (lung impairment, for example) might become typical in a population for a time, in a pandemic or due to environmental disaster or degradation.

The relevant notion of function also allows that not all effects of even properly func-tioning traits are their functions. Consider an artifact case for a moment: a belt buckle deflects a bullet and saves the life of the soldier wearing it. The buckle has the function to buckle the belt and help hold up the trousers. It does not have the function to stop a bullet, though it might do that too. It might *serve the function* of stopping a bullet or *act as* a bullet stopper, so to speak, without having the function to do so. Similarly, some effects of biological traits are not their functions. Hearts have the function to pump blood rather than make lub-dub sounds, but they do both. At least intuitively, hearts might even make frequent adaptive contributions to the survival and reproduction of the individuals whose hearts they are without those contributions being their functions. For instance, by making lub-dub sounds, they might also assist doctors in diagnosing treatable disease.

On the etiological theory, the function/dysfunction and function/non-function distinctions are due to functions being grounded in the past selection of ancestral traits, rather than in the present dispositions of current instances, as well as what there was selection *for* (what past adaptive effects contributed to the selection) as opposed to merely whatever was done by what was selected. In what follows I assume that the teleonomic notion is sensitive to the function/dysfunction and function/non-function distinctions. The proponents of different theories of this notion of function draw these distinctions in somewhat different ways in line with their preferred theories, but none reject these distinctions altogether,[7] nor could they without changing the subject.

The question of whether a teleonomic notion of function plays a significant scientific role in biology is in large part the question of what role the function/dysfunction and function/non-function distinctions play.

FUNCTIONAL EXPLANATION

This section considers the role of these distinctions in explanation. For the sake of brevity of exposition, I assume the etiological theory to do so, leaving it as an exercise for readers to reflect on the implications of other theories. One kind of functional explanation has already been mentioned, but without any discussion of its significance for biology. Moreover there are two kinds of functional explanation to consider.

The kind of functional explanation mentioned in earlier sections ascribes a function to something to answer a why-question concerning its origin, presence, or persistence. Why is there chlorophyll? Why do veins have valves? On an etiological theory of functions, to be told that the function of chlorophyll in plants is to perform photosynthesis is to be told why chlorophyll was selected, and to be told that the function of veins in valves is to prevent blood from flowing backward and help return it to the heart is to be told why veins in valves were selected.

Asking and answering why-questions can be illuminating, not only for understanding the evolutionary history of some trait but also for understanding how complex organic systems operate. In the 17th century, Harvey's initially surprising and controversial discovery that the blood circulates is in part attributed to his asking and answering a series of quite specific why-questions (e.g., why do veins have more valves than arteries?). Harvey's 1628 book, which describes his discoveries, often speaks of ends and purposes. He might have imagined that the "Nature, who does nothing in vain" was the Christian God, since he was a Christian, but (setting aside the Panglossian implications, as well as the idea of divine design) analogous thinking about natural selection may play a similar role in discovery. For instance, asking why we have eyelashes more recently led researchers to test the hypothesis that mammalian eyelashes, which are about a third of the length of the eye, are the best length for reducing airflow toward the eye and protecting it from particle deposition and excessive evaporation. While not in the same league as Harvey's discovery, this helps illustrate the point that large and small discoveries can be facilitated by asking and answering why-questions. Both the function/dysfunction and function/ accident distinction are relevant for this type of reverse engineering, given that a trait's function(s) and not its pathological effects or other non-function effects are why traits of the type were selected.

A second kind of functional explanation answers how-questions. How does photosynthesis occur? How does the circulatory system circulate blood? Biologists answer such questions by conceptually decomposing a system into its component parts and ascribing diverse functions to them. This can be done at multiple *levels* of analysis. That is, a system that is conceptually decomposed into its main parts can be further decomposed into sub-parts, and the sub-parts further decomposed into yet simpler sub-sub-parts, and so on. In a componential analysis of a system, the simpler and simpler parts are ascribed simpler and simpler causal roles, and thus the circulation of blood, for instance, can be explained at various levels of analysis.

It might at first glance seem that this latter type of explanation does not involve a teleonomic notion of function, but instead a mere notion of a causal contribution. Other scientists also give componential analyses of complex systems, decomposing mechanisms into parts at different levels of analysis, and without ascribing teleonomic functions to the components. The formation of a planetary system is a complex process that cosmologists try to explain at multiple levels of analysis, from supermassive black holes and the galaxies surrounding them to sub-atomic particles. Yet contemporary cosmologists do not claim that stars have the function to send heavy elements into the interstellar medium to help form vast clouds of molecular matter, that pre-planetary clumps of matter have the function to collide and accrete into larger clumps, or that the different elements and compounds have functions to do what they must if a solar system like ours is to form. They do not, anyway, ascribe malfunction-permitting functions. Stars do not malfunction.

It is true that cosmologists and biologists both give componential or mechanistic explanations (in the sense elucidated by Craver & Darden 2013). Biologists, however, ascribe teleonomic functions when explaining how complex living systems operate. But why use the teleonomic notion of function in this kind of context? One might think that it ought not to be used because physiological outcomes, just like cosmological ones, depend on the causal contributions qua causal contributions of the parts involved (Cummins 1975; Godfrey-Smith 1993). Physiological outcomes depend on whether certain causal contributions are made, not on whether these contributions are functions, pathological effects, or non-function effects. So one might think that the teleonomic notion of function can have no scientific significance in this type of componential (aka mechanistic) explanation, even if one wants to allow that it could have some other sort of significance, such as a moral or social significance in clinical medicine. This opinion is sometimes accompanied by the claim that functional norms are interest-laden (e.g., see Cummins & Roth 2010).

Against this is the view (supported by Boorse 1977 and Neander 2015 among others) that a (teleonomic) notion of function has a role in generalization. Consider the problems biologists face when trying to give useful general descriptions of how complex living systems *of a type or kind* operate. Describing the operation of a single cell is a huge challenge by itself. Describing the operation of a single multicelled organism is vastly more so. Add to this that each individual changes over time. Now add that there are usually billions of individuals in a species. Consider how the multiplicity of individuals, combined with their complexity, creates a momentous challenge with respect to providing useful general explanations of how living systems of a type or kind operate. The more complex a system, the more variables there are that can vary from one individual to the next. Sui generis genomes and complex developmental and maturational interactions with the environment ensure that a great deal of this potential variation is realized in a species. And, in addition to normal variation, there are countless ways in which complex multicelled organisms can malfunction.

Describing the actual causal contributions qua mere causal contributions of each component part of each individual in a species is of course not a remotely viable option in practice. The experimental work of biologists often focuses day to day on discovering the causal roles of a few features of a few individuals in a few controlled circumstances. But this work is almost always intended to further the larger collective enterprise of giving useful general descriptions of some type or kind of system as well—of the normal human immune system or of the normal human visual system, for example.

If the etiological theory of function is correct, a so-called "normal system" is, in the first instance, one in which each part that was selected to do something is disposed to do what it was selected to do. I say "in the first instance" because the description of the normal system could also include a description of other aspects of the system.[8] For instance, it could include a description of universal features that cannot be changed due to developmental or architectural constraints, along with any adaptive or for that matter maladaptive effects they have.

This style of idealization to normal or proper functioning might be useful for a number of reasons. Natural selection tends to drive adaptive traits to fixation, and most organisms are mostly normal most of the time, and so the composite portrait of the system that functions "as designed" (in the neo-Darwinian sense) has useful generality. But there are different ways to be normal in a species (e.g., there are sex- and age-related differences in adaptations, adaptations to local environments, polymorphisms that have resulted from distribution-sensitive effects on fitness, and so on), and the relevant notion of function can accommodate these differences. A description of the system that functions "as designed" also captures the way in which diverse parts of the system are, to a first approximation, co-adapted to work together. The description of the normal system is not a description of mere heterogeneity, but of *organized complexity*. While the description of the normal system abstracts from malfunction, the practice of describing normal systems does not simply ignore malfunction. The function/dysfunction distinction is a useful tool for understanding how normal systems operate, because to understand what happens when something goes wrong, and learn which capacities associate and dissociate when they do, is a useful tool for probing normal functional dependencies. Pathology is also efficiently described and understood against a background description and understanding of normal functioning, as specific deviations from normal functioning. The practice of describing normal systems, in this sense of "normal," also gives biologists across different laboratories, generations, and continents a stable descriptive target. What is normal will change with time, but at the slow pace of evolution (at least on the etiological theory, and when the selection is phylogenetic). It will not change with mere changes in researcher experimental techniques, lab conditions, or ways of recording and reporting data.

Again, this idealization strategy does not commit physiologists to thinking that *every* trait of a living system has a function, that traits that have functions were *optimally* designed for them, or that selection *as opposed to* other things (such as the random production of alternatives, drift, or developmental or architectural constraints) are responsible for preserving traits in populations. Selection operates within constraints, alongside drift, on alternatives that randomly arise, and it can certainly result in less than optimal designs. Descriptions of normal systems (even in the etiologist's sense) are consistent with this.

To understand whether the relevant notion of function plays a useful role, we also need to think about how the generalization problem might be solved without it. An alternative idea is that biologists specify *ceteris paribus* laws in this kind of context. These are "laws" or at least generalizations concerning how certain types or kinds of systems behave in certain circumstances, when "all else is equal" or (roughly) in the absence of interfering factors. But this approach invariably allows malfunction to count as an important source of interference; when a human immune system or human visual system malfunctions all else is not equal. So this reintroduces the relevant notion of function, even though it does not give it a central place. In my opinion, by lumping the possibility of malfunction together with the possibility of meteor strikes and the like, those who support this idea fail to appreciate how central is the role of the relevant notion of function is in componential analyses of living systems in biology.

A second alternative one finds in the literature is a hand wave at statistics. In relation to this, note that an ahistorical-statistical theory of the relevant notion of function is a genuine alternative to an etiological theory of that notion. Thus I am distinguishing mere hand waving at statistics from the attempt to offer a serious, detailed ahistorical-statistical theory of function. Mere hand waving at statistics is no real answer since there are many ways to collate statistics with respect to complex organic systems. Even if functional norms were basically statistical, we would need an account of how the statistics are sorted and reported to capture the function/dysfunction and function/non-function distinctions. Serious ahistorical-statistical theories of function try to address this issue. Merely gesturing at the use of statistics to provide some sort of idealization does not.

Since it is easy to misunderstand my aim here, I repeat that it is not to argue for the etiological theory against other theories of the relevant notion of function (such as the ahistorical-statistical, propensity, or cybernetic theories). The topic is whether biology needs a *teleonomic* notion of function, one that at least intuitively has a teleological flavor and in any event respects the function/dysfunction and function/non-function distinctions. The etiological theory of this notion is popular and in my opinion is correct. But readers might disagree with this, and yet agree that the relevant notion has a significant scientific role to play.

FUNCTIONAL CLASSIFICATIONS

A brief discussion of the possible role of teleonomic functions in typing traits is included in this penultimate section. Its most obvious role in this respect concerns analogous categories of traits. Analogous categories are contrasted with homologous categories. Standardly, two structurally similar traits in two separate species are said to be homologous if they were inherited from a common ancestor, regardless of whether they have the same function. For example, vertebrate forelimbs (e.g., human arms, horse forelegs, bird wings) are homologous, as are some of the bones involved in hearing in mammals (the malleus and incus) to certain jawbones in reptiles. In contrast, two traits are analogous if they share the same function as a result of independent evolution. Thus avian wings, bat wings, and insect wings are analogous, since these wings evolved independently as an adaptation for flight in several separate lineages. Analogous categories are always functional categories. Homologous categories need not be.

Analogous categories are of interest in understanding the extent to which similar selection pressures or ecological opportunities lead to similar strategies. But they might not be considered genuine natural kinds, or not especially interesting ones, since the similarities among members of an analogous category might be few or superficial, making the role of the concept of the category inferentially poor as opposed to rich (Amundson & Lauder 1994).

It is, however, a mistake to equate functional categories with analogous categories, since trait classifications can use multiple criteria. There can be cross-classification involving function and homology (e.g., pectoral fins), involving function and taxon (e.g., Pterodactyl wings), or involving function, morphology, and taxon (e.g., low aspect ratio versus high aspect ratio avian wings), to mention a few ways in which function could combine with other criteria to determine non-analogous classifications of traits. The significance of the teleonomic notion of function for trait classification therefore does not rest on how useful analogous categories are.

Many categories of traits, as with many categories of artifacts, are malfunction-inclusive. Must a mousetrap be able to catch a mouse? Or does a broken mousetrap still count as a mousetrap? I take the answer to be affirmative. Similarly, a broken, deformed, diseased or paralyzed Pterodactyl wing is still a wing, even if it does not enable flight. *If* having a certain function is *required* for membership in such a category, the notion of function is malfunction-permitting (and not simply a notion of a causal disposition) since being able to perform the function is not required. The issues hereabouts are admittedly complicated by the fact that trait types can be malfunction-inclusive without directly involving a teleonomic notion of function. For example, Pterodactyl forelimbs can be broken and yet still count as Pterodactyl forelimbs. But the claim that teleonomic function often plays a role alongside other criteria (homology, morphology, molecular signature, the mechanisms of inheritance and development, taxon, and so on) is plausible. It is plausible, for instance, that the individuation (as opposed to the identification) of distinct mechanisms in an organism relies on functional considerations. It is also plausible that, even when the functional criteria are replaced with other criteria in an operational definition (such as a specific molecular signature), the motivation for individuating the mechanism in that way, as opposed to in another way, often involves functional considerations (Garson 2013).

One objection raised against the claim that functional considerations play a role in trait classification is that circularity threatens. If a token trait's function depends on what type of trait it is, and how it is typed in turn depends on its function, this seems circular. And almost every theory of functions assigns functions in the first instance to traits *of a type*. For example, the etiological theory says that the function of a token trait x is to do z only if traits *of x's type* were selected for z-ing. How, then, can x being of the x-type in turn depend on it having the function to z? There might appear to be just two options here: deny that an item's being of the x-type depends on its having the function to z, or deny that its having the function to z depends on its being of the x-type. Unusually, Bence

Nanay (2010) argues in favor of the second option, thus allowing for functional categorizations. But I believe there is also a third option, which is that being a trait of a certain type and having a certain function can co-supervene. In other words, the circularity is not vicious. According to the etiological theory, the trait type and the function can supervene on the same selection history (Neander & Rosenberg 2013). A token trait's being a wing *and* having the function to enable flight, for instance, co-supervenes on the token's location in a lineage (in terms of its relations of ancestry and descent), and on whether selection for flight operated on the part of the lineage to which it belongs. If we draw lines in the lineage with respect to when selection for flight started and/or stopped, a (non-vestigial) wing that has the function of flight belongs to a part of the lineage during which there was selection for flight.

So it remains plausible that the teleonomic notion of function plays a significant role in trait classification, especially given its role in explanation. But since there are many different trait classifications, which may use different criteria in different combinations in delineating them, as well as a number of different philosophical analyses of the teleonomic notion of function, the role of this notion in trait classification is not something that can easily be settled.

CONCLUDING REMARKS

Contemporary philosophers who maintain that biology still needs "teleology" rarely mean that it needs the idea of intelligent or divine design. What they usually mean is that biology needs a teleonomic notion of function, which is sensitive to the function/dysfunction and function/non-function distinctions, and which can be given a naturalistic analysis. One analysis of this notion is given by the etiological theory, which tells us (roughly) that the function of an item is to do what it was (or items of the type were) selected to do by natural selection, but there are other competing analyses on offer. This notion seems to play a role in two different kinds of functional explanations. One is explicitly teleonomic. Such functional explanations explain items (in part) in terms of their functions; on the etiological theory, in terms of what they were selected to do. While these kinds of explanations are derivative of selectional explanations, there is an important role in discovery for asking and answering why-questions. The other kind of explanation is an operational explanation of how a living system operates when it functions normally or properly, which is a componential analysis or mechanistic explanation of a normal system. This provides for some useful generality in the face of variation. Further, the teleonomic notion of function might also play a role in trait classification, although the issues here are not easily disentangled. In assessing the significance of its classificatory role, we need to keep in mind that analogous categories of traits might only be one kind of functional category, since teleonomic function might play a role along with other criteria, such as homology or morphology.

NOTES

1. Baddeley & Hitch (1974: 86).
2. This is the title of Amador et al. (2015).
3. The nature of the everyday concept of an artifact is in part an empirical question. See, for example, Kelemen & Carey (2007) for some empirical backing.

4. For details, see Garson 2012 and 2016.

5. There are strands of discussion sympathetic to this idea in Nagel 1977, Wimsatt 1972, Boorse 1976, McShea 2012, and Trestman 2012.

6. Christopher Boorse (2002) prefers to say of a token x that's dysfunctional with respect to z that normal xs have the function to z, and that token x would have the function to z if it were normal. For present purposes, however, we can treat this way of speaking as a terminological variant.

7. For example, Boorse offers responses to the possibility of typical dysfunction in his 1977 and 2002. In the latter he relies on functions pertaining to typicality in a time-extended population.

8. Boorse (1977: 557) aptly calls this the "composite portrait" of a species.

REFERENCES

Amador, G., Mao, W., DeMercurio, P., Montero, C., Clewis, J., Alexeev, A., & Hu, D. 2015. "Eyelashes divert airflow to protect the eye." *Journal of the Royal Society Interface* 12: 2014–1294.

Amundson, R. & Lauder, G. 1994 "Function without purpose." *Biology and Philosophy* 9: 443–469.

Baddeley, A. & Hitch, G. 1974. "Working memory." *The Psychology of Learning and Motivation* 8: 47–89.

Beckner, M. 1959. *The Biological Way of Thought* (University of California Press).

Bigelow, J. & Pargetter, R. 1978. "Functions." *Journal of Philosophy* 84: 181–196.

Boorse, C. 1976. "Wright on functions." *The Philosophical Review* 82: 70–86.

Boorse, C. 1977. "Health as a theoretical concept." *Philosophy of Science* 44: 542–573.

Boorse, C. 2002. "A rebuttal on functions." In A. Ariew, R. Cummins, & M. Perlman (eds.), *Functions: New Readings in the Philosophy of Psychology and Biology* (Oxford University Press) 63–112.

Craver, C. & Darden, L. 2013. *In Search of Mechanisms* (University of Chicago Press).

Cummins, R. 1975. "Functional analysis." *Journal of Philosophy* 72: 741–765.

Cummins, R. 2002. "Neo-teleology." In A. Ariew, R. Cummins, & M. Perlman (eds.), *Functions: New Readings in the Philosophy of Psychology and Biology* (Oxford University Press) 164–174.

Cummins, R. & Roth, M. 2010. "Traits have not evolved to function the way they do because of a past advantage." In F. Ayala & R. Arp (eds.), *Contemporary Debates in Philosophy of Biology* (Wiley-Blackwell) 72–86.

Garson, J. 2012. "Function, selection, and construction in the brain." *Synthese: Special Issue on Neuroscience and its Philosophy* 189: 451–481.

Garson, J. 2013. "The functional sense of mechanism." *Philosophy of Science* 80: 317–333.

Garson, J. 2016. *A Critical Overview of Biological Functions* (Springer).

Godfrey-Smith, P. 1993. "Functions: Consensus without unity." *Pacific Philosophical Quarterly* 74: 196–208.

Harvey, W. [1628] 2005. *On the Motion of the Heart and Blood in Animals* (Kessinger).

Kelemen, D. & Carey, S. 2007. "The essence of artifacts: Developing the design stance." In E. Margolis & S. Laurence (eds.), *Creations of the Mind: Theories of Artifacts and Their Representation* (Oxford University Press) 212–230.

McShea, D. 2012. "Upper-directed systems: a new approach to teleology in biology." *Biology and Philosophy* 27: 663–684.

Millikan, R. 1989. "In defense of proper functions." *Philosophy of Science* 56: 288–302.

Nagel, E. 1977. "Teleology revisited: Goal-directed processes in biology." *Journal of Philosophy* 74: 261–279.

Nanay, B. 2010. "A modal theory of function." *Journal of Philosophy* 107: 412–431.

Neander, K. 1991. "The teleological notion of 'function.'" *Australasian Journal of Philosophy* 69: 454–468.

Neander, K. 1995. "Pruning the tree of life." *British Journal for the Philosophy of Science* 46: 59–80.

Neander, K. 2015. "Functional analysis and the species design." *Synthese: Special Issue on Teleological Organization.* doi: 10.1007/s11229-015-0940-9.

Neander, K. & Rosenberg, A. 2013. "Solving the circularity problem for functions." *Journal of Philosophy* 109: 613–622.

Sober, E. 1984. *The Nature of Selection: Evolutionary Theory in Philosophical Focus* (University of Chicago Press).

Trestman, M. 2012. "Implicit and explicit goal directedness." *Erkenntnis* 77: 207–236.

Wimsatt, W. 1972. "Teleology and the logical structure of function statements." *Studies in History and Philosophy of Science* 3: 1–80.

Wright, L. 1973. "Functions." *Philosophical Review* 82: 139–168.

Wright, L. 1976. *Teleological Explanation* (University of California Press).

II

Evolution and Information

6

Evolution and Information: An Overview

Ulrich Stegmann

Many biological processes appear to involve the storing, transmission, and processing of information. Heredity is often conceptualized as the transfer of information from parents to offspring and development as the expression of that information. Information also figures at higher levels of the biological hierarchy; for example, in animal communication and individual decision-making. Evolutionary biology, too, has seen its share of "informational thinking."

Yet, information is a notoriously slippery concept that often generates confusion and cross-talk. When reading a newspaper we hope to acquire information about an event of interest, in which case information is a piece of knowledge or, perhaps, a proposition. In information theory, a branch of mathematics that can be traced back to Claude Shannon's (1948) theory of communication, information is concerned with a set of quantifiable, probabilistic features. And then there is the sort of information that one thing can carry about another; for example, the information that the growth rings of trees carry about the age of trees. This kind of information is known as "natural information." Traditionally, natural information is construed as a mind-independent feature of the world, something that organisms can exploit in order to guide their behavior. It is therefore also likely to be important for evolutionary processes.

Natural information has received considerable philosophical attention since the 1970s. Fred Dretske's (1981) book *Knowledge and the Flow of Information* was an early landmark, which shaped much subsequent thinking in the field. The present chapter surveys some of this work, delineating what has been achieved and identifying some of the recurrent challenges. A specific exploration of the evolutionary significance of natural information, however, is a matter for future research and hence beyond the scope of this chapter.

EVOLUTION AND INFORMATION

Before surveying the philosophical work on natural information it is useful to mention some of the points of contact between evolution and information. One of the earliest connections was made by George Williams (1966), who combined natural selection with information in order to define genes. Genes, for Williams, are units of hereditary information that are subject to a certain amount of selection, where "information" is intended to mean something quite distinct from matter and energy. Williams's gene concept was criticized for implying a mysterious domain of information (Godfrey-Smith & Sterelny 2016) as well as for privileging genes as the only object of evolution (Griesemer 2005).

Another point of contact was elaborated in John Maynard Smith and Eörs Szathmáry's (1995) influential book *The Major Transitions in Evolution*. Major transitions are important evolutionary innovations, such as the transition from uni- to multicellular organisms. Maynard Smith and Szathmáry argued that the major evolutionary transitions are intimately connected with changes in the way in which information is stored and transmitted. Among these changes is the emergence of non-genetic (epigenetic and cultural) inheritance systems. Another important change is the transition from "limited" to "unlimited" inheritance systems, which essentially differ in the amount of information they can transfer. Maynard Smith and Szathmáry also maintained that the amount of genetic information increases over evolutionary time. The latter claim was later assessed with the help of information theory. In computer simulations, a population of nucleic acids with random base sequences was subjected to mutations and selection for efficient protein-binding sites. After 700+ generations, the nucleic acids had converged on non-random base sequences at the binding sites (Schneider 2000). This convergence amounted to an increase in (quantitative) information at the binding sites because the base sequences at these sites had become more and more predictable. The study also undermined a contention popular among defenders of Intelligent Design—namely, that natural processes like mutation and selection are insufficient for increasing information over evolutionary time.

More recently it has been argued that information theory can improve the theoretical basis of evolutionary biology. The theory of natural selection is the formal centerpiece of evolutionary biology, and it is exemplified by Ronald Fisher's (1930) "fundamental theorem of natural selection." The theorem describes how the fitness of individuals affects the rate of change in a population's average fitness. The rate of change is closely related to a population's growth rate, and it was found that a population's growth rate can be described with a quantity that has an information theoretic interpretation (Jeffrey's divergence). This fact, in turn, allows an information theoretic articulation of the theory of natural selection as a whole. According to Steven Frank (2012), such an articulation is preferable to the statistical formalisms of the classical theory (but see Sarkar 2014; Crawford 2015).

INTRODUCING NATURAL INFORMATION

Natural information was introduced above with the example of growth rings of trees. Growth rings are informative in the sense that they can reveal tree age. Dretske (1981) captured this idea as follows:

When a scientist tells us that we can use the pupil of the eye as a source of information about another person's feelings or attitudes, that a thunder signature (sound) contains information about the lightning channel that produced it, that the dance of a honeybee contains information as to the whereabouts of the nectar or that the light from a star carries information about the chemical constitution of that body, the scientist is clearly referring to information as something capable of yielding knowledge. A state of affairs contains information about X to just that extent to which a suitably placed observer could learn something about X by consulting it.   (1981: 45)

In short, the basic intuition behind information is that it may yield knowledge. But *how* is it possible that one event can tell us about another? Exactly when is an observer "suitably placed"? Answering these questions is the task of a philosophical *theory* of natural information. In other words, theories of information aim to identify the specific conditions under which one event or state of affairs carries information about another.

Some terminological conventions will be useful in what follows. First, let's distinguish the information-carrying state of affairs (r) from the state about which it carries information (s). Second, let's refer to the information content, which r carries about s, with the predicate ". . . is F." Hence, growth rings (r) carry information about tree age (s), and the information they carry is, say, that the tree is three years old (is F), rather than some other age. Similarly, a fingerprint (r) carries information about the murderer (s), and the information it carries is that the murderer is Moriarty (is F), rather than someone else. The task of a philosophical theory of information is then to fill in the dots in the proposition "r carries information that s is F if and only if . . . ." The dots are placeholders for individually necessary and jointly sufficient conditions.

DRETSKE'S ACCOUNT

Dretske's (1981) work is a good starting point for surveying philosophical theories of natural information. Dretske argued that r's ability to carry information consists in a probabilistic relation between r, s's being F, and one's background knowledge k. He articulated the relevant probabilistic relation in terms of conditional probabilities, that is, the probability of an event *given* that some other event has occurred. Here is Dretske's analysis of (natural) information:

A signal r carries the information that s is F = the conditional probability of s's being F, given r (and k), is 1 (but, given k alone, less than 1).   (1981: 65)

This analysis involves two conditional probabilities. The first is the probability of s's being F given *both* the signal r and background knowledge k; the second is the probability of s's being F given only the background knowledge k, in the absence of signal r. Signal r carries the information that s is F just in case the first conditional probability is unity and the second is smaller than unity. Plausibly, both conditions are satisfied in the case of light emitted from distant stars. The probability that a given star (s) has a certain chemical composition (F) is less than 1 in the absence of knowing what light (r) it emits. But the star cannot fail to have that composition if it emits light of a certain quality. Hence, the light carries the information that the star has a certain chemical composition.

Three features of Dretske's (1981) account are worth highlighting. One feature is the reliance on laws of nature. According to Dretske, only states connected by laws of nature, or logical principles, can have conditional probabilities of 1. Stars "must" have a certain chemical composition if they emit light of a certain quality because a star's composition is nomologically related to the emitted light. Accidental coincidences, by contrast, cannot guarantee conditional probabilities of 1. Suppose my neighbor drives to work in the morning shortly before the local bus arrives. Assuming my neighbor drove off just now, there is a high conditional probability that the bus will arrive shortly. But since my neighbor is not connected to the bus by a law of nature, a change to the bus schedule will diminish the conditional probability of the bus arriving.

The second feature of Dretske's account is that natural information is always correct. Information is veridical. According to the first condition, r carries the information that s is F only if the occurrence of r raises to 1 the conditional probability that s is F. In other words, the first condition excludes the possibility that r carries the information that s is F in case s is actually not F.

The third feature is the role of background knowledge. Dretske illustrates the significance of background knowledge with a "shell game." There are several shells, and players must guess which one covers a nut. A player choses one shell and lifts it to determine whether or not she guessed correctly. If not, she lowers the shell back into place and another player gets his turn. The game continues until the nut is found. Now suppose there are four shells in the game, and shells 1 and 2 are found empty. At that point a new player joins but is not told about the empty shells. The game resumes with shell 3 being lifted and found empty. At this stage, shell 3's being empty carries different information for the different players. For the original players it indicates that the nut is beneath shell 4, because this is the only remaining shell. However, for the new player it does not carry this information, because his background knowledge does not include the fact that shells 1 and 2 had been turned over already and found empty. So, a given state of affairs can carry information for some observers but not others, and whether or not it does depends on what they know.

All three features proved to be contentious. For instance, the dependence of natural information on a subject's epistemic situation undermines the view that natural information is an observer-independent fact "out there" in the world. It also blocks analyzing doxastic states in terms of natural information, because the appeal to natural information then presupposes a notion of, say, knowledge. Dretske tried to minimize these worries by assuming shared background knowledge among observers, in which case no explicit reference to background knowledge is needed.

Many authors agree that Dretske's probabilistic condition is too demanding (e.g., Loewer 1983; Suppes 1983). Recall that for r to carry the natural information that s is F it must be nomologically impossible for r to occur if s is not F. But many states appear to carry information about others without satisfying this condition. For example, extreme weather events like cold spells can cause trees to form extra growth rings, in addition to the ones reflecting their age. Nevertheless, the growth rings of trees are taken to carry information about their age; an entire discipline (dendrochronology) has emerged on that assumption and proven successful. Furthermore, animals gain information from features that do not guarantee the occurrence of the indicated state (e.g., Godfrey-Smith 1989). Vibrations in a spider's net, for instance, carry information that there is prey without guaranteeing its presence.

Another persistent worry is that there is no suitable notion of probability (Loewer 1983). For example, if probabilities are understood as subjective degrees of belief, then doxastic states cannot be analyzed in terms of natural information, as this would presuppose a notion of belief. Other interpretations also run into difficulties. On a propensity interpretation, probabilities are dispositions of some states to bring about others. However, some states carry the information that others obtain while lacking any disposition to cause them (Demir 2008). For instance, animal alarm signals indicate an approaching predator without in any way causing the predator's approach.

A natural response is to identify and then eliminate the sources of these difficulties. Two such sources are the reliance on background knowledge and probabilities. The counterfactual theory (Cohen & Meskin 2006) therefore seeks to exclude both from a theory of natural information.

COUNTERFACTUAL THEORY

Suppose a doorbell is set up in such a way that it only ever rings if someone is at the door. The ringing of such a bell *implies* that someone is at the door. So, we can learn from its ringing that someone is there and, intuitively, the bell's ring carries the information that someone is at the door. According to the counterfactual theory (Cohen & Meskin 2006), the bell's ring carries this information because the ringing is counterfactually related in the right way with people at the door. For if the doorbell only ever rings if someone is at the door, then the following counterfactual is true: if nobody were at the door, then the bell would not ring. The obtaining of this counterfactual relation is seen as constitutive of natural information:

> x's being F carries information about y's being G if and only if the counterfactual conditional 'if y were not G, then x would not have been F' is non-vacuously true.
>
> (2006: 3)

Here, x designates the information-carrying state, y the state about which information is being carried, and ". . . is G" denotes the information content. Stipulating that the counterfactual must be non-vacuously true avoids problems resulting from the standard semantics of counterfactuals. Suppose it is metaphysically *necessary* that y is G. Then the counterfactual "If y were not G, then x would not have been F" has an impossible antecedent: by hypothesis, it is necessary that y is G; that is, there exists no possible world in which y is not G. Now, counterfactuals with impossible antecedents are vacuously true, according to the standard semantics of counterfactuals (the Lewis–Stalnaker theories). If vacuously true counterfactuals could ground natural information, then x's being F would carry the information that y is G, even though y could never fail to be G, irrespective of whether or not x is F.

Though Cohen and Meskin (2006) aim to put much distance between their account and Dretske's, they are very similar in one crucial respect. Both imply that natural information is veridical. That is, if x's being F carries the information that y is G, then y is G. It cannot be the case that x's being F carries the information that y is G when in fact y is not G.

Since the counterfactual theory does not rely on probabilities, it circumvents the difficulties related to probabilities. And in virtue of identifying natural information with certain counterfactual relations in the world, information is independent of the epistemic situation of observers. For this reason, natural information can be employed, as Dretske intended, to analyze notions in epistemology and philosophy of mind, such as knowledge and perceptual content.

However, the independence of information from an observer's background knowledge has its downside. Recall the stage in Dretske's shell game when the original players have witnessed shells 1 and 2 being empty and a new player, who does not know about the empty shells, joins the game. The game resumes with the lifting of shell 3, which also turns out to be empty. Intuitively, shell 3's being empty carries the information that the nut is under shell 4 for the original players, but not for the new player, and the difference is due to their diverging background knowledge regarding shells 1 and 2. Dretske's account reproduces this intuition, but the counterfactual theory does not (Scarantino 2008). In order for shell 3's being empty to carry the information that the nut is under shell 4, the following must hold: if the nut were not under shell 4, then shell 3 would not be empty. But this counterfactual is false, because if the nut weren't hidden under shell 4, then it may be hidden under any other shell, including 1 and 2. Put differently, among the possible worlds in which the nut is not under shell 4, worlds in which it is hidden under shell 1 or 2 may be just as close or closer than worlds in which it is hidden under shell 3. Consequently, on the counterfactual theory, shell 3's being empty does not carry the information that the nut is under shell 4, not even for the original players.

The counterfactual theory gives up on Dretske's idea that natural information is a probabilistic relation. But it retains his view that if a state carries natural information about another, then the latter always obtains. Probabilistic theories of natural information take the opposite route.

PROBABILISTIC INFORMATION

Dretske's (1981) theory was criticized for being too demanding in requiring that natural information renders a state *certain*. Probabilistic accounts maintain that even imperfectly related events carry natural information under certain conditions, and they then aim to identify these conditions.

Two probabilistic relations should be distinguished from the outset, that is, degrees of coincidence and probability-changing. The degree of coincidence between one event or state of affairs (A) and another (B) is the degree to which As co-occur with Bs. If 20 percent of As co-occur with Bs, then A's degree of coincidence with B is 20 percent. For example, if 20 percent of alarm calls co-occur with predators, then the correlation between alarm calls (A) and predators (B) is 20 percent. Put in probabilistic terms, the conditional probability of an approaching predator given an alarm call is $p(B|A) = 0.2$.

Alternatively, *A* can correlate with *B* in the sense of "changing" *B*'s probability. Suppose again that the conditional probability of an approaching predator is $p(B|A) = 0.2$. In addition, suppose that predators *always* elicit alarm calls, so that the conditional probability of an approaching predator without an alarm call is $p(B|\neg A) = 0$. In this case, the call increases the predator's probability from 0 percent to 20 percent [$p(B|A) - p(B|\neg A) = 0.2$]. Note that in the examples considered so far, coincidence and probability change had the same value (0.2). But this need not be the case. If, for instance, 20 percent of predators slip through without eliciting alarm calls [$p(B|\neg A) = 0.2$], then an alarm call does not make a predator more likely than it was before. Despite a positive degree of coincidence, there is no change in the predator's probability.

Probabilistic accounts of natural information come in several versions (Millikan 2000, 2004; Shea 2007; Piccinini & Scarantino 2010; Scarantino & Piccinini 2010; Skyrms 2010; Scarantino 2015). The differences mainly concern the type of probabilistic relation that is deemed constitutive of information; for example, whether it consists in coincidence or probability-changing, and whether or not the probabilistic relation must obtain for non-accidental reasons. The most sophisticated account is Andrea Scarantino's (2015) "probabilistic difference maker theory."

Scarantino (2015) seeks to capture the central idea of probabilistic accounts, that even imperfectly correlated events can carry natural information, by appealing to probability-changing:

> *Incremental Natural Information* (INI): *r*'s being *G* carries *incremental natural information* about *s*'s being *F* relative to background data *d* if and only if $p(s \text{ is } F | r \text{ is } G \ \& \ d) \neq p(s \text{ is } F | d)$.
>
> (2015: 423)

According to this account, carrying natural information is a matter of one state's changing (increasing or decreasing) the probability of another. For example, the occurrence of an alarm call (*r*'s being *G*) carries incremental information about a predator approaching (*s*'s being *F*) if the call makes a predator more (or less) likely than in the absence of an alarm call. On the other hand, if the call makes no difference to the probability of a predator, then the call does not carry information about it. Note that the change in probabilities is relative to background data. Background data are sets of propositions which an observer brings to bear and which may or may not be true. In the case of alarm calls it includes the listeners' assumption that alarm calls make an approaching predator more likely. In the shell game, the background data include knowing that shells 1 and 2 are empty.

Relativizing probabilities on background data turns natural information into a receiver-dependent, three-term relation. It ceases to be something "out there" in the world that organisms simply encounter. Nevertheless, once the signal and background data are fixed, changes to probabilities ensue independently of an observer's hopes or opinions. Furthermore, relativizing probabilities affords a response to the reference problem. Correlations exist only relative to some reference class, and it has been argued that determining reference classes is always arbitrary (Millikan 2013). But relativizing probabilities systematically on receivers' background data allows a non-arbitrary delineation of the reference class.

Increasing the probability of an event does not necessarily make it *probable* to occur (similarly, for decreasing an event's probability). In the example above, an alarm call increases the probability of an approaching predator from 0 percent to 20 percent.

This increase to 20 percent does not make it probable that a predator is approaching; that would require the probability to be higher than chance (e.g., 70 percent). The increase to 20 percent just means that an approaching predator is now more likely than it was before (without the alarm call). Nevertheless, Scarantino (2015) argues that the overall conditional probability of an event is an important aspect of natural information and has been neglected by other probabilistic accounts. This feature is captured as follows:

> *Degree of Overall Support* (DOS): the degree of overall support provided by a signal r's being G carrying incremental natural information about s's being F relative to background data d is equal to $p(s$ is $F|r$ is G & $d)$.

> (2015: 423)

In the example above, r's being G raises the probability of s's being F to 20 percent; its overall support for s's being F is thus $p(s$ is $F|r$ is G & $d) = 0.2$.

So far we have looked at the conditions under which one thing carries natural information about another. But what is the content of that information? What does r's being G "tell" us? Any given state of affairs changes the probabilities of many states simultaneously, not merely of one other state. For example, an alarm call increases the probability of an approaching predator, but also of the predator having been detected, the receivers taking evasive action, and so on. At the same time, the call decreases the probability of the predator being successful, of the caller continuing to forage, and so on. Furthermore, the call provides some degree of overall support for all these states. According to Scarantino, they are all part of the alarm call's information content. The formal definition of information content is too complex to reproduce here, suffice it to say that it includes three features: the identity of the states whose probabilities are changed, the amount of change, and the probabilities after the change.

One attraction of probabilistic accounts is that natural information can be shown to do genuine explanatory work (Stegmann 2015). Suppose an organism responds with a type of behavior B to events of type A. Since organisms exhibit B specifically in response to experiencing A, B is part of a dispositional property of these organisms. Suppose we wish to explain, in the first instance, why the organism responds to a particular event a with a manifestation of behavior b. The explanation will proceed as follows:

[E 1]
1. Organism o has the disposition to respond to A with B
2. There was an A-token (a)
Therefore, o responded to a with behavior b

In other words, perceiving a together with a disposition to respond to A-tokens explains why the organism responded with b. This is an ordinary causal explanation and does not appeal to the A-token raising the probability of some other event. It therefore does not seem to involve natural (probabilistic) information. However, the role of information becomes apparent once we push our inquiry a step further, specifically by asking why the organism has the disposition in the first place. Let's assume that the disposition is acquired through associative learning. Acquiring the disposition can then be explained as follows:

[E 2]

1. In the past, the conditioned stimulus A increased the probability of the unconditioned stimulus C.

2. Other conditions necessary for associative learning were satisfied (e.g., surprise and belongingness).

Therefore, the organisms acquired the disposition to respond to A with B.

Since stimulus A increases the probability of C (premise 1), A carries natural information about C's occurrence. And this fact partly explains why the organism acquired the disposition, as per [E 2]. Natural information is therefore explanatory. Ultimately, natural information is also explanatory of an organism's token behavior b, because having the disposition partly explains b, as per [E 1]. Probabilistic theories therefore show how natural information can play a well-delineated and significant explanatory role.

Of course, probabilistic accounts have their difficulties. Most obviously, the lack of an adequate interpretation of probability, which plagued Drestke's theory, resurfaces here. It remains to be seen whether appeals to objective interpretations of probability (e.g., objective Bayesianism) can fill this gap.

Proponents of probabilistic accounts often believe that they successfully capture the role of information in the brain and behavioral sciences. Scarantino (2015), for instance, maintains that the information carried by vervet monkey alarm calls is probabilistic information about predators. While alarm calls do carry such information, it is not obvious that this is the kind of information which scientists actually attribute to alarm calls and which figures in their explanations and predictions (Stegmann 2013). This can be seen, for instance, by comparing the information content attributed by scientists with the content predicted by probabilistic accounts. The information content scientists attribute to alarm calls can be fairly narrow; for example, a vervet's snake alarm call is supposed to signal the presence of a dangerous type of snake, mostly pythons (Seyfarth & Cheney 2003). But on probabilistic accounts, including Scarantino's, the information carried by alarm calls is about much more than approaching predators (see above).

Another challenge concerns "wild tokens." When r's being G correlates with s's being F, then it is possible that some rs are G without there being ss that are F. Do such "wild" r-tokens carry the natural information that s is F? If they do, then natural information can be false. This implication is problematic, because it undermines the entrenched view that natural information is always true and, in addition, it requires an alternative criterion for distinguishing between natural information and representational content. On the other hand, if wild tokens are not informative, then carrying natural information cannot merely be a matter of being an instance of a probabilistically related type. And this outcome contradicts the main tenet of probabilistic accounts (Stegmann 2015).

NATURAL INFORMATION AS CAUSATION

This section considers a third set of theories of natural information, which analyze information in terms of causation. Causation is usually understood as a relation between things in the world, which either does or does not hold between any particular pair of things. The causal relation that obtains between a token cause and its token effect is known as singular causation.

Singular causation is the essence of natural information, according to Karen Neander's (2013) account of "singular causal information." Where r and e are particulars, "r carries the natural indicative information that e if e is a cause of r." Thus, any given effect carries the information that its cause has occurred. And it does so simply in virtue of the fact that it has been caused by that particular. If a red tomato causes a certain activation pattern (RED) in a sensory neuron, then RED carries the information that there is a red thing. Neander maintains, in addition, that singular causal information can be "forward-looking," that is, token causes can carry (prescriptive) information about their token effects. This ensures, for instance, that "motor instructions carry information about the movements they cause" (Neander 2013: 27).

An attraction of singular causal information is its ability to rule out misinformation. By definition, a RED token carries the information that there is something red only if it was actually caused by something red. Furthermore, Neander (2013) maintains that our intuitions about natural information match her account. Recall that extreme weather events may cause trees to form additional growth rings; for example, a severe cold spell may cause a five-year-old tree to add a sixth ring. Neander argues that our intuitive judgment in this case would be that the sixth ring does not mean or indicate that the tree is six years old. This intuition would be in line with her account, because the tree's age does not cause the sixth ring and therefore does not carry information about age.

A potential worry is that biologists and cognitive scientists attribute information to many fewer entities than this account does. For instance, the light-dependent reduction in a light receptor's firing rate is taken to signal the presence of light, whereas the detachment of part of the retina is not regarded as indicating the presence of the tiny breaks that caused it (except for an ophthalmologist). Yet both events carry singular causal information. The worry is not decisive, however. Probabilistic accounts face the same worry and, furthermore, the cognitive scientists' more selective use of "information" may be due to pragmatic considerations that are layered on top of a causal notion of information, as Neander suggests.

Another notion built on a causal analysis of information is "mechanistic information" (Bogen & Machamer 2010). Consider the chain of events in mechanisms like protein synthesis. The components occurring at one stage of protein synthesis exert a causal influence on its subsequent stages. The central idea is to identify information with this causal influence, subject to two conditions. One condition is that the components serve an organism's usual needs, by advancing the mechanism to its end state. The second condition is that the components have long "reach" (roughly, their causal influence extends far downstream and consists in strongly reducing the number of possible downstream consequences). In Jim Bogen and Peter Machamer's words: "Mechanistic information is the causal influence that entities and activities at one step in the operation of a mechanism exert to select teleologically appropriate results for production in one or more subsequent steps" (2010: 858).[1]

The account draws interesting distinctions between types of causal influence. And, in contrast to Neander's theory, it appears to construe information as a matter of degree because features like "reach" come in degrees. A number of key issues remain open, however. Does the intended content of mechanistic information concern upstream causes, downstream consequences, or both? Is mechanistic information veridical? There are also questions about the theory's intended scope. While it is not promoted as a general theory of natural information, it is meant to account for phenomena that usually qualify as natural information; for example, the information that the activation pattern of a worm's sensory neurons carries about pressure applied to its skin. Perhaps the theory's scope is more specific, an account of biological information or even just of informational mechanisms.

Ultimately, causal theories of natural information face a common problem. Natural information is often carried by states that are not causes of one another. The fall in a barometer's mercury column indicates a storm, but neither causes the other. Instead they are the effects of a common cause, the drop in atmospheric pressure. Causal accounts rule out such stock examples and, consequently, the possibility of animals extracting information from the myriads of non-causal correlations they encounter.

ACKNOWLEDGMENTS

I thank Gerry Hough, Federico Luzzi, and Stephan Torre for their helpful comments on the draft.

NOTE

1. Griffiths & Stotz (2013) also provide a causal account of information, specifically of genetic information. However, they place different constraints on the kind of causal relations that qualify as carrying information. Note that the converse project, explaining causation in terms of information, has also been attempted (e.g., McKay Illari 2011).

REFERENCES

Bogen, J. & Machamer, P. 2010. "Mechanistic information and causal continuity." In P. McKay Illari, F. Russo, & J. Williamson (eds.), *Causality in the Sciences* (Oxford University Press) 845–864.

Cohen, J. & Meskin, A. 2006. "An objective counterfactual theory of information." *Australasian Journal of Philosophy* 84: 333–352.

Crawford, D. 2015. "Sarkar on Frank." *Philosophy of Science* 82: 122–128.

Demir, H. 2008. "Counterfactuals vs. conditional probabilities: A critical examination of the counterfactual theory of information." *Australasian Journal of Philosophy* 86: 45–60.

Dretske, F. 1981. *Knowledge and the Flow of Information* (MIT Press).

Fisher, R. 1930. *The Genetical Theory of Natural Selection* (Clarendon Press).

Frank, S. 2012. "Natural selection: How to read the fundamental equations of evolutionary change in terms of information theory." *Journal of Evolutionary Biology* 25: 2377–2396.

Godfrey-Smith, P. 1989. "Misinformation." *Canadian Journal of Philosophy* 19: 533–550.

Godfrey-Smith, P. & Sterelny, K. 2016. "Biological information." In E. Zalta (ed.), *Stanford Encyclopedia of Philosophy*. http://plato.stanford.edu/entries/information-biological

Griesemer, J. 2005. "The informational gene and the substantial body: On the generalization of evolutionary theory by abstraction." In M. Jones & N. Cartwright (eds.), *Idealization XII: Correcting the Model, Idealization and Abstraction in the Sciences* (Rodopi) 59–115.

Griffiths, P. E. & Stotz, K. 2013. *Genetics and Philosophy: An Introduction* (Cambridge University Press).

Loewer, B. 1983. "Information and belief." *Behavioral and Brain Sciences* 6: 75–76.

Maynard Smith, J. & Szathmáry, E. 1995. *The Major Transitions in Evolution* (Oxford University Press).

McKay Illari, P. 2011. "Why theories of causality need production: An information transmission account." *Philosophy & Technology* 24: 95–114.

Millikan, R. 2000. *On Clear and Confused Ideas: An Essay about Substance Concepts* (Cambridge University Press).

Millikan, R. 2004. *The Varieties of Meaning* (MIT Press).

Millikan, R. 2013. "Natural information, intentional signs and animal communication." In U. Stegmann (ed.), *Animal Communication Theory: Information and Influence* (Cambridge University Press) 133–146.

Neander, K. 2013. "Toward an informational teleosemantics." In D. Ryder, J. Kingsbury, & K. Williford (eds.), *Millikan and her Critics* (Blackwell) 21–36.

Piccinini, G. & Scarantino, A. 2010. "Computation vs. information processing: Why their difference matters to cognitive science." *Studies in History and Philosophy of Science* 41: 237–246.

Sarkar, S. 2014. "Does 'information' provide a compelling framework for a theory of natural selection? Grounds for caution." *Philosophy of Science* 81: 22–30.

Scarantino, A. 2008. "Shell games, information, and counterfactuals." *Australasian Journal of Philosophy* 86: 629–634.

Scarantino, A. 2015 "Information as a probabilistic difference maker." *Australasian Journal of Philosophy* 93: 419–443.

Scarantino, A. & Piccinini, G. 2010. "Information without truth." *Metaphilosophy* 41: 313–330.

Schneider, T. 2000. "Evolution of biological information." *Nucleic Acids Research* 28: 2794–2799.

Seyfarth, R. & Cheney, D. 2003. "Signalers and receivers in animal communication." *Annual Review of Psychology* 54: 145–173.

Shannon, C. 1948. "A mathematical theory of communication." *Bell System Technical Journal* 27: 379–423, 623–656.

Shea, N. 2007. "Consumers need information: Supplementing teleosemantics with an input condition." *Philosophy and Phenomenological Research* 75: 404–435.

Skyrms, B. 2010. *Signals: Evolution, Learning, and Information* (Oxford University Press).

Stegmann, U. 2013. "A primer on information and influence in animal communication." In U. Stegmann (ed.), *Animal Communication Theory: Information and Influence* (Cambridge University Press) 1–39.

Stegmann, U. 2015. "Prospects for probabilistic theories of natural information." *Erkenntnis* 80: 869–893.

Suppes, P. 1983. "Probability and information." *Behavioral and Brain Sciences* 6: 81.

Williams, G. 1966. *Adaptation and Natural Selection* (Princeton University Press).

7

The Construction of Learned Information through Selection Processes

Nir Fresco, Eva Jablonka, and Simona Ginsburg

INTRODUCTION

The aim of this chapter is to offer an analysis of information in the context of learning and communication. Information has been previously analyzed in the context of phylogenetic evolution, suggesting a receiver-centered, functional notion of information (Jablonka 2002). This analysis is complemented here by addressing the question of how an individual can learn and develop during ontogeny through exchanging and processing information. To this aim, we present a broad notion of *functional information* that can be applied to both phylogenetic and ontogenetic learning processes. We focus on ontogenetic learning, and claim that learned, functional information is produced through exploration processes and selective stabilization in a receiver.

Two basic notions should be characterized from the outset: "function" and "functional information." "Function" refers to the causal role of a structure, process, act, or strategy that has contributed to the goal-directed behavior of an encompassing system (based on Cummins 1975; Wright 1973; Kitcher 1993; for a discussion see Jablonka 2002). Importantly, the level at which the causal role is described may be different for different systems and contexts, and this may determine the level to which goal-directedness may be attributed. For example, although most of the specific neural circuits that are activated during trial-and-error learning do not have a direct beneficial effect (since most lead to errors), learning through trial-and-error *as a strategy* is highly functional. Hence, although specific neural circuits played a causal role in learning, this is not the level at which natural selection acted. Hence, a direct and straightforward relationship, between the input and the response to it, brought about by natural selection, is unlikely for complex multicomponent systems such as those involved in learning.[1]

"Functional information" refers to any difference in the (external or internal) environment of a system that has made a systematic difference to the system's goal-directed (teleological) behavior (extending Gregory Bateson's definition of a "bit of information" as "a difference which makes a difference" (1972: 315)). For a biological structure or process to have a *systematic functional effect* it must have been produced through ontogenetic and/or natural selection processes.

We understand the notion of selection in the broad, Pricean sense. According to George Price, selection is a sampling process that may or may not involve replication and multiplication, such as the selection involved in trial-and-error learning (1995). Thus understood, we interpret some examples of learning in Pricean terms, suggesting that information and selection are connected through exploration processes followed by selective stabilization processes that do not necessitate multiplication and replication.

We begin by defining key notions for a broad information taxonomy. In the next section, we discuss general selection and the Price equation, and suggest that learned, functional information can be measured in Pricean terms and the final section concludes the chapter.

A TAXONOMY OF INFORMATION

As the basis of our account of functional information, we define four key notions: "datum," "sign," "signal," and "symbol," so as to allow a broad taxonomy of information. One consequence of these definitions is that "information" is a graded concept.

> *Definition 1—"Datum."* A datum is either any act, event, process, or structure, or its absence,[2] to which a receiver can, but does not yet, functionally respond by being sensitive to variations in its form.[3]

> *Definition 2—"Sign."* A sign is a datum the receiver evolved either to overtly respond to or to acquire an altered disposition to respond to through past natural, ontogenetic, or cultural selection.

> *Definition 3—"Signal."* A signal is a sign that may have a learned component and is sent by a sender that evolved, through past natural, ontogenetic, or cultural selection, to emit it as a sign for particular receiver types.

> *Definition 4—"Symbol."* A symbol is a signal that is part of a systematic, rule-governed, self-referential-signaling system.

According to Definition 1, the classification of something as a datum is relative to a *receiver*, and, the system/organism has to be able to receive an input from the environment, which is broadly understood to include also the internal milieu, not only what is *external* to the receiver. The definition implies the receiver's sensitivity to the Batesonian difference mentioned above. (Quarks, to which an organism is insensitive, do not qualify as sensory data.) A physical substance can be replaced with another and still convey the same information so long as its form remains the same. The receiver, thus, has to be sensitive to variations in the *form* (i.e., organization), rather than the particular physical makeup, of the input. For example, nightingales (*Luscinia megarhynchos*) are sensitive not only to the phrase types of a particular male song,[4] but also to their transition pattern complexity (Weiss et al. 2014).

To be considered a datum, while the receiver should be able to detect the event, act, process, or structure concerned, the receiver does not yet functionally respond to it. Thus, a cloud qualifies as a datum for an organism that perceives it but neither responds to it nor changes the organism's future disposition to respond to it, thereby making it informationally neutral. This means that a datum cannot be what experimental psychologists call a "US" (unconditioned stimulus), something which activates an innate neural circuit.

For a datum to qualify more narrowly as a sign, it should elicit on average an adaptive response by the receiver. Every US (e.g., electric shock for mammals) is, therefore, a sign. An adaptive response should be broadly construed so as to include not only overt behavioral action, but also a change in a disposition to respond (e.g., predictions or beliefs).[5] This suggests that there is an intimate relation between information and (phylogenetic or ontogenetic) memory, for learning requires memory. Moreover, a neutral datum can become a sign through a process of learning: a dark cloud is a sign for an ape that has learned its significance (in psychological jargon, the cloud is a conditioned stimulus, CS). For not only does the ape see dark clouds in the sky, but it can also take appropriate action when needed (e.g., seek shelter from the rain) on the basis of the predictive association between dark clouds and rain. By contrast, for a baby ape, a dark cloud is only a datum, given its lack of functional importance for the ape at this stage of development.

When a datum becomes a sign it bears non-zero information for the receiver. An innate sign can have positive, often maximum, information for the receiver (it may require only a single trial by the receiver). For example, the Leatherback turtle, *Dermochelys coriacea*, has an irregular pink area on the crown of its head (its "pink spot") that is used to respond to light. When day-length shortens, the evolved pink spot with its neural underpinnings "interprets" the light (by initiating a complex cascade of reactions) as a decrease in day-length (the informational sign), making the turtle leave its foraging areas and migrate (Davenport et al. 2014).

A signal is a special case of a sign that is transmitted by a sender, and that evolved in the sender to convey information for the receiver. According to definition 3, a signal implies a *sender* that evolved to emit it.[6] While data and signs may be environmental, a signal originates in a sender that was selected to emit functionally informational signs for receivers. Selection should be broadly construed so as to include not only phylogenetic, but also cultural processes of selection and ontogenetic processes of selective (differential) stabilization. That is, selection here includes all processes that can contribute to adaptation via variation and selective retention. Importantly, signals, unlike signs, often have a sending cost. Senders invest in producing reliable signals, and receivers, who benefit from responding to reliable signals, exert pressure on senders to invest in the reliability of the signal by responding favorably to reliable signals while ignoring unreliable ones (Zahavi 2008).[7]

Two concrete examples of ontogenetic selection should help clarify the notions of datum, sign, and signal. First, consider the learning process of a rat in a Richard Morris water maze (1984). A rat is placed in a small pool filled with water and milk (to make the water murky thereby hiding a small, submerged platform in the pool (see Figure 7.1a)). The rat tries to escape from the water by whipping its tail. Once the platform is found, it offers the rat a temporary escape from swimming. The rat undergoes various training trials in the pool, typically, having access to either distal or proximal environmental indicators that can be used to spatially navigate (it can also learn about its own actions, but here we focus on learning from external indicators). Spatial learning requires some basic abilities, such as intact eyesight and swimming, as well as basic strategies, such as learning to climb on the platform and to swim away from the wall (Vorhees & Williams 2006).

How do data and signs come into play in such a setting? The rat is exposed to many *data* in the environment by perceiving multiple environmental stimuli, yet not all of them result in a particular behavioral response. Some of these environmental stimuli are not only perceived by the rat, but also trigger a positive functional response, because they become associated with a desired outcome. (The rat starts processing some data as *signs*.) The rat's learning may be viewed as a selection process of eliminating possible trajectories to that platform until the rat is satisfied with a minimal set of "optimal" trajectories it has settled on. (Figure 7.1b graphically shows the decrease in the lengths of trajectories explored.) In the next section, we revisit this experiment to show how the relevant information in the perceived sign(s) can be measured.

Communication about predators in vervet monkeys (*Chlorocebus aethiops*) is a good example of signals. Vervets give acoustically different alarm calls for leopards, eagles, and snakes, where each type of call elicits a different, adaptive response. Individuals run into trees when they hear a leopard alarm, look up in the air when they hear an eagle alarm, and stand up on their hind legs and peer into the grass around them when they hear a snake alarm call (Cheney & Seyfarth 1990). The calls of adults are selective and precise: "leopard alarms" are given to mammalian carnivores (mainly leopards), "eagle alarms" primarily to martial eagles and crowned eagles, and "snake alarms" almost only to pythons.

Although there is an innate basis for the production and usage of these calls, there is much fine-tuning through individual and social learning in the response to the alarm calls. Infants and juveniles make many mistakes, and give alarm calls to species that pose no danger to them. However, with time and experience they enhance the relation

A The watermaze

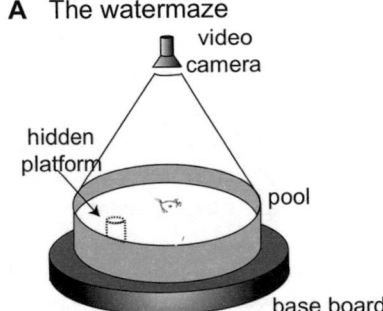

B Paths and latency during place navigation

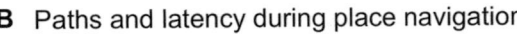
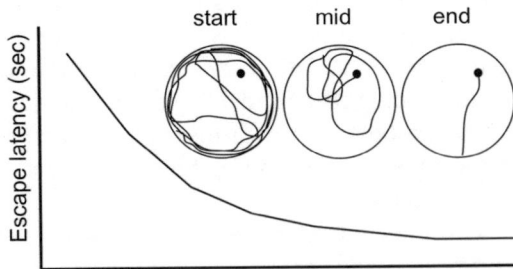

Figure 7.1 a) A typical water-maze set-up including a videocamera and a hidden platform; b) representative escape latency graph and swim paths across various stages of training: initial swimming at the side-walls, then circuitous paths across the area of the pool, and finally directed path-navigation. (Figure reproduced with permission from Scholarpedia and Richard Morris.)

between each type of alarm call and the corresponding stimulus (Seyfarth & Cheney 2010: 94). The calls, clearly, evolved to communicate different types of danger to conspecifics, though the monkeys seem not to have a psychological intention to warn others (Cheney & Seyfarth 1990; Seyfarth & Cheney 2010). The decrease in the irrelevant calls is a function of learning, and the amount of information in the signals increases with time and learning trials.

Our provisional definition of a symbol (following Deacon 1997, 2011; Jablonka & Lamb 2014: 196), emphasizes symbols' reference not only to things in the world, but also to other symbols within the symbolic system. Symbolic language allows reference to things that do not exist, because it need not depend on references to the world. In the case of formal logic, symbols may even refer *only* to other symbols. Strong semantic-symbolic information does not seem to exist in non-human animal communication, because animal signals do not form a self-referential system (Jablonka & Lamb 2014: 201). Symbols need not be confined to linguistic constructs, since pictures, paintings, diagrams, replicas, and maps can also convey symbolic information.

Since the usage of informational terms in the literature is variable, we briefly compare our definitions to some related ones (a detailed comparison exceeds the scope of this chapter). A *cue* is defined in animal communication theory as "a feature of the world, animate or inanimate, which can be used by an animal to guide future actions" (Maynard Smith & Harper 2003: 3; following Hasson 1994). Our comparable definition of a datum is different in that we see it as a difference maker only *in potentia*. When this potential is manifest (e.g., through learning) the datum becomes a sign. Luciano Floridi defines a datum as a difference relation between two variables that is both binary and symmetric: "nothing is a datum *per se*, without its counterpart" (2011: 356). Our definition, in contrast, explicitly presupposes both a receiver and its ability to detect the act, event, process, or structure (thereby making a datum less ubiquitous in nature compared with a Floridian datum). Our definition also presupposes some learnable correlation in the environment in the sense required by Andrea Scarantino's (2015) receiver-dependent natural information.

Our definition of a sign resembles Charles Peirce's definition, but is more restricted. In his later work, a sign is defined "as anything which is so determined by something else, called its Object, and so determines an effect upon a person, . . . [an effect called] its Interpretant" (Peirce 1998: 478). According to Peirce, a sign is whatever signifies—be that a written symbol or a picture. The interpretant is whatever effect the sign/object relation has on a person—or more generally, a receiver. On our definition, too, a sign signifies a relation to something in the receiver's environment that can (or does) elicit an effect on the receiver. However, when senders send phylogenetically, ontogenetically, or culturally evolved signs, these signs qualify as signals, and symbolic signifiers qualify as symbols (see definitions 3 and 4).

Importantly, definition 2 marks a crucial distinction from Peter Corning's account of functional information. What he calls "control information" "is always defined in terms of the functional relationship between the source and the user" (2007: 308). But, on Corning's account, functional information is, by definition, cybernetic: it is about the capacity to control the acquisition, disposition, and utilization of matter or energy in cybernetic processes (2007: 302). Definition 2 makes an implicit distinction between what may be called "descriptive" and "prescriptive" information. The latter—Corning's control information—is reflected in the receiver's overtly responding to a datum. The former is reflected in the receiver's acquiring an altered disposition to respond. Descriptive information is not unique to humans. Corvids, for example, have been shown to be capable of recalling "specific what-where-and-when memories about the past" that can *later* be used for planning (Clayton 2015: 209). Ruth Millikan calls the most primitive (information-based) representations in simple organisms "pushmi-pullyus" (2004: x, 77). She argues that their descriptive and prescriptive functions are undifferentiated. The vervet's alarm calls tell what kind of predator is near (*descriptive* information) and directs the response appropriate to that predator-type (*prescriptive* information) (Millikan 2004: 81).

Lastly, a signal is standardly defined in animal communication theory as "any act or structure which alters the behavior of other organisms, which evolved because of that effect, and which is effective because the receiver's response has also evolved" (Maynard Smith & Harper 2003: 3). Definition 3 is compatible with this standard definition, but the evolved aspect can be a general, rather than a specifically evolved communicative pre-disposition. A signal, as in some cases of animal communication and in human language, can evolve through cultural evolution and requires ontogenetic learning for its acquisition and use.

INFORMATION AND GENERAL SELECTION

We now return to the relation between functional information and general selection. Price developed the most general formalization of selection, often referred to as the "Price equation" (a mathematical statement describing how the average value of any character changes in a population from one generation to the next; see below). His broad notion of selection to which we alluded earlier, is clearly expressed in the following passage.

> Selection has been studied mainly in genetics, but of course there is much more to selection than just genetical selection. In psychology, for example, trial-and-error learning is simply learning by selection. . . . In linguistics, selection unceasingly shapes and reshapes phonetics, grammar, and vocabulary. In history we see political selection in the rise of Macedonia, Rome, and Muscovy. Similarly, economic selection in private enterprise systems causes the rise and fall of firms and products. And science itself is shaped in part by selection, with experimental tests and other criteria selecting among rival hypotheses.
>
> (Price 1995: 389)

Price's notion of selection includes *both* the familiar concept of Darwinian selection—selection among multiplying, replicating entities (D-selection)—and Sample selection—a process of selecting a subset from a set according to some value criterion (S-selection). S-selection *does not* require reproduction or replication, and even the selection of radio stations with the turning of a dial exemplifies it (an example Price used to explain this process). At the most general level, he was interested in the relation between selection and information, and hoped that a link would be forged between information and selection theories.

Such a relation has been actively explored among theoretical evolutionary biologists (cf. Donaldson-Matasci et al. 2010; Lachmann 2013). In fluctuating environments, there is a measurable relation between developmental responses at the organism level and Shannon's mutual information. It has been shown that, under the appropriate conditions, the fitness benefit of an organism's ability to identify and respond to a relevant environmental input equals the mutual information between the input and the environment (Donaldson-Matasci et al. 2010). Organisms can choose, within a generation, the right phenotype for the environment based on an environmental input, rather than following a strict proportional-betting strategy for choosing among possible phenotypes.

Our analysis broadens the scope of the relation between information and selection, adding learned, functional information constructed by S-selection. Adopting the Pricean view of selection, we formulate a definition of general selection in terms of information as a decrease in uncertainty:

> *Definition 5—"General Selection."* The exploration, sampling process in the receiver that leads to the decrease in uncertainty of approaching or reaching an attractor (i.e., usually, a more stable, adaptive homeostatic state).

D-selection can be seen as a special case of the exploration-stabilization principle involving differential reproduction through multiplication and replication. S-selection, on the other hand, is an exploration-stabilization process that does not involve multiplication or replication.

S-selection applies to everyday observations of coping with novel challenges. It is implicitly based on plasticity, the ability of organisms to display flexible responses to altered, predictable, and unpredictable environmental conditions. When organisms are faced with challenges to which they do not have a specific evolved solution, they mobilize internal processes of random or semi-random search, exploring the environment. Internal states that lead to the alleviation or resolution of the problem are stabilized. When a datum upgrades to a sign through a learning process, the organism's interpretation system undergoes a stabilization process in response to the environmental condition. The closer the organism gets to a stable, adaptive state (e.g., following learning), the more informative the relevant perceived sign is. A rat that learns to swim toward the platform is less uncertain about how to reach the platform and it gains increasing information from visual signs by learning from its own experience and (potentially) from others. A vervet learns to emit and respond to others' alarm calls more accurately and with more certainty with experience (Cheney & Seyfarth 1990: 128–138).

Exploration and selective stabilization mechanisms are all based on a similar principle: the generation of a large set of local variations from which only a subset is eventually stabilized and manifested. Dynamically stabilized states are called by system biologists "attractors." In physics, an attractor is considered a state toward which a system tends to evolve for a wide range of initial conditions. Physical states that get close enough to the attractor remain close even if slightly perturbed. Biological attractors are usually functional: the mechanisms that enable them to be reliably reached (despite different initial conditions) evolved or are maintained by natural selection. For example, when a cell, in which growth had been arrested, can find through exploration an internal state that allows it to divide, the networks that produce this functional state are stabilized (e.g., Stern et al. 2007).

Biology is full of examples of phenotypic adjustment (or phenotypic accommodation) through exploration-stabilization processes. Consider foraging in army ants, for example: ants move in random directions, marking outgoing and returning paths with pheromones while doing so. A path that is productive (i.e., leads to food) is used more frequently than others and, hence, is more heavily marked and leads to further use, thereby stabilizing this particular usage. The persistence of the more frequently used path through this type of positive feedback is, therefore, an example of selective stabilization (Deneubourg et al. 1989). Other examples include spindle-formation within cells, exploration of metabolic networks in highly stressed cells, exploratory search of roots for water during drought, and exploratory cultural learning in humans.

Exploration and selective stabilization is a strategy that is employed across the board (this principle is illustrated in Figure 7.2). At the phylogenetic level, it is the only strategy that can adaptively cope with the unknown. This strategy is central to what happens in the nervous system during development and learning, and occurs at several levels of the nervous system's development and function. First, nerve cells are overproduced and most of them die unless they are stabilized by "survival factors" during development (Levi-Montalcini et al. 1996). Second, the synapses in surviving neurons are also overproduced, and most are pruned: connections that have the highest functional efficacy persist according to Hebb's rule: "neurons that fire together wire together." The third level of connectivity between neuronal groups is more contentious. Both Jean-Pierre Changeux (Changeux et al. 1973; Changeux & Dehaene 1989) and Gerald Edelman (1989) suggested that S-selection occurs in the nervous system during ontogenetic development and learning. S-selection is based on both selection for neuronal survival and selection for synapse stability. For Edelman, it is reentry between neuronal groups (maps) that constitutes the third level of stabilization. For Changeux, the third level is constituted by higher levels of exploration and stabilization between neural networks that occur in the brain during complex learning; for example, when a songbird learns its species-typical song.

It should be noted that many imaging and electrophysiological studies have shown that during learning (especially skill learning) changes occur in the activation patterns of neurons and in the structural organization of brain areas subserving particular activities (see, e.g., Bernardi et al. 2013; for reviews see Hill & Schneider 2006; Chang 2014). Groups of neurons that are highly active when a child learns to ride a bike, for example, are less active or no longer active after she acquired the skill. Such findings are compatible with the ideas of stabilization and neural selection in learning.

Others have challenged the idea that S-selection can account for complex forms of neural learning (Fernando et al. 2012). Selection without replication, on their view, can lead, for example, to learning a *particular* behavior, but not a strategy. They conjecture that in order to account for complex cognition, D-selection must occur in complex brains through replication-like processes allowing the copying of network topology (including local circuitry and not just receptive fields) from one brain region to another (Fernando et al. 2008).

However, even if D-selection occurs in brains, the existence of S-selection through differential stabilization of synaptic connections, for example, is generally accepted.

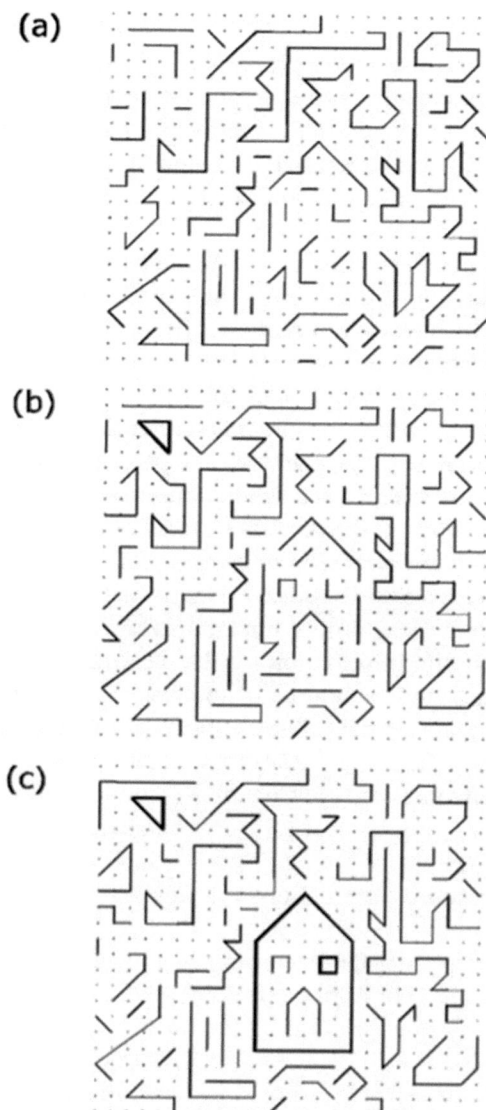

Figure 7.2 An imaginary system and its rules. a) Lines between points continuously form, disappear, or extend to join neighboring points (exploration). b) The longer the line, the less it changes, but increased stability is achieved only when a closed shape (e.g., the triangle) is formed. c) A functional closed struc-ture (e.g., the house) has enhanced stability, and increases the likelihood of closure of related structures, while inhibiting the generation of lines within the closed structure. (Adapted with permission from Figure 11.8 in Jablonka & Lamb 2014.)

Our view is that functional, learned information should be understood in terms of exploratory neural activity through which the organism selects those stimuli whose effects match its internal states and lead to perception or goal-directed behavior. When the exploratory neural activity leads to stabilization, the information born by the stimuli for the organism is greater than that born during the exploration process. It is also quite plausible that such selection processes in the brain contribute to the formation of neural representations. For example, the rat navigating in the water maze uses some cognitive map of its trajectory to the platform. As a function of its learning, some neural relations form in the rat's brain that correspond to external world relations (e.g., the actual trajectory used to navigate), and they persist long enough to support learning of these external relations.[8]

Can functional information be measured? We suggest that learned, functional information relative to a receiver at a given time can be measured using the complement of Prediction Error (PE), which we call *Predictive Fit* (PF = (1 − PE)). The dynamics of informational change during ontogenetic learning can be measured using the Price equation, which can accommodate both S- and D-selection.

PE reflects the degree to which an outcome consequent to perceiving the relevant sign/signal/symbol is *surprising*. When acting on a perceived sign yields a different outcome from what the receiver expects, a prediction error occurs. This error is followed by neural processes that "update" the receiver's model of the world and lead to a usually more adaptive response. The decrease in uncertainty of reaching the attractor state can be seen as a decrease in PE and increase in PF. Signs that accurately predict an adaptive response *without further learning* reduce uncertainty to zero (PE = 0), thereby implying that the sign bears maximum information for the receiver in that context (PF = 1). The functional content of the sign is measured by focusing on the surprise of the receiver as indicated by its learning rate (in the rat's case, locating a submerged platform by associating it with the indicator). In discrete trial-level models of conditioning, where animals learn to predict what outcomes are contingent on which events, the Rescorla–Wagner learning rule (Equation 1) can be used (Niv & Schoenbaum 2008: 265–66).

$$Vnew = Vold + n(\text{outcome} - \text{prediction}) = Vold + n(R - Vold)$$

Equation 1. The Rescorla–Wagner learning rule to be used in discrete trial-level models of conditioning. (For term definitions, see Equation 2.)

$$PE = Vnew - Vold = n(R - Vold)$$

Equation 2. The Rescorla–Wagner learning rule assumes that the increment in the associative strength of a sign/signal on a learning episode is directly determined by PE. n is the learning rate that depends on the particulars of the CS (the sign/signal used) and US (e.g., safety in the case of the rat). R is a scalar quantity denoting the absolute goodness of the outcome, which can be normalized to 1. *Vnew* and *Vold* stand for the new and old predictions associated with the observed stimulus, respectively (in units of predicted reward).

In the Morris water maze setting, one can design an experiment where the increasing gain of information from the sign over time is measured using the Rescorla–Wagner learning rule. PE gets nearer to zero as the number of "shortest" trajectories to the platform stabilizes. To evaluate PE, we divide the total number of trials into trial-sets for consecutive time points—days. Accordingly, 100 trials, for example, are divided into ten sets occurring on ten consecutive days. Each trial-set has an average and a distribution of the lengths of trajectories taken with the different distributions partially overlapping so that the shorter trajectories on day one are, on average, the more common trajectories on day two, and so on. On each trial, the change in the distribution of trajectories (i.e., the average trajectory and the variance for a trail set) is examined as a function of the rat's learning. We assume that the environment and the general state of the animal (except its learning-success) do not change during the ten trial days. The uncertainty reaches zero (and PF reaches one, or close to one) when the rat's path to the platform stabilizes on the shortest trajectory without further improvement (cf. the circle labeled "end" in Figure 7.1b).

PE is measured by $Vnew - Vold$ (see Equation 2). Finding safety by reaching the platform (the US) has the maximal functional value (PF) when the rat is safe. V is the prediction associated with the observed flag carrying information about the platform's location. R is a scalar quantity specifying the functional value of the outcome (e.g., a temporary escape from swimming, and stress relief). We analyze the experiment in terms of Pavlovian conditioning with the CS being the sight of the flag (the CS precedes the US, since the sight of the flag predicts the location of the platform). Before learning starts, V (i.e., the strength of prediction or the associative strength) equals 0. $Vold$ here is the rat's expected value of reward: how little effort the rat expects it would take to get to the platform (i.e., what the shortest path from the starting point to the platform is, and how large the variance in that set is). n is the learning rate ($0 \leq n \leq 1$), an empirical parameter, that determines how much each specific experience affects future prediction and depends on the properties of the US and CS. Although it can be derived experimentally depending on the rate of shortening the trajectories throughout the trials, Equation 3 below does not use it. Instead, Equation 3 is based on $Vnew - Vold$ (see Equation 2), and it can be used to calculate the functional information gained from the flag at any given trial-set. By including the variance of the trial-set, the rat's expectation of the trajectory is considered (the variance grows smaller from one trial-set to another). Both $Vnew$ and $Vold$ are constrained by considering the minimal trajectory length to the platform throughout the whole trial.

Equation 3. Applying the Rescorla–Wagner learning rule to the rat's learning in the water maze. PE ranges between 0 and 1. The average trajectory length is calculated by considering all the trajectories the rat swims in a given trial-set. The variance is normalized to [0,1] range, for example, by using minmax function linear transformation. The minimal trajectory length can be constrained by specifying the physically shortest path relative to the rat's location in the numerator.

The Rescorla-Wagner measure of PE, however, does not apply to all learning types. For example, when learning is not divisible into discrete trials, temporal difference (TD) learning measures (Niv & Schoenbaum 2008: 266–267) can be used. According to the TD learning model, PE occurs as soon as the sign is perceived. This model also allows measuring second- or higher-order conditioning. There are other types of learning to which neither the Rescorla–Wagner nor the TD model applies; for example, learning of relationships between events that do not have an affective component, such as stimulus–stimulus learning (when there is no reward). Here, unsupervised learning models can be used to measure PE (Niv & Schoenbaum 2008: 269).

However PE is measured, there is a clear intimate, dynamic relation between information and selection. This relation can be described by applying Price's equation to the S-selection processes that reduce the receiver's uncertainty relative to a particular goal or state of affairs. The Price equation (a mathematical statement describing how the average value of any character changes in a population from one generation to the next, see below)[9] consists of a covariance term, which provides an abstract expression of selection, and an expectation term, which includes all other evolutionary processes, such as changes in transmission or inheritance.

$$w_{av} \bullet \Delta z_{av} = Cov(w, z) + E(w \bullet \Delta z)$$

Equation 4. The Price equation.

For the case of evolution by natural selection, the individual's character value is z, its fitness is w, and the discrepancy between the character values of itself and its offspring is Δz; the subscript av denotes average. $Cov(w, z)$ describes the change in the frequency of a trait (or gene) following selection, a covariance between individuals' character values and their relative abundance or fitness. $E(w \bullet \Delta z)$ describes changes in the trait (or gene) value due to the process of transformation from one "generation" to another, which in the genetical case would be the result of changes due to recombination or mutation (this term is often neglected).

Because of its generality, Price's equation has to be qualified when applied to specific cases. For the case of learning, we replace w (fitness) with $(1-PE)$ or PF as a measure of the information gained from an observed sign/signal/symbol. When the animal has finished learning, the information gained from the perceived sign/signal/symbol is maximal ($PE = 0$ and $PF = 1$).

For present purposes, the Price equation is not used to measure change in the trait (or performance) value in a biological population from one generation to the next. Rather, the actual mean value of an individual receiver's characteristic is measured across successive performances of a particular task through learning, such as the rat's learning based on its ability to associate the flag with the platform. We assume that prior experiments with rats can establish that the flag is the only relevant sign in this experimental setting.[10] The rat's trait value (z) can be of *any trait*. The *relevant traits* in our case are traits that contribute to its performance in the pool (e.g., the concentration of stress hormone in the rat's blood during trials). The equivalent of Price's fitness (w) here is the rat's reduction in uncertainty or the increase in information ($1-PE$ or PF) about the platform's location. The closer PF is to 1, the higher the information acquired. Δz is the change in the trait value during learning across successive trials. The Price equation explains that the change in the trait value occurs through selective stabilization process. Accordingly, we substitute PF (or $1 - PE$) for w in the Price equation. The second term in this equation can be neglected, if we assume that besides learning nothing else changes the rat's measured trait.

$$\Delta z_{av} = (Cov\,[(1 - PE), z)] + E((1 - PE) \bullet \Delta z)) / (1 - PE)_{av}$$

Equation 5. The Price equation describing the dynamic relation between learning and information by using the relevant PE measure.

We are aware that the observed changes between trial-sets reflect changes in the rat's brain – that is, changes in some neural associations that get reinforced, diminished or eliminated across trial-sets. However, since we presently have little access to these correlates, we measure their behavioral reflections—the change in the lengths of the trajectories traversed as a function of learning across successive trial-sets.

CONCLUSION

Our analysis has focused on the question of how an individual learns and develops through the exchange and processing of information. We have defended a notion of *functional information* that is applicable to both phylogenetic and ontogenetic processes. Functional information is construed as a difference that has made a difference to the goal-directed behavior of a receiver.

Information turns out to be a graded concept. We have distinguished among sign, signal, and symbol (with a datum being the base type). Our claim is that functional information is, in all cases, intimately connected to selection processes, and that learned, functional information can be quantified using Prediction Error measures. Since signs, signals, and symbols are constructed during selection-learning processes, the Price equation can be used for quantifying over information and selection at the level of individual learning by substituting fitness for the relevant success-measure (for the learning case we described it as the predictive fit, PF $(1 - PE)$). Our analysis provides further evidence for the intimate relation between information and general selection, thereby vindicating one of Price's important intuitions.

NOTES

1. The same arguments are valid for gene networks within a cell that are activated as the cell responds to changes in environmental conditions.
2. The absence of something can also be informational; for example, when it is expected but missing. This accords with the Batesonian difference that makes a difference.
3. Note that recognizing something as a *datum* for an organism can mostly be done only a posteriori (e.g., after recognizing it being a *sign* for that organism).
4. These songs qualify as signals by our account.
5. It is important to distinguish between behavioral and neural responses. Although, in some sense, all neural responses are "overt," there are levels of organization in biological systems that one should always bear in mind.
6. While focusing on inter-organism communication, our definitions do not preclude intra-organism communication, such as the exchange of signals among neurons.
7. A useful distinction is between efficacy cost that is necessary to transmit the specific information unambiguously, and strategic cost that is paid at the equilibrium, in addition to the former, to ensure the reliability of the signal (Maynard Smith & Harper 2003: 7). Production cost is the sum of the efficacy and strategic costs. It is debatable whether the production cost has to include a strategic cost component to maintain honesty under sender's conflict of interest (Számadó 2011: 4).
8. This does not imply that correspondence is sufficient for the formation of representation.
9. For an accessible introduction to the Price equation see Gardner 2008.
10. This is not a strong assumption. For rats do not have the capacity to orientate through echolocation; for example, and they cannot see the submerged platform in the murky water.

REFERENCES

Bateson, G. 1972. *Steps to an Ecology of Mind* (Ballantine).

Bernardi, G., Ricciardi, E., Sani, L., Gaglianese, A., Papasogli, A., Ceccarelli, R., Franzoni, F., et al. 2013. "How skill expertise shapes the brain functional architecture: An fMRI study of visuo-spatial and motor processing in professional racing-car and naïve drivers." *PLoS ONE* 8: e77764.

Chang, Y. 2014. "Reorganization and plastic changes of the human brain associated with skill learning and expertise." *Frontiers in Human Neuroscience* 8: doi:10.3389/fnhum.2014.00035.

Changeux, J.-P. & Dehaene, S. 1989. "Neuronal models of cognitive functions." *Cognition* 33: 63–109.

Changeux, J.-P., Courrege, P., & Danchin, A. 1973. "A theory of the epigenesis of neuronal networks by selective stabilization of synapses." *Proceedings of the National Academy of Sciences of the United States of America* 70: 2974–2978.

Cheney, D. & Seyfarth, R. 1990. *How Monkeys see the World: Inside the Mind of Another Species* (University of Chicago Press).

Clayton, N. 2015. "Ways of thinking: From crows to children and back again." *Quarterly Journal of Experimental Psychology* 68: 209–241.

Corning, P. 2007. "Control information theory: The 'missing link' in the science of cybernetics." *Systems Research and Behavioral Science* 24: 297–311.

Cummins, R. 1975. "Functional analysis." *Journal of Philosophy* 72: 741–765.

Davenport, J., Jones, T., Work, T., & Balazs, G. 2014. "Pink spot, white spot: The pineal skylight of the leatherback turtle (*Dermochelys coriacea* Vandelli 1761) skull and its possible role in the phenology of feeding migrations." *Journal of Experimental Marine Biology and Ecology* 461: 1–6.

Deacon, T. 1997. *The Symbolic Species: The Co-evolution of Language and the Brain* (W.W. Norton).

Deacon, T. 2011. "The symbol concept." In K. Gibson & M. Tallerman (eds.), *The Oxford Handbook of Language Evolution* (Oxford University Press) 393–405.

Deneubourg, J., Goss, S., Franks, N., & Pasteels, J. 1989. "The blind leading the blind: Modeling chemically mediated army ant raid patterns." *Journal of Insect Behavior* 2: 719–725.

Donaldson-Matasci, M., Bergstrom, C., & Lachmann, M. 2010. "The fitness value of information." *Oikos* 119: 219–230.

Edelman, G. 1989. *Neural Darwinism: The Theory of Neuronal Group Selection* (Oxford University Press).

Fernando, C., Karishma, K., & Szathmáry, E. 2008. "Copying and evolution of neuronal topology." *PLoS ONE* 3: e3775.

Fernando, C., Szathmáry, E., & Husbands, P. 2012. "Selectionist and evolutionary approaches to brain function: A critical appraisal." *Frontiers in Computational Neuroscience* 6: doi:10.3389/fncom.2012.00024.

Floridi, L. 2011. *The Philosophy of Information* (Oxford University Press).

Gardner, A. 2008. "The Price equation." *Current Biology* 18: R198–R202.

Hasson, O. 1994. "Cheating signals." *Journal of Theoretical Biology* 167: 223–238.

Hill, N. & Schneider, W. 2006. "Brain changes in the development of expertise: Neuroanatomical and neurophysiological evidence about skill-based adaptations." In K. Ericsson, N. Charness, P. Feltovich, & R. Hoffman (eds.), *The Cambridge Handbook of Expertise and Expert Performance* (Cambridge University Press) 653–682.

Jablonka, E. 2002. "Information: Its interpretation, its inheritance, and its sharing." *Philosophy of Science* 69: 578–605.

Jablonka, E. & Lamb, M. 2014. *Evolution in Four Dimensions: Genetic, Epigenetic, Behavioral, and Symbolic Variation in the History of Life* (MIT Press).

Kitcher, P. 1993. "Function and design." *Midwest Studies in Philosophy* 18: 379–397.

Lachmann, M. 2013. "The value of information in biology." In U. Stegmann (ed.), *Animal Communication Theory: Information and Influence* (Cambridge University Press) 357–375.

Levi-Montalcini, R., Skaper, S., Dal Toso, R., Petrelli, L., & Leon, A. 1996. "Nerve growth factor: From neurotrophin to neurokine." *Trends in Neurosciences* 19: 514–520.

Maynard Smith, J. & Harper, D. 2003. *Animal Signals* (Oxford University Press).

Millikan, R. 2004. *Varieties of Meaning: The 2002 Jean Nicod Lectures* (MIT Press).

Morris, R. 1984. "Developments of a water-maze procedure for studying spatial learning in the rat." *Journal of Neuroscience Methods* 11: 47–60.

Niv, Y. & Schoenbaum, G. 2008. "Dialogues on prediction errors." *Trends in Cognitive Sciences* 12: 265–272.

Peirce, C. 1998. *The Essential Peirce: Selected Philosophical Writings*, vol. 2 (Indiana University Press).

Price, G. 1995. "The nature of selection." *Journal of Theoretical Biology* 175: 389–396.

Scarantino, A. 2015. "Information as a probabilistic difference maker." *Australasian Journal of Philosophy* 93: 419–443.

Seyfarth, R. & Cheney, D. 2010. "Production, usage, and comprehension in animal vocalizations." *Brain and Language* 115: 92–100.

Stern, S., Dror, T., Stolovicki, E., Brenner, N., & Braun, E. 2007. "Genome-wide transcriptional plasticity underlies cellular adaptation to novel challenge." *Molecular Systems Biology* 3: doi:10.1038/msb4100147.

Számadó, S. 2011. "The cost of honesty and the fallacy of the handicap principle." *Animal Behavior* 81: 3–10.

Vorhees, C. & Williams, M. 2006. "Morris water maze: Procedures for assessing spatial and related forms of learning and memory." *Nature Protocols* 1: 848–858.

Weiss, M., Hultsch, H., Adam, I., Scharff, C., & Kipper, S. 2014. "The use of network analysis to study complex animal communication systems: A study on nightingale song." *Proceedings of the Royal Society B: Biological Sciences* 281: doi: 10.1098/rspb.2014.0460.

Wright, L. 1973. "Functions." *Philosophical Review* 82: 139–168.

Zahavi, A. 2008. "The handicap principle and signalling in collaborative systems." In P. d'Ettorre & D. Hughes (eds.), *Sociobiology of Communication* (Oxford University Press) 1–10.

8

Genetic, Epigenetic, and Exogenetic Information

Karola Stotz and Paul Griffiths

INTRODUCTION

The most popular account of genetic information in contemporary philosophy is "teleo-semantics" (Millikan 1984; Maynard Smith 2000; Shea 2007; Kingsbury, Chapter 20 this volume). This yields a semantic notion of information—the information in a gene is a description or an instruction and as such genetic information can be true or false, obeyed or disobeyed. It also defines the information content of a gene in terms of the evolution-ary history of that gene. Physically identical genes can have entirely different information content if they evolved due to different selective pressures. This way of thinking about genetic information corresponds closely to the image of genes in popular science. Genes are coded messages instructing organisms to develop in one way or another, and those instructions were written by evolution.

But the teleosemantic approach fails to meet two obvious desiderata for an account of genetic information. First, genetic information ought to feature in causal explanations of development. But the historical nature of teleosemantic information means that the information content of a gene can be changed or removed altogether without any effect on how the organism develops (Griffiths 2013): teleosemantic information is causally inert. Second, heredity is usually thought to be a precondition of evolution by natural selection, and heredity is widely supposed to involve the transmission of genetic infor-mation. Teleosemantics, in contrast, implies that the evolution of any novel character begins without any genetic information about that character being present. The genes that influence a character only carry information about it *after* that character has evolved by natural selection acting on those genes. We return to teleosemantics at the end of the chapter and discuss how it may play a role in "ultimate biology" that complements other notions of genetic information in "proximate biology." For now, however, our focus will be on a strictly proximate sense of genetic information—a sense in which that informa-tion plays a substantive causal role in the operation of living systems.

GENETIC INFORMATION: CRICK AND INFORMATIONAL SPECIFICITY

If there is one thing all philosophers who have written about genetic information agree on it is that genetic information is not merely the application of Claude Shannon's information theory to biological systems. Many components of any physical system will contain mutual information about one another merely in virtue of the fact that they form part of a causal network. But the fact that genes are "informational molecules" is meant to be distinctive, something that marks systems with a genome out from other chemical systems and also marks genes out from many other components of living systems (Maynard Smith 2000; Godfrey-Smith 2000; Bergstrom & Rosvall 2009; Stegmann 2014). In the philosophical literature, claiming that genetic information is merely Shannon information has been a way to minimize its theoretical importance (Sterelny & Griffiths 1999; Griffiths 2001).

In our view there *is* something distinctive about genetic information, but it is a distinctive property best captured using the Shannon formalism. It is a property first clearly identified by Francis Crick in the period of the mid-20th century when, it is generally agreed, molecular biology became an informational science (Kay 2000).

Although Watson and Crick had used the term "genetic information" earlier, the first substantial use is in Crick's 1958 statement of the Sequence Hypothesis and Central Dogma, some of the most influential ideas in the history of molecular biology:

The Sequence Hypothesis
In its simplest form it assumes that the specificity of a piece of nucleic acid is expressed solely by the sequence of its bases, and that this sequence is a (simple) code for the amino acid sequence of a particular protein.

The Central Dogma
This states that once "information" has passed into protein *it cannot get out again.* In more detail, the transfer of information from nucleic acid to protein may be possible, but transfer from protein to protein, or from protein to nucleic acid is impossible. Information means here the *precise* determination of sequence, either of bases in the nucleic acid or of amino-acid residues in the protein.

(Crick 1958: 152–153 italics in original)

Here Crick identifies the specificity of a gene for its product with the information coded in the sequence of the gene. By doing so, he links the idea of information very closely to one of the fundamental organizing concepts of biology. Biological specificity is nothing less than the "orderly patterns of metabolic and developmental reactions giving rise to the unique characteristics of the individual and of its species" (Kleinsmith 2014). From the second half of the 19th to the first half of the 20th century specificity was "the thematic thread running through all the life sciences" (Kay 2000: 41), starting with botany, bacteriology, immunology, and serology. Specificity came to be understood in terms of the complementary three-dimensional shapes of biomolecules that exhibit specificity for one another. By mid-century, quantum mechanics had provided the necessary insight to explain the observed structural complementarity between molecules in terms of the quantum-physical forces that allow biomolecules to form weak hydrogen bonds with one another.

Crick introduced a new, more abstract conception of specificity in terms of how one molecule can precisely specify the linear structure of another. The information that specifies the product is no longer carried by a three-dimensional structure but instead by the linear, one-dimensional order of elements in each sequence—their *colinearity*. Amongst other consequences, this means that specificity becomes independent of the medium in which this order is expressed (i.e., DNA, RNA, or amino acid chain) and of the kind of reaction by which the specificity is transmitted (i.e., transcription or translation). The same information/specificity flows continuously through these three media and through both processes.

According to Crick, the process of protein synthesis involves "the flow of energy, the flow of matter, and the flow of information." While he noted the importance of the "exact chemical steps," he clearly separated this flow of matter and energy from what he regarded as "the essence of the problem"—namely, the problem of how to join the amino acids in the right order. The flow of "hereditary information," defined as "the specification of the amino acid sequence of the protein," solved this critical problem of "sequentialization" (Crick 1958: 143–144). In "Central dogma of molecular biology" Crick clarified his earlier position:

> The two central concepts which had been produced . . . were those of sequential information and of defined alphabets. Neither of these steps was trivial. . . . This temporarily reduced the central problem from a three dimensional one to a one dimensional one. . . . The principal problem could then be stated as the formulation of the general rules for information transfer from one polymer with a defined alphabet to another. (Crick 1970: 561)

The philosopher Gregory Morgan[1] corresponded with Crick late in his career about the original inspiration for using the term "information." This 1998 correspondence shows the consistency of Crick's view over forty years. He states that "information" was "merely a convenient shorthand for the underlying causal effect"—namely, the "precise determination of sequence." Information for him meant only "detailed residue-by-residue determination."

Amongst the many virtues of Crick's conception of information as the encoding of specificity through the precise determination of the sequence of gene products is that it can be made precise using information-theoretic tools. In the next section we show how this is done. In the subsequent sections we show that the resulting measure of "Crick information" can be applied to a wider range of components of biological systems than Crick himself supposed.

INFORMATION, BIOLOGICAL SPECIFICITY, AND CAUSAL SPECIFICITY

Rather than rest with an intuitive notion of specificity we can make the idea precise. Sahotra Sarkar proposed that one variable is a biologically specific cause of another if there is a bijective mapping between the values of the variables. Each value of the first variable corresponds to one and only one value of the other variable, and vice versa (Sarkar 2005: 267).

This analysis of biological specificity is identical to the analysis of a broader notion, "causal specificity," given independently by James Woodward (2010).

Woodward's account of causal specificity is part of a larger program to analyze the idea of causation as it figures in the practice of the special sciences, including biology (Woodward 2003, 2014). According to this "interventionist" or "manipulationist" view of causation, two variables are causally related if it is possible to manipulate the value of one by intervening on the other. In the limiting case, variables are causally related if there is a single pair of values of each variable that are related in this way, even if the two are unrelated across the rest of their ranges. Clearly, then, the interventionist theory of causation needs to differentiate between causes in various ways—to identify causes that "are likely to be more useful for many purposes associated with manipulation and control" (Woodward 2010: 315). A number of different ways to distinguish types of causes have been suggested, and one of these is causal specificity.

The intuitive idea is that interventions on a highly specific causal variable C can be used to produce any one of a large number of values of an effect variable E, providing what Woodward terms "fine-grained influence" over the effect variable (Woodward 2010: 302). In earlier work we and our collaborators have developed an information-theoretic framework in which to measure the specificity of causal relationships within the interventionist account (Griffiths et al. 2015). Our proposal formalizes the simple idea that the more specific the relationship between a cause variable and an effect variable, the more information we will have about the effect after we perform an intervention on the cause. This led us to propose a simple measure of specificity:

SPEC: the specificity of a causal variable is obtained by measuring how much mutual information interventions on that variable carry about the effect variable

Specificity measures the mutual information between *interventions on* C and the variable E. This is not a symmetrical measure because the fact that interventions on C change E does not imply that interventions on E will change C. Formally, $I(\hat{C}; E) \neq I(\hat{E};C)$, where I is mutual information and $\hat{C}$ is read "do C" and means that the value of C results from an intervention on C (Pearl 2009).

This measure adds precision to several aspects of the interventionist account of causation. Any two variables that satisfy the interventionist criterion of causation will show some degree of mutual information between interventions and effects. This criterion is sometimes called "minimal invariance"—there are at least two values of C such that a manipulation of C from one value to the other changes the value of E. If the relationship $C{\rightarrow}E$ is minimally invariant—that is, invariant under at least one intervention on C—then C has some specificity for E, that is, $I(\hat{C}; E) > 0$ (Griffiths et al. 2015).

We propose that causal relationships in biological systems can be regarded as informational when they are highly causally specific. Biological specificity, whether stereochemical or informational, seems to be simply the application of the idea of causal specificity to biological systems. The remarkable specificity of reactions in living systems that biology has sought to explain since the late 19th century can equally be described as the fact that living systems exercise "fine-grained control" over many variables within those systems. Organisms exercise fine-grained control over which substances provoke an immune response through varying the stereochemistry of rec-ognition sites on antibodies for antigens. They catalyze very specific reactions through varying the stereochemical affinity of enzymes for their substrates, or of receptors for their ligands. Organisms reproduce with a high degree of fidelity through the informa-tional specificity of nucleic acids for proteins and functional RNAs. Genes are regulated in a highly specific manner across time and tissue through the regulated recruitment of trans-acting factors and the combinatorial control by these factors and the cis-acting sites to which they bind. These are all important aspects of why living systems appear to be "informed" systems, and what is distinctive about all these processes is that they are highly causally specific.

GENETICS AND EPIGENETICS: TWO SOURCES OF CAUSAL SPECIFICITY

Elsewhere we have termed the encoding of specificity for the linear sequence of biomol-ecules "Crick information" (Griffiths & Stotz 2013). If a cause makes a specific difference to the linear sequence, it contains Crick information for that molecule. This definition embodies the essential idea of Crick's sequence hypothesis, without in principle limit-ing the location of information to nucleic acid sequences as Crick does. Our definition of Crick information can be applied to other causal factors that affect the sequence of biomolecules.

Crick's Central Dogma and Sequence Hypothesis were based on a very simple pic-ture of how the specificity of biomolecules is encoded in living cells. They assume that the sequence of the gene not only *precisely* determines the sequence of the product, but also *completely* determines it. We now know that in eukaryotes additional information is needed to regulate gene expression. The large discrepancy between the number of cod-ing sequences and the number of gene products led to the insight that the informational specificity in coding regions of DNA must be *amplified* by other biomolecules in order to specify the whole range of products. Additional specificity of a kind not captured by the original sequence hypothesis is required.

Different mechanisms of gene regulation co-specify the final linear product of the gene in question, first by *activating* the gene so it can get transcribed, second by *selecting* a chosen subset of the entire coding sequence (e.g., alternative splicing), and third by *creating* new sequence information through the insertion, deletion, or exchange of sin-gle nucleotide letters of the RNA (e.g., RNA editing). Thus specificity, and hence Crick information, is distributed between a range of factors in addition to the original cod-ing sequence: DNA sequences with regulatory functions, diverse gene products such as transcription, splicing, and editing factors (usually proteins), and non-coding RNAs (Stotz 2006).

Absolute specificity turns out to be not inherent in any single biomolecule in these molecular networks but induced by regulated recruitment of many molecules and com-binatorial control of transcription and post-transcriptional processing by those mole-cules (Ptashne & Gann 2002). And it is here that we will find that the networks cannot be reduced to DNA sequences plus gene products. The recruitment, activation, or transpor-tation of transcription, splicing, and editing factors renders them functional and allows the environment to have *specific* effects on gene expression. Not just morphogenesis at higher levels of organization, but even the determination of the primary sequence of gene products is a process of "molecular epigenesis" that cannot be reduced to the information encoded in the genome alone (Stotz 2006; Griffiths & Stotz 2013).

Eva Jablonka and Marion Lamb were amongst the first biological theorists to see the full significance of epigenetics for understanding biological information. They wrote that "DNA is not just a passive information carrier, it is also a responsive system" (Jablonka & Lamb 1995: 2). But this insight has older roots. In an immediate response to Crick's new picture of sequential information coded in DNA, the ciliate biologist David L. Nanney pointed out that:

> This view of the nature of the genetic material . . . permits, moreover, a clearer conceptual distinction than has previously been possible between two types of cellular control systems. On the one hand, the maintenance of a "library of specificities," both expressed and unexpressed, is accomplished by a template replicating mechanism. On the other hand, auxiliary mechanisms with different principles of operation are involved in determining which specificities are to be expressed in any particular cell. . . . these two types of systems . . . will be referred to as "genetic systems" and "epigenetic systems."   (Nanney 1958: 712)

What is remarkable about Nanney's discussion is the rapidity with which he realized that the new view of the gene proposed by Crick implied the necessity for epigenetic mechanisms that have since been uncovered. Nanney hypothesized that the utility of epigenetic control systems "lies precisely in their ability to respond specifically to altered environmental conditions" and suggested that the influence of these systems should be understood in terms of their "specificity of induction" of developmental effects (Nanney 1958: 713, 715). Elsewhere Nanney likened them to "signal interpreting devices, yielding predictable results in response to specific stimuli from inside and outside the cell" (Nanney 1959: 333). Very much in line with this idea, Griffiths and Stotz (2013) have argued that epigenetic factors relay environmental signals to the genome.

The existence of two sources of developmental information implies that heredity needs to provide both genetic and epigenetic information if it is to reproduce living systems. In the following sections we make some distinctions between different senses of "epigenetic" and explore the *complementary* roles of the different forms of biological information.

EPIGENETIC AND EXOGENETIC INHERITANCE

"Epigenesis" is the ancient idea that the outcomes of biological development are *created* during the process of development, not *preformed* in the inputs to development. In earlier centuries, epigenesis was contrasted to the "preformationist" view that organisms already exist in miniature within sperm or ova (Roe 1981). Evolutionary developmental biologist Brian Hall notes: "As a continuation of the concept that development unfolds and is not preformed (or ordained), epigenetics is the latest expression of epigenesis" (Hall 2011: 12).

The related term "epigenetics" was introduced by Conrad H. Waddington in a broad sense that is almost synonymous with "development" (Waddington 1940, 1942). Epigenetics was the study of the causal processes by which many genes interact with one another and with many environmental factors to produce an organism. Today, however, most biologists understand the term in a narrower sense, ultimately derived from the work of Nanney quoted above (Nanney 1958; Haig 2004). In this narrower sense, epigenetics is the study of the mechanisms that determine which genome sequences will be expressed in a cell and how they will be expressed. Epigenetic mechanisms control cell differentiation in multicellular organisms, or the life cycle of unicellular organisms. Epigenetic mechanisms give cells their identity as cells of a particular type. When epigenetic modifications are maintained through mitosis this produces cell-line heredity.

This ambiguity in the term "epigenetic" itself is relatively unproblematic, but it produces a genuinely confusing ambiguity when biologists talk of "epigenetic inheritance." In the narrow sense of "epigenetic" there is epigenetic inheritance only in cases when a methylation pattern, chromatin modification, or the like is transmitted through the germline from one generation to the next. That is to say, when the mechanisms that make cell identity mitotically heritable, between cell generations, also make some aspect of cell identity meiotically heritable, between generations of whole organisms. However, the term is often meant far more broadly, to include every mechanism by which parents can influence the phenotypes of their offspring other than through the inheritance of nuclear DNA. To avoid confusion we have suggested referring to this broader class of mechanisms as "exogenetic inheritance" (Griffiths & Stotz 2013: ch. 5). This leaves us with three categories of inheritance—genetic, epigenetic, and exogenetic— each of which is a source of information for the developing organism. Our category of exogenetic comprises both Jablonka and Lamb's (2005) behavioral and symbolic inheritance systems.

To add to the confusion, many forms of exogenetic inheritance are *mediated* by epigenetic inheritance. The epigenetic modifications in question do not pass through the germline but are reconstructed anew in each generation. Some call these "transgenerational epigenetic effects" (Youngson & Whitelaw 2008) or "experience-dependent epigenetic inheritance" (Danchin et al. 2011). Despite involving epigenetic mechanisms, these are examples of *exogenetic* heredity because the epigenetic marks are not inherited by one cell or organism from another, but reestablished via an environmental influence (often reliably produced by a parent). In one well-studied example, epigenetic mechanisms have been shown to mediate the behavioral inheritance of stress reactivity and maternal care behavior in rats (Meaney 2001). Maternal behavior in the form of licking and grooming establishes stable patterns of methylation in certain genes in the pups' hippocampus. These affect gene expression and therefore brain development, with downstream effects on the behavior of the next generation of mother rats. The behavior of these second-generation mothers reestablishes the patterns of methylation in her pups without the actual patterns of methylation being inherited via the germline. Her epigenetic gene expression pattern predisposes the new mother to recreate the behavior of her mother. Interestingly, the parent–offspring correlations created through these transgenerational epigenetic effects are often much higher than those observed in narrow-sense epigenetic inheritance, like the famous case of the agouti mouse (Wolff et al. 1998).

THE EVOLUTIONARY SIGNIFICANCE OF GENETIC, EPIGENETIC, AND EXOGENETIC INHERITANCE

Evolutionary biologist Adam S. Wilkins gives a clear statement of a conventional view about the evolutionary significance of epigenetic inheritance:

> If an epimutation is to have evolutionary importance, it must persist. . . . This matter is central to whether epimutations can be treated as equivalent to conventional mutations or whether, if they have some degree of stability, some new population genetic theory is needed.   (Wilkins 2011: 391)

Some cases certainly meet this criterion. In a comprehensive review of transgenerational epigenetic inheritance, Eva Jablonka and Gal Raz (2009) conclude that epigenetic inheritance is ubiquitous, and has been shown to persist for up to three generations in humans and up to eight generations in other animal taxa. In plants, which lack comprehensive reprogramming of DNA in each generation, the stability of epigenetic inheritance can rival genetic inheritance. Many cases of true epigenetic inheritance, however, particularly in mammals, would not meet the criterion of multigenerational stability. Such cases may also disappoint with respect to their efficiency or fidelity of transmission, resulting in much lower parent–offspring correlation than genetic inheritance.

However, it is simply not correct that epigenetic change will affect evolution only if the changes themselves persist for more than one generation. In conventional quantitative genetics, the importance of genetics is that Mendelian assumptions let us work out what phenotypes (and hence fitnesses) will appear in the next generation as a function of the phenotypes in the last generation. Epigenetic and exogenetic inheritance both change this mapping from parental phenotype to offspring phenotype, and therefore affect evolution. Both epigenetic and exogenetic inheritance appear in quantitative genetics as "parental effects": correlations between parent and offspring phenotypes above and beyond correlations between parent and offspring genotypes, which are also not the result of a shared environment independently influencing both parent and offspring. It has long been understood that one-generation parental effects can substantially alter the dynamics of evolutionary models, and change which state a population will evolve to as an equilibrium (Lande & Price 1989; Wade 1998). Wilkins's argument appears to be a non sequitur.

Although all three forms of heredity have evolutionary significance, this does not mean that they have *the same* evolutionary significance. Epigenetic marks are sensitive to environmental factors in that they are first "established by transiently expressed or transiently activated factors that respond to environmental stimuli, developmental cues, or internal events" (Bonasio et al. 2010: 613). Hence epigenetic variations may, indeed, be less stable than genetic ones, because they are in principle reversible by the same mechanisms that induced them. This may make them more adaptive in variable environmental conditions than genetic variation (Jablonka & Lamb 1995; Holliday 2006). Many hypotheses about the evolutionary origins of epigenetic inheritance stress its value in spatially and temporally heterogenous environments, where it allows rapid responses to change. This kind of rapid heritable response is sometimes referred to as "transgenerational adaptive plasticity" (Sultan 2015).

All three forms of heredity provide information for development in the precise sense outlined above. But the widely held view that genetic inheritance is somehow more basic than epigenetic or exogenetic inheritance, and that nucleic acids are distinctively "informational molecules," is not without foundation. With the exception of some forms of structural heredity such as the membrane inheritance system, both epigenetic and some exogenetic inheritance can reasonably be thought of as systems for the hereditary regulation of genome expression. At the heart of these heredity systems, then, is the ability of nucleic acids to provide templates for the synthesis of biomolecules. The evolutionary breakthrough that came with nucleic acid heredity was the provision of "a sequestered molecular template used by cells to transfer specificities to subsequent (cellular) generations" (Sarkar 2005: 94). Moreover, the way in which Crick information is encoded in nucleic acid templates bears many of the hallmarks of a well-designed Shannon information channel (Bergstrom & Rosvall 2009). Nucleic acid heredity is a key innovation in the history of life that allowed the highly efficient transfer of large quantities of Crick information from one cell to the next.

The mistake made by many authors who have tried to identify the special role of nucleic acid in heredity, it seems to us, has been to focus on the relative importance of different heredity systems in enabling future evolution, rather than this foundational role of nucleic acid heredity in the history of life, a point not threatened by the parity thesis (see below).

A related error is to identify the special role of nucleic acids in heredity with the idea that a greater proportion of developmental information flows through the genetic heredity channel than through epigenetic or exogenetic channels. This idea is associated with attacks on the "parity thesis," "according to which the roles played by the many causal factors that affect development do not fall neatly into two kinds, one exclusively played by DNA elements the other exclusively played by non-DNA elements" (Griffiths & Gray 2005: 420; see also Griffiths & Knight 1998). One consequence of parity is that "Any defensible definition of information in developmental biology is equally applicable to genetic and non-genetic causal factors in development" (Griffiths 2001: 396; see also Griffiths & Gray 1994). Our approach to information in this chapter clearly meets this constraint, which was inspired by Susan Oyama's calls for "parity of reasoning" in nature/nurture disputes (Oyama 2000 and elsewhere). Critics have responded to the parity idea by admitting that non-genetic causes can *in principle* carry information, but insisting that *far more* information is carried by genes (e.g., Sterelny et al. 1996). The approach to information described above allows a quantitative assessment of this claim and reveals that it is more plausible as a claim about the sources of specificity for evolutionary change than as a claim about the sources of developmental specificity (Griffiths et al. 2015: 543–550).

In conclusion, all three heredity systems have evolutionary significance. The genetic heredity system may well have been optimized for the ability of organisms to transfer biological specificity between cells. This ability, perhaps the key innovation in the history of life, was dependent on the invention of nucleic acid-based heredity. However, the addition of epigenetic and exogenetic heredity systems amplifies that information, allowing a greater range of products to be specified by the same template resources. Epigenetic mechanisms also provide the control engineering that enables the flexible expression of those template resources during development, and in response to different environmental demands. At least some epi- and exogenetic mechanisms of inheritance have been optimized to allow organisms to respond to environmental demands on timescales intermediate between individual development and genetic selection. In modern organisms the amount of information transmitted between the generations through epi- and exogenetic heredity may exceed the amount transmitted in the underlying nucleic acid template resources considered in isolation, although nothing important turns on which form of heredity "wins" in this comparison, as each clearly plays a significant role.

PROXIMATE AND ULTIMATE INFORMATION

Our concern in the previous sections has been with proximate information, information that does causal work in living systems. In this section we turn to ultimate information, to information defined in terms of evolutionary purposes, and its relationship to proximate information.

Teleosemantic accounts define the information in a biological object in terms of the effect that that object was designed to produce by natural selection. This is "ultimate information" because it is derived from ultimate biology—the study of why organisms evolved to their current state. The most thoroughly developed account is Nicholas Shea's theory of "inherited information" (2011).

Shea accepts the parity thesis, and argues that inherited information is found in all genetic and certain special environmental causes of development. Our concern is that the presence of inherited information, whether in genes or environments, cannot contribute to proximate explanations of biological development. If two organisms contain the same allele, one by inheritance and the other as a *de novo* mutation, then according to Shea the first allele contains inherited information and the second does not. But they will, of course, affect the organism that carries them in exactly the same way, all else being equal. The presence or absence of "inherited information" in Shea's sense makes no difference in development (Griffiths 2013, and see above). This highlights the urgency to identify the link between ultimate and proximate information.

Finding the link is made easier by the fact that Shea's formulation, unusually amongst teleosemantic theories, makes a connection with the Shannon formalism. According to Griffiths (2016), the essence of Shea's proposal is that a developmental cause contains "inherited information" if (a) manipulating that cause affects how the organism develops, (b) the causal variable contains mutual information about the environment, and (c) the whole system evolved because it was fitness enhancing by matching appropriate developmental outcomes to different environments. We can apply this proposal to a typical case of adaptive phenotypic plasticity, such as the development of defensive armor in water fleas when developing fleas are exposed to chemical traces of predators (Lüning 1992). The mechanism that produces this effect contains inherited information about the need for armor. Griffiths goes on to point out that we can remove the claim (c) about history from Shea's definition of information to get a corresponding proximal notion of information, which Griffiths terms "adaptive information." A developmental cause contains adaptive information if it contains mutual information about an environmental variable and affects development so that developmental outcomes and environments correlate in a way that enhances fitness. The relationship between "adaptive information" and "inherited information" is exactly the same as the relationship between an adaptive trait and a trait that is an adaptation. Every adaptation was, by definition, an adaptive trait in the past, and an adaptive trait will become an adaptation in future generations if it is successful enough. The ideas of adaptiveness and adaptation are complementary and both are needed to describe the process of natural selection. In the same way, both adaptive information and inherited information are needed to describe the natural selection of information systems.

It is now possible to state the relationship between ultimate and proximate information. Adaptive information is a special case of proximate biological information, in the sense defined in earlier sections, where the presence of that information enhances the fitness of the organism. Ultimate—or "inherited"—information is proximate biological information that exists in current organisms because earlier instances of the same information led to the evolutionary success of ancestral organisms. The presence of DNA sequences that contain Crick information about the structure of biomolecules is explained by the adaptive advantages of having been able to produce those molecules in the past (pseudogenes that can no longer be transcribed might be said to carry "vestigial Crick information"). The presence of proximate information in epigenetic marks and exogenetically inherited factors is explained by the adaptive advantages of being able to regulate genome expression in the past, and of the ability to adaptively match the environment on shorter timescales, as explained above.

The mistake in much of the existing literature is to try to use teleosemantics to *define* the proximate information that is a substantive causal factor in the operation of living systems. Proximate information can—indeed it must—exist before inherited information in order to be selected and thereby get to count as inherited information. Furthermore, not all the proximate information we can measure in living organisms is also ultimate information, just as not every feature of an organism is an adaptation, but we can expect that much of it will be.

CONCLUSION

In 1958 Crick introduced the idea that biological specificity could take the form of information coded in the order of bases in amino acids. Here and elsewhere we have argued that the idea of biological information is best understood in a way inspired by his work (Griffiths & Stotz 2013). Crick's definition of information as the precise determination of the order of elements in a biomolecule can be analyzed using the Shannon formalism and the interventionist approach to causation. The presence of information in Crick's sense corresponds to a highly specific causal relationship between the order of elements in a biomolecule and some cause of that biomolecule, such as the nucleic acid from which it was transcribed or translated. Specificity can be measured as the mutual information between interventions on the cause variable and the effect variable (the order of elements in the downstream biomolecule).

This definition of information can be generalized. First, other causes of the structure of biomolecules can also be highly specific, and thus contain Crick information. Second, we can apply the same measures to other causal relationships within organisms, where the effect variable is not simply the order of elements in a biomolecule. For example, enzymes are highly specific for their substrates, and this relationship can be measured in the same way that we measure the specificity of a gene for its product. Biological information, we propose, is the very same thing as biological specificity. Living systems are "informed systems" because they exercise a high degree of specificity over their internal processes. Crick information is a special case of biological information, just as the specificity of nucleic acids for their products is a special case of biological specificity.

Both Crick information and biological information more broadly can be inherited via genetic, epigenetic, and exogenetic mechanisms. But there is an important sense in which the genetic heredity system stands out from the others: nucleic acid heredity was the key innovation in evolution that allowed the highly efficient transfer of large quantities of Crick information from one cell to the next. Many other heredity systems exploit this system to achieve the transfer of specificity to the next cell or organism generation. But the special status of genetic heredity in this respect does not support some of the other claims about it in the philosophical literature. For example, it does not support the widely held view that genes contribute much more information to development than other factors. In modern organisms a small number of coding sequences produce orders of magnitude more transcripts through epigenetic regulation of gene expression, and do so in a highly regulated manner across space and time. When we look for the source of the specificity manifested in this regulated genome expression, much of it will be found in epigenetic and exogenetic sources. This is not to say that all three heredity systems play *the same* role in development, but the differences between them are not captured by saying that one contains all or most of the information expressed in development.

Nor is it the case that only genetic information is inherited. Epigenetic and exogenetic heredity also transmit information across generations. Because of this, all three forms of heredity have evolutionary significance: they all affect which phenotypes will be seen in the next generation as the result of the differential success of competing phenotypes in the previous generation. Therefore they all affect the dynamics of evolution, as can be seen in quantitative genetic models that incorporate these other forms of heredity. The idea that a heredity system affects the course of evolution in proportion to how stable individual hereditary marks are across evolutionary time is a non sequitur. Again, this is not to say that all three heredity systems play *the same* role in evolution. On the contrary, we have argued that they play complementary roles.

Finally, the biological information found in current organisms has an evolutionary history. It is possible to look at organisms either from the perspective of proximal biology, characterizing their structure and function, or from an ultimate perspective, characterizing the teleological function of those structures and functions. In the same way, biological information can be viewed proximally, as a causal factor in the operation of living systems, or ultimately, looking at its teleological significance.

NOTE

1. Personal communication. We are extremely grateful to Morgan for making Crick's replies of March 20 and April 3, 1998 available to us.

REFERENCES

Bergstrom, C. & Rosvall, M. 2009. "The transmission sense of information." *Biology and Philosophy* 26: 159–176.
Bonasio, R., Tu, S., & Reinberg, D. 2010. "Molecular signals of epigenetic states." *Science* 330: 612–616.

Crick, F. 1958. "On protein synthesis." *Symposia of the Society for Experimental Biology* 12: 138–163.

Crick, F. 1970. "Central dogma of molecular biology." *Nature* 227: 561–563.

Danchin, É., Charmantier, A., Champagne, F., Mesoudi, A., Pujol, B., & Blanchet, S. 2011. "Beyond DNA: Integrating inclusive inheritance into an extended theory of evolution." *Nature Reviews Genetics* 12: 475–486.

Godfrey-Smith, P. 2000. "On the theoretical role of 'genetic coding.'" *Philosophy of Science* 67: 26–44.

Griffiths, P. 2001. "Genetic information: A metaphor in search of a theory." *Philosophy of Science* 68: 394–412.

Griffiths, P. 2013. "Lehrman's dictum: Information and explanation in developmental biology." *Developmental Psychobiology* 55: 22–32.

Griffiths, P. 2016. "Proximate and ultimate information in biology." In M. Couch & J. Pfeifer (eds.), *The Philosophy of Philip Kitcher* (Oxford University Press) 74–91.

Griffiths, P. & Gray, R. 1994. "Developmental systems and evolutionary explanation." *Journal of Philosophy* 91: 277–304.

Griffiths, P. & Gray, R. 2005. "Three ways to misunderstand developmental systems theory." *Biology and Philosophy* 20: 417–425.

Griffiths, P. & Knight, R. 1998. "What is the developmentalist challenge?" *Philosophy of Science* 65: 253–258.

Griffiths, P. & Stotz, K. 2013. *Genetics and Philosophy: An Introduction* (Cambridge University Press).

Griffiths, P., Pocheville, A., Calcott, B., Stotz, K., Kim, H., & Knight, R. 2015. "Measuring causal specificity." *Philosophy of Science* 82: 529–555.

Haig, D. 2004. "The (dual) origin of epigenetics." *Cold Spring Harbor Symposia on Quantitative Biology* 69: 67–70.

Hall, B. 2011. "A brief history of the term and concept of epigenetics." In B. Hallgrimsson & B. Hall (eds.), *Epigenetics: Linking Genotype and Phenotype in Development and Evolution* (University of California Press) 9–13.

Holliday, R. 2006. "Epigenetics: A historical overview." *Epigenetics* 1: 76–80.

Jablonka, E. & Lamb, M. 1995. *Epigenetic Inheritance and Evolution: The Lamarkian Dimension* (Oxford University Press).

Jablonka, E. & Lamb, M. 2005. *Evolution in Four Dimensions: Genetic, Epigenetic, Behavioral, and Symbolic Variation in the History of Life* (MIT Press).

Jablonka, E. & Raz, G. 2009. "Transgenerational epigenetic inheritance: Prevalence, mechanisms, and implications for the study of heredity and evolution." *Quarterly Review of Biology* 84: 131–176.

Kay, L. 2000. *Who Wrote the Book of Life: A History of the Genetic Code* (Stanford University Press).

Kleinsmith, L. 2014. "Biological specificity." doi:10.1036/1097–8542.082900.

Lande, R. & Price, T. 1989. "Genetic correlations and maternal effect coefficients obtained from offspring-parent regression." *Genetics* 122: 915–922.

Lüning, J. 1992. "Phenotypic plasticity of *Daphnia pulex* in the presence of invertebrate predators: Morphological and life history responses." *Oecologia* 92: 383–390.

Maynard Smith, J. 2000. "The concept of information in biology." *Philosophy of Science* 67: 177–194.

Meaney, M. 2001. "Maternal care, gene expression, and the transmission of individual differences in stress reactivity across generations." *Annual Review of Neuroscience* 24: 1161–1192.

Millikan, R. 1984. *Language, Thought & Other Biological Categories* (MIT Press).

Nanney, D. 1958. "Epigenetic control systems." *Proceedings of the National Academy Sciences USA* 44: 712–717.

Nanney, D. 1959. "Epigenetic factors affecting mating type expression in certain ciliates." *Cold Spring Harbor Symposia on Quantitative Biology* 23: 327–335.

Oyama, S. 2000. *The Ontogeny of Information: Developmental Systems and Evolution* (Duke University Press).

Pearl, J. 2009. *Causality: Models, Reasoning, and Inference* (Cambridge University Press).

Ptashne, M. & Gann, A. 2002. *Genes and Signals* (Cold Spring Harbor Laboratory Press).

Roe, S. 1981. *Matter, Life, and Generation: Eighteenth-Century Embryology and the Haller-Wolff Debate* (Cambridge University Press).

Sarkar, S. 2005. *Molecular Models of Life: Philosophical Papers on Molecular Biology* (MIT Press).

Shea, N. 2007. "Consumers need information: Supplementing teleosemantics with an input condition." *Philosophy and Phenomenological Research* 75: 404–435.

Shea, N. 2011. "What's transmitted? Inherited information." *Biology and Philosophy* 26: 183–189.

Stegmann, U. 2014. "Causal control and genetic causation." *Noûs* 48: 450–465.

Sterelny, K. & Griffiths, P. 1999. *Sex and Death: An Introduction to the Philosophy of Biology* (University of Chicago Press).

Sterelny, K., Dickison, M., & Smith, K. 1996. "The extended replicator." *Biology and Philosophy* 11: 377–403.

Stotz, K. 2006. "Molecular epigenesis: Distributed specificity as a break in the central dogma." *History and Philosophy of the Life Sciences* 28: 527–544.

Sultan, S. 2015. *Organism and Environment: Ecological Development, Niche Construction, and Adaptation* (Oxford University Press).

Waddington, C. 1940. *Organisers and Genes* (Cambridge University Press).

Waddington, C. 1942. "The epigenotype." *Endeavour* 1: 18–20.

Wade, Michael J. 1998. "The evolutionary genetics of maternal effects." In T. Mousseau & C. Fox (eds.), *Maternal Effects as Adaptations* (Oxford University Press) 5–21.

Wilkins, A. 2011. "Epigenetic inheritance: Where does the field stand today? What do we still need to know?" In S. Gissis & E. Jablonka (eds.), *Transformations of Lamarckism: From Subtle Fluids to Molecular Biology* (MIT Press) 389–393.

Wolff, G., Kodell, R., Moore, S., & Cooney, C. 1998. "Maternal epigenetics and methyl supplements affect agouti gene expression in Avy/a Mice." *FSAB Journal* 12: 949–957.

Woodward, J. 2003. *Making Things Happen: A Theory of Causal Explanation* (Oxford University Press).

Woodward, J. 2010. "Causation in biology: Stability, specificity, and the choice of levels of explanation." *Biology and Philosophy* 25: 287–318.

Woodward, J. 2014. "A functional account of causation; or, A defense of the legitimacy of causal thinking by reference to the only standard that matters—usefulness (as opposed to metaphysics or agreement with intuitive judgment)." *Philosophy of Science* 81: 691–713.

Youngson, N. & Whitelaw, E. 2008. "Transgenerational epigenetic effects." *Annual Review of Genomics and Human Genetics* 9: 233–257.

9

Language: From How-Possibly to How-Probably?

Kim Sterelny

LANGUAGE AND LINEAGE

Language confronts the human evolution community with an inescapable challenge. The challenge is inescapable because language is both a unique feature of human social life—no other animal has anything like it—and a fundamental feature of human life and cognition. Only the expressive power of human language makes possible our forms of social life. For that form of life is dependent on a large reservoir of collectively built, publically available information; it is dependent on complex, error-intolerant coordination, often over considerable time depth and spatial extent; it is infused with norms and narratives about who we are, and who we are not; it depends not just on social learning but active teaching. Precursors of these aspects of social life, very likely, antedated the emergence of language as we now know it. But coordination over time and space with differentiation of roles; the narratives of social identity; the kinship systems that structure social interaction in many communities; all this and more depends on language. Likewise, our ability to use this system, and live in the social world that it makes possible, has shaped our mind (see, for example, Dennett 1991; Tomasello 2014). So an evolutionary theory of human nature—of how we became such an unusual great ape—must include an account of language and its emergence.

The challenge is made difficult by the fact that the nature of language remains profoundly controversial, despite its familiarity. One of those controversies—its relationship to great ape communication and the nature of precursor systems—will come into focus in the section "Codes, Speaker Meaning, and Communicative Intentions." In this respect language contrasts with other unique and critical features of human sociality; for example, with large-scale cooperation. Here the explanatory target is in clear view. Human social life depends on cooperation within groups which are too large for the individuals cooperating to be close relatives, and too large for each individual to be sure that they will interact with every other individual repeatedly. So cooperation in these groups cannot be explained by kin selection or reciprocal altruism.[1] So how and why did large-scale cooperation emerge; why is it stable?

Famously, explaining the emergence of language is also made difficult by evidential considerations. As those working on the evolution of language endlessly grumble, language does not fossilize. It is true that in favorable circumstances, there is fossil evidence of voluntary control of complex vocalization. But though charting the emergence of voluntary and precise control of vocalization is certainly relevant to the evolution of language, language may be gestural, and vocalization may be musical. There have also been some recent suggestions that there are *genetic fossils* and these give us a direct historical signal about the antiquity of language. There is now direct evidence about the genome of two extinct human species: Neanderthals and Denisovians (and very recent hints of a third (Prüfer et al. 2014)). These genomes can be compared to each other and to ours, enabling us to make reasonable estimates of the genome of *Homo Heidelbergensis*, the presumptive common ancestor of *sapiens*, Neanderthals, and Denisovians. To the extent that we can identify genes especially relevant to our possession of language, we can estimate the presence of these genes in our sibling species, and our common ancestor. Dan Dediu and Stephen Levinson (2013) have recently argued on these grounds that *sapiens*, Neanderthals, and Denisovians all had language by inheritance from our common ancestor. While this line of thought is innovative, these genetic fossils at best show that these other human species had the *genetic resources* needed for language.[2] That is by no means the same thing as having language. Our deep time ancestors very likely had the genetic resources needed for formal quantitative reasoning, but without the cultural invention of numerals and a number line, those resources could not be exploited. Arguably, the same is true of language. It depends on cultural scaffolds, not just appropriate genetic potential. In any case, this line of evidence cannot tell us anything about the incremental sequence of language evolution.

The aim of this chapter is to propose a set of constraints specifying an empirically and theoretically constrained model of the evolution of language, and to present a partial and skeletal view of the evolution of language that fits that framework as a "proof of concept." In the final section, I take up one substantive issue in developing that skeletal view, our conception of the baseline from which the distinctive features of human language emerged. The framework is organized around Brett Calcott's idea of a lineage explanation (Calcott 2008). Evolutionary theory is incrementalist: new structures and capacities emerge through a series of small changes. If the communicative capacities of the last common ancestor of the hominin/pan clade were approximately those of the living chimps, there is a huge phenotypic gap between the LCA and living humans, and that gap was not crossed in a single, or a few, steps. Calcott's proposal makes this idea precise. A lineage explanation of a novel trait is the specification of an incremental sequence that meets three conditions: (i) the explanation identifies the organization of the (nascent) structure or capacity at each stage of the sequence, showing how the structure or system, thus organized, makes a contribution to the overall phenotype of the agent; (ii) each stage of the sequence varies from its predecessor and successor in a relatively minor way; the members of the sequence are just one plausible evolutionary step apart; (iii) the step from one stage to its successor is selectable.[3] In the case of language, this condition has an extra twist. In explaining the evolution of social, interactive traits, we need to explain why a variant is advantageous, or at least not penalized, when rare; to explain how successive variants invade ancestral populations.

A classic morphological example of a lineage explanation is the model of eye evolution in Dane-Erik. Nilsson and Susanne Pelger's work (1994). In their explanation of eye evolution, the incremental changes were in the structural components of the eye: the lens, retina, iris, and the like. In a model of the evolution of language, the components will be the psychological, morphological, and social factors that make communication and enhanced communication possible. By the time something like full language has arrived, many suppose that very distinctive and specialized cognitive capacities are essential (Hauser et al. 2002). But on any plausible view of the incremental growth of our communicative capacities in general, and language more specifically, there needed to be other changes in our cognitive kit, in our emotional and motivation dispositions, and in our social environment. A lineage explanation, ideally, would present an explicit account of these capacities, and show how incremental changes in one result both in accommodating changes in other capacities, and in changes in overall hominin phenotypes.

FILTERING CANDIDATE EXPLANATIONS

Communication depends on a complex mosaic of interacting capacities, so a picture of the evolution of language will necessarily be coevolutionary, with changes in one communication-relevant capacity selecting for changes in others. These will include memory and processing capacities. Language imposes heavy demands on short-term memory and processing capacities (Christiansen & Chater 2016). The lexicon (and the importance of common knowledge to conversational interaction) ensures that language imposes huge demands on long-term memory as well. Theory of mind is crucial. Interpreting others, and successfully signaling to them, will often depend on understanding their point of view. Indeed, we shall see in the section "Codes, Speaker Meaning, and Communicative Intentions" that on one view it is the single most central cognitive capacity we need. Communication, once it becomes elaborated and context specific, requires voluntary control of increasingly precise, elaborate, and error intolerant action sequences.[4] It requires motivational changes: social tolerance, turn taking, attentional focus, active listening, sharing information. For these reasons, expansions in hominin communication required, and coevolved with, changes in the social environment: to greater tolerance, cooperation and collaboration (Tomasello 2008, 2014; Sterelny 2012a). Even if the language faculty in the narrow sense emerged in a single step (Berwick et al. 2013), we would still need an incrementalist, lineage explanation of the evolution of language, for the computational machinery that runs a recursive syntax and phonology would not enable us to use language without a raft of supporting competences, none of which are present in suitable form in living great apes. Conversation is a phenomenally impressive capacity, as agents produce long and exact sequences of phonemes or gestures, while monitoring and interpreting others' sequences, and while remaining sensitive to the social and physical environment, and to the common knowledge that makes conversation work.

There is no consensus on the nature of language. But there is a rough consensus on four features that any decent theory of language should explain (for a similar set of criteria, see Odling-Smee & Laland 2009). The most widely discussed of these is syntax. Sentences are structured sequences of words; moreover, the meaning of the sentence is determined by the meaning of the individual items and how they are organized. Furthermore, on the standard view,[5] it is a compositional system with no upper bound, as structures can be recursively embedded or conjoined. Non-human communication systems do not have this compositional, structured organization, and so a critical problem is to explain how and why a compositional, unbounded system could evolve out of an unstructured system.[6]

Second, words are not calls, even when calls are "referential," as vervet calls famously are. Calls carry information (or misinformation) about the time and place of the interaction, not about the elsewhere and the elsewhen, let alone about the merely possible, the imagined, the forbidden, the required. It is not too hard to understand how systems of signal and response evolved, for signals can begin as functional or involuntary responses to (what will become) the target of the signal. Bared teeth can morph from preparing to attack to a warning to stay away. As the target of the signal is in the here and now, and salient to the agents involved in an interaction, agents can learn to associate the signal with its target. Words (and structures of words) do not typically have their referents in the here and now, so it is much harder to see how producing and understanding such symbols evolved, and what they evolved from. It is far from obvious that associatively understood signals could morph into symbols with displaced reference by any incremental process.[7] There are no involuntary responses to, or preparations for, the elsewhere, the elsewhen, or the merely possible, to morph into a symbol for (say) "The river may be dangerous tomorrow." Likewise, agents cannot learn symbol meaning, whatever that might be, by association, precisely because symbols have displaced reference.

Third, we need to explain the incremental emergence of the extraordinary levels of cooperation manifest in conversation, despite the fact conversations take place between agents with opposed, or only partially overlapping interests; for example, in bargaining and negotiation. Why would we expect such conversations to be honest, or mostly honest? Without such an expectation, why would anyone talk or listen? Moreover, the very features of language that make it so flexible and useful also make deception possible, and apparently easy. The arbitrariness of words, the capacity to represent the elsewhere and the elsewhen, the fact that talk is cheap—all combine to ensure that utterances carry no intrinsic guarantee of their veracity. Almost all human social worlds have conflicts of interest, and so a theory of the evolution of language should explain why conversations are honest enough to make talking and listening worthwhile.

Finally, a striking feature of language is that it emerged only in our lineage. Why would that be, given its apparent power and benefit? One answer to that question links cooperation, honest communication, and hominin mentalizing skills. One potential explanation of the stability of honest communication is that humans and only humans have advanced theory of mind skills (and more generally, metarepresentational skills) and these skills allow us to detect deception; if not perfectly then reliably enough to stabilize conversation (Sperber et al. 2010; Cloud 2014). One version of this idea is the suggestion that only such agents can police (and then punish) deception reliably enough to keep conversations honest enough to be useful. Another is that utterances are very special forms of intentional acts: their intended effects are on the minds of other agents, and their intended means is that the audience recognize that intention and realize that they are intended to recognize that intention. The uniqueness problem, on this view, becomes that of explaining how and why such rich forms of mind reading evolved in our lineage, but not in others.

A PATHWAY TO LANGUAGE

In recent work I have begun to develop a lineage explanation that satisfies the criteria set out above (Sterelny 2012b, 2016a, 2016b). It is a sketch; what follows is a sketch of a sketch. The crucial point is that though the scenario is speculative in places, it is empirically constrained. Though hominin communicative capacities do not leave direct historical traces, specific capacities have prerequisites, co-requisites, and consequences which do leave traces. For expository convenience, I will set out this scenario as a three-phase model. I begin with the initial transformation of hominin lifeways.

The Erectine Revolution

Beginning with the ancient emergence of bipedal lifeways, there were changes in hominin morphology and lifeways that selected for an expanded role for communication, and for the cognitive and motivational mechanisms that make possible such an expanded role. Bipedalism itself is important, as it leads to much larger range sizes. Historically known foragers have local ranges much larger than chimp territories; their territory sizes are like those of top predators rather than scrounging omnivores (Layton et al. 2012). Larger territory sizes select for enhanced skills in memory and navigation. In fission-fusion social organizations, bipedalism also selects for enhanced communication and planning (Jeffares 2014), since fusion can no longer be left to chance encounter and homing in on the sounds of social interaction.

Bipedalism was followed, or accompanied, by a revolution in hominin diets, which in turn had great social and cognitive consequences. Sometime beginning about 2.5 mya a dietary revolution began in our lineage, with meat and other high value foods becoming more important, allowing a reduction in gut size, and expansion in brain size, and a reorganization and reduction of tooth and jaw mass. Meat might not have been the only component in this change,[8] but it was one. By the erectines, around 1.7 mya, there is quite persuasive evidence that organized and targeted large to medium game ambush hunting had become important (Bunn 2007, 2012; Pickering 2013), presumably emerging via various forms of scavenging and less systematic hunting. Ambush hunting has cognitive, social, and communicative implications:

- It involves a shift toward a more technologically intensive lifeway, which selects for greater executive control and planning, both because tools must be made before they are needed and then carried in anticipation of need, and because (for Acheulian-grade technology) tool-making is a planned, inner-template guided activity, with a sequence of strikes and retouches thought through in advance (Stout 2010, 2011).
- It involves a shift toward a more information-intense lifeway, as successful ambush hunting requires intimate knowledge of both the target animals and the local environment.

- A more skill- and information-intensive lifeway selects for improved social learning. In particular, Peter Hiscock (2014) has argued persuasively that learning stone working skills by unguided trial and error is quite dangerous.
- Ambush hunting selects for planning and coordination, hence communication, not just cooperation. Ambush hunters must select targets, role division, and setting in advance. Moreover, since they had only short-range, low-velocity weapons, cooperation at the kill site would have been essential for effective and relatively safe dispatch of the targets.

The suggestion, then, is that erectine communicative capacities were much richer than those of great apes, and these capacities evolved both through direct selection for enhanced communication, and selection on social and cognitive capacities that themselves enhanced communicative ability.

In my view, those expanded capacities came through gesture rather than voice.[9] This suggestion is supported by the following considerations:

- Great apes have significant gestural repertoires, and their use of gesture is context sensitive, under voluntary control, and shaped by learning. Great ape vocalizations are probably not learned, and may not be fully under voluntary control. Coordinating via vocalization would require a new capacity for controlled, context specific vocalization.
- Gestures and mimes, in contrast to vocalizations, are naturally structured. That tends to be true even of very simple and abbreviated gestures. For while chimps do not informationally point, they indicatively beg, and gestures are oriented at both social partners and the object of desire. Chimps also use attention-getting signals. It is not clear whether they concatenate these signals with other gestures like directed begging; if so, even those sequences would be structured (Moore 2016).
- The use of an inner template to guide a motor sequence decouples the execution of that sequence from the physical substrate to which it is normally directed. Thus a template-guided, precise, error-intolerant, action sequence can be redeployed as a signal whose target need not be part of the immediate environment. Going through the motions of knapping a core can become a signal suggesting a trip to a local flint site; going through the motions of emerging from hiding and spearing can become a signal suggesting a hunt. Evolving artisan capacities can be coupled to improving theory of mind and to communicative intentions to yield a form of displaced reference; no further piece of mental equipment is needed.

By the evolution of the erectines, or perhaps shortly thereafter, a gesture-mime system was established that enabled agents to communicate about, and coordinate on short-run, spatially nearby, forward planning, in addition to mediating social interactions in the here and now. These enhanced communicative capacities rest only on incremental advances on great ape capacities for (i) planning and coordination (evidence from foraging and diet); (ii) template-guided control of action sequence (evidence from task analysis of Acheulian technology (Stout 2010, 2011)); (iii) enhanced cooperation and hence enhanced impulse control (evidence of cooperative hunting); (iv) enhanced social learning (evidence of ambush hunting (Pickering 2013) and an analysis of the costs of self-taught stone working skills (Hiscock 2014)). Of course the evidence for this scenario is suggestive, not decisive. But it goes beyond mere speculation.

The Vocal Turn: From Erectus to Heidelbergensis

If hominins did not begin as vocal communicators, they became vocal communicators. On the framework presented here, that transition began with the initial emergence of a cooperative environment, which in turn selected for the inhibition of emotional response. As Sarah Hrdy points out, human social life depends on our having vastly better impulse control than the chimps (Hrdy 2009). Impulse control includes the inhibition of emotional response, including the vocalization of emotion. Vocalization became less reflexive. Second, as Robin Dunbar has long argued, larger and more complex hominin social organization both required conflict defusing capacities and put stress on the primitive great ape peacemaking and affiliation mechanism—one-on-one grooming. In larger and more intense social worlds, there simply is not time for one-on-one grooming (Dunbar 1996; Gamble et al. 2015). Music and song are affectively powerful human universals, so Dunbar has suggested that some form of proto-musical vocal grooming emerged to supplement physical grooming (Gamble et al. 2015). Selection for song-like peacemaking and bonding would select both for precise control[10] of vocal performance and for performance in response to specific social situations. Third, once hominins evolved precise top-down control of their vocal production, it would naturally be incorporated into mime and gesture, and communication would come to be a hybrid of manual, bodily, and vocal elements. Much human communication still is such a hybrid. Finally, after these changes, economic considerations would drive a shift in the center of gravity from gesture to voice: (i) as Liz Irvine has pointed out (personal communication), once there are three or more conversational partners, gesture is very hard to track visually; (ii) a gesture-based system makes it impossible to act and communicate (and hence coordinate[11]) at the same time. Once vocal control had evolved for other reasons, efficiency considerations could and would make voice central. One important effect of this shift is that iconicity plays a diminishing role in communication. Unlike gesture and mime, vocal signals are typically arbitrary, with little resemblance between signal and target.

As far as one can tell from the fossil data, the Heidelbergensians had evolved near-modern vocal control. So the admittedly more speculative suggestion made here is that by the emergence of the Heidelbergensians, there had been a shift to voice as the dominant vocal mode and a modest expansion over the erectines in Heidelbergensian expressive power, to support the social learning needed by their enhanced technology and the richer mapping of their local environment. They had increasing control of fire (Gowlett & Wrangham 2013; Atwell et al. 2015) and perhaps also the first glimmerings of composite technologies (Barham 2013).

Toward Language

According to the framework presented here, late mid-Pleistocene hominins had significant communicative capacities, initially based on a gestural system, but becoming supplemented and enriched by vocal elements. But these capacities do not much resemble full language, or even the lexically rich protolanguages documented from language contact situations. Mid-Pleistocene hominins were technically adept cooperative foragers, but there is:

- no sign in the physical record of a diverse or regionally varied technical repertoire, or of the capacity to reliably accumulate innovations. There are some ambiguous signals of the control of fire in the period 1.5 mya–800 kya, becoming more regular and convincing after that period, but not unmistakable until about 400 kya.

- little sign of trade and exchange networks; raw materials were typically sourced relatively close to the places artifacts were used.
- no sign yet of Clive Gamble's (1999) "release from proximity": of social relations mediated by such cultural tools as kinship systems, shared ritual experience, or exchange networks rather than face to face interaction.
- no sign yet of lives mediated by ritual or material symbols.
- no sign of planning or coordination over weeks or months, rather than a day or two. Mid-Pleistocene lifeways had begun the hominin shift toward a delayed return economic life, but as far as we can tell from the physical record, the planning horizon was still close.

Given these considerations, the communicative challenges posed by Acheulian cultures (about 1.7 mya–400 kya) were much less complex than those facing historically known people. But between 400 kya and 100 kya, hominin lifeways were transformed in at least three and perhaps four ways:

1. There was take-off of cumulative culture, especially from 100 kya on.
2. The economy of cooperation shifted to one based on indirect reciprocation, reputation, and partner choice rather than on immediate return mutualism. As a consequence, agents had to track costs and benefits over longer time frames and for less easily compared resources.
3. Social ties depended less on intimate daily interaction, and more on symbolically mediated connections. For example, Gamble has argued that the final out-of-Africa movement involved intentional migration rather than mere passive diffusion into suitable habitat, and that these mini-migrations depended on social relations that were stable across time and space (Gamble 2008).
4. If Gamble and others are right about the emergence of spatially extended groups, this period probably also saw the emergence of more vertically complex social organizations. In historically known forager societies, bands are nested in communities, and these in turn are networked with other communities to form larger units (sometimes called "tribes"), which in turn are themselves typically part of larger ethno-linguistic units. At around 100 kya, we do see longer raw material transport, and that is the signal of some kind of change in network shape; perhaps of trading or gifting; perhaps of resource-gathering work teams having the information and the security to travel significant distances to gather high-quality materials.

Material culture became more diverse and elaborate; foraging became more demanding, as the resource envelope expanded. Most especially, social life became much more challenging. With this, the demands on hominin communicative capacities increased greatly:

- Social networks were mediated by such cultural inventions as kinship systems and gift rituals.

- Cumulative culture depends on high fidelity, high volume social learning, and this in turn probably depends on some explicit teaching (e.g., of the medical, technological, and culinary uses of the local flora).
- Cooperation mediated by indirect reciprocity is stable only if reputation is reliably tracked, and in a spatially dispersed group this depends on gossip. Agents must be able to report accurately and precisely on the words, deeds, and intentions of others.
- Coordination over larger spatial and temporal scales requires agents to refer to time and place combinations, again with some precision.
- Cumulative culture increases the community's technical and informational repertoire and hence technical vocabulary.
- A more dispersed social world increases the information gradient within the community, which increases the value of communication but also requires making contextual background explicit.
- Larger and more vertically complex social worlds, combined with a reciprocation-based economy, make the management of conflict more difficult, and select for the invention of cultural tools of conflict management. Plausibly these include:
  - a larger role for norms of division and sharing;
  - norms and customs for managing multiple social identities (band members will typically have one set of obligations in virtue of their local network of daily interaction, coexisting with distinct kinship ties and reciprocal obligations beyond the band);
  - a local narrative life of shared stories that express the group's identity and connection to place;
  - an expanded role for ritual and proto-religion. While ritual might depend only on music, dance, and material symbols, these other social technologies depend on language-like capacities.

The picture suggested here is structurally similar to that suggested for the erectine revolution. There was direct selection for improved communicative capacities. But there were also concurrent changes which made such selection more likely to be effective:

- A rich communication system depends on high-fidelity, high-volume cross-generation cultural learning (think of the lexicon), and there are signs in the record from around 200 kya that cumulative cultural evolution, also dependent on high-fidelity and high-volume learning, was flickering into play, presumably through some mix of genetic and cultural changes.
- One distinctive feature of the human mind is the extent to which it both depends on, and can take advantage of, external scaffolds (Clark 2008). The gradual emergence of a richer material culture over the last 300 kya is an increase in hominin informational resources, not just their material resources.
- Michael Tomasello has argued that this period saw further advances in our social intelligence, in particular the evolution of joint intentionality. This involves motivational factors, but also the capacity to map complex task structures in an agent-neutral way, and to appreciate what is common knowledge, and what is not, amongst an interacting group (Tomasello 2014). Common knowledge, in turn, is critical to planning and interpreting communicative acts.

- An essential element of full linguistic competence is of course mastery of syntax. The exact nature of this mastery, and the extent to which the computational processes are unique to language, remain obscure. But it has been argued that elaborate motor skills have the same fundamental computational organization, with procedures embedded within other more global procedures—procedures that can be reapplied within the overall organization of action (Stout & Chaminade 2012). If this is right, one of the core competences of language might be a modified version of those involved in elaborate motor skills.

The ideas just sketched are at best a fragment of a lineage explanation. But even so, the sketch is not just a how-possibly story. It is empirically constrained. Increasingly rich forms of communication have increasingly elaborate cognitive, motivational, and social prerequisites and corequisites, and those forms of communication also make possible new forms of social life (e.g., long-distance material transport (see Marwick 2003)). We have direct information about the technology, the geographic distribution, the ecological breadth, and the foraging behavior of ancient hominins. Morphology encodes information about physical capacities (gait, dexterity, power), diet, life history, mating patterns, and perhaps social complexity.[12] A fluke trace fossil—the Laetoli footprints—is a trace of impressive executive control, for it shows that 3.6 mya Laetolians could not just track another's footprints through fresh ash, they could consistently and precisely step into those prints (Shaw-Williams 2014).

Collectively, these sources impose significant and increasing constraints on our reconstructions of ancient hominin social and ecological lives, and these in turn impose constraints on our reconstruction of their communicative capacities. For example, the evolution of language is a special case of a quite demanding form of cumulative cultural evolution. As language became the phonologically, morphologically, lexically complex system it is today, an increasingly large volume of information had to pass from one generation to the next.[13] The evolution of language, as language approached its current levels of richness and complexity, depended on high-volume, high-fidelity cultural learning, and that is true even if strong versions of linguistic nativism are correct. There is persuasive evidence from the archaeological record, supported by formal models of the interaction between reliable social transmission and demographic structure (Powell et al. 2009), that high fidelity cultural learning stabilized relatively late in human evolution, perhaps around 100 kya. If so, human languages could not have reached their current levels of richness and complexity earlier.

CODES, SPEAKER MEANING, AND COMMUNICATIVE INTENTIONS

I shall finish this chapter by taking up in more detail one of the problems of constructing a lineage explanation: our characterization of the baseline, and of the role of advanced theory of mind in the evolution of language. The scenario outlined in the previous section saw hominin communicative capacities as evolving initially as a richer and more flexible version of pre-existing great ape capacities, with elements we think of as distinctive to language becoming part of the mix gradually, and not as a package deal. Displaced reference and structured signals arrive relatively early, piggybacking on

the ecologically driven expansion of top-down control and on selection for coordination. The cooperative nature of language thus arises as initially a side-effect of ecological cooperation, as that cooperation expands in scope, and then changes its economic
basis and socio-communicative organization with the shift from immediate return
mutualism to indirect reciprocation. The notorious arbitrariness of language is mostly
the product of a shift from the gestural to the vocal mode, though no doubt habitually
used gestures become abbreviated and conventionalized. The stock of atomic constituents comes on stream gradually, reflecting the slow and unstable expansion of technical
and natural history knowledge, and social worlds that long depended on shared history
and daily interaction for the stability of relationships, rather than social-symbolic categories. I would bet long odds against erectines having a concept of *father's brother's
son*. On this model, the functional richness of language is late: its role in myth, ritual,
norm, and gossip is a response to the challenges of managing cooperation and social
peace in social environments that had become more vertically complex, more dependent on indirect reciprocation, more dependent on individual bet-hedging mediated
by horizontal connections between groups. Syntax, too, on this view, is late, whether
as a de novo cognitive adaptation or as an exaptation. On any view like this, there is
no historical moment when hominins first had language. Critical elements arise incrementally and not as a package deal. On this view, language is genealogically connected
to, and built on top of, great ape gestural communication, while being a great deal more
than that earlier system.

There is an influential alternative that rejects the genealogical link to great ape communication. This view derives from an analysis of speaker meaning by H. P. Grice,
and has recently been developed by Dan Sperber, Thom Scott-Phillips, and Michael
Tomasello.[14] The core claim is that genuinely meaningful utterances—the bedrock
phenomena of language—are "ostensive intentional" acts: acts committed with overt
communicative intentions and requiring sophisticated theory of mind. On this view,
language evolved incrementally not from great ape communication but from social
intelligence and mind reading. Animal communication systems, including chimp gestures, are *codes*, and codes are coordinating devices made possible through association.
Senders associatively build signals as natural signs of aspects of a joint environment, and
receivers associatively learn to link signal, state, and response. These systems have very
limited flexibility, for association can only shape existing, pre-potent responses to situations and signals, and this sharply constrains the recruitment and use of new signals.
Human communication, the analysis goes, is not a code. It is *an evidence-inference interaction*, mediated on both sides by advanced theory of mind and common knowledge,
and so has none of the intrinsic limits of code systems. With the right stage-setting, my
pointing to my tooth can let you know that next week's speaker is out for your blood.
Codes cannot gradually morph into ostensive-intentional systems, as the underlying
cognitive mechanisms are entirely distinct. But great apes could incrementally add theory of mind capacities until a threshold is reached, at which point ostensive-intentional
communication becomes possible. Once it is possible, then habitual, then central to
the lives of the agents in question, ostensive intentional signaling becomes regularized,
systematized, expanded, made more precise, domesticated (Cloud 2014) through a rich
and systematically used lexicon, and through the syntactic structuring of that lexicon

into phrases and sentences. But meaning began with the initial emergence of ostensive intentional communication, once a theory of mind threshold was crossed, rather than built from the foundations provided by great ape gesture.

Richard Moore gives a clean depiction of the essential structure of this view of ostensive-intentional communication (Moore 2016). A sender S means something by a signal u if and only if S sends u to R intending:

1. R to produce a particular response r, and
2. R to recognize that S intends (1).[15]

The intended response r determines what u means. The fact that R's intention is overt— R wants the target audience to know what he/she is doing (via clause 2)—is what makes u a meaningful, communicative act, rather than just an attempt to manipulate or influence R. As a cruel joke I can drop an emetic in R's drink intending him to spend most of the night on the toilet. But my doing so does not count as an act of communication. If I ostentatiously wave a bottle of wine at friend at a party, the implicit question is a genuine communication. The high church version of this view insists that r *is itself* a cognitive response; a representation in the audience of the speaker's state of mind. In the case above, the intended response is that she recognize that I am wondering whether she would like me to pour her a drink (or something similar).

A full discussion of this view of language and meaning is well beyond the scope of this chapter. My aim here is to resist its dichotomized conception of communication systems in two ways. First, I shall suggest that much routine human conversation does not fully fit the ostensive-intentional framework, while obviously not being a form of association-driven signaling, and so the framework appears to import the dichotomy into the human sphere. Second, following up a suggestion of Richard Moore, there does seem to be an incremental track from implicitly overt, cognitively undemanding signals to the full-blown metarepresentational intentional acts described by high church Griceans.

The evidence-inference view of meaning depicts conversation as an extraordinarily cognitively demanding interaction. On this model, a speaker has a complex high-order intentional state, and chooses linguistically and contextually appropriate clues to that intention. These clues enable the audience to recognize these intentions. None of this can be habitual, for the choice of clue is sensitive to linguistic context and common knowledge, both of which change as the conversation progresses. Yet these choices are made very rapidly, apparently effortlessly, with a low error rate. This is surely too complex a picture to be taken literally as a description of the cognitive mechanisms involved in routine communication, especially as quite young children are conversationally competent, yet seem not to have the metarepresentational competence that this model requires, for they do not pass false belief tests. Scott-Phillips takes up this challenge, arguing that young children do indeed have these mentalizing capacities. For very young children pass non-linguistic versions of the false belief test (Scott-Phillips 2015b: 71–75).[16] He suggests that these experiments show the necessary mentalizing capacities are in place very early, and that they show that high-order intentionality is not intrinsically cognitively demanding. However, the Gricean analysis of routine linguistic communication depends on very fluent and skilled high-order mentalizing. It requires the capacity in real time, under the temporal constraints imposed by the speech stream, to smoothly integrate high-order mentalizing, registration of the context of interaction, and the speaker's awareness of common knowledge and its limits. For only thus (given the analysis) can the speaker select highly relevant linguistic clues to her communicative and informational intentions. On this analysis, language processing had better not interfere with or disrupt mentalizing, as it is necessary for those capacities to be smoothly integrated with speaking and interpreting. A pragmatic, processing-load explanation of three-year-old failure to pass the false belief test is inconsistent with the basic Gricean model of human communication, given their conversational competence.

I think that it is indeed likely that some form of high-order mentalizing is like perception: fast, automatic, effortless for the agent, even though it seems cognitively complex to the theorists.[17] Children, perhaps even very young children, probably have some kind of automatic, subdoxastic, relatively encapsulated quasi-perceptual, quasi-modular system that registers the perceptual field of those that they interact with, and which enables them to anticipate behavior in the light of another's perceptual field (even when their own perceptual field is saliently different). Such a system would be automatic and relatively effortless; it would not, for example, impose heavy demands on attention. Encapsulated mutual tracking could then be integrated smoothly with coordinated or interactive behavior which is less routinized, and more demanding of cognitive resources. This idea, however, supports a quite different picture from the one offered by the Griceans. First, an encapsulated perceptual tracking-action anticipation system is quite different from pragmatic competence required by evidence-inference interaction. Providing and reading clues to communicative and informational intentions goes far beyond registering one another's perceptual field and taking into account the constraints of perceptual field on the control of action. Social context, previous history of interaction, non-perceptual common knowledge—all of these are important to the choice and interpretation of clues. Pragmatic competence cannot depend on an encapsulated system (as Fodor noted long ago (Fodor 1983)). So toddler success at false belief tasks does not show that somewhat older children have the basic cognitive tools for pragmatic competence, despite failing the classic, linguistically encoded versions of the false belief task. Second, and importantly, it gives a plausible option for a form of social cognition that might have sufficed for the gestural and protolanguage competence of hominins that lived after the Last Common Ancestor and before our recent, large-brained ancestors. Moreover, it suggests an incremental model of the emergence of Gricean meaning from less cognitive-rich signaling.

Ostensive intentional signaling is overt signaling, and there can be an increasingly rich account of what it is for S's production of u to be overt. In the initial stage of the transition to Gricean meaning, the overt production of u is just the fact that S's production of u is not deceptive. S's production is public information, and the probability that R will respond to u with r would not be reduced, were R to be aware that S produced u (and aware that S wanted R to r). In the second stage of the transition, the transaction between S and R is *explicitly cooperative*: S expects R's recognition of S's production of u to boost the probability of r. So

1. *S* produces *u* intending *R* to *r*.
2. *S* signals to *R* his/her production of *u*.

This is Moore's deflationary account of an overt communication intention (Moore 2016). In the final stage of this transition, *S* expects *r* to depend on *R*'s recognition of *S*'s intention. So

1. *S* produces *u* intending *R* to produce *r*, and
2. *S* expects *R*'s production of *r* to depend on *R*'s recognition that *S* produced *u* intending that *R* produce *r*.

The idea that *S*'s intention is overt has transitioned from one in which *S*'s goals would not be undermined by the audience's recognition of their presence to one in which it critically depends on that recognition. It is also probably true that the character of *r* changes through this trajectory: from an initial state in which *r* is typically an action to states in which *r* is typically a representation in the audience of a cognitive state of the speaker. Perhaps early in this transition, *S*'s acts count as nudges rather than utterances. No matter: there is a relatively smooth pathway from intentional, signal-like acts that do not depend on rich mentalizing capacities to fully Gricean speaker meaning.

NOTES

1. This is not a feature of the recent: many forager and small-scale farming societies are large enough to be large in this sense.
2. We cannot currently identify the genes relevant to language, so Dediu and Levinson's argument was based on the close overall similarity of the genomes of the three species and the thought that there would have to be a very significant genetic difference between a language-capable and a languageless hominin.
3. More precisely, no step is blocked by selection. Drift and populations with polymorphisms may well play a role in the evolutionary trajectory through which novelties appear.
4. To recycle an example of Dan Dennett's illustrating the error-intolerance of language as it now is, consider the great social and semantic differences between the following two messages, differing only by a single phoneme: "Call me" rather than "Ball me."
5. Though there are dissenters; see, for example, Everett 2005; Evans & Levinson 2009.
6. It is tempting to describe animal signaling systems as fixed and finite, having a very limited range of independent signals. That is roughly right, except that many animal signals are continuous rather than discrete; the intensity and duration of a call carries information about (say) the level of a threat.
7. Terrence Deacon (1997) presses this point powerfully.
8. Cooked or otherwise processed plant food might also have been important (Wrangham 2009).
9. An increasingly popular view: see, for example, Tomasello 2008; Corballis 2011.
10. Selection for precise control might well be reinforced by the value of vocal imitation to foragers; call imitation can be used to manipulate prey behavior.
11. Or teach: an expert knapper could not, for example, both demonstrate a technique and simultaneously comment on it, directing attention to the most salient elements.
12. This depends on one's view of Dunbar's attempts to correlate relative neocortical size to group size; I am myself skeptical.
13. No one thinks the lexicon or morphology of a language is innate.
14. See Grice 1957; Sperber & Deidre 1986; Scott-Phillips 2015a, 2015b. The analysis that follows relies heavily on Scott-Phillips's recent defenses of this viewpoint.

15. There is also a no-deception condition: S does not intend to conceal or deceive R about those intentions.
16. The experimental paradigm relies on the fact that agents look longer at scenes that surprise them. It is a surprise if someone is looking for a hidden object "in the wrong place." But what counts as the wrong place will vary for the infant, depending on whether they can appreciate the fact that another agent has a mistaken belief about where a doll is.
17. Thus Scott-Phillips argues: "More likely, it is, like simple mindreading, something we do habitually and subconsciously, as part of our everyday, low-level perceptions of the world around us" (Scott-Phillips 2015b: 73).

REFERENCES

Atwell, L., Kovarovic, K., & Kendal, J. 2015. "Fire in the Plio-Pleistocene: The functions of hominin fire use, and the mechanistic, developmental and evolutionary consequences." *Journal of Anthropological Sciences* 93: 1–20.

Barham, L. 2013. *From Hand to Handle: The First Industrial Revolution* (Oxford University Press).

Berwick, R., Friederici, A., Chomsky, N., & Bolhuis, J. 2013. "Evolution, brain, and the nature of language." *Trends in Cognitive Sciences* 17: 89–98.

Bunn, H. 2007. "Meat made us human." In P. Ungar (ed.), *Evolution of the Human Diet: The Known, the Unknown, and the Unknowable* (Oxford University Press) 191–211.

Bunn, H. 2012. "Bovid mortality profiles and early hominin meat: Foraging capabilities at Olduvai Gorge, Tanzania." *European Society for the Study of Human Evolution*, Bordeaux, France.

Calcott, B. 2008. "Lineage explanations: Explaining how biological mechanisms change." *British Journal for the Philosophy of Science* 60: 51–78.

Christiansen, M. & Chater, N. 2016. "The now-or-never bottleneck: A fundamental constraint on language." *Behavioral and Brain Sciences*: doi:10.1017/S0140525X1500031X, e62.

Clark, A. 2008. *Supersizing the Mind: Embodiment, Action, and Cognitive Extension* (Oxford University Press).

Cloud, D. 2014. *The Domestication of Language: Cultural Evolution & the Uniqueness of the Human Animal* (Columbia University Press).

Corballis, M. 2011. *The Recursive Mind: The Origins of Human Language, Thought, and Civilization* (Princeton University Press).

Deacon, T. 1997. *The Symbolic Species: The Co-evolution of Language and the Brain* (W.W. Norton).

Dediu, D. & Levinson, S. 2013. "On the antiquity of language: The reinterpretation of Neandertal linguistic capacities and its consequences." *Frontiers in Psychology* 4: doi.org/10.3389/fpsyg.2013.00397.

Dennett, D. 1991. *Consciousness Explained* (Little, Brown and Company).

Dunbar, R. 1996. *Grooming, Gossip, and the Evolution of Language* (Harvard University Press).

Evans, N. & Levinson, S. 2009. "The myth of language universals: Language diversity and its importance for cognitive science." *Behavioral and Brain Sciences* 32: 429–448.

Everett, D. 2005. "Cultural constraints on grammar and cognition in Piraha: Another look at the design features of human language." *Current Anthropology* 46: 621–646.

Fodor, J. 1983. *The Modularity of Mind* (MIT Press).

Gamble, C. 1999. *The Palaeolithic Societies of Europe* (Cambridge University Press).

Gamble, C. 2008. "Kinship and material culture: Archaeological implications of the human global diaspora." In N. Allen, H. Callan, R. Dunbar, & W. James (eds.), *Early Human Kinship: From Sex to Social Reproduction* (Blackwell) 27–40.

Gamble, C., Gowlett, J., & Dunbar, R. 2015. *Thinking Big: How the Evolution of Social Life Shaped the Human Mind* (Thames and Hudson).

Gowlett, J. & Wrangham, R. 2013. "Earliest fire in Africa: Towards the convergence of archaeological evidence and the cooking hypothesis." *Azania: Archaeological Research in Africa* 48: 5–30.

Grice, H. 1957. "Meaning." *Philosophical Review* 66: 377–388.

Hauser, M., Chomsky, N., & Fitch, W.T. 2002. "The faculty of language: What is it, who has it, and how does it evolve?" *Science* 298: 1569–1579.

Hiscock, P. 2014. "Learning in lithic landscapes: A reconsideration of the hominid 'tool-using' niche." *Biological Theory* 9: 27–41.

Hrdy, S. 2009. *Mothers and Others: The Evolutionary Origins of Mutual Understanding* (Harvard University Press).

Jeffares, B. 2014. "Back to Australopithecus: Utilizing new theories of cognition to understand the Pliocene hominins." *Biological Theory* 9: 4–15.

Layton, R., O'Hara, S., & Bilsborough, A. 2012. "Antiquity and social function of multilevel social organization among human hunter-gatherers." *International Journal of Primatology* 33: 1215–1245.

Marwick, B. 2003. "Pleistocene exchange networks as evidence for the evolution of language." *Cambridge Archaeological Journal* 13: 67–81.

Moore, R. 2016. "Meaning and ostension in great ape gestural communication." *Animal Cognition* 19: 223–231.

Nilsson, D-E. & Pelger, S. 1994. "A pessimistic estimate of the time required for an eye to evolve." *Proceedings of the Royal Society of London, Series B* 256: 53–58.

Odling-Smee, J. & Laland, K. 2009. "Cultural niche construction: Evolution's cradle of language." In R. Botha & C. Knight (eds.), *The Prehistory of Language* (Oxford University Press) 99–121.

Pickering, T. 2013. *Rough and Tumble: Aggression, Hunting and Human Evolution* (University of California Press).

Powell, A., Shennan, S., & Thomas, M. 2009. "Late Pleistocene demography and the appearance of modern human behavior." *Science* 324: 1298–1301.

Prüfer, K., Racimo, F., et al. 2014. "The complete genome sequence of a Neanderthal from the Altai Mountains." *Nature* 505: 43–49.

Scott-Phillips, T. 2015a. "Nonhuman primate communication, pragmatics, and the origins of language." *Current Anthropology* 56: 56–80.

Scott-Phillips, T. 2015b. *Speaking Our Minds* (Palgrave-Macmillan).

Shaw-Williams, K. 2014. "The social trackways theory of the evolution of human cognition." *Biological Theory* 9: 16–26.

Sperber, D. & Deidre, W. 1986. *Relevance, Communication, and Cognition* (Blackwell).

Sperber, D., Clément, F., Heintz, C., Mascaro, O., Mercier, H., Origgi, G., & Wilson, D. 2010. "Epistemic vigilance." *Mind and Language* 25: 359–393.

Sterelny, K. 2012a. *The Evolved Apprentice* (MIT Press).

Sterelny, K. 2012b. "Language, gesture, skill: The coevolutionary foundations of language." *Philosophical Transactions of the Royal Society, Series B* 367: 2141–2151.

Sterelny, K. 2016a. "Deacon's challenge: From calls to words." *Topoi* 35: 271–282.

Sterelny, K. 2016b. "Cumulative cultural evolution and the origins of language." *Biological Theory* 11: 173–186.

Stout, D. 2010. "The evolution of cognitive control." *Topics in Cognitive Science* 2: 614–630.

Stout, D. 2011. "Stone toolmaking and the evolution of human culture and cognition." *Philosophical Transactions of the Royal Society, Series B* 366: 1050–1059.

Stout, D. & Chaminade, T. 2012. "Stone tools, language and the brain in human evolution." *Philosophical Transactions of the Royal Society, Series B* 367: 75–87.

Tomasello, M. 2008. *Origins of Human Communication* (MIT Press).

Tomasello, M. 2014. *A Natural History of Human Thinking* (Harvard University Press).

Wrangham, R. 2009. *Catching Fire: How Cooking Made us Human* (Profile Books).

10

Acquiring Knowledge on Species-Specific Biorealities: The Applied Evolutionary Epistemological Approach

Nathalie Gontier and Michael Bradie

Reflection on what it is like to be a bat seems to lead us, therefore, to the conclusion that there are facts that do not consist in the truth of propositions expressible in a human language. We can be compelled to recognize the existence of such facts without being able to state or comprehend them.   (Nagel 1974)

INTRODUCTION

Evolutionary epistemology is an inter- and transdisciplinary research area that associates both with philosophy of biology and with the evolutionary sciences. It understands knowledge as an evolved phenomenon displayed by *all* biological species (Campbell 1974; Wuketits 1989; Bradie 1986; Gontier 2006a). Evolutionary epistemologists investigate how species acquire and transmit information and knowledge about the world, how and to what extent the evolved systems of knowledge of biological species in turn inform us of the ontological state of the universe, and how knowledge itself evolves over the course of evolutionary time.

In this chapter, we outline how, by making use of the evolutionary sciences, evolutionary epistemology differs from traditional epistemological fields, and we demonstrate how evolutionary epistemology fits into the broader field of philosophy of biology. Besides by means of natural selection, evolution can occur by a myriad of evolutionary mechanisms and we briefly outline how this plurality results in various evolutionary epistemologies. While early evolutionary epistemologists favored a hypothetical realist position, today scholars favor constructivist approaches to knowledge. This means that scholars no longer adhere to the view that organisms re-present an outer world through the process of adaptation, but that organisms actively participate in constructing the world, by building species-specific biorealities. Rather than present an encompassing evolutionary epistemology, in this chapter we provide a research program on how to study these biorealities.

KNOWLEDGE IS AN EVOLVED PHENOMENON DISPLAYED
BY ALL BIOLOGICAL ORGANISMS

From the ancient Greeks onwards, philosophers have assumed that only humans can obtain true knowledge ("*episteme*") of the world as it is in itself. True knowledge was understood to involve a relation between an individual human knower and the outer world (Pinxten 1997), a relation that was thought to be expressed exclusively in linguistic propositions (Figure 10.1).

At the dawn of the 20th century, however, the early Wittgenstein (1922) demonstrated that we cannot prove by making use of logic that our linguistic propositions relate to the world. The later Wittgenstein (1953) understood language not as an epistemic tool whereby we represent the world, but as a system that results from sociocultural interactions.

Subsequently, two new schools developed concerning how we are to understand knowledge, one proclaiming that knowledge is a sociological phenomenon, the other that it is an evolved phenomenon. Following Wittgenstein, scholars understood "paradigms" (Kuhn 1962) and "epistemic fields" (Foucault 1969) as "regimes of truth" (Foucault 1971) or "scientific research programs" (Lakatos 1978) that are defined by human actors who are part of sociocultural and political communities (Figure 10.2). From within such a Sociology of Knowledge, knowledge becomes redefined as a relation between different knowers (Munz 1993).

Figure 10.1 Classic view of knowledge

For classic philosophers, epistemology goes hand in hand with solving the reference problem, that is, the problem of how our language, that is considered an expression of our thoughts and empirical observations, corresponds to the matters of fact of the outer world. For empiricists, the world gives us impressions that we transform into language that we use to speak about the world. For rationalists, our mind possesses the right mental, linguistic categories to understand our senses and the objects they perceive in the world. Knowledge is therefore defined as a somewhat direct relation between humans and the world that becomes expressed in language.

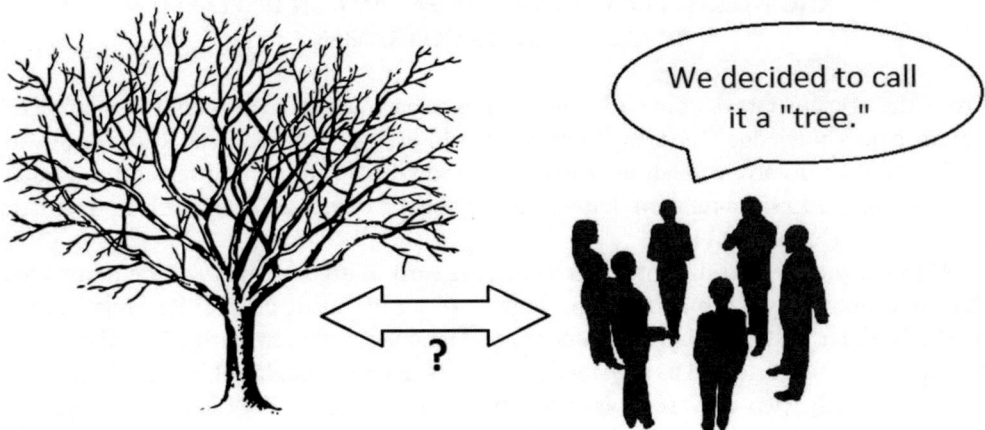

Figure 10.2 Sociology of knowledge

Knowledge is understood as a consensus-based relation entertained between different humans. How our knowledge relates to the physical environment is secondary to understanding how power, hegemony, and sociocultural traditions are established and how they found worldviews.

By adopting an evolutionary stance, evolutionary epistemologists reject the idea that knowledge is *solely* the product of sociocultural traditions where humans develop consensus-views on what is true or false; and they reject the idea that knowledge is exclusively located in the human mind or the consequence of empirical observation. Instead, all organisms are recognized to have knowledge and knowledge is redefined as a relation between the evolved organism and the outer world (Wuketits 2006). The mind, our senses, language, culture, and society are recognized as evolved phenomena (Campbell 1959), and evolutionary epistemology therefore tries to provide an evolutionary foundation for the sociological and cultural phenomena that are associated with knowledge formation. The cognitive capacity to learn (Piaget 1971) and the sociocultural capacity to transmit learned knowledge across generations (Skinner 1986) are recognized as outcomes of evolutionary processes. Because knowledge is defined as a relation between all organisms and the world, for evolutionary epistemologists the question of how our evolved knowledge relates to the outer, physical world remains a valid research question (Figure 10.3).

Organisms have evolved anatomical adaptations to their environments, as well as cognitive schemes of reaction and behavioral patterns that allow them to survive and reproduce in the world. Evolutionary epistemologists (Lorenz 1941; Campbell 1959) understand the evolved anatomy, cognition, and behavior of organisms as systems of information that embody knowledge about the world. It is embodied because it neither comes in the form of language, nor do species need to be conscious about the knowledge they have—their bodies literally embody knowledge.

Many biological individuals also demonstrate knowledge that surpasses their individual anatomy, cognition, and behavior. Either the knowledge is somehow "carried" by the group instead of by each organism individually (which is something that happens especially in sociocultural systems), or the group-knowledge externalizes and often materializes (in termite mounds, for example, or colonies, or shared technological tool complexes that become part of the species' environment). Several primates and most hominids, for

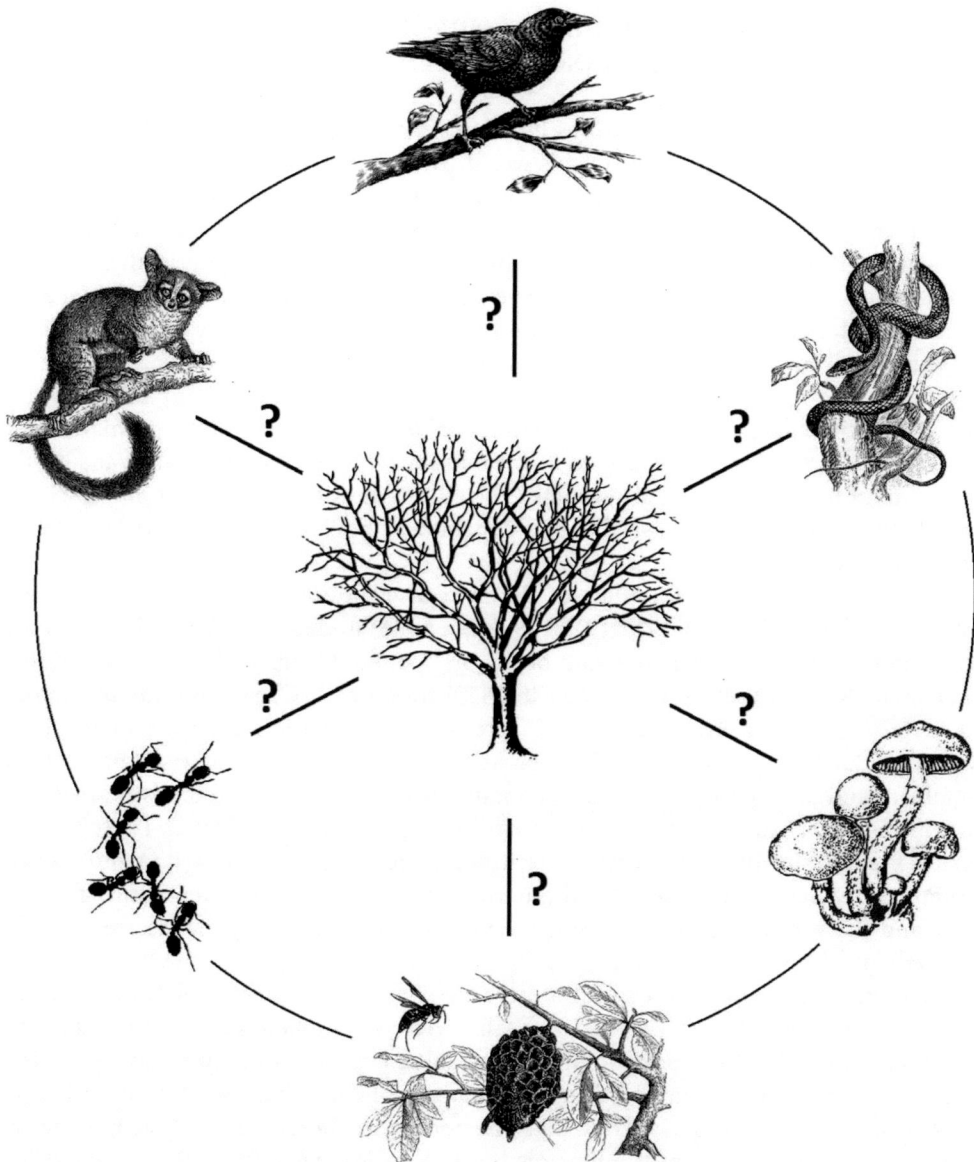

Figure 10.3 Evolutionary epistemology

Knowledge is understood as an evolved relation between the entire organism and its environment. Organisms have evolved anatomical features, cognitive capacities, and behavioral patterns that provide them with means to respond to and anticipate their biotic and abiotic environment, enabling them to survive and reproduce. These features are understood as informational systems that enable the acquisition of knowledge about the environment.

example, accumulate information on how to use and manufacture tools, knowledge that is culturally transmitted over generations through time. And humans have a biologically evolved capacity to learn language, but they learn the specific language they speak from their caregivers. Societies, cultures, or languages function as "extended minds" (Clark & Chalmers 1998; Clark 1999) that are studied from within a general evolutionary framework, as evolved traits.

DIFFERENT EVOLUTIONARY THEORIES RESULT IN DIFFERENT EVOLUTIONARY EPISTEMOLOGIES

It is a fascinating fact that the various life forms that exist today have evolved different means to survive and reproduce in this world. These different means represent evolved information, but the question remains if, how, and to what extent this information references the outer world. Bats and dolphins, for example, have evolved particular anatomical, cognitive, and behavioral traits that enable them to navigate in their specific environment. Bats are well suited to survive and reproduce in a terrestrial or aerial environment, but, unlike dolphins, they do not possess knowledge on how to survive within a marine environment. Each organism connects with certain but not other aspects of the living and non-living world, up to the point that one can argue that organisms live in species-specific biologically informed realities or "*biorealities*." We return to this concept later in the chapter. Here, we focus on how evolutionary epistemology tries to answer the questions of how organismal knowledge evolved, and how it lends insight into the outer world. The answer given is variable and dependent upon the evolutionary framework one adheres to.

Following Darwin, the founders of the field of evolutionary epistemology (Lorenz 1941; Campbell 1959, 1974; Piaget 1971; Skinner 1986) understood the evolved anatomical features, cognitive capacities, and behavioral patterns from within an adaptationist (Lamarck 1809) and selectionist (Darwin 1859) framework. Organisms are understood to adapt to the outer environment, which results in species-specific ways of surviving and reproducing; and natural and sexual selection occurs at the interface between an organism and its environment (which contains both biotic and abiotic components). If organisms display anatomical features, cognitive capacities, and behavioral patterns that enable them to survive long enough to reproduce, then in accordance with natural selection theory, it may be assumed that these traits are adapted to the outer world. If organisms and the various traits they display are not adaptive, then organisms are naturally weeded out.

Such an adaptationist and selectionist view enabled the early evolutionary epistemologists to maintain a hypothetical realist view (Campbell 1960). Organisms are considered to "fit" to the world when they are able to survive and reproduce, and adaptations are understood as evolved re-presentations of the outer world. Konrad Lorenz (1941), for example, famously wrote that "the hoof of the horse is already adapted to the ground of the steppe before the horse is born." Karl Popper (1963, 1972) saw a direct analogy with natural selection theory when he assumed organisms to be like evolved conjectures about the world that somewhat approximate reality or are "corroborated" by reality when they are adapted, and that are "refuted" or "falsified" by the external environment when they are maladaptive. Donald Campbell (1974) drew strong parallels between natural selection theory and Popper's theory of conjectures and refutations, as well as B.F. Skinner's (1986) theory of trial and error learning, and suggested that our bodies constantly test our hypothetical ideas about the outer world. And following Popper (1972: 145; 1984: 19–31) who defined "organs" as embodied "theories" and theories as exosomatic organs, Peter Munz (1993) called organisms "embodied theories" and theories (understood as both scientific thoughts and sociocultural ideas) were dubbed "disembodied organisms." One can, for example, state that a fish provides an as yet unfalsified theory of the water, and by studying the anatomy and behavior of a fish, one can acquire knowledge of the water. A scientific theory is a disembodied organism because it evolves through time, it develops new features, and unfit ideas are eliminated by the environment.

From the viewpoint of hypothetical realism, natural selection becomes nature's way to conjecture, test, falsify, or confirm evolved hypotheses or theories about the world, hypotheses that are embodied in the form of evolved organisms. Knowledge accumulates or grows over evolutionary time, and biological species can even be understood as instruments to acquire knowledge of the outer world (Munz 1993). We humans, for example, cannot see ultraviolet light, but we can point toward scorpions and butterflies that can. Besides the technical instruments that humans have developed over the course of history, life's biodiversity provides a further means to uncover the various layers of reality.

The hypothetical realism adhered to by these scholars (Campbell 1960; Hooker 1989) is nonetheless already tempered through the acknowledgment that there remains a discrepancy between the biological and the purely physical world. Organisms are not literal re-presentations of the outer world because (1) living organisms simply differ from the abiotic world even though they are made up of inorganic matter; (2) adaptive organisms differ from one another; and (3) successful survival and reproduction does not necessarily imply that organisms are adaptive, it merely means that they are not maladaptive. The fact that some organisms survive and reproduce might be due to an abundance of resources, and if a scarcity would arise, competing or predator species might cause the former to go extinct. Non-falsified organisms or theories are not immediately representative of the outer world, they might simply not have been put to the test yet.

Today, the early hypothetical realist views are treated more cautiously. Since the foundation of the Modern Synthesis, numerous fields have questioned several tenets of the core neo-Darwinian framework. New research avenues have opened up including, amongst others:

1. Evo-devo: a field that studies niche construction (Lewontin 1982, 2000; Odling-Smee 1988), phenotypic plasticity (West-Eberhard 2003), developmental systems (Griffiths & Grey 1994; Oyama 2000), and epigenetic mechanisms (Jablonka & Lamb 1989, 2006).
2. Ecology (Van Valen 1973; Odum 1994; Lewontin 2000): a field that divides "the environment" into a biotic and abiotic component and investigates the various interactions amongst them.
3. Macroevolution (Serrelli & Gontier 2015): a research endeavor that studies evolution at the grand scale, beyond populations.
4. Reticulate evolution (Gontier 2015): a vernacular term for evolution as it occurs through hybridization (Arnold 1997), horizontal gene transfer, infectious heredity, symbiosis, and symbiogenesis (Margulis & Sagan 2000; Rosenberg et al. 2007; Gilbert et al. 2012).

These various schools have further specified the exact nature of the organism, the environment, and the relation that exists between these entities. The founders of the Modern

Synthesis treated the organism and the environment as somewhat homogenous entities, and the relation between them was considered unidirectional from the environment to the organism. An organism was either adapted to its external environment or it was not, and it was the environment that, through selection, molded which organisms survived and reproduced.

Today, we recognize that organisms are heterogeneous entities that besides displaying adaptive traits, often simultaneously display maladaptive traits, or neutral traits (Kimura 1968), and existing traits can become exapted for other functions (Gould & Lewontin 1979). Our laryngeal tract, for example, enables a rich vocal palate that allows for differentiated speech, but it also facilitates choking.

Ecological and macroevolutionary perspectives have further diversified "the environment." For one, the environment where the hypothesized struggle for existence occurs is by and large made up of other organisms (Van Valen 1973). Second, these different organisms group together into complexes that form populations, communities, ecosystems, and ultimately the biosphere (see Figure 10.4). And third, the abiotic environment also consists of a multilayered nested hierarchy, where abiotic processes can influence the further course of evolution (Salthe 1985; Tëmkin & Eldredge 2015).

Constructivist approaches necessitate rethinking the classic relation between organisms and the environment and provide more reciprocal and dynamic views. There does not merely exist a struggle for existence over scarce resources, and organisms are not merely passive vehicles that adapt to their environment due to outer selection processes. They actively construct their environment (Gould & Lewontin 1979; Lewontin 1982) in such a way that even inhospitable environments become inhabitable. Numerous organisms anticipate winter and store food. On an ecological level, many species not merely compete, but also interact in such a way that new metabolic cycles, tissues, or biological individuals are formed (Margulis & Sagan 2000). Anaerobic bacteria, for example, that get poisoned by oxygen, symbiotically populate the oxygen-low gut environment of many mammals where they contribute to the good digestion of their host; and lichens are symbiotic organisms made up of distinct species—algae that partner up with cyanobacteria and/or fungi.

Rather than demonstrate adaptation toward an external environment, niche construction and symbiosis demonstrate adaptability (Warburton 1956), or an organism's ability to actively modify and build the environment in such a way that it becomes adapted to it. Such traits do not necessarily result in a one-to-one correspondence with the physical world that exists outside the organism; rather, organisms are less representative or referential of an "independent outer world." Species form an active part in building the biosphere, an influence that extends well into the abiotic world. The earth's atmosphere, for example, is oxygen-rich, and 90 percent of that oxygen was created as a waste-product by early cyanobacteria (Margulis & Sagan 2002).

Constructivist approaches (Vollmer 1984; Riedl 1987; Hooker 1989), and even non-adaptationist views (Wuketits 2006) have been put forward to explain how organisms and the species they group into embody knowledge, not merely by "re-presenting" the "outer" world, but by actively constructing it. While hypothetical realism assumes a fit between the organism and the environment, within constructivist views it is recognized that over evolutionary time distinct biorealities have emerged. As such, there is no constant "outer world," but an ever-changing sequential series of emerging biorealities.

ORGANISMS CONSTRUCT SPECIES-SPECIFIC BIOREALITIES

Ancient philosophers investigated how human knowledge (epistemology) provides insight into the nature of the earth and the larger cosmos (ontology), but they thought the cosmos and its components were stable entities and any change that occurred was assumed to be repeated in a cycle of coming and becoming. Ancient philosophers therefore merely sought to find the right order of the cosmic hierarchy, a hierarchy and order they assumed to be eternal. Today, we know that the universe originated approximately 13.7 billion years ago, and life originated on earth some 4 billion years ago. Every abiotic and biotic entity must therefore have come into being, and what exists as "real" is variable over time. With the origin of the universe, various types of matter and energy originated, and with the origin of living beings, various forms of knowledge evolved. The knowledge that species embody and materialize is diverse, and results from organism-environmental interactions. The relation between the biotic and abiotic world is dynamic in kind, and the knowledge that accumulates from this interaction is variable in time and space. Consequently, there is no essential and given knowledge system, nor does there exist a fixed or invariant truth that is waiting to be discovered by these various knowledge systems.

Whether and how the various knowledges that the living world acquired are in one-to-one correspondence with an "outer world," understood as a stable and purely physical world, are the wrong questions to ask. Rather, over time, multiple species-specific and species-bounded biological realities, or biorealities have evolved, realities that are also bounded by physical and chemical laws. What counts as "real" or "true" for one species might not count as "real" for another. We humans, for example, have especially adapted to a mesocosm (Vollmer 1984), a world of middle-sized objects. We cannot observe the bacteria that occupy our skin or gut, or the molecules that form a table. Bacteria, on the other hand, even without possessing vision or brains, can easily overcome these biophysical boundaries and establish a biochemical communication with our skin, or penetrate deep inside table wood.

Because biological realities are species-specific and because species evolve, the various biorealities that are formed over time are not stable. There is not one homogenous outer world "out there." Rather, over evolutionary time multiple and varying biorealities emerged. When species go extinct their specific bioreality often ceases to exist while new biorealities become constructed when new species evolve. Biorealities expand and contract in time, in congruence with the species that build them, but boundaries are fuzzy. Though species-specific, different biorealities often overlap, especially when different species occupy the same niche where they are dependent upon the same resources; or when species share common descent, and therefore share common traits that enable them to modify and construct their environment in similar ways. A purely solipsistic view, for example, extrapolated toward organisms or species, is impossible just because we can prove that at least all eukaryotic organisms are related to one another by common descent.

In sum, constructivist approaches necessitate a more dynamic and emergent view of the world. In so far as there existed an "outer world" before life originated, earth has significantly altered in association and perhaps even in correspondence with the life forms that have evolved over time. There does not exist an external relation between the organism and the world, the world has changed inside out, because of the organisms that have evolved, and therefore the world itself becomes an emerging and changing entity. A valid question then becomes how these biorealities are ordered over time, that is, if and how these different biorealities match together into a nested hierarchy (or multiple nested hierarchies), and how they together form a reality (or multiple realities). On a higher level, and to some extent, the various species-specific "biorealities" give proof of the existence of such larger hierarchically structured and nested entities.

APPLIED EVOLUTIONARY EPISTEMOLOGY AND THE WIDER FIELD OF PHILOSOPHY OF BIOLOGY

In this part we present a research program for how to study emerging biorealities. Applied evolutionary epistemology covers the following five research areas:

1. What aspects or traits of biological individuals count as information or knowledge?
2. How can evolutionary theories explain the origin and evolution of these information and knowledge systems?
3. Where do these knowledge systems evolve?
4. How do the evolved knowledge systems underlie the construction of various biorealities and how can the latter in turn lend insight into the ontological layers of the world?
5. Can evolutionary mechanisms themselves be regarded as knowledge-acquiring systems?

The first research endeavor roughly coincides with research on the units of evolution (what evolves, Table 10.1), the second with research on the various evolutionary mechanisms that explain how these units evolve (Table 10.2); and the third coincides with research on the levels of evolution (where in "reality" or "the environment" these units evolve according to certain mechanisms, Figure 10.4). These three research endeavors are tightly related to one another, because identifying units of evolution always coincides with identifying the locus or level where this unit evolves as well as with the mechanism whereby the unit evolves at a certain level (Gontier 2010).

Darwin identified the organism (what we today designate as the phenotype) as the unit of evolution, and he argued that the organism evolves at the level of the environment by means of natural selection. Following Darwin, philosophers (Brandon 1982) and evolutionary

Table 10.1 Examples of units of selection

Unit	Characterization
Replicator	any entity able to "create copies of itself" (Dawkins 1976: 15)
	something that demonstrates "fecundity, longevity, and copying-fidelity" (Dawkins 1976: 18)
Meme	"a unit of imitation" (Dawkins 1976: 192)
	"brain structures whose 'phenotypic' manifestation as behaviour or artefact is the basis of their selection" (Dawkins 1982: 164)
Interactor	"an entity that directly interacts as a cohesive whole with its environment in such a way that replication is differential" (Hull 1980: 318)
Culturgen	"that which generates culture" (Lumsden & Wilson 1981: 26)
	"a relatively homogeneous set of artifacts, behaviors, or mentifacts (mental constructs having little or no direct correspondence with reality) that either share without exception one or more attribute states selected for their functional importance, or at least share a consistently recurrent range of such attribute states within a given polythetic set" (Lumsden & Wilson 1981: 27)

biologists (Williams 1966; Dawkins 1976, 1982; Lewontin 1970) originally focused exclusively on identifying the units and levels of natural selection, by attempting to "universalize" the theory in approaches that became known as "universal Darwinism" (Dawkins 1982, 1983) or "universal selectionism" (Cziko 1995). Besides anatomical, phenotypic form, these scholars also understand cognition or individual and even group behavior to be units that evolve by means of natural selection. And from within evolutionary epistemology, any and all biologically evolved traits can be understood as systems of information.

Universalization implies the following. When it is argued that science (Toulmin 1972; Hull 1988) or culture (Cavalli-Sforza & Feldman 1981; Lumsden & Wilson 1981; Boyd and Richerson 1985; Laland et al. 1995; Mesoudi 2015) evolve by means of natural selection, it needs to be demonstrated how Darwin's theory of natural selection can be extrapolated toward sociocultural phenomena such as scientific or cultural knowledge, and also the units of sociocultural selection (e.g., ideas, rituals, or practices), as well as the levels where these entities evolve, need to be specified. In this regard, Michael Bradie (1986) distinguished between the evolution of epistemological mechanisms (EEM) and the evolutionary epistemology of theories (EET) program: the former investigates the evolution of the biological organs and systems employed in the acquisition of knowledge, while the latter investigates the evolution of the knowledge corpuses that are constructed by knowers.

To explain how natural selection theory can be applied to the sociocultural domain, pioneering evolutionary epistemologists and evolutionary biologists (Table 10.2) have developed several "heuristics" (Campbell 1959, 1960), "Darwinian principles" (Lewontin 1970), or "universal selection formulas" (Hull 1980; Plotkin 1994).

Theorizing on the levels of selection (Figure 10.4) used to be associated with theorizing on how the "superorganic" (Hutton 1788; Spencer 1876; Sapir 1917) relates to the inorganic and organic layers of reality. Today, it is associated more with multilevel selection theory (Okasha 2005) as well as with ecologically and macroevolutionary-oriented fields where scholars actively build hierarchies of the biotic and abiotic world. We return to this when we discuss the fourth point.

First, it is important to emphasize that the classic units and levels of selection debate is currently more accurately defined as the units and levels of evolution debate (Gontier 2010, 2012). Scholars nowadays recognize the existence of a myriad of units that evolve at various levels (Table 10.3), by a multitude of evolutionary mechanisms beyond natural

Table 10.2 Examples of universal selection formulas

Blind variation and selective survival (Campbell 1959)

Blind variation and selective retention (Campbell 1960)

Phenotypic variation, differential fitness (because of different environments), and heritability of that fitness (Lewontin 1970)

Conjectures and refutations (Popper 1963, 1972)

"A process in which the differential extinction and proliferation of interactors cause the differential perpetuation of the replicators that produced them" (Hull 1980: 381)

Replication, variation, and environmental interaction (Replicator, interactor, lineage) (Hull 1980; Hull et al. 2001)

Blind trial and error learning (Skinner 1986)

Generate, test, regenerate schema (Plotkin 1994)

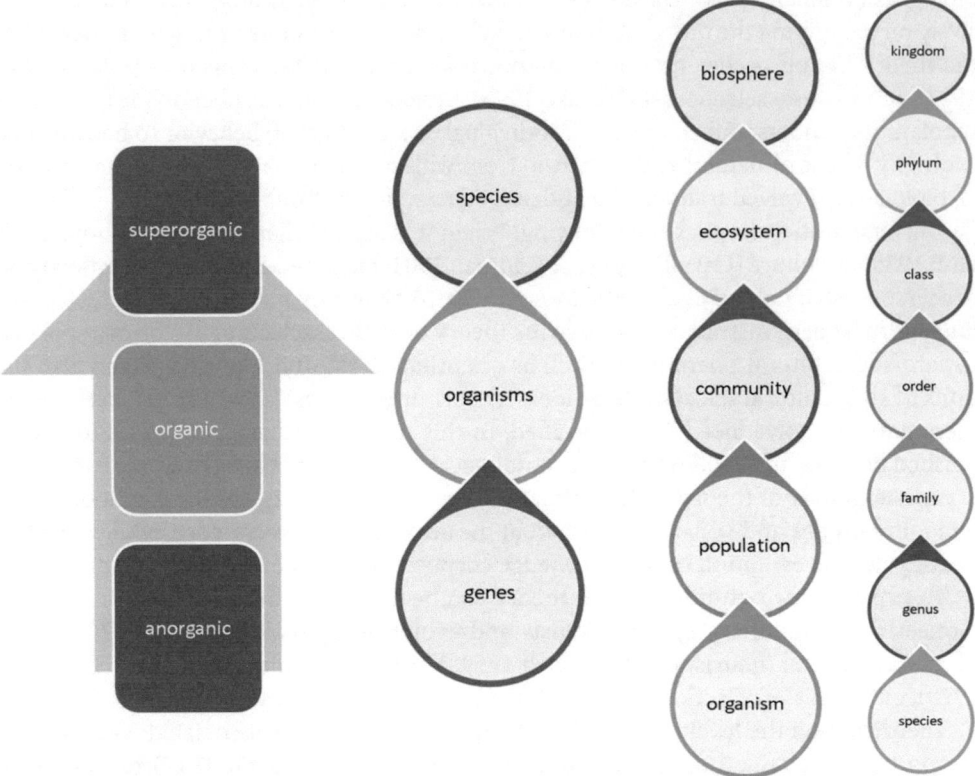

Figure 10.4 Examples of hierarchies of the biotic and abiotic world

From left to right, Hutton's geological and Spencer's sociocultural distinction between the anorganic or non-living, organic or living and superorganic or sociocultural world; the biological hierarchy of what are considered real entities; the ecological hierarchy; and the theoretical genealogical hierarchy.

Table 10.3 Units of non-selectionist mechanisms

Unit	Characterization
Developmental systems theory	
Reproducer	"The pieces of development that are passed on in progeneration [that] confer the capacity to develop on the offspring" (Griesemer 2000: 361)
	"Special or developmental progeneration is multiplication with material overlap of mechanisms conferring the capacity to develop. Development is acquisition of the capacity to reproduce. Reproduction, therefore, is progeneration of entities that develop." (Griesemer 2000: 361)
Symbiosis and Symbiogenesis theory	
Holobiont	"The symbiotic complex" consisting of an individual biont and its (multiple) symbionts (Margulis 1991: 2–4)
Hologenome	"Host genome + microbiome, including viriome" (Rosenberg & Zilber-Rosenberg 2014: 111)

selection. Evolution can be reticulate, or it can occur by means of genetic and ecological drift, niche construction, epigenetic mechanisms, and so on.

However, much work remains to be done in these new fields. The mechanisms that underlie reticulate evolution or the evolutionary emergence of (epigenetic) developmental systems are still not fully understood, and scholars still need to "universalize" the units whereupon and the levels whereat such evolution occurs. Eco-evo-devo fields (Griffiths & Gray 1994; Hahlweg 1989; Jablonka & Lamb 1989, 2006; Oyama 2000) are currently investigating how organismal knowledge is often not reducible to genes. Internal structures such as bodily organs, hormonal and vascular systems, or complex biochemical gene-protein interactions are complex adaptive systems (Hooker 1989) that possess "information" that can become transmitted across generations. And reticulate evolution (Margulis & Sagan 2000, 2002; Gontier 2015) can occur by means of symbiosis, symbiogenesis, infectious heredity, hybridization, and lateral gene transfer. Sometimes, such reticulate evolution requires the horizontal transmission of genetic material, sometimes it requires genetic exchange through sex, and sometimes it involves the merging of different bodies into one another, such as bacteria that infect an organism's airways.

Non-genetic and epigenetic information is also horizontally exchanged and shared between individuals at a sociocultural level. Strong parallels can be drawn between language and culture borrowing and symbiogenesis, for example (Gontier 2006b, 2007), but the mechanisms whereby reticulate evolution occurs are yet to be universalized.

Turning to the fourth point, the recognition that numerous units, levels, and mechanisms of evolution exist necessitates us to investigate how these units and levels interact as well as how the various mechanisms interact to bring forth evolution (Gontier 2010). Units and levels form hierarchically nested realities (Salthe 1985; Tëmkin & Eldredge 2015; Pievani 2015), and the nested ontological layers increase in complexity in what concerns their organization. Studying this increasing complexity relates to theorizing on the major transitions (Table 10.4) that life has gone through since its evolutionary emergence, transitions that are often defined in terms of new means to store and pass on information (Maynard Smith & Száthmary 1995; Calcott & Sterelny 2011).

Table 10.4 Maynard Smith & Száthmary's (1995) major transitions in evolution

Major Transitions in Evolution
Replicating molecules to populations of molecules
Independent replicators to chromosomes
RNA to DNA
Prokaryotes to Eukaryotes
Asexual clones to sexual populations
Protists to animals, plants and fungi
Solitary individuals to colonies
Primate societies to human societies

These ideas also find their roots in the works of early evolutionary epistemologists, who shared the aim to build a hierarchy or "taxonomy of behavior" (Lorenz 1941, 1958; Tinbergen 1963) that would map onto the tree of life and portray the evolution of knowledge. Campbell (1959, 1974), for example, distinguished between ten and twelve stages of "inductively achieved knowledge," and his stages provided an evolutionary line-up of the various informational systems that evolved.

Such theorizing therefore connects with the theorizing on how evolutionary mechanisms themselves are knowledge-acquiring devices, which brings us to the fifth point of the program. Lorenz (1941), the founder of ethology, Jean Piaget (1971), the founder of developmental psychology, Skinner (1986), one of the founders of behaviorism and operant conditioning, were all also evolutionary epistemologists; and Campbell (1959) had a background in comparative psychology. These scholars commenced the investigation of cognition and knowledge as an organismal trait that establishes itself during development (ontogeny), through trial and error, observational and other kinds of learning, and this development in turn is driven both by our evolved anatomical and neurocognitive constitution and by our equally evolved sociocultural environment, over the course of phylogeny. Skinner (1986) investigated operant conditioning in various animals, and he developed a theory of language learning that was based upon such operant conditioning. Behaviorist schools became accompanied with cognitive schools of thought (Piaget 1971) that attempted to enter the "black box" that had been the mind up until then.

Lorenz (1941) wrote what has become a classic paper in the field wherein he reevaluated Kant's synthetic a priori claims from within an evolutionary perspective. Most eukaryotic organisms are born with "instincts" or "fixed action patterns" that can be understood as inborn knowledge because they comprise a set of biological expectations about and responses to a specific environment. Newborn ducks, for example, instinctively follow the first thing they see once they hatch. Under normal circumstances, they first see their mother, and it is adaptive to follow her around because she provides food and protection. The newborn ducks do not see the mother as a "mother," "caregiver," or "protector," but they evolved to instinctively follow the first thing they see. Under experimental conditions, Lorenz was able to demonstrate that they would also follow him, or even mobile toys such as trains. Such behavior is called "imprinting," and it demonstrates a specific type of inborn knowledge about the environment. Before the duck was able to learn through experience what its mother can do for it, it "knows" to follow her around. Similarly, animal courtship and mating or fighting are highly ritualized behaviors, called "fixed action patterns," that enable organisms to respond adequately to certain environmental cues. Experiments show that many organisms know how to behave sexually or violently before having observed or learned the behavior or having acted accordingly.

Such behavior is "known" from birth onward (and thus a priori given), but the reason organisms possess this inborn knowledge is that it evolved over the course of evolution. As such, Lorenz (1941), and later also Campbell (1959, 1974), reinterpreted Kant's synthetic a priori (innate) knowledge as knowledge that was obtained synthetically (inductively or a posteriori), by our ancestors over the course of phylogeny. The justification for such claims was found in the theory of natural selection. Such a view furthermore implies that knowledge accumulates over time, and early ethologists assumed that over the course of evolution, learned habits turned into inborn instincts, and thus somehow became conserved and transmitted through generations over time. This was a somewhat neo-Lamarckian claim, and current eco-evo-devo schools are studying the possibility of such inheritance of acquired characteristics, within and beyond the genome.

CONCLUDING REMARKS

Evolutionary epistemology is a field particularly concerned with solving the problem of the development and transmission of information and knowledge through time, across all domains of life. Knowledge is thereby broadly conceived, and includes anatomical, cognitive, behavioral, and sociocultural traits displayed by organisms. As such, it has identified a fundamental question that remained unanswered by the founders of the Modern Synthesis—namely, the question of how new information becomes introduced, stored, and transmitted over generations through time.

Natural selection *strictu sensu* is a theory that explains how existing information is selected over time and how maladaptive information is deleted. To explain how novel information becomes introduced, Darwin made use of Lamarckian inheritance theory, but the founders of the Modern Synthesis conjectured that new traits resulted solely from small random genetic mutations. Today, we know that much more counts as information than what is stored in genes, and this non-genetic and epigenetic information can become stored and transmitted through epigenetic mechanisms, drift, and mechanisms that underlie reticulate evolution.

Philosophy is most successful when it launches new research fields or even entire new scientific disciplines. Many of the founding fathers of evolutionary epistemology were also founding fathers of fields such as ethology, comparative psychology, behaviorism, and cognitive psychology. These fields have now evolved into the new evolutionary sciences that include evolutionary psychology, evolutionary linguistics, evolutionary anthropology, evolutionary sociology, and so on. Nonetheless, many of the questions first raised by evolutionary epistemologists have not been answered satisfactorily by the new evolutionary sciences. And because the research program has to some degree been incorporated into other disciplines, the major philosophical issues raised by the early evolutionary epistemologists have unfortunately been somewhat abandoned by philosophers. The problem of inductively acquired knowledge, or how ontogenetically acquired traits become players in phylogeny remains crucial for both philosophers and evolutionary scholars. The above outlined applied evolutionary epistemological research program provides a means whereby these different fields can interact to find answers.

ACKNOWLEDGMENTS

Gontier kindly acknowledges the support of the Portuguese Fund for Scientific Research (grant ID SFRH/BPD/89195/2012 and project number UID/FIL/00678/2013) and the Centre for Philosophy of Science of the University of Lisbon.

REFERENCES

Arnold, M. 1997. *Natural Hybridization and Evolution* (Oxford University Press).

Boyd, R. & Richerson, P. 1985. *Culture and the Evolutionary Process* (University of Chicago Press).

Bradie, M. 1986. "Assessing evolutionary epistemology." *Biology and Philosophy* 1: 401–459.

Brandon, R. 1982. "The levels of selection." *Proceedings of the Philosophy of Science Association* 1: 315–323.

Calcott, B. & Sterelny, K. (eds.). 2011. *The Major Transitions in Evolution Revisited* (MIT Press).

Campbell, D. 1959. "Methodological suggestions from a comparative psychology of knowledge processes." *Inquiry* 2: 152–183.

Campbell, D. 1960. "Blind variation and selective retention in creative thought as in other knowledge processes." *Psychological Review* 67: 380–400.

Campbell, D. 1974. "Evolutionary epistemology." In P. Schilpp (ed.), *The Philosophy of Karl Popper Vol. 1* (La Salle) 413–459.

Cavalli-Sforza, L. & Feldman, M. 1981. *Cultural Transmission and Evolution* (Princeton University Press).

Clark, A. 1999. "An embodied cognitive science?" *Trends in Cognitive Sciences* 3: 345–351.

Clark, A. & Chalmers, D. 1998. "The extended mind." *Analysis* 58: 7–19.

Cziko, G. 1995. *Without Miracles* (MIT Press).

Darwin, C. 1859. *The Origin of Species* (Murray).

Dawkins, R. 1976. *The Selfish Gene* (Oxford University Press).

Dawkins, R. 1982. "Replicators and vehicles," In King's College Sociobiology Group (ed.), *Current Problems in Sociobiology* (Cambridge University Press) 45–64.

Dawkins, R. 1983. "Universal Darwinism." In D. Hull & M. Ruse (eds.), *The Philosophy of Biology* (Oxford University Press) 15–35.

Foucault, M. 1969. *L'Archéologie du Savoir* (Gallimard).

Foucault, M. 1971. *L'Ordre du Discours* (Gallimard).

Gilbert, S., Sapp, J., & Tauber, A. 2012. "A symbiotic view of life: We have never been individuals." *Quarterly Review of Biology* 87: 325–341.

Gontier, N. 2006a. "Introduction to evolutionary epistemology, language and culture." In N. Gontier, J. van Bendegem, & D. Aerts (eds.), *Evolutionary Epistemology, Language and Culture* (Springer) 1–26.

Gontier, N. 2006b. "Evolutionary epistemology and the origin of language: Taking symbiogenesis seriously." In N. Gontier, J. van Bendegem, & D. Aerts (eds.), *Evolutionary Epistemology, Language and Culture* (Springer) 195–226.

Gontier, N. 2007. "Universal symbiogenesis: An alternative to universal selectionist accounts of evolution." *Symbiosis* 44: 167–181.

Gontier, N. 2010. "Evolutionary epistemology as a scientific method: A new look upon the units and levels of evolution debate." *Theory in Biosciences* 129: 167–182.

Gontier, N. 2012. "Applied evolutionary epistemology: A new methodology to enhance interdisciplinary research between the human and natural sciences." *Kairos* 4: 7–49.

Gontier, N. 2015. "Reticulate evolution everywhere." In N. Gontier (ed.), *Reticulate Evolution* (Springer) 1–40.

Gould, S. J. & Lewontin, R. 1979. "The spandrels of San Marco and the Panglossian paradigm: A critique of the adaptationist programme." *Proceedings of the Royal Society London, Series B: Biological Sciences* 205: 581–598.

Griesemer, J. 2000. "Development, culture and the units of inheritance." *Philosophy of Science* 67: S348–S368.

Griffiths, P. & Gray, R. 1994. "Developmental systems and evolutionary explanation." *Journal of Philosophy* 91: 277–304.

Hahlweg, K. 1989. "A systems view of evolution and evolutionary epistemology." In K. Hahlweg & C. Hooker (eds.), *Issues in Evolutionary Epistemology* (State University of New York Press) 45–78.

Hooker, C. 1989. "Evolutionary epistemology and natural realism," In K. Hahlweg & C. Hooker (eds.), *Issues in Evolutionary Epistemology* (State University of New York Press) 101–150.

Hull, D. 1980. "Individuality and selection." *Annual Review of Ecology and Systematics* 11: 311–332.

Hull, D. 1988. *Science as a Process* (University of Chicago Press).

Hull, D., Langman, R., & Glenn, S. 2001. "A general account of selection: Biology, immunology, and behavior." *Behavioral and Brain Sciences* 24: 511–573.

Hutton, J. 1788. "Theory of the Earth." *Transactions of the Royal Society of Edinburgh* 1: 209–304.

Jablonka, E. & Lamb, M. 1989. "The inheritance of acquired epigenetic variations." *Journal of Theoretical Biology* 139: 69–83.

Jablonka, E. & Lamb, M. 2006. *Evolution in Four Dimensions* (MIT Press).

Kimura, M. 1968. "Evolutionary rate at the molecular level." *Nature* 217: 624–626.

Kuhn, T. 1962. *The Structure of Scientific Revolutions* (Chicago University Press).

Lakatos, I. 1978. *The Methodology of Scientific Research Programmes* (Cambridge University Press).

Laland, K., Kumm, J., & Feldman, M. 1995. "Gene-culture co-evolutionary theory." *Current Anthropology* 36: 131–146.

Lamarck, J. 1809. *Philosophie Zoologique* (Dentu Libraire, Museum d'Histoire Naturelle).

Lewontin, R. 1970. "The levels of selection." *Annual Review of Ecological Systems* 1: 1–18.

Lewontin, R. 1982. "Organism and environment." In H. Plotkin (ed.), *Learning, Development and Culture* (Wiley) 151–170.

Lewontin, R. 2000. *The Triple Helix* (Harvard University Press).

Lorenz, K. 1941. "Kant's Lehre vom Apriorischen im Lichte gegenwärtiger Biologie." *Blätter für Deutsche Philosophie* 15: 94–125.

Lorenz, K. 1958. "The evolution of behavior." *Scientific American* 199: 67–78.

Lumsden, C. & Wilson, E. 1981. *Genes, Mind, and Culture* (World Scientific Publishing).

Margulis, L. 1991. "Symbiogenesis and symbionticism." In L. Margulis & R. Fester (eds.), *Symbiosis as a Source of Evolutionary Innovation* (MIT Press) 1–14.

Margulis, L. & Sagan, D. 2000. *What is Life?* (University of California Press).

Margulis, L. & Sagan, D. 2002. *Acquiring Genomes* (Basic Books).

Maynard Smith, J. & Szathmáry, E. 1995. *The Major Transitions in Evolution* (Oxford University Press).

Mesoudi, A. 2015. "Cultural evolution: A review of theory, findings and controversies." *Evolutionary Biology*. doi:10.1007/s11692-015-9320-0.

Munz, P. 1993. *Philosophical Darwinism* (Routledge).

Nagel, T. 1974. "What is it like to be a bat?" *Philosophical Review* 83: 435–450.

Odling-Smee, F. 1988. "Niche constructing phenotypes." In H. Plotkin (ed.), *The Role of Behavior in Evolution* (MIT Press) 73–132.

Odum, H. 1994. *Ecological and General Systems* (Colorado University Press).

Okasha, S. 2005. "Multilevel selection and the major transitions in evolution." *Philosophical Biology* 72: 1013–1025.

Oyama, S. 2000. *The Ontogeny of Information* (Duke University Press).

Piaget, J. 1971. *Genetic Epistemology* (W.W. Norton).

Pievani, T. 2015. "How to rethink evolutionary theory: A plurality of evolutionary patterns." *Evolutionary Biology*. doi:10.1007/s11692-015-9338-3.

Pinxten, R. 1997. *When the Day Breaks* (Peter Lang).

Plotkin, H. 1994. *Darwin Machines and the Nature of Knowledge* (Penguin Books).

Popper, K. 1963. *Conjectures and Refutations* (Routledge & Kegan Paul).

Popper, K. 1972. *Objective Knowledge* (Clarendon Press).

Popper, K. 1984. "Critical remarks on the knowledge of lower and higher organisms, the so-called motor systems." In O. Creutzfeldt, R. Schmidt, & W. Willis (eds.), *Sensory Motor Integration in the Nervous System* (Springer-Verlag) 19–31.

Riedl, R. 1987. *Begriff und Welt* (Parey).

Rosenberg, E. & Zilber-Rosenberg, I. 2014. *The Hologenome Concept* (Springer).

Rosenberg, E., Koren, O., Reshef, L., Efrony, R., & Zilber-Rosenberg, I. 2007. "The role of microorganisms in coral health, disease and evolution." *Nature Reviews Microbiology* 5: 355–362.

Salthe, S. 1985. *Evolving Hierarchical Systems* (Columbia University Press).

Sapir, E. 1917. "Do we need a 'superorganic'?" *American Anthropologist* 19: 441–447.

Serrelli, E. & Gontier, N. (eds.). 2015 *Macroevolution* (Springer).

Skinner, B.F. 1986. "The evolution of verbal behavior." *Journal of the Experimental Analysis of Behavior* 45: 115–122.

Spencer, H. 1876. *The Principles of Sociology, Vol. 1* (Williams and Norgate).

Tëmkin, I. & Eldredge, N. 2015. "Networks and hierarchies: Approaching complexity in evolutionary theory." In N. Gontier & E. Serrelli. (eds.), *Macroevolution* (Springer) 227–275.

Tinbergen, N. 1963. "On aims and methods of ethology." *Zeitschrift für Tierpsychologie* 20: 410–433.

Toulmin, S. 1972. *Human Understanding* (Princeton University Press).

Van Valen, L. 1973. "A new evolutionary law." *Evolutionary Theory* 1: 1–30.

Vollmer, G. 1984. "Mesocosm and objective knowledge: On problems solved by evolutionary epistemology." In F. Wuketits (ed.), *Concepts and Approaches in Evolutionary Epistemology* (D. Reidel) 69–121.

Warburton, F. 1956. "Genetic assimilation: Adaptation versus adaptability." *Evolution* 10: 337–339.

West-Eberhard, M. 2003. *Developmental Plasticity and Evolution* (Oxford University Press).

Williams, G. 1966. *Adaptation and Natural Selection* (Princeton University Press).

Wittgenstein, L. 1922. "*Tractatus Logico-Philosophicus* (trans. C. Ogden) (Kegan Paul).

Wittgenstein, L. 1953. *Philosophical Investigations* (trans. G. E. M. Anscombe & R. Rhees) (Blackwell).

Wuketits, F. 1989. "Cognition: A non-adaptationist view." *La Nuova Critica* 9–10: 5–15.

Wuketits, F. 2006. "Evolutionary epistemology: The non-adaptationist approach." In N. Gontier, J. van Bendegem, & D. Aerts (eds.), *Evolutionary Epistemology, Language and Culture* (Springer) 33–46.

III

Human Nature

11

Human Nature: An Overview

Stephen M. Downes

INTRODUCTION

Debates about human nature inform every philosophical tradition from their inception (see Stevenson 2000 for many examples). Evolutionarily based criticisms of human nature are of much more recent origin. Ironically, most evolutionarily based criticisms of human nature are directed at work whose avowed goal is to biologicize human nature and even to place human nature within an evolutionary frame. Here I will focus on accounts of human nature that begin with and come after E.O. Wilson's sociobiology. I will also focus on criticisms of human nature that arose first as responses to sociobiology. There are some more recent approaches to human nature that have much in common with the sociobiological approach and I will show that critical arguments developed to target sociobiology have purchase on related recent approaches to human nature. In what follows I will briefly outline some well-known accounts of human nature. Next I will briefly outline some key evolutionarily based arguments against such accounts of human nature. I conclude by summarizing the evolutionary case against biological accounts of human nature and endorsing it.

Some evolutionary arguments against human nature arise from debates about species and the issue of whether or not essentialism is appropriately applied in the context of species delineation. I will only briefly introduce these issues as they are dealt with in more detail in John Wilkins and Kevin LaPorte's chapters in this section. Other evolutionary arguments against human nature center around the question of normality—is there a coherent concept of a normal human? Finally, some arguments against human nature focus on whether we can cleanly divide nature and culture. I briefly outline examples of these kinds of arguments below but the delineation of nature from culture is dealt with in much more detail in Louise Barrett and Maria Kronfeldner's chapters in this section.

BIOLOGICALLY BASED HUMAN NATURE

In the final chapter of his large work on sociobiology, Wilson (1975) argues that human social behavior has a biological basis just as animal social behavior does. He provides a more detailed defense of this view in his follow-up book, *On Human Nature* (1978). Here he argues, first, that human nature is best characterized by the collection of distinctive behaviors that are universally distributed throughout all cultures and, second, that these behaviors are best understood as having been shaped by natural selection. In other words, the genes underlying our social behavior are more highly represented as a result of selection. Wilson proposes that the goal of "human sociobiology is to learn whether the evolution of human nature conforms to conventional evolutionary theory" (2013: 16). He expressed optimism that we will soon identify the "genes that influence behavior" (2013: 21). Wilson acknowledges that there is genetic diversity in humans just as in any species, saying:

> we are a single species . . . one great breeding system through which genes flow and mix in each generation. Because of that flux, mankind viewed over many generations shares a single human nature within which relatively minor hereditary influences recycle through ever changing patterns, between the sexes and across families and entire populations.   (2013: 23)

Here he foreshadows later accounts of human nature (discussed below) that aim to account for variation as part of our nature. This approach is more liberal than the view that there is a collection of distinct behaviors, determined by a collection of genes that characterize our nature. This more restrictive view (or disciplined view as Lewens (2015) puts it) is the one usually associated with Wilson and the view that is the focus of much critical scrutiny.

Evolutionary psychologists propose that human nature is not a collection of universal human behaviors but rather a collection of universal psychological mechanisms underlying these behaviors. This view retains some of the structure of the sociobiological view but relocates the focus of explanatory work. The adaptations—products of natural selection—that evolutionary psychologists focus on are underlying psychological mechanisms. John Tooby and Leda Cosmides pithily sum up the view as follows: "the concept of human nature" is "based on a species-typical collection of complex *psychological adaptations*" (1990: 17). Here is David Buller's characterization of the evolutionary psychologists' view: "human nature consists of a set of psychological adaptations that are presumed to be universal among, and unique to, human beings" (2005: 423). Donald Symons clearly expresses the evolutionary psychologists' approach as follows: "all accounts of human action . . . *imply* a human nature" and this nature is "a diverse array of complex, specialized brain/mind mechanisms" (1987: 89). Symons says that Darwinians are in a better position than others in the social sciences to account for our nature thus construed. Evolutionary psychologists acknowledge the wide variety of human behavior and cultures but argue that the best way to explain this variation is in terms of the underlying mechanisms we have in common. On their account, the selective pressures that shaped our psychological mechanisms were active thousands of years ago and so these mechanisms are adaptations that helped our ancestors in their environments. Like

Wilson, evolutionary psychologists propose that human nature consists of traits that no longer vary. Evolutionary biologists say that such traits are at fixation. Selection can act so that a trait, or allele, becomes fixed in a population but selection can also result in alternative traits or alternative alleles being present in a population. Evolutionary psychologists account for some of the manifest variation in our behaviors in terms of a mismatch between our ancient psychological mechanisms and more recent challenging environments. We will see below that some evolutionary psychologists challenge this view of human nature but the view summarized here clearly captures the target of several evolutionary criticisms of human nature.

Edouard Machery (2008) presents and defends a notion of human nature that he claims is "an important notion of human nature [that] is compatible with evolutionary biology" (322). Machery calls his notion of human nature the "nomological notion." The nomological notion states that "human nature is the set of properties that humans tend to possess as a result of the evolution of their species" (Machery 2008: 323). On this account, bipedalism is part of human nature but supporting Liverpool Football Club is not. According to Machery, the nomological notion of human nature rules out certain kinds of explanations for a trait if it is part of human nature. Specifically, "any explanation to the effect that [a trait's] occurrence is exclusively due to enculturation or to social learning" (2008: 326) is ruled out. Machery adds that this constraint does not rule out that social learning could be part of the explanation of the trait but if a trait is part of human nature, it cannot be completely explained by appeal to culture or social learning. Machery proposes that traits arising purely as a result of local cultural circumstances are very unlikely to be common among humans. On his account, the idea that a trait is common among humans is a necessary condition for that trait being part of human nature. Machery refers to this as the universality proposal. So his account contains two central proposals, the evolutionary proposal and the universality proposal, summed up here: "humans have many properties in common as a result of the evolution of their species" (2008: 328). Machery's account of human nature shares features of both the sociobiological account and the evolutionary psychologists' account already introduced. This is by design, as Machery holds that his nomological notion of human nature is also important "because this notion of human nature is probably the relevant one for understanding sociobiologists', such as E.O. Wilson, and evolutionary psychologists' interest in human nature" (2008: 328). Elsewhere Machery says "the current attempt to reconceptualize human nature aims in part at explicating the notion of human nature that is used in the human behavioral sciences," adding that his universality proposal and evolutionary proposal are "necessary for this task" (2012: 478). The relevant human behavioral sciences for Machery are sociobiology and evolutionary psychology.

Richard Samuels proposes and defends "causal essentialism" about human nature. He says "human nature is a suite of mechanisms that underlie the manifestation of species-typical cognitive and behavioral regularities" (2012: 2). Samuels says that human nature picks out a "set of phenomena that will form a focus of empirical enquiry for some region of science" (2012: 4), thus his account is very similar to Machery's. However Samuels develops his view via criticisms of Machery. Samuels argues that "causal essentialism" provides a causal and explanatory function for human nature while the nomological notion is merely descriptive. Samuels says this about the nomological notion: "if human nature just is the set of human-typical regularities, then it clearly cannot be the cause of these regularities, underlying or otherwise" (2012: 18). Samuels does not claim that the nomological notion plays no explanatory role at all but he says "Natures are supposed to be underlying structures that play a central role in the explanation of an entity's more superficial properties; and this is not something that the nomological conception can give us" (2012: 18). There is not as much distance between the two views on this issue as Samuels thinks. Both present a set of common human traits to comprise human nature and both argue that human nature has an explanatory role to play in the behavioral sciences. Where the two views do differ is that they each highlight different behavioral scientists. Where Machery points to sociobiologists and evolutionary psychologists, Samuels says that his causal essentialist notion is more in line with the proximal mechanisms proposed by neuroscientists and cognitive psychologists.

The accounts of human nature outlined so far all are couched in terms of traits we have in common. There are several biologically based accounts of human nature that emphasize human variation and aim to treat variation as part of human nature. The accounts introduced so far either exclude highly varying traits from human nature or aim to explain variation in terms of traits in common that comprise human nature or both. Next I introduce some of the accounts of human nature that encompass variation.

Elizabeth Cashdan, like many in her field of evolutionary anthropology, is well aware of the human diversity that Wilson notes but she rejects the assumption, held by Wilson and many in her field, that "human nature is found solely in its universals—in the traits found in every society" (2013: 71). Those who hold this assumption (e.g., Brown 1991) say that traits found in some cultures but not others are "culturally constructed and without an evolutionary foundation" (2013: 71). We saw this distinction preserved in Machery's account of human nature above. In contrast Cashdan assumes that we evolved to be flexible and we exhibit phenotypic plasticity. She says that examining human nature should start with the question of how natural selection shaped our flexibility. She says that "we cannot understand our universal human nature without understanding the variability in its expression" (2013: 71). Cashdan argues that our nature is found in patterns of variation. She proposes to reveal these patterns in variation by appealing to norms of reaction, which are "the pattern of expression of a genotype across a range of environments" (2013: 71). Evolutionary biologists present norms of reaction by plotting the relation between a trait value and an environmental factor for specific genotypes (see Figure 11.1). Norms of reaction reveal variation in traits produced by the same gene expressed in changing environments. According to Cashdan, all the reaction norms for all our genes in all environments taken together constitute our nature.

Evolutionary psychologist H. Clark Barrett (2015) shares Cashdan's view of how we should approach human nature and so differs from many evolutionary psychologists in emphasizing variation. He says "both variation and lack thereof (if any) should be of interest to those studying humans" (Barrett 2015: 324). For Barrett our "species is a thing," and he takes species and nature to be equivalent here, "a big wobbly cloud that is different from the population clouds of squirrels and palm trees. To understand human minds and behaviors, we need to understand the properties of our own cloud, as messy as it might be" (2015: 332). For Cashdan, reaction norms, taken together, are the patterns in our variation that constitute our nature. Barrett has a slightly different take on the situation, for him: "Even on a probabilistic, population-minded, reaction-norm-based view

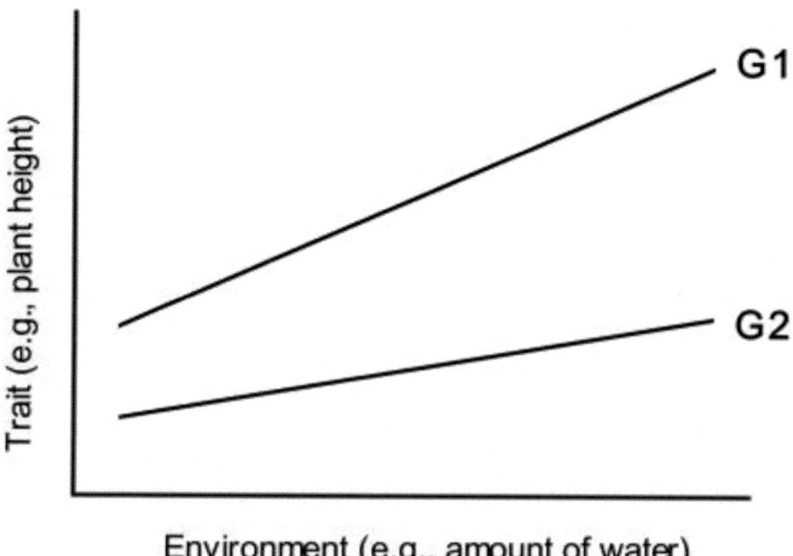

Figure 11.1 Relation between a trait value and an environmental factor for specific genotypes

of human nature, we eventually want answers about how variation is structured across humans" (2015: 324). Here Barrett indicates that there is more to the patterns in human variation than just reaction norms and this is what makes up our nature.

Grant Ramsey's (2013) account of human nature is set up in contrast with Machery's nomological account of human nature. Ramsey asks "why should we presume that it is the *sameness* across individuals that is of interest to scientists, and not their variation?" (2013: 986). Like Cashdan he thinks "it is a mistake to hold that only traits universal (or nearly universal) in the human species are of scientific interest and should be included within human nature" (2013: 986). Also, like Cashdan, Ramsey draws on evolutionary biology for help in characterizing the relevant variation. He appeals to life history theory evolution (see, e.g., Roff 2002). Life history theorists emphasize the variation at different life stages of each individual and the different selection pressures that impact upon different life stages. Frogs' drastically different morphologies at different life stages exemplify the need for this evolutionary approach. For Ramsey "individual nature is defined as the pattern of trait clusters within the individual's set of possible life histories" and "human nature is defined as the pattern of trait clusters within the totality of extant human possible life histories" (2013: 987). He calls this the "life-history trait cluster" (LTC) account of human nature. Different possible life histories for organisms result from the range of possible developmental responses organisms make to differing environmental circumstances. He also proposes that "characterizations of features of human nature are merely descriptions of patterns within the collective set of human life histories" (2013: 988). According to Ramsey, the LTC framework reveals "patterns within and across human heterogeneity" (2013: 992).

Paul Griffiths (2011) also emphasizes variation but he proposes an account of human nature that appeals to a different evolutionary perspective than the accounts by Barrett, Cashdan, or Ramsey. He echoes Barrett, Cashdan, and Ramsey when he says

"The search for a shared human nature cannot be a search for human universals; it must instead be a way to interpret and make sense of human diversity" (2011: 326). Instead of appealing to norms of reaction or life histories, Griffiths adopts the Developmental Systems Theory (DST) perspective (see, e.g., Griffiths & Gray 1994, 2001). He says that the DST perspective shows that to find out what an organism is like we need to look to "external" influences, such as the organism's environment. According to DST, organisms' developmental processes are not solely a product of internal factors, of which the usual candidate is the genome. Griffiths says "If an animal's nature is what explains its species-typical development, then its nature includes many of the environmental influences with which 'nature' has traditionally been contrasted" (2011: 326). One implication here is that you cannot find out what humans are like by ignoring society or culture (2011). For Griffiths, society and culture are potent causal determinants of our developmental trajectories and so make up our nature. Griffiths's account does not allow for Machery's principled distinction between products of evolution and products of culture or learning. On this account it is important to note that an organism's environment is not understood as merely a constraint or influence on development. Rather an organism's environment is as important a contributor to development as genes or any other relevant causal factor (cf. Griffiths & Gray 1994, 2001). This implies that his account also invokes a wider range of traits to characterize our nature than either Cashdan's or Ramsey's accounts. Cultural artifacts, for example, can be part of our nature on this account. Griffiths takes his notion of human nature to serve an explanatory function, saying "human nature in the causal sense includes the causes of difference as well as of uniformity" (2011: 319). Griffiths agrees with Cashdan that evolution can favor phenotypic plasticity and he points out that this means that adaptations need not necessarily be species typical (2011: 325) or need not be at fixation in the population, to use the terminology introduced above (cf. Buller 2005).

EVOLUTIONARY CHALLENGES TO HUMAN NATURE

As we have seen, biologically influenced accounts of human nature either emphasize traits in common, or universal traits, or include differing numbers of human trait variants. These accounts all converge on the ideas that human nature is something that can be characterized in biological terms and is a proper object of study. Philosophers and social scientists direct general skeptical arguments at both these central ideas. (See Prinz 2012 for a sustained skeptical attack on both these ideas.) This debate has similar contours to debates about race. In debates about race some defend eliminativism, arguing simply that there is no such thing as race, and some defend constructivism, arguing that while there is such a thing as race, it is a social construction. Here I will not discuss broader skeptical attacks on human nature. (See Kronfeldner, Roughley, & Toepfer 2014 for a discussion of a broad range of arguments about human nature.) Rather, I focus on a smaller subset of criticisms of human nature all of which draw on evolutionary thinking. Throughout this section I take "human nature" to stand for "biologically based accounts of human nature" unless I indicate otherwise.

We can distinguish four types of evolutionarily based criticisms of human nature. They are that defenders of human nature (i) presuppose an untenable species concept,

(ii) cannot adequately account for human variation, (iii) require a notion of normal or natural that is not supported by evolutionary biology, or (iv) presuppose an untenable distinction between the impacts of biology and culture. These types of criticisms are related and the details of criticisms under each type differ. I will go through these types of criticism in turn.

Barrett speaks for all defenders of human nature when he says "Whatever human nature is, it's a biological phenomenon, with all that implies" (2015: 321). Evolutionarily based criticisms of human nature capitalize on Barrett's final clause. The first criticism is that defenders of human nature presuppose an untenable species concept. The untenable species concept in question is an essentialist (Buller 2005; Hull 1978, 1986; Sober 1980) or typological species concept (Mayr 1994; Sober 1980). The problematic essentialist species concept is roughly that a particular trait or specific collection of traits characterizes each species (cf. Buller 2005; Hull 1986; Sober 1980). Sober emphasizes that the essentialist species concept plays an explanatory role, the characteristic(s) shared by all members of a species explain(s) "why they are the way they are" (1980: 354). The problem is that the essentialist notion of species is not supported by evolutionary biology. Rather, evolutionary biologists treat species as historical entities, lineages, with beginnings and, in the case of extinction, ends. There are several distinct species concepts defended by evolutionary biologists but all are consistent with this rough outline. (John Wilkins in Chapter 12 and Joseph LaPorte in Chapter 13 in this section of the volume treat essentialism and the various species concepts in more detail.)

Buller (2005: 441–442) provides a quick and dirty way of distinguishing between essentialist and lineage style species concepts. Say we became extinct but that some time after our extinction a new species arises with exactly the same cluster of characteristics in common as our species. On essentialist grounds, these would be the same species but evolutionary biologists cannot recognize them as such. Individual organisms in a species are connected by their place in a lineage and that lineage, the whole historical entity, is the species. Buller explains how holding the lineage-style species concept can be turned into an argument against human nature by David Hull: "As Hull says, if species are individuals, 'then particular organisms belong in a particular species because they are part of that genealogical nexus, not because they possess any essential traits. No species has an essence in this sense. Hence there is no such thing as human nature' [Hull 1978: 358]" (Buller 2005: 450).

The question now is whether this criticism holds against any of the defenders of human nature presented here. Buller argues that evolutionary psychologists succumb to this criticism and Hull argues that sociobiologists do also. Both sociobiologists and evolutionary psychologists (but not Barrett) say that we are characterized by a set of traits that we have in common and, further, both argue that our wide variety of traits can be explained in terms of this set of traits. For sociobiologists we are characterized by genes we have in common and for evolutionary psychologists we are characterized by psychological mechanisms we have in common. Buller's and Hull's criticisms appear to be on target in these cases. Machery and Samuels both explicitly deny that they adopt an essentialist approach to species. Machery says that his nomological account is not intended to delineate our species and Samuels says that his account does not serve a "taxonomic function." Machery says that his account does serve an explanatory function but he says he does not claim "the fact that humans have the same nature explains why generalizations can be made about them" (2008: 323). Perhaps the case could be pushed against Machery and Samuels (Lewis, forthcoming, does so against Machery) but the evolutionarily based case against their accounts of human nature is made better by adopting one of the other types of criticism discussed below. Similarly, it would be difficult to make the case that Cashdan, Barrett, Ramsey, or Griffiths presupposes an essentialist species concept (Louise Barrett does push this claim against H. Clark Barrett in Chapter 14 in this section). As presented here, this type of criticism is successful against only two of our accounts of human nature.

The second type of criticism holds that defenders of human nature cannot adequately account for human variation. Most criticisms of this type can be expressed as follows: defenders of human nature do not account for variation in a manner consistent with evolutionary thinking (cf. Lewens 2015). Hull ties his version of this criticism tightly to his version of the above criticism concerning essentialism. Elliott Sober's (1980) version of this criticism is also tied closely to an anti-essentialist line and I will outline this first. While Sober's line of argument is not directed squarely at human nature, it nicely frames relevant issues about accounting for variation. Further, Buller (2005) applies Sober's arguments to good effect in this context. Finally, Sober's arguments can be cleanly separated from the issue of competing species concepts.

Sober distinguishes between two ways of confronting and accounting for variation. Ernst Mayr puts the distinction this way: "For the typologist the type is real and the variation is an illusion, while for the populationist, the type is an abstraction and only the variation is real. No two ways of looking at nature could be more different" (1994: 158). Sober says that Mayr's typologists, or typological thinkers, account for variation within a species or population as interference with the natural state or prototype for that population. Sober says that this approach appeals to the "Natural State Model," which distinguishes between the natural state of a kind of object and its state resulting from an interfering force (1980: 360). Sober traces the Natural State Model to Aristotle but notes that it is the underlying explanatory model in Newtonian mechanics and other scientific fields. Newton's first law clearly illustrates this approach. Applied to biological contexts, the Natural State Model accounts for variation in a population of organisms as arising from interference with the organisms' natural state. Sober also stresses that according to the Natural State Model "the invariance underlying . . . diversity is the possession of a particular natural tendency by each individual organism" (1980: 370). Variation in a population is accounted for in terms of each individual's deviation from natural type. In contrast, for evolutionary biologists variation is characterized at the level of populations. Evolutionary biologists—populationists or population thinkers—account for variation in a population in terms of variation in previous generations of the population. Evolutionary biologists capture variation in a population via norms of reaction (introduced above). From this perspective there are no "natural" states and states that result from interference. All of the phenotypes expressed in the population are natural and the variation is a property of the population.

Buller takes evolutionary psychologists to task for invoking the Natural State Model in their account of human nature. He goes on to argue that the Natural State Model "can't be founded in evolutionary biology" and so evolutionary psychologists' proposal that underlying psychological mechanisms constitute our nature is not an evolutionary view (Buller 2005: 432). This version of the criticism tells against evolutionary psychology (excluding Barrett) but what about the other defenders of human nature? Samuels's is the other approach outlined here that most clearly adheres to the Natural State Model but he explicitly states that his account of human nature is not an evolutionary account. As a result, his view cannot fail for not being consistent with evolutionary thought. Further, Samuels does not take his task to be accounting for variation. This implies that there is no variation in cognition or neuroanatomy to account for but there is massive variation in human cognition (see, e.g., Henrich et al. 2010) and neuroanatomy (see, e.g., Amundson 2000). The Natural State Model is not only an inappropriate approach to accounting for variation, its adherents are predisposed to downplay or ignore variation. Evolutionary thinkers also criticize defenders of human nature for ignoring or not accounting for variation, which is not consistent with evolutionary thinking.

Ignoring or downplaying variation can be a consequence of focusing on traits that do not vary, or are fixed. We have seen that several human nature defenders invoke universal traits or traits in common or species typical traits. Evolution does result in the fixation of traits but also sustains variation in populations (cf. Buller 2005; Hull 1986; Lewens 2015). This point applies at both the phenotypic and the genetic level. There is a high degree of heterozygosity in human populations with some populations having higher percentages than others (Buller 2005; Hull 1986; Sober 1980). Also, there is enormous variation in genotypes associated with particular clusters of traits; for example, the immune system. Machery says that his account of human nature is designed to account for traits we have in common rather than traits that vary. He also says that traits that are part of human nature result from evolution. There are many traits that result from evolution that we do not share and many are very rare. Lewens says that Machery draws an arbitrary line around some evolved traits by counting only those we share as part of our nature. Drawing this line is not supported on evolutionary grounds and to do so results in an impoverished evolutionary account (Lewens 2015: 67). So to ignore variation or deliberately rule it out is not consistent with evolutionary thought. Also, while not perhaps strictly sticking to the Natural State Model, ignoring variation indicates typological rather than population thinking. So accounts of human nature that cannot account for variation or simply ignore variation are not evolutionary and also fail to confront phenomena that need accounting for.

Barrett, Cashdan, Ramsey, and Griffiths all acknowledge human variation and all agree that this variation results from evolution. They all reject accounts of human nature that adopt the Natural State Model, accounts that appeal to human universals and accounts that downplay or ignore variation. Yet these accounts have still been subject to criticism from other evolutionary thinkers. The criticism is that these views "have no theoretical meaning" (Buller 2005: 420) or constitute "simply a collection of informative truths about humans" (Lewens 2015: 77) (cf. Hull 1986). Further, these accounts amount to no more than the proposal that we consistently apply the various evolutionary methods we use to study all organisms to the study of humans (Downes 2016). Here is Buller's version of this point: "One possibility is that the concept of human nature could refer to the totality of human behavior and psychology." He goes on to say that his version of human nature "has no particular theoretical meaning; it is merely an abbreviation for talking about the rich tapestry of human existence" (2005: 420). Lewens, discussing Ramsey but making a point applicable to any of the people under discussion here, says: "For Ramsey, a description of human nature is simply a collection of informative truths about humans. His account demonstrates that extreme liberality is the price of defensibility in this domain" (2015: 77). Lewens later says: "Once an account of human nature is loosened up so as to make room for variation and learning, there is no way to gain control of it" (2015: 79). Lewens calls these views "libertine" accounts of human nature. So accounts of human nature that ignore variation or account for it by appealing to the Natural State Model are not evolutionary, and accounts that include variation do no special work over and above the work done by the evolutionary tools they invoke.

The third type of criticism says that human nature requires a notion of normal or natural that is not supported by evolutionary biology. This type of criticism is closely connected to considerations about variation. Hull makes this connection by considering norms of reaction. He says that when confronted with norms of reaction "the conviction is sure to remain that in most cases there must be some normal developmental pathway through which most organisms develop or would develop if presented with the appropriate environment" (1986: 8). On this conception, the normal or natural pathway for humans characterizes human nature. There is no normal or natural slice of the norm of reaction countenanced by evolutionary theory. When we look at a coarse grained trait like adult height in humans, we can assess average height and we can assign upper and lower bounds to observed height but there is no normal or natural height to be discovered. Douglas Futuyma says that if "human nature is our behavioral norm of reaction, which includes everything that people do," we can't say that any of this is not natural or not normal, it is certainly all natural from an evolutionary standpoint (1998: 743). Many people have made the point that characterizing what is normal for humans is not supported on evolutionary grounds (Buller 2005; Dupré 1998; Hull 1986; Lewens 2015; Sober 1980). Of the accounts presented here, other than sociobiology and evolutionary psychology, Machery's account comes the closest to endorsing a notion of normality when he says that traits are species typical, but he does not explicitly link his nomological account of human nature with normality. So this criticism has bite against sociobiologists and evolutionary psychologists but it would take work to make it stick against Machery or others presented here.

Here is a really clear statement of the distinction between culture and biology: "Human evolution has biological and cultural components. Man's biological evolution changes his nature; cultural evolution changes his nurture" (Dobzhansky 1962: 23). Several of the accounts of human nature presented here share this distinction, including sociobiology, evolutionary psychology, and the nomological account. The final type of criticism of human nature is that such a clear distinction between the biological and the cultural is not defensible on evolutionary grounds and, as a result, human nature concepts presupposing this distinction fail.

One way the distinction between biology and culture is defended is to claim that social behavioral patterns arising only in a small number of human populations are due to culture (see, e.g., Machery 2008). But we have seen that all manner of variation is sustained by evolution; for example, differing proportions of heterozygosity in sets of alleles in different human populations. The fact that there is variation in social behavior between populations does not rule out evolutionary factors contributing to this variation. Even if the relevant behavior is extremely rare, it could still be a consequence of evolutionary processes. Rare alleles are present in the total species genome and most of them are likely not present as a result of cultural processes. Even this last claim requires more care because, as we shall now see, the best explanation for the presence of some alleles in some human populations often appeals to cultural phenomena.

As Theodosius Dobzhansky presents the biology/culture distinction, biological processes produce biological traits and cultural processes produce cultural traits. However, there are examples of gene-culture co-evolution (cf. Lewens 2015). Adult humans in some populations are capable of digesting lactose and hence able to consume milk as part of their diet. Clare Holden and Ruth Mace assess all of the possible explanations for lactose digestion capacity and provide very strong support for the claim that the capacity is an "adaptation to dairying" (Holden & Mace 1997). Other environmental hypotheses do not account for the presence of the capacity in the relevant populations. The mechanism supporting lactose digestion capacity involves an enzyme that supports breakdown of lactose. This enzyme is present in people who have alleles that code for the enzyme. In this case, dairy farming, a cultural phenomenon on most accounts, is key to accounting for the presence of certain alleles in a population.

Evolutionary anthropologists provide numerous examples of human traits whose evolutionary explanations blur the cultural/biological distinction. Kim Sterelny makes a very strong case that many human traits arise as a result of intricate co-evolutionary processes and that to understand the evolution of human traits we must take niche construction into account (see his 2012). Many animals structure components of their environment: rabbits dig burrows, termites construct city like mounds, and so on. Humans can be thought of as the niche constructors par excellence. We are surrounded by what we build. We build not just concrete artifacts but also language and other prototypically cultural products. According to Kim Sterelny (2012), to account for the evolution of human traits, we must acknowledge the dynamic selective impact of all of our environments, including the environments we have constructed. Doing this puts a great deal of pressure on any clear distinction between biology and culture as independent determinants of distinct clusters of our traits. So, claiming that human nature is a product of biological evolution as opposed to cultural evolution, presupposes a distinction that is not supported by evolutionary thought. There is plenty of variation that results from selection, and prototypically biological traits arise as the result of prototypically cultural processes and vice versa. Both Louise Barrett and Maria Kronfeldner pursue different aspects of the nature culture divide (or lack thereof) in their chapters in this section of the current volume.

CONCLUSION

Wilson challenged sociobiology to "learn whether the evolution of human nature conforms to conventional evolutionary theory" (1978: 34). What we can conclude from the discussion above is that biological conceptions of human nature do not conform to evolutionary theory. Such human nature concepts presuppose an untenable species concept; ignore or cannot adequately account for human variation; require a notion of normal or natural that is not supported by evolutionary biology; or presuppose an untenable distinction between the impacts of biology and culture. Evolutionary thought supports none of these presuppositions. There is great promise for providing evolutionary explanations for human traits of many kinds but little promise for biologically based accounts of human nature. Those who provide evolutionary accounts of the origin of human social behaviors, such as Barrett, Cashdan, Griffiths, and Ramsey, could do so more profitably without invoking human nature.

REFERENCES

Amundson, R. 2000. "Against normal function." *Studies in History and Philosophy of Science Part C: Studies in History and Philosophy of Biological and Biomedical Sciences* 31: 33–53.

Barrett, H.C. 2015. *The Shape of Thought: How Mental Adaptations Evolve* (Oxford University Press).

Brown, D. 1991. *Human Universals* (McGraw-Hill).

Buller, D. 2005. *Adapting Minds: Evolutionary Psychology and the Persistent Quest for Human Nature* (MIT Press).

Cashdan, E. 2013. "What is a human universal? Human behavioral ecology and human nature." In S.M. Downes & E. Machery (eds.), *Arguing About Human Nature* (Routledge) 71–80.

Dobzhansky, T. 1962. *Mankind Evolving: The Evolution of the Human Species* (Yale University Press).

Downes, S.M. 2016. "Confronting variation in the social and behavioral sciences." *Philosophy of Science* 83: 909–920.

Dupré, J. 1998. "Normal people." *Social Research* 65: 221–248.

Futuyma, D. 1998. *Evolutionary Biology* (Sinauer).

Griffiths, P. 2011. "Our plastic nature." In S. Gissis & E. Jablonka (eds.), *Transformations of Lamarckism: From Subtle Fluids to Molecular Biology* (MIT Press) 319–330.

Griffiths, P. & Gray, R. 1994. "Developmental systems and evolutionary explanation." *Journal of Philosophy* 91: 277–304.

Griffiths, P. & Gray, R. 2001. "Darwinism and developmental systems." In R. Gray, P. Griffiths, & S. Oyama (eds.), *Cycles of Contingency: Developmental Systems and Evolution* (MIT Press) 195–218.

Henrich, J., Heine, S., & Norenzayan, A. 2010. "The weirdest people in the world." *Behavioral and Brain Sciences* 33: 61–83.

Holden, C. & Mace, R. 1997. "Phylogenetic analysis of the evolution of lactose digestion in adults." *Human Biology* 69: 605–628.

Hull, D. 1978. "A matter of individuality." *Philosophy of Science* 45: 335–360.

Hull, D. 1986. "On human nature." *PSA: Proceedings of the Biennial Meeting of the Philosophy of Science Association* 2: 3–13.

Kronfeldner, M., Roughley, N., & Toepfer, G. 2014. "Recent work on human nature: Beyond traditional essences." *Philosophy Compass* 9: 642–652.

Lewens, T. 2015. *Cultural Evolution* (Oxford University Press).

Lewis, C. Forthcoming. "A response to Machery's 'A plea for human nature.'"

Machery, E. 2008. "A plea for human nature." *Philosophical Psychology* 21: 321–329.

Machery, E. 2012. "Reconceptualizing human nature: Response to Lewens." *Philosophy & Technology* 25: 475–478.

Mayr, E. 1994. "Typological versus population thinking." In E. Sober (ed.), *Conceptual Issues in Evolutionary Biology* (2nd ed. MIT Press) 157–160.

Prinz, J. 2012. *Beyond Human Nature: How Culture and Experience Shape the Human Mind* (W.W. Norton).

Ramsey, G. 2013. "Human nature in a post-essentialist world." *Philosophy of Science* 80: 983–993.

Roff, D. 2002. *Life History Evolution* (Sinauer).

Samuels, R. 2012. "Science and human nature." *Royal Institute of Philosophy Supplement* 70: 1–28.

Sober, E. 1980. "Evolution, population thinking and essentialism." *Philosophy of Science* 47: 350–383.

Sterelny, K. 2012. *The Evolved Apprentice: How Evolution Made Humans Unique* (MIT Press).

Stevenson, L. (ed.). 2000. *The Study of Human Nature* (Oxford University Press).

Symons, D. 1987. "Darwin and human nature." *Behavioral and Brain Sciences* 10: 89.

Tooby, J. & Cosmides, L. 1990. "On the universality of human nature and the uniqueness of the individual." *Journal of Personality* 58: 17–67.

Wilson, E.O. 1975. *Sociobiology: The New Synthesis* (Harvard University Press).

Wilson, E.O. 1978. *On Human Nature* (Harvard University Press).

Wilson, E.O. 2013. *On Human Nature* (excerpts). In S.M. Downes & E. Machery (eds.), *Arguing About Human Nature* (Routledge) 7–23.

12

The Reality of Species: Real Phenomena Not Theoretical Objects

John Wilkins

INTRODUCTION

Many scientists and philosophers have said something like "species are the units of evolution" or "the units of biodiversity"; it is even the title of some well-known books on the subject (Ereshefsky 1992; Claridge et al. 1997). What does this even mean? In this chapter I argue that species are real, phenomenal objects rather than objects of any biological theory, let alone of evolutionary theory. *Species* is a term which everybody thinks they understand, but which nobody agrees upon, to denote the "basic units" of groups of biological organisms. Philosophically, there have been three stances taken in the past fifty years: the *species essentialist* view that all organisms of a species must share some (usually genetic) properties (Devitt 2008); the *individualist* view that species are historical objects without essences (Hull 1976, 1978); and the *homeostatic* view that species are coherently maintained objects due to some causal processes (Boyd 2010).

In addressing the impact of evolution upon the concept of species, we must first ask what *are* the units of evolution? The ontology depends a lot on what theory is being employed. When talking about population genetics, the basic units, of course, are the *allele* and the *locus* (Griffiths & Stotz 2007). When talking about development, then the unit is the *organism*, as it also is when you are talking about ecological interactions, although "species" is used here as the term for a class of ecologically exchangeable organisms, that is, organisms that play the same role in the local ecosystem (Wilkins 2007a, 2007b). Although organisms are pretty well all different (which is the point of population genetics), for the purpose of trophic webs (the food webs of ecology), conspecifics are treated as being interchangeable elemental units. Then there are the larger units of evolution: *populations* (Godfrey-Smith 2009), and the particular revision of that concept, the "deme" (Winsor 2000). A deme is basically the population that can interbreed—the term in the equations of population genetics is "N_e," the number of effective, or reproductive, individuals. However, non-breeders also play a role in many species in contributing to the fitness of their kin, by helping raise them, or finding food, so the ontology here depends solely upon what the research issues are. While "population" is itself somewhat fuzzy (Gannett 2003)—the more migration there is between two (sexual) populations, the more they start to look like a single population—it is a theoretical object (Millstein 2006).

But species? No biological theory *requires* them. True, ecologists and conservation biologists use the word *species*, but what they are doing is using field guides as a surrogate for the ecological roles played in an ecosystem by individuals of the species who are more or less normal—the "wild type." Likewise, medical and biological researchers do the same thing with their model organisms. *Mus musculus*, or the common mouse, is used as a model because it is assumed that each individual member of that species shares the same properties (developmental cycles, phenotypes). But in practice they use "strains" that are specially bred to see the effects of gene knockouts, for example. The "objects" here are the genetic strains and the organisms (Ankeny 2000). Systematists use the word *species*, but the explanations given of species *being* species are manifold. The notion of a "gene pool" or "metapopulation" is the foundation of one such explanation (de Queiroz 2005, 2007). But the theories used, the explanations, are not theories of species; they are theories of gene exchange, reproduction, fitness, adaptation, and so on. Species are being *explained*; they do no work in explaining.[1]

So, there are two ways we might go if species are not theoretical objects. One is that we may deny that species exist, and a lot of people do this, and have done since the *Origin* and even before, Lamarck being the most prominent. I call them *species deniers*, although the usual term is *species conventionalists*, or *nominalists* (both philosophically and historically misleading terms). A version of species denial is to replace the term *species* with some "neutral term." *Deme* was one of these, but it got subverted by population geneticists for the meaning given above. Other examples include *operational taxonomic units* (Sneath & Sokal 1973; Sokal & Sneath 1963), *least inclusive taxonomic units* (Pleijel & Rouse 2000), *evolutionarily significant groups* (Hey 2001), and so on. In each case, the term *species* came or is coming back into use. Why is *species* so durable?

The second alternative is *not* taking the term and concept as a theoretical term. *Species* is a useful term because species are real phenomena.[2] That is, they are things observed that call for explanation, they are *explicanda*. Theories of biology explain why there are species, although the same theories do not apply in the same way for all particular species (Wilkins 2003). Biology is not that neat. Some species are explained in textbook fashion through the acquisition of reproductive isolating mechanisms formed in geographical isolation; some are not. There are species formed by hybridization,[3] by sexual selection, and of course asexual or mostly asexual species that are maintained, as I argue elsewhere (2007a, 2007b), by adaptation to niches.

If there is no general theoretical account of species, why do we have this category? It might be because we tend to name things that look similar to us. This is what species deniers think: it is all about us and our cognitive dispositions, not the things themselves. However there *are* some general features of species that license us calling them all species: *species are salient phenomenal objects.* They are salient not because of our perceptual tendencies alone but because they do exist. They are a bit like mountains. Each particular mountain is caused by definite processes, but every mountain is not caused by the same processes. We *identify* mountains because they are there. We *explain* them with theories of tectonics, vulcanism, or (if they are dunes) wind. Analogously, species are clusters of genomes, phenotypes, and organismic lineages. We explain them because they need explaining. A species is (roughly) where the lineages of genes, genomes, parent–child relationships, haplotypes, and ecological roles all tend to coincide. Not all of these need to coincide in every case, but so long as most of them do, they are species, and we must give an account of them.

PHENOMENAL OBJECTS

Perhaps the most crucial practical aspect of the species concept debate lies in its relevance to conservation, but it is not the most theoretically interesting. Biology, like most sciences, has a need for units of measurement which are grounded in the real world. So *species*—the "rank" of biology that is agreed on most sides as the most or only natural one in the Linnaean hierarchy—determines many measures of biology in fields from genetics to ecology. If, as a significant number of specialists think, the rank is a mere convention (Mishler 1999), then those measures become arbitrary and meaningless. Therefore we need to consider what sort of "unit" a species might be. I can think of three alternatives.

The first is that species are, in fact, simply *a matter of convention*, which is to say, something that makes things convenient for us in communication, just as John Locke said in the *Essay* (Bk III, ch. V, §9; although that was about *logical* species, not biological species).[4] Instead, claim researchers like polychaete specialist Frederick Pleijel (1999) and geneticist Jody Hey (2001), we need to replace the notion *species* with something like a "least inclusive taxonomic unit" (LITU; Pleijel) or "evolutionary group" (Hey).[5] There are other replacement concepts in the offing. And the so-called "phylogenetic species concept" is not really a concept of species, at least in one of the versions under that name, so much as something very like a LITU that gets *called* a species (Wheeler & Platnick 2000).

The second alternative is that *species* plays a theoretical role in biology, and this seems intuitively right: we talk about species as the units of evolution, so they are supposed to be required by evolutionary biology; and likewise in ecology, species are the unit that is crucial in defining the biodiversity of a region or ecosystem. But if species are theoretical objects, we ought to find them as a *consequence* of theory, not as a "unit" that we feed into theoretical or operational processes, and this is not the case. Population genetics and evolutionary theory have populations, haplotypes, alleles, trophic nodes, niches and so on, but what they do not have are species. In every case where species are *used* in theory, they are primitives, or stand as surrogate terms for the other things mentioned. Theory does not define species. This might be challenged by adherents of Ernst Mayr's biological species concept, or one of the derivative or related conceptions—a species is a protected gene pool, as Mayr said (1970: 13). However, the vast bulk of life would not be in species if that were the case, and, anyway, species were well described and identified long before genetics was developed, up to two centuries before. So they must at least be things that can be observed in the absence of theory.

The third alternative is that species are not theoretical objects at all; they are *objects that have phenomenal salience*. That is, we do not *define* species, we *see* them. Consider mountains again. Mountains are hard to define, and they have a multitude of geological causes, ranging from uplift, subduction, vulcanism, differential erosion, and so forth. "Mountain" is not a theoretical object of geology—subduction zones, tectonic plates, and volcanoes are. A mountain is just something you *see*, although there are no necessary sets of properties (or heights) that mountains have to have, and it is often vague when differentiating between them. A mountain calls for an explanation, and the explanation relies on theory, but equally so do mesas, land bridges, and caves.

So the proposed answer to the question "What is a species?" is that a species is something one sees when one realizes that two organisms are in some relevant manner the same. They are natural objects, not mere conveniences, but they are not derived from explanations, they call for them.

THEORY-DEPENDENCE AND DERIVATION

Traditionally, a theoretical object was something that theory required or employed but which was not empirically ascertainable—"electron" *c.*1920, "gene" prior to 1952, and perhaps still "Higgs boson." But this is a positivist sense of *theory*—a formal system in which objects are either observational or theoretical. Whether or not one is now a logical empiricist instead of a logical positivist, objects are much more nuanced than that. There is a school of thought that treats scientific ontology, the set of objects that one thinks exists in a domain, as basically the bound variables of the best theory of that domain, that is, Quine's "to be is to be the value of a variable" (Quine 1948: 32). In this case, species would be theoretical objects if they were such variables of a theory. But if they are not, we need to establish what sort of ontological status they, and other phenomenal non-theoretic objects, may have.

Consider planetary orbits. Observed and debated for a very long time before Newton proposed a general physics that accounted for them (and made predictions about them), orbits were demoted by Newton from theoretically important objects (heavenly spheres) to special cases of larger and more universal physics. "Planetary orbit" is thus a special instantiation of astrophysical dynamics, which aperiodic comets, star systems, and even entire galaxies obey. Even if no orbits actually existed (and we can perhaps envisage this in some world) under this physics, the movements of objects would be still covered by Newtonian dynamics. Likewise species. They obey the biology of populations, interbreeding, selection, drift, and so on, but they are not theoretical objects, any more than planetary orbits are in physics.

This chapter's general characterization of species is that they are the nexus of the coalescence of genes, haplotypes, parent–child lineages and so on, at or about the same level. In abstract terms, species are these coalescences that are distinct from other such coalescences, something I have called the *synapomorphic* conception of species (Wilkins 2003), and each and every one has a general set of properties and modes of speciation, and a unique set of these that only they have (the synapomorphies, or shared characters, which are causally active in maintaining separation). Because each species is a unique historical object, this makes its *modality* (as I have called it) something that they tend to share only with those taxa they are closely related to. As a result, the modality of a species is as much an evolved trait as having a vertebral column or a nuclear membrane. The causal process whereby a species evolved and is maintained depends upon shared ances-tral traits such as developmental machinery, genetic sequences, and ecological resources. While these may be very similar between related species, they cannot be expected to be the same in each case—not all Rhagoletids will speciate by host race transfer, for instance—and so each species will have a special set of causes. No theory will capture all and only these causes (not every sexual species will be caused by allopatric isolation, and not every asexual species will be caused by a single niche adaptation) except at a level of generality that is so vague as to preclude explanation in terms of mechanisms. As a take-home exercise to the reader, try to imagine under which conditions organisms like ours would not form species at all.

There have been several proposals for what makes an object "theoretical."[6] The first is that of Quine: something exists just to the extent that our best theory of a given domain requires them (i.e., binds them with a quantifier). On this view, species are simply not theoretical, and indeed do not exist, because if I am right that no theory of biology *requires* species, then they are never the value of a bound variable in any model of biology.

Another similar but not so restrictive view is "Ramseyfication," according to which what a theory requires is based on a formalization of the theory—a "Ramsey sentence" (Psillos 2006; also called a "Carnap-Ramsey sentence" or a "Ramsey-Lewis sentence," see Koslow 2006). Objects exist so long as they are represented either by primitive terms (values of variables, or constants) of the theory or combinations or derivations of those. A primitive here might be empirical, so that species might be primitives of biology, but are not themselves explained by it. This is not the case with species, though, because in every such case of which I am aware, one can replace "the species $X y$" with something like "a local population of $X y$" or "organisms that behave in a way that is typical of $X y$" for functional accounts such as ecological ones. In other words, the species $X y$ is replace-able with objects that the theory actually employs. The Ramsey approach, recently also called the "Canberra Plan" (Braddon-Mitchell & Nola 2009; Jackson 1998), treats these objects as non-objects. Sometimes this is played out under "Structural Realism" in which a theory structure is true (Psillos 2006), but the objects it poses which are "unobservable" may or may not be real, so long as the theory is empirically adequate in other ways. This is irrelevant here.

What makes an object theoretical, and are there other roles objects and their repre-sentation play in science? For our view to work, it must be that there are objects that are described by the theory, which in the domain of that theory have a certain coher-ence or unity as objects. Mechanisms like tectonic drift are obviously theoretical in that sense. But mountains are more difficult. Mountains are real things, but the category as a whole lacks theoretical coherence. That is, a mountain has no theoretical place qua "mountain," but as a particular mountain, say, Mount St Helens or the Matterhorn, it calls for explanation. If you have to map them, travel around them, or climb over them, they are as real as anything can be, but the choice of demarcation between peaks can be conventional or even just something that perception hands to us on a plate. Nothing in theory demands that this particular mountain exists, or even that there are mountains. On a planet with an atmosphere and no tectonics, after a reasonable period, there may be none. Species are like that. They are real facts about the world, which we perceive rather than define. Of course, this makes identifying them relative in a way to the rules and capacities of perception. If we had poor vision, we might not "perceive" mountains until we had telescopic surveyor's sights. Once we have that technology, though (which, note, does not rely upon the theories of geology), we *do* see mountains. Similarly, we may need to use all kinds of assay techniques to see species, but when we have them they are seen.

An instructive example is the discovery by Murray Littlejohn and his advisor of the different species that had previously been called *Rana pipiens*, the "leopard frog" of the southern United States (Littlejohn & Oldham 1968). The leopard frog is widespread, and Littlejohn was using a new piece of equipment designed for speech therapy, the sonograph, to graph the mating calls of these frogs. He discovered that there were up to six distinct mating calls (although new species have since been identified: Platz 1993). Since mating calls in amphibians are highly species-particular, Littlejohn proposed that *R. pipiens* was a species complex in which morphology and ecology were indistinguishable, but that mating was restricted within the mating call groups, the species. Subsequent work proved this to be the case. The differentiation was always there, but you needed the right assay technique. This is not species being "constructed" or any other "postmodern" nonsense. While the *concept* we have of those species is being constructed (and reconstructed as new evidence comes in), the concept denotes realities, either of classes of things that are theoretical, such as populations, haplotypes, genes, developmental sequences or cycles, and so on, or of things that are not required by the theory. When we construct a concept, we are *learning* about the things we describe. It is like finding that Everest has a hitherto-hidden peak that is even higher. Our concept of Everest changes, but the thing itself was already as it is.

WHAT ARE SPECIES?

First of all we must take issue with the entire way the debate has been framed over the past 150 years or so and assert: *There is only one species concept* (Wilkins 2011). That is to say, there is only one concept that we are all trying to define in many ways, according to both our preferred theories of how species come into being and maintain themselves over evolutionary time and what happens to be the general case for the particular group of organisms we have in our minds when we attempt our definitions. The former case is what we might call *theoretical* conceptions of species, where a "conception" is a definition of the word and concept of species. The latter are the *prototypical* conceptions of species. If you work in, say, ichthyology, then your conception of species has to deal with the usual facts about fishes (Rosen 1978). If you are a fern botanist, then those organisms set up your prototype (Wagner 1983). The debate over what species are has been driven by differing prototypes as much as by different theories of speciation. Elsewhere (Wilkins 2009), I list some twenty-six conceptions[7] of species in the modern (post-Synthesis) literature. I am going to focus now on the few basic ideas that underlie nearly all of these.

The first conception is based on *reproductive isolation*. Since the Synthesis of genetics and Darwinian evolution was formed, the ruling notion of species generation (speciation) was based on the criterion of sexual populations that are isolated from each other, so that they evolve in divergent ways, leading to populations that when they meet, if they do, in the same range, they no longer tend to interbreed, and their gene pools are now distinct over evolutionary time scales. The conception of species that the Synthesis adopted as a result of this genetic-evolutionary view is sometimes misleadingly called the *biological species concept* (or BSC). It is called this because it was contrasted to the practices of museum taxonomists, who identified species based on differences in the morphology of captured or collected specimens. This was held to be a sterile methodology where the data were more in the heads of the taxonomists than in the real world. Hence, the BSC was biological, while the museum approach was conventional (due to the conveniences of the taxonomists). But the leading idea of the BSC is not that things live in messy populations, although that is part of it, but rather that these populations are *reproductively isolated* from each other. So I prefer to call this conception the *Reproductive Isolation Species Conception* (RISC), or "isolationist" conception for short.[8] There are several versions of it, but the basic idea—that something inhibits interbreeding when they meet—is common to them all.

Criticisms of the RISC began early (Ehrlich 1961). For a start, it was observed that there was a disconnection between the theoretical justification for the RISC and the ways in which taxonomists who adopted it did their taxonomy. To be confident that you *have* a RISC taxon, you really need to do breeding experiments. Many quite diverse morphs in, say, butterflies, that were identified as distinct species in the 19th century, turned out to be different genders of the same species. "Aha!" said the isolationists, "this is a failure of morphology." But when similar cases occurred and were found to be different genders *before* the Synthesis, these so-called "morphologists" had no problem seeing them as the same species on that ground. It was understood that form was only a guide to the underlying biological reality, not an end in itself. Worse, isolationists themselves use morphology to identify their species. Breeding experiments, even when technically possible, take enormous time and resources. So while *theoretically* isolationists are basing their work on reproductive isolation, *practically* they are doing just what their supposed mistaken predecessors did. This might lead us to think that the older workers were not so silly after all.

The second of our broad conceptions of species is *ecological isolation*, and is often called the *Ecological Species Conception*, which goes back in one form or another to Linnaeus. However, it acquired currency in modern times when another Swedish botanist named Göte Turesson did some studies during the 1920s of plant morphologies in different ecological conditions (Turesson 1922, 1925, 1929). Turesson coined the term *ecotype* to describe these differing morphologies. He distinguished between ecotypes and *ecospecies*, which were populations prevented by adaptation to a particular ecological niche from interbreeding. In the 1970s, Leigh Van Valen offered a new version, based on the fact that American oaks will freely interbreed, but that the ecological types remain constant (Van Valen 1976). In these cases, the "species" is effectively maintained by the ecological niche. Similar cases are common in plants and single-celled organisms, though less so among multicellular animals. Bacteria and other single-celled organisms which do not often exchange genes may be entirely maintained by this. Lacking sex, they cannot be RISC species, so Turesson coined another term for them, *agamospecies* ("sexless species"). However, in animals asexual reproduction has evolved from sexual species many times (*parthenogens* "virgin origins"), while in plants it is much more common (*apomicts* "apart from mixing").

The third kind of species conception is known variously as *Morphological, Typological,* or *Essentialist,* but these labels are misleading. Sometimes it is called *the Linnaean Conception,* because it is supposed to be the default view before genetics and evolution were discovered, and hence the view of Linnaean taxonomy. This is a bit unfair— Linnaeus never clearly defined a species concept,[9] and the standard view at the time was that of John Ray, in which a species was twofold: a *form,* which is *reproduced.* This conception was never isolated from normal reproduction by parents. (Moreover, Linnaean and Rayesque species were not defined by essences either, but that is for another time.) The important thing was that it was the overall organization of the organisms that defined them as a species, so long as it was reproduced. Ray's own "definition" was "the distinguishing features that perpetuate themselves in propagation from seed" (quoted in Mayr 1982: 256). Ray's definition was designed to cover plants, but he explicitly extended it to animals, and it was the first time any naturalist had ever given a purely *biological* definition of "species." It was not based on Aristotle or any logical system, but on observation. This earlier definition remained the standard view at the time Darwin began his work, via the authority of Baron Cuvier.

A fourth general conception is based on the convenience of biological work, including mutual communication between specialists: *species conventionalism,* the view that, as Locke had said, species are made for communication, and nothing else. For Darwin, species were real but temporary things, and he believed there was no special rank or level in biology that was unique to species. Contrary to common opinion since the turn of the 20th century (and earlier, vide Agassiz), Darwin was not a conventionalist, but evolutionary thinking made it harder to be exact about species.

This leads us to our final conception. Based on *evolutionary history,* it has two main versions: the *phylogenetic* species conceptions based on cladistics, and the so-called *evolutionary* species concepts, which are often a mixture of the RISC, the ecological species conception, and phylogenetic accounts of reconstructed history. The former are often more like the RISC, because they rely on there being separation of lineages over a large time as defined by their sharing or not evolved traits, and this implies genetic isolation. The latter do not rely on RISC, but only that after the fact the lineages remain distinct for whatever reason. These conceptions are process-based, and are equally as non-operational as the RISC, but cladistics at least has a large number of mathematical and formal techniques for drawing up their cladograms. The problem is that, without some way of saying what the level of separation is for species, cladistics can divide lineages up to a very small level (such as haplotype groups), leading to "taxonomic inflation" (Isaac et al. 2004; Padial & de la Riva 2006). Phylogenetic species can run to as much as nine or ten times in number compared to the ordinary ("Linnaean") kind. The debate rages through the modern systematics community.

After all that, what is a species? Any *universal* concept of species has to range over the entire evolutionary tree, but the modes of being a species will depend on what ways they have evolved to remain distinct from each other. Hence, none of the particular conceptions is sufficient or necessary to cover being a species in all organisms. *However, each conception only tells us what species sometimes are. They do not tell us why the things so defined should even be* called *"species."* For example, RISC proponents will often say that asexual organisms (*agamospecies*) are not really species at all, because they lack the defining properties of species which is, of course, reproductive isolation

(Fisher 1930: 135; Dobzhansky 1951: 275; Mayr 1942: 122). So we should call them something else—agamospecies, quasispecies, pseudospecies, paraspecies, etc. This has the unwanted consequence that the bulk of life does not exist in species, but only those few clades that happened to evolve sex. I think we should say that *all* organisms come in "kinds," some of which are sexual kinds. Others come in genetic bundles or are clustered for ecological reasons, and many are a mixture.

So here is a "definition" of the word *species*:

> A species is any lineage of organisms that is distinct from other lineages because of differences in some shared biological property set.

It has to be a *lineage* in order to distinguish biological species (but not just RISC species) from species of chemical compounds, minerals, and symptomatic diseases. However, while all species are lineages, not all lineages are species, not even the monophyletic ones. It has to be a *causal* definition, because formal approaches do no explanatory work (in short, the formalist definition merely restates that there are differences). And it has to be based on *biological properties*, because non-biological properties like range or geography are not enough to include or exclude populations and organisms. All the various conceptions try to give the differences in shared biological properties some detail—and when we look at them that way, it becomes clear why none of them are sufficient or necessary for *all* species: the mechanisms that keep lineages distinct evolved uniquely in every case, and so generalizations cover only some, not all, of life.

PATTERN RECOGNITION AND ABDUCTION

We tend to classify epistemic activities into two kinds: induction (regarding which controversies over its warrant are well known) and deduction (with many arguments about its applicability). But there is something else that we do to learn about what exists in the world: *classification* (Wilkins & Ebach 2013). Classification is typically regarded as either of the other two kinds of inference, but it is, instead, a third kind, similar to abduction, or inference to the best explanation.

When we classify in a theory-lacking domain we are not yet inductively constructing theory, and we are not able to deduce from theory (since there is not any yet) the classes of objects in the domain we are investigating. What is happening here is *pattern recognition* (Bishop 1995). It is one of the distinguishing features of neural network (NN) systems, such as those between our ears, that they will classify patterns. They do so in an interesting fashion. Rather than being cued by theory or explanatory goals, NNs are cued by stereotypical "training sets." In effect, in order to see patterns, you need to have prior patterns to train your NN.

Where do these training sets come from? There are several sources. One is evolution itself: we are observer/classifier systems of a certain kind. This gives us a host of cue types to which we respond by training our stereotype classifier system. For example, we respond to movement of large objects, to differences in color and shade, and so on, in our optical system. The problem is that so long as our survival and reproductive success is ensured, evolution cannot guarantee us access to the way things "really" are. At best it gives us a good balance between false positives and false negatives. It is good enough, as it were, for government work (Godfrey-Smith 1991); but is it good enough for science and metaphysics?

Science proceeds by refining its categories of what exists in the world based on two main sources: evidence and explanatory force. In the case of a domain of investigation for which there is as yet no explanation, all we have is evidence; but apart from our evolved dispositions to respond to certain stimuli (called our *Umwelten* by Jakob von Uexkull (1926)), how can we identify the salient aspects of evidence? There is an almost infinite amount of possible information we might use, and so we must glean the *right* sources of information.

One source is economic necessity. Over time, farmers and hunters will tend to respond to the features of the things they are engaged in acquiring and using that are more or less important for success, because those features which are not salient will impose a cost of time and effort that tends to reduce success. This is a process very like natural selection, and has been the basis for what came to be known as *evolutionary epistemology*, in which a parallel process to biological evolution occurs in the domain of knowledge. Cognitive traditions become better at acquiring reliable knowledge because ideas and approaches that do not aid this goal are costly and are abandoned.

However, we have a superfluity of cognitive and conceptual resources. We can retain ideas and practices that are not really natural for social reasons, such as rituals and "explanations" that have no counterpart with the reality being dealt with. The fact that a particular culture is successful at farming by relying upon a ritual calendar (as in pre-Indonesian Bali) does not warrant belief in Hindu gods. The functional aspects of the rituals act to transmit the information even if nobody in the culture fully understands why those rituals make farming successful (Lansing 2007). When a classifier recognizes patterns in economic circumstances, what counts is not the conceptual superstructure, the theories, and the ideologies, but the categories of what matters—in this case of water, soil, and landscapes. How might this explain the success of science? Taxonomists are classifiers in a particular economic context: that of professional science. When a taxonomist encounters organisms in the wild, they are in the same situation as a hunter who hunts in that ecology. To succeed at taxonomy, as to succeed at hunting, the agent must know the *right* things about the target objects. A hunter who does not know what different species of bird look like or how they behave or where they live is in exactly the same economic conditions as a taxonomist who also lacks knowledge. Neither will end up with dinner on the plate (qua hunter or taxonomist). In the case of the taxonomist, the gap between failure and hunger is somewhat more distal than for the hunter (but hunters typically get most of their food from foraging rather than hunting anyway, courtesy of the non-hunters, mostly women, in their village), but ultimately economic success depends directly upon correct pattern recognition. Mayr was fond of telling the story of how when he visited Papua in the 1930s, he and the local hunters identified the same species of bird, with an exception where Western ornithologists also disagreed, and he used this as justification for the reality of those (and all) species. He inferred that science was able to discover kinds of things that were real in the world. However, when E. O. Wilson tried the same experiment about ants, a subject he knows intimately, instead of the locals counting the same species he did (several dozen), he found they did not discriminate them (Wilson 1992: 39). Why did Mayr's informants know their birds while Wilson's did not know their ants? The answer is that birds were of economic importance to locals and ornithologists, while ants were of economic importance only to Wilson and other myrmecologists.

By "economic" I do not mean fiscal, but the acquisition of resources, success at which gives the person investigating a living. What distinguishes scientific success is a unique socioeconomic system of professionalism, credit in society, and access to funds and resources like labs, students, and equipment. The motivations of the individuals concerned are several, often (but not always) based on personal curiosity, but curiosity is not enough if you do not get the resources to do the work. So we are very good at turning our perceptual pattern recognition systems to scientific work. What evolution provides, science refines. It happens that pattern recognition and the subsequent classificatory activities can deliver reliable knowledge of the world when it matters. But given that it is parasitic upon those evolved capacities, and given that scientists are social organisms, this is not without its failures. Social influences, particularly the inherited traditions of ritual and conception that history bequeaths, can skew and bias our categories about the world. This is where theory and experiment come in.

Science, by way of its historical accidents, also seeks to explain things in ways that can be tested. Here the ordinary philosophical issues come into play—we inductively generalize based on the patterns we have recognized, and form hypotheses, and from those hypotheses we derive deductive consequences, which we can test in ways that are not circular, which do not rely upon our original observations. And so we can eliminate hypotheses that do not fit the facts, more or less. This is what Karl Popper and the evolutionary epistemologists built their views upon. What evolutionary epistemology never explained, nor Popper, was how we came up with those hypotheses in the first place. Pattern recognition does.

CONSEQUENCES

This approach makes sense of several facts about biological science.

It explains why we recognized species well in advance of there being anything remotely like a theoretical explanation of them, from the 16th century onwards. Ray formally defined biological species for the first time in 1686, but his view was implicit in the work of natural historians going back to Aristotle and Theophrastus. Genetic and developmental accounts of species did not arise until around 1900.

It explains why when replacement terms are proposed for *species*, they tend to settle on the same sorts of phenomena, and eventually *species* makes a comeback. It also explains why it is that when autochthonous peoples employ organisms economically, say by hunting or raising them, they recognize the same sorts we do for scientific reasons (Atran 1987). Things are phenomenally salient if you have to interact with them.

But most of all it explains something about science. I would like to briefly sketch what I think are the implications of phenomenal objects in the ontology of a science. In the traditional view of science, observation is theory-dependent and objects are theoretical. I am proposing that some objects are *not* theory-dependent in the domain under investigation, and in doing so I can explain why it is that so much of biology is what Rutherford sneeringly called "stamp collecting." Before you can begin to formulate theories, you have to gather together the objects under explanation and organize this information into a taxonomy, otherwise it is not even clear what the domain of the theory *is*. The traditional view of science of the 20th century ignored classificatory activities as uninteresting; I am suggesting it is one of the crucial and essential aspects of a science.

One might object that of *course* these objects are theoretical: to observe them is to identify a difference by measurement, and that implies an assay or methodological protocol. This is usually true, although species and mountains do not need much if any theoretical ancillary assumptions. But the point is that they do not need the *theory under investigation* in order to be phenomenal objects. That is, if they are theory-dependent, they are dependent on theories outside the domain in question. Moreover, they are often tokens of a class of phenomenal objects that call for explanation in *those* theories as well (consider optical theories, or genomic clusters in genetic theories).

Since the dependence here is a general kind (such as for optics), the theory-dependence is benign. With respect to our theory *T*, there is no special dependence on which the observations are being made, so the phenomena are *T*-independent. This does not mean there is such a thing as *completely* naive observation—nobody ever starts from total naivety or from a *tabula rasa*. Even observers in the mountains of Papua New Guinea are informed by prior ideas and experience. But we can say they observe species, and do not thereby need to define them.

To summarize:

1. A species is something that forms phenomenal, salient, lineages of populations of organisms and genes.
2. A species can have a particular mode based on evolved biological properties.
3. The species *conception* applied in each case depends on whether that species meets the conditions for that conception.
4. Each species is a phenomenon that calls for a conception and an explanation.

So we do not need to have a monistic or singular definition of *species*, because species are things to be explained, they are *explicanda*, not an a priori category or rank into which every biological organism must be fitted.

ACKNOWLEDGMENTS

Thanks to Gal Kober, Marc Ereshefsky, Quentin Wheeler, and Brent Mishler for pushing me on this. I must also acknowledge all the specialists with whom I have had long, often frustrating for us both, arguments on this topic leading to my 2009 book and the PhD it was based upon. In particular I must thank David Williams of the NHM, Malte Ebach, Gareth Nelson, and Brent Mishler, none of whom can be fairly thought to agree with me. Ingo Brigandt first made the claim that species are not theoretical objects (Brigandt 2003), to my knowledge.

NOTES

1. One possible exception is the work species do in "species selection" theories (Jablonski 2008; Rice 1995; Gould & Lloyd 1999), but it is arguable whether these are actually theories as such, and equally arguable whether the properties that are "theoretical," which play a role in causal explanations, are those of the

species, populations within the species, or the individuals or kin groups. If species selection is taken to mean that species whose members have a particular property (like eurytopy) tend to speciate more often, then "species" in this sense is merely a mass noun (Grandy 2008).

2. This is not to deny that *species* is also maintained by conventional and social practices. If entire volumes are dedicated to describing species, anyone who wishes to be taken seriously in that field has to refer to those described objects.

3. For example: Knobloch 1959, Mallet 2007, 2008; Arnold & Meyer 2006; Wagner 1983.

4. This was the view of John Maynard Smith as well (1958).

5. Brent Mishler (pers. comm.) has suggested that a cladistic taxonomy requires only the *smallest named and recognized clades* (SNaRCs). This is not a replacement term for *species* however, as there is no rank beyond what has been recognized as monophyletic.

6. See Brittan 1986; Ladyman 1998; French & Ladyman 2003.

7. A conception is a variant definition of a concept. This in turn is distinct from the various formulations of conceptions of the concept. For example, other researchers have given several hundred formulations (Lherminier & Solignac 2000), but mostly these are comprised of a much lower number of conceptions.

8. The historical account of this terminology, which derives from Ernst Mayr, is given in my 2009 book.

9. Although he did use an interbreeding criterion (Müller-Wille & Orel 2007), and so may be charitably interpreted as holding a RISC.

REFERENCES

Ankeny, R. 2000. "Fashioning descriptive models in biology: Of worms and wiring diagrams." *Philosophy of Science* 67 (suppl. vol.): S260–S272.

Arnold, M. & Meyer, A. 2006. "Natural hybridization in primates: One evolutionary mechanism." *Zoology* 109: 261–276.

Atran, S. 1987. "The early history of the species concept: An anthropological reading." In S. Atran et al. (eds.), *Histoire du Concept D'Espèce dans les Sciences de la Vie* (Fondation Singer-Polignac) 1–36.

Bishop, C. 1995. *Neural Networks for Pattern Recognition* (Oxford University Press).

Boyd, R. 2010. "Homeostasis, higher taxa, and monophyly." *Philosophy of Science* 77: 686–701.

Braddon-Mitchell, D. & Nola, R. 2009. *Conceptual Analysis and Philosophical Naturalism* (MIT Press).

Brigandt, I. 2003. "Species pluralism does not imply species eliminativism." *Philosophy of Science* 70: 1305–1316.

Brittan, G. 1986. "Towards a theory of theoretical objects." *PSA Proceedings of the Biennial Meeting of the Philosophy of Science Association* 1986: 384–393.

Claridge, M., Dawah, H., & Wilson, M. 1997. *Species: The Units of Biodiversity* (Chapman & Hall).

de Queiroz, K. 2005. "Different species problems and their resolution." *BioEssays* 27: 1263–1269.

de Queiroz, K. 2007. "Species concepts and species delimitation." *Systematic Biology* 56: 879–886.

Devitt, M. 2008. "Resurrecting biological essentialism." *Philosophy of Science* 75: 344–382.

Dobzhansky, T. 1951. *Genetics and the Origin of Species* (Columbia University Press).

Dowling, T. & Secor, C. 1997. "The role of hybridization and introgression in the diversification of animals." *Annual Review of Ecology and Systematics* 28: 593–619.

Ehrlich, P. 1961. "Has the biological species concept outlived its usefulness?" *Systematic Zoology* 10: 167–176.

Ereshefsky, M. (ed.). 1992. *The Units of Evolution: Essays on the Nature of Species* (MIT Press).

Fisher, R. 1930. *The Genetical Theory of Natural Selection* (Clarendon Press).

French, S. & Ladyman, J. 2003. "Remodelling structural realism: Quantum physics and the metaphysics of structure." *Synthese* 136: 31–56.

Gannett, L. 2003. "Making populations: Bounding genes in space and in time." *Philosophy of Science* 70: 989–1001.

Godfrey-Smith, P. 1991. "Signal, decision, action." *Journal of Philosophy* 88: 709–722.

Godfrey-Smith, P. 2009. *Darwinian Populations and Natural Selection* (Oxford University Press).

Gould, S. J. & Lloyd, E. 1999. "Individuality and adaptation across levels of selection: How shall we name and generalize the unit of Darwinism?" *Proceedings of the National Academy of Sciences USA* 96: 11904–11909.

Grandy, R. 2008. "Sortals." In E. Zalta (ed.), *The Stanford Encyclopedia of Philosophy*. http://plato.stanford.edu/entries/sortals.

Griffiths, P. & Stotz, K. 2007. "Gene." In M. Ruse & D. Hull (eds.), *Cambridge Companion to Philosophy of Biology* (Cambridge University Press) 85–102.

Hey, J. 2001. "The mind of the species problem." *Trends in Ecology & Evolution* 16: 326–329.

Hull, D. 1976. "Are species really individuals?" *Systematic Zoology* 25: 174–191.

Hull, D. 1978. "A matter of individuality." *Philosophy of Science* 45: 335–360.

Isaac, N., Mallet, J., & Mace, G. 2004. "Taxonomic inflation: Its influence on macroecology and conservation." *Trends in Ecology & Evolution* 19: 464–469.

Jablonski, D. 2008. "Species selection: Theory and data." *Annual Review of Ecology, Evolution, and Systematics* 39: 501–524.

Jackson, F. 1998. *From Metaphysics to Ethics: A Defence of Conceptual Analysis* (Clarendon Press).

Knobloch, I. 1959. "A preliminary estimate of the importance of hybridization in speciation." *Bulletin of the Torrey Botanical Club* 86: 296–299.

Koslow, A. 2006. "The representational inadequacy of Ramsey sentences." *Theoria* 72: 100–125.

Ladyman, J. 1998. "What is structural realism?" *Studies in History and Philosophy of Science Part A* 29: 409–424.

Lansing, J. 2007. *Priests and Programmers: Technologies of Power in the Engineered Landscape of Bali* (Princeton University Press).

Lherminier, P. & Solignac, M. 2000. "L'espèce: Définitions d'auteurs." *Sciences de la Vie* 323: 153–165.

Littlejohn, M. & Oldham, R. 1968. "*Rana pipiens* complex: Mating call structure and taxonomy." *Science* 162: 1003–1005.

Mallet, J. 2007. "Hybrid speciation." *Nature* 446: 279–283.

Mallet, J. 2008. "Hybridization, ecological races and the nature of species: Empirical evidence for the ease of speciation." *Philosophical Transactions of the Royal Society, Series B: Biological Sciences* 363: 2971–2986.

Maynard Smith, J. 1958. *The Theory of Evolution* (Penguin).

Mayr, E. 1942. *Systematics and the Origin of Species from the Viewpoint of a Zoologist* (Columbia University Press).

Mayr, E. 1970. *Populations, Species, and Evolution* (Harvard University Press).

Mayr, E. 1982. *The Growth of Biological Thought: Diversity, Evolution, and Inheritance* (Harvard University Press).

Millstein, R. 2006. "Natural selection as a population-level causal process." *British Journal for the Philosophy of Science* 57: 627–653.

Mishler, B. 1999. "Getting rid of species?" In R. Wilson (ed.), *Species, New Interdisciplinary Essays* (MIT Press) 307–315.

Müller-Wille, S. & Orel, V. 2007. "From Linnaean species to Mendelian factors: Elements of hybridism, 1751–1870." *Annals of Science* 64: 171–215.

Padial, J. & de la Riva, I. 2006. "Taxonomic inflation and the stability of species lists: The perils of ostrich's behavior." *Systematic Biology* 55: 859–867.

Platz, J. 1993. "*Rana subaquavocalis*, a remarkable new species of leopard frog (*Rana pipiens* complex) from Southeastern Arizona that calls under water." *Journal of Herpetology* 27: 154–162.

Pleijel, F. 1999. "Phylogenetic taxonomy, a farewell to species, and a revision of *Heteropodarke* (*Hesionidae, Polychaeta, Annelida*)." *Systematic Biology* 48: 755–789.

Pleijel, F. & Rouse, G. 2000. "Least-inclusive taxonomic unit: A new taxonomic concept for biology." *Proceedings of the Royal Society of London, Series B: Biological Sciences* 267: 627–630.

Psillos, S. 2006. "Ramsey's Ramsey-sentences." In M. Galavotti (ed.), *Cambridge and Vienna: Frank P. Ramsey and the Vienna Circle* (*Vienna Circle Institute Yearbook* 12) 67–90.

Quine, W.V. 1948. "On what there is." *Review of Metaphysics* 2: 21–38.

Rice, S. 1995. "A genetical theory of species selection." *Journal of Theoretical Biology* 177: 237–245.

Rosen, D. 1978. "Vicariant patterns and historical explanation in biogeography." *Systematic Zoology* 27: 159–188.

Sneath, P. & Sokal, R. 1973. *Numerical Taxonomy: The Principles and Practice of Numerical Classification* (W.H. Freeman).

Sokal, R. & Sneath, P. 1963. *Principles of Numerical Taxonomy* (W.H. Freeman).

Turesson, G. 1922. "The species and variety as ecological units." *Hereditas* 3: 10–113.

Turesson, G. 1925. "The plant species in relation to habitat and climate." *Hereditas* 6: 147–236.

Turesson, G. 1929. "Zur Natur und Begrenzung der Artenheiten." *Hereditas* 12: 323–334.

Uexkull, J. von. 1926. *Theoretical Biology* (Kegan Paul, Trench, Trubner).

Van Valen, L. 1976. "Ecological species, multispecies, and oaks." *Taxon* 25: 233–239.

Wagner, W. 1983. "Reticulistics: The recognition of hybrids and their role in cladistics and classification." In N. Platnick & V. Funk (eds.), *Advances in Cladistics* (Columbia University Press) 63–79.

Wheeler, Q. & Platnick, N. 2000. "The phylogenetic species concept (*sensu* Wheeler and Platnick)." In Q. Wheeler & R. Meier (ed.), *Species Concepts and Phylogenetic Theory: A Debate* (Columbia University Press) 55–69.

Wilkins, J. 2003. "How to be a chaste species pluralist-realist: The origins of species modes and the synapomorphic species concept." *Biology and Philosophy* 18: 621–638.

Wilkins, J. 2007a. "The concept and causes of microbial species." *Studies in History and Philosophy of the Life Sciences* 28: 389–408.

Wilkins, J. 2007b. "The dimensions, modes and definitions of species and speciation." *Biology and Philosophy* 22: 247–266.

Wilkins, J. 2009. *Species: A History of the Idea, Species and Systematics* (University of California Press).

Wilkins, J. 2011. "Philosophically speaking, how many species concepts are there?" *Zootaxa* 2765: 58–60.

Wilkins, J. & Ebach, M. 2013. *The Nature of Classification: Kinds and Relationships in the Natural Sciences* (Palgrave Macmillan).

Wilson, E. 1992. *The Diversity of Life* (Penguin).

Winsor, M. 2000. "Species, demes, and the omega taxonomy: Gilmour and *The New Systematics*." *Biology and Philosophy* 15: 349–388.

13

Modern Essentialism for Species and Its Animadversions

Joseph LaPorte

We often hear about the death of essentialism in the philosophy of biology. Biological species are supposed to have no essences and no necessary characteristics. So according to Gillian Barker and Philip Kitcher, "There are no properties essential to *Drosophila melanogaster* (the famous fruit fly) or to *Homo sapiens*" (2014: 40). Before Darwin, the familiar story goes, species appeared to be definable in terms of their essences. But biologists now get by without these, "identifying species without supposing that they have essences" (Barker & Kitcher 2014: 41). Michael Ghiselin agrees: species "cannot be defined, in the sense of listing properties they simply must have" (1987: 129). For Michael Hardimon, "the concept species can be articulated in either essentialist or nonessentialist ways. Contemporary biology has vindicated the nonessentialist alternative as true of real existing species" (2013: 5).

As you would suppose, there is a whole nest of views associated with the offending essentialist doctrine. Some of the views are mistaken; there is no clean tradition. But there is a central strand of essentialism that enjoys some promise. It is helpful to sort through the different strands. The conclusion here will be that essentialism survives objections in an important, relevant form. However, not *all* work associated with species' essences can be supported by essence. A potential concern is whether species *essences*, coming to what they come to, will have lost their significance; or whether *species*, coming to what they come to, essentially, will have lost their significance. These concerns will be addressed in their turn.

A ROBUST ESSENTIALISM FOR SPECIES

Biologists today accept that all species they look at come from others, in an evolutionary tree: and they generally delimit species, or characterize them, according to their place in that tree. In this respect the mainstream "species concepts," as they are called, are more or less uniformly historical or phylogenetic. Species tend to be characterized as a historically connected group of organisms with this or that origin in *particular* organisms on the family tree.

There are other things to take into account besides historical connections, in delimiting one species from another species; but history counts. What else counts, by way of delimiting species? Here informed answers vary. A classic answer as to what makes a historically connected group of interacting organisms into a species is that there is *interbreeding* or potential interbreeding between its constituent organisms, to the exclusion of outsiders to that species. This tells *which* of the historically demarcated organismal groups to which an organism like you belongs is its *species*—H. sapiens, say, as opposed to the mammals as a whole: your species is the group united to you by interbreeding potential, or "reproductive isolation." There are other proposals for what else binds a species besides historical roots (see LaPorte 2004 for elaboration of competitors). But let us take the most prominent approach for granted at the moment.

The interbreeding approach accounts for the Darwinian news that species come from other species in a historical lineage (and again, so do other live approaches). That is supposed to be a problem for essentialism. Here is Makmiller Pedroso: "Organisms are conspecific not because they necessarily share a certain property, the taxon's essence, but because they are (spatiotemporally restricted) parts of the same species," which is a historical lineage (Pedroso 2012: 182; see also Andreasen 2005: 105).

Let us be a little more specific about the central essentialist tenet at issue. Call it the "invariance" of a species' essence:

> INVARIANCE: There is a property that all and only members of any species share, namely its essence. It is a necessary truth that all members of the species share that property, because the property captures just what it is to be that species and not some other.

Criticisms of the proposal that species are characterized by any such essentialist invariance are legion. For Marc Ereshefsky, in order "to see the failure of essentialism we need only consider" this "first tenet," namely the central essentialist idea "that all and only the members of a kind have a common essence" (2010b).

The trouble is supposed to be that species have no genetic or otherwise qualitative characteristics that could serve as the essence: they vary too much genetically. Variability is at the very heart of Darwinism. Since *invariance* is central to essentialism, and *variability* is central to Darwinism, the two are thought to be irreconcilable. As Jody Hey puts it, the "idea, that variation among organisms is the crucial stuff of changing life and of life's progress, is devastating to essentialism" (2001: 62). By articulating a convincing evolutionary theory that accords a key role to variation, "Darwin delivered a one–two punch that pretty much obliterated essentialism" (Hey 2001: 61).

We should pause here to observe a controversial premise of the argument. In the past several years, explicit emphasis on the importance of a species' shared structure has grown, making it controversial to say that there are no genetic or structural characteristics common to a species. Indeed, even before this articulation found a voice relatively recently with the burgeoning of "HPC" ("homeostatic property cluster") accounts of species and other similar accounts, there had been celebrated species concepts that allowed for genetic or structural characterizations of species; but informed discussion was more attentive to historical aspects of such species concepts than structural ones. (Joel Cracraft's favored species concept is an example of a species concept that was a major player long before people talked as much about genetic or qualitative characterizations of species, yet one that countenances genetically delimited species (see LaPorte 2004 for discussion). Cracraft also recognizes that species are delimited by *historical* pedigree, as do more recent champions of genetic essences (see Brigandt 2009: 80n; Devitt 2008: 346; Dumsday 2012).) Emphasis, not content, seems to have changed: there has been and continues to be a recognition, on the part of many, that species are delimited by qualitative characteristics like genetic or structural traits, *as well as historical* pedigree. For all of these workers, structure has a role, perhaps along with *reproductive isolation*, as one delimiting factor to distinguish a species. For all of these workers, history is another delimiting factor.

Let us ignore this complication concerning renewed attention to qualitative characteristics for species: let us suppose that a species has no genetic or qualitative characteristics shared by its members. Why is this supposed to be a problem for essentialism? That is not obvious. What *is* obvious, if biologists cannot identify any *genetic* or qualitative characteristics that could be essential to a historical species, is that the *essence* is not genetic or qualitative. A species might then instead be delimited essentially by its historical *origin* and by connecting relations between members. Such historical essences are familiar in the essentialist tradition. For example, Saul Kripke argues in classic works (for further discussion, see LaPorte 2004) that individuals essentially come from their own original stock, and not some other just like it. The idea is intuitive. You are an organism whose parts share a similar genetic makeup. But that is not what makes them parts of your body. Somatic mutations do not remove tissue from you or make it part of some other organism with genes more like those of the mutant tissue.

Organisms have historical essences. So it might be for species. So it *is* for species, if indeed species are properly delimited by origin and, say, reproductive connections: the respective origin and connections would then be what, roughly speaking, "accounts" for something being the species *Citrus limon* or *Homo sapiens* respectively, instead of anything else we might care to discuss. Nor is it problematic that species might or might not all be delimited by *reproduction*, so far as historical essentialism is concerned. Reproductive isolation is a popular standard but there are competitors—and a similar essentialist story applies to species as they are properly characterized by competitors, too. Perhaps species are lineages delineated by origins and *genetic* structure, or origins plus *ecological role*, say. If species are individuated *that* way, then essentialism is simply reformulated.[1]

This proposal that species have historical essences, now well known, is viewed with considerable suspicion in the informed community. Mark Ellis (2011: 356) complains that so-called historical essentialists "creatively revive the illusion of species essentialism." Many suggest that the proposal breaks with what tradition has *all along* been calling "essentialism," so that the proposal "robs essentialism of its essence" as Olivier Rieppel puts it (2010: 670); whether such historical essentialism "should be considered as essentialistic in some novel sense of the term, or liberated entirely from essentialism" is moot, he says. Ereshefsky goes further: historical "essentialism is not essentialism," because it does not uphold what is "core" to the tradition (2010b). Others voice similar qualms (see below).

Is so-called "historical essentialism" really just a new doctrine called by a venerable name? If so, then this defense of what is supposed to be essentialism—historical essentialism—enjoys all the benefits of theft over honest toil (apologies to Bertrand Russell).

Fortunately, historical essentialism is the real thing; it is not a cheat. There should be little question that essence, as outlined under the rubric "INVARIANCE," is deeply rooted in the essentialist tradition. *Whatever* essence is supposed to be, it is surely supposed to be *what* or *how* some entity has to be *in order to be* what it *is* rather than some other entity. Very roughly you could think of essence as what "defines" the item. This is the main thing essence is supposed to be all about. And this basic idea of essence would surely permit essences historical as well as qualitative.

Recall the essence of the lemon. Some have maintained, mistakenly, that the essence of *lemon*, or of this particular lemon tree, is purely genetic. But there is nothing about the nature of essences to preclude answers that accord better with recent biology, answers like, "the lemon's essence is to be what has such and such origin." The essentialist question is merely, "What makes the lemon to be *this* instead of that other citrus?" (I point to one and then another tree), or, to vary the question, "What is 'defining' of the lemon, in the respect that nothing else could qualify as fitting the definition?" Such a question arises however we delimit species; so "clearly that question will not go away," as Samir Okasha says, "if we think that *Citrus Limon*," being a biological species, has a history that differentiates it essentially (Okasha 2002: 194; Devitt is also especially clear on this point (2008: 347–348).

Historical essences are essences. They meet the requirements: they explain what it is to be something—namely, a historically delimited thing. Even so, you could still object that historical essences are not the *sort* of essences that essentialists and anti-essentialists have traditionally been talking about: you could argue this point on the grounds that context, say, reveals otherwise. Context can sometimes in this way limit the range of what is under discussion: thus, even though dust and air are "stuff," strictly speaking, context would rule out counting them as "stuff" remaining in a warehouse we are emptying of inventory. After the last of the inventory is gone, it is okay to say, "there's no more stuff there." If a neophyte objected, "Wait, there's still dust and air in there—so there's stuff after all!" we would conclude he did not know what we were talking about. In the same way, the contexts presupposed in traditional debates between essentialists and anti-essentialists might make it okay to say, "species have no essences." To say, "Wait, there are *historical* essences" might indicate a failure to understand what people have been talking about in the relevant contexts.

Have traditional contexts in fact suggested that historical origins are irrelevant to whether there are "essences"? On the contrary. First, as we have seen, recent generations of philosophers have recognized historical essences for other sorts of entities like individual *organisms*. Historical essences thus featured prominently in past generations of philosophical discussion well before the question of any application to species dawned—or rather re-dawned, because the application to *species* is not really new either. On the contrary, the alleged essentiality of a *species' historical origins* to that species is an old idea. Hence, long ago John Locke canvassed the position. As Locke characterizes him, the historical essentialist holds that a species has "a real essence, which he thinks certainly conveyed by generation." Locke himself worried that the position would mire us in skepticism: "If the species of animals and plants are to be distinguished only by propagation, must I go to the Indies to see the sire and dam of the one, and the plant from which the seed was gathered that produced the other, to know whether this be a tiger or that tea?" (1995: III.vi.23). Historical essences counted as bona fide essences in the context of that debate over whether species have essences.

So there is nothing wrong-headed about being open to historical origins as essential to a species. Historical essences accord with the essentialist demands of INVARIANCE. They qualify as essences. And context has honored their relevance. They count as what essentialists and anti-essentialists have been talking about, traditionally.

TOO MANY DEMANDS ON ONE NOTION OF ESSENCE: ESSENCES FOR SPECIES AND FOR SPECIMENS

The primary job of essence, summarized in the foregoing rubric INVARIANCE, is to explain what it is to be this or that. That is what essence is about most fundamentally. But essences have been associated with less fundamental work, too. Essences have been associated with diverse work *other* than that limned by INVARIANCE, as well as work limned by INVARIANCE. Because diverse work has been attributed to essences, it is understandable that confusion has arisen about what essences are supposed to do, or whether self-declared "essentialists" today deserve the name. Let us consider some of the other work associated with essence, besides that most fundamental work limned by INVARIANCE.

An ancient line would require the essence of a species S not only to explain what it is to be S *itself*, but also what it is to be this or that individual organism *belonging* to S, in the following respect. Each organism belonging to S is sometimes thought *itself* to belong essentially: to have essentially the properties defining of S.

The history of ideas indicates repeatedly that when manifold demands are made upon a single notion, usually without proper distinctions, the confusing welter of demands must be sorted out; often enough, no single notion turns out to be able to bear the burden of all the different sorts of work attributed to one thing. So it may well be here. The doctrine that their species is essential to the separate *organisms* in the species seems much more problematic than the essentialist doctrine originally limned above with INVARIANCE, according to which there are distinguishing boundaries essential to a species itself, boundaries to which organisms have to conform if they are to belong to that species and not some other. So there is not much reason to think that essence, as articulated by INVARIANCE, also honors the demand that organisms essentially belong to their species. (We might call this doctrine "Essential Membership.")

The difficulty with Essential Membership is this: that current characterizations of species, those of the experts, do not seem amenable. Return to the best-established view according to which reproductive isolation does the job of delimiting. If that is so, then it would seem possible for an organism to belong to one species rather than another contingently. So suppose that a population of frogs becomes isolated on an island. Geographical isolation alone is not reproductive isolation. But such an isolate might develop slightly different reproductive tendencies that would discourage interbreeding with the parental stock on the mainland. Then the members of the isolate would thereby qualify as belonging to a new species, but that could easily be by virtue of events *contingent* to *individual* frogs. Tendencies of an organism can be contingent to it, including reproductive habits.

You might object that dispositions in this case are not contingent to individuals because, you might say, *genetic* constitution would be responsible for a disposition or tendency to reproduce with the organisms of one lineage rather than another, which disposition qualifies that organism to be of the one species (which is a lineage) rather than the other; you might add that genetic constitution is essential to an organism and you might infer that reproductive isolation would therefore also be essential to individual frogs. But the objection would not seem to meet its mark. Let us assume, controversially (Ghiselin 1987: 137), that reproductive isolation must arise by way of genetic difference between the lineages isolated one from another. A new isolate of a parent lineage might indeed undergo a slight genetic change, causing its reproductive call to differ in tone, or causing the frogs to call at a different time. Distinctive *tones* or *calling times* are the sorts of things that serve as isolating mechanisms. In the foregoing case, even though a slight genetic variation causes a new isolate to qualify as a new species, distinct from the parental species by virtue of a genetically prompted mechanism that would prevent inter-breeding between the two lineages, the reproductive isolation would still seem clearly to be contingent to any given frog—*not*, or not only, because the genetic change responsible for the distinctive reproductive routine is so slight that it would seem inessential to any given frog (though that seems reasonable to say), but also because other variables besides that slight genetic difference will be necessary for the distinctive features of the call that produce the isolation. The acoustic properties of a frog's call depend on variables such as the temperature of the surrounding environment, or other noises within earshot. These variables, which are surely not essential to any frog's being itself instead of some other individual, can cause the same frog, with its very same genetic tendencies, to sound a call that works a little differently. Because that difference is the sort of thing that causes reproductive isolation, that difference is therefore the sort of thing causing the frog to be a member of a new species instead of the one it would otherwise be a member of. There are many similar examples, but this one will serve to illustrate the basic problem in principle. Let us develop it a bit.

Suppose that a population S2 takes to an island, from the mainland stock species S. The island periodically swings to slightly higher temperatures than the mainland. So the island individuals of S2 adapt by becoming more flexible about the mating call: depending on the temperature, the ritual can sound a little shriller. Meanwhile, the mainland undergoes an influx of interfering noises in places. The stock species S adapts accordingly a lower call, depending on whether interference is present. Now, the isolate and the stock organisms are not reproductively isolated so long as they would, in general, cross. And conditions might typically allow for that—typically an island specimen of S2 would reproduce were it transplanted to the mainland—but not in the occasional special circumstances of the interfering noises. By parity a typical organism from the mainland species S might not be challenged at reproduction, were it sent to the island, except for the occasional season or part of a season during which temperatures rise. The specimen would not integrate with the shriller ritual.

So far there is no speciation. But if noise pollution were to become the norm on the mainland and a rise in temperature on the island, there *would* be reproductive isolation—a settled, stronger-than-geographical barrier to reproductive integration. Now members of the two lineages would not belong to the same species. Instead, there would be two new species. There is the mainland species, S1, and the island species, S2 (plus the parent, S). Yet clearly the individual organisms belonging to S1 and S2 *could* be conspecific with members of S, for all their individual identities required. Were it not for noise and temperature-related pollution that does not compromise the individual identity of any individual specimen, the organisms in S1 and S2 would be members of S.[2]

TOO MANY DEMANDS ON ONE NOTION OF ESSENCE: ESSENCES AND THE LAW-LIKE SECURING OF PROPERTIES

Tradition has assigned too many roles to the same notion called "essence," when tradition has suggested that a species' essence must be essential to individuals. What other doctrines commonly associated with essence must be distinguished from the basic notion of *species essence* (from the rubric INVARIANCE)? Something often mentioned is the important theoretical role essence plays in supporting powerful *generalizations*.

Clusters of theoretically important properties are sometimes supposed to arise by nomological necessity from essence. Laws governing a species or other lineage might necessitate its child-rearing practices, metabolism, camouflage, social interaction, communication, culture, or other higher functions of intelligence, and so on.

Can properties supporting law-like generalizations be tightly associated with a species in this way? It is doubtful. We can illustrate briefly just for the reproductive approach. A single genetic mutation could switch a transition in breeding habits, and thus distinguish a new species. The bulk of the genetic heritage, with the best connection to non-historical properties, might divide saliently along other lines than reproductive ones. No cluster of important properties has to follow species divisions.

Consider by way of illustration a stock species that sends out colonies in a tripartite geographical dispersal. A colonial population B might become reproductively isolated by way of a minor genetic change from the stock species A1, while a second colony, A2, becomes geographically isolated from A1 but not reproductively isolated from A1—so A2 remains conspecific with A1. Yet A2, although conspecific with A1, might be the outsider with respect to salient law-like generalizations. A2 might lose law-like properties belonging to B and A1, having to do with camouflage or vision or whatnot, and A2 might acquire unique new such properties not shared by B and A1, which undergo less change. The salient law-like generalizations might be stronger between A1 and B than between A1 and A2, though species-lines break differently. Reproductive isolation is not the only candidate here for what counts as a division into species, but something similar could be said about other candidates, such as a switch of ecological role.

So species' essences are not to be tied too closely to non-historical properties or to law-like generalizations supported by clusters of properties. Often because of the organisms making up the lineage, there *will* be such a correlation; but not necessarily. Sometimes striking similarity is attended by rifts in interbreeding; other times, markedly different strains, so far as non-historical properties are concerned, can be counted as one species reproductively. There are plenty of observed cases of this (e.g., in marine corals), and the theoretical point is clear in any case: historical essences of the sort species have could hardly be expected in any hard and fast way to assure better correlation to non-historical properties.

All this reflection on essence's connection with laws might make us uneasy: what we recognize and call "species" should have scientific significance. So if our characterizations of "species" fail to uphold law-like generalizations, should we conclude that "species" as we have characterized them fail to serve biological science, by failing to mark theoretically important groups? No.

If the species in fact delimited by the reproductive account and other accounts accepted by biologists *were* to turn out to be of little biological significance, then we should be sent back to the drawing board. We should be forced to discard the foregoing characterizations of "species" as *mis*characterizations. But that threat does not materialize because the species named by biologists are in fact of biological significance. Species as biologists characterize them do tolerably well at serving to frame theoretically important biological inquiry.

To be sure, law-like generalizations are important to some sorts of scientific inquiry, including some biological inquiry. But such generalizations might be unavailable or inappropriate in many salient biological contexts. Other biological contexts, which feature other sorts of inquiry and other sorts of answers, have their own value and salience in biology, as Stephen Jay Gould for one has emphasized: "Answers that invoke general laws of nature rather than particular contingencies of history" are not unimportant, then; but even so a good part of what biologists do is to look for explanations "rooted in history, pure and simple" (Gould 1991: 121). Sometimes a question has a historical answer: e.g., Why is QWERTY the standard? Answer: Because the keyboard started that way. Why do kiwis have oddities like a big egg? Answer: "Kiwis are as they are because they were as they were" (Gould 1991: 121).

Tight general laws of the sort associated with chemical elements' essences do not seem to characterize biological species' essences, and certainly would not seem to arise by metaphysical necessity from species' defining essences. Even so, theoretically interesting generalities tend to hold. Species are often observed to be causally integrated, as well as integrated genetically, morphologically, ecologically, behaviorally, and so on. They do not have to be; but they often are. Law-like connections to properties do not obtain by way of the necessity of essence; but for all that, they *tend* to hold.

WHEN CONSILIENCE FAILS: SPECIES' ESSENCES, SPECIESHOOD'S *ESSENCE* AND WHY THESE MATTER

Because consilience is not perfect, ecological and morphological distinctness can obtain without reproductive isolation, say, as it does in the case of oaks, the classic example. This is why there are competing species concepts, that is, different ways to characterize species. Thus, you might recognize one or another species of oaks as that *lineage* with such and such original population along with all and only the organisms united to it by *shared ecological role* (instead of reproductive ties). You might so characterize your species because ecology and not reproductive isolation seems to obtain and more importantly to explain what is salient about the oak lineage.

Reflection on this sort of discrepancy between species concepts might engender skepticism about whether species are one natural *category* of lineage. Perhaps the category of species is disjunctive: it embraces a motley mix of lineage types, important for different reasons and united by different sorts of mechanisms.

If some species are united reproductively and others ecologically, that might seem to threaten the integrity or theoretical interest of species essences. But that threat does not materialize. Suppose that the species category is disjunctive, because some species are reproductively united and others are ecologically united. This would not indicate that any one species—a species of oaks, say—has a disjunctive essence. It would not. That species' essence would be ecological, not *ecological OR reproductive*. Individual lineages like the different oaks or like *H. sapiens* would remain integrated and natural biologically, united by one measure or the other for species.

You might question whether the category *species* has any essence or at least any scientifically significant essence: its essence might turn out to be motley, and not worth recognizing. But for all that, there would be no reason to doubt that individual species' essences matter. After all, if the species themselves matter, then their several essences matter, because if the individual species matter, then it matters just what makes these species what they are.

Does the species *category* have a theoretically interesting essence? That is, is there a theoretical property that is defining of species (as opposed to a theoretical property that is defining of *Q. rubra* or of *H. sapiens*)? Or is the species category, rather, disjunctive or even gerrymandered? Some have even suggested that the category should be retired and that we should stop talking about "species" under a common label altogether. Again, there is no need to commit to much here by way of these issues, in order to establish an interesting species essentialism. Even if we were to retire—or, better, qualify by further specification—the common label "species," it would not undermine an important essentialism concerning the different species. However, I will suggest all too briefly that the category itself *is* of theoretical interest and that essentialism is plausible here too. First, there is more to be said concerning the significance of different species' essences.

An apparent threat to the significance of species' essences is implicit in the suggestion that individual species are not worth the biological attention accorded to them. Something like this can occasionally come across in the vibes of certain systematists who suggest that there is nothing distinguished about species as opposed to any other lineage (Velasco 2013: 10–14); yet even if this were so, it takes little away from the importance of species and their essences, because different levels of lineage *all* have importance. Think of the canines, say, a genus—or chihuahuas, an ancient breed from Asian stock, that coevolved with Native American humans. The canines and the chihuahuas are theoretically interesting lineages each with its own distinctive heritage.

And species may be especially important: they have certainly received special attention and recognition from biologists on the whole. As Kevin de Queiroz says:

> Biologists commonly assert that species are fundamental biological units, comparable in importance to fundamental units at lower levels of organization, such as organisms and cells.   (2011: 29)

Such special attention is sometimes criticized; but my own sense, for what that is worth, is that biologists know by acquaintance that species are a good rough marker of about the level on which they ought to focus—even if there are different kinds of species and even if it is hard to articulate what is important about them. Biologists themselves certainly have a general sense that they have that tacit understanding, as even critics of the species concept(s) acknowledge (e.g., Hey 2001: especially 8–9, 24, 190).

The power of species comes into relief when we compare alternatives. Some workers seem to prefer *populations* of organisms within a species, say, instead of species, as a focal point (see, e.g., Velasco 2013: 10, 13, 14). But consider that no one cares about a population of Greeks colonizing a lonely island like Delos, however low their contact with the mainland, until something more important happens. When something more important does happen on an island, so that speciation occurs, *then* we care: the relatively recently evolved human "hobbit" species *Homo floresiensis* is headlines!

It is hard to focus on a rough level of distinctiveness for lineages without that level being around where one or another species concept would put it anyway; thus, for example, *populations* with distinctions marked enough to tempt a systematist to accord them recognition are captured in the nets of species "splitters." There are complications, to be sure (hence, the recognition of "metaspecies"). And species are not the only lineages worth recognition. Still, what seems most often most germane in the broad contexts where species are invoked—say, in conversations about biodiversity—is a salient measure that takes into account, roughly at least, both *distinctiveness* and also the history of *splitting* events marking permanent divisions (speciation). Divisions between species tend to provide such a measure, to a decent approximation, in the world as it happens to be (recall that there is *approximate* consilience). Consider this question, important for comparing the Anthropocene period in which we live to earlier periods, like the Cretaceous: "Were all of the periods with mass extinction events, like the extinction event that took the dinosaurs, caused by accelerated extinctions—or did some occur as a result of the *slowing down of speciation*?" It is a question that is closely, perhaps inextricably, tied to the idea of species: distinctive, permanent lineages that have separated stably. We need *some* handy measure to relate information like this quickly. Systematics is about conveying relevant information efficiently. It seems hard to do better than reference to species, which is in fact something like the settled standard.[3]

All of this talk about a handy measure or a common measure—specieshood—or, alternatively, of different measures—eco-specieshood, bio-specieshood, etc.—suggests a higher-order essentialism about the species category or categories. If there are such categories, then presumably they have certain characteristics integral to what it is to *be* the relevant category and not others. So although *species essentialism* can be defended in both fact and significance without settling these higher-order matters of category, we might still observe a plausible case for essentialism here, too, at the level of *category*.

The simplest approach to these matters is by way of assuming that different species concepts have their own interest and relevant application. Each has an essence. Eco-specieshood is essentially characterized as the category to which belong, or the property possessed by, those lineages that qualify as eco-species (like the different oak species). Lineages that amount to eco-species, like *Quercus Rubra*, qualify as eco-species by virtue of *playing a certain ecological role*, such as occupying a certain habitat, providing such and such services like the removal of waste, the transformation of nutrients, or the maintenance of shelter. That is what it *is* to belong to that category; it specifies the essence of the category. Eco-specieshood, then, would seem itself to be an entity with an essence. Bio-specieshood is the correlate that makes appeal to reproductive isolation instead of ecology. And there are other species concepts, some of them cross-classifying. These categories would all have their own essence and they would all be interesting. Perhaps there is a species concept unifying all of these: a Unified Species Concept. If so, then this is its own higher-level species category whose essence explains how eco-species, bio-species, and so on, qualify to be *species-simpliciter*, according to a unified understanding of *species*.

CONCLUSION

In conclusion, interesting essentialism about species seems tenable. Species taken severally have historical essences. We might plausibly maintain essentialism even regarding the species category(/categories), though that conclusion has been but briefly discussed here.

NOTES

1. Because genetic characterizations tend not to hold hard and fast, some workers who suppose that genes play a role in essence do not realize that the essences they recognize, in effect, comply with INVARI-ANCE. Thus, in a nice elaboration of Richard Boyd's position that species have "HPC" essences, Michael Rubin explains that "traditionally, the essence of a natural kind is thought of as a property or a collection of properties whose exemplification by a given individual is both necessary and sufficient for that individual to count as a member of the kind." Nevertheless, says Rubin, "Boyd relaxes this requirement so that an individual may belong to an HPC kind even if it fails to instantiate some of the properties in the kind's HPC essence" (2008: 500–501). The error of such a move is exposed by considering the properties that tend to accompany a species. Are these part of its essence? Not if they merely *tend* to accompany the species. If there can be exceptions to the rule so far as possession of these properties goes (a single specimen might behave and look differently from other organisms in its species because of its particular genetic mutations), then there would need to be some account for why the mutant is still a member of the species; and the answer will invoke the species' *essence*—that is, what it *is* to be of that species. *That* remains invariant. It will generally come down to history, which indicates that the historical component of the relevant accounts of species is what is really doing all the work even though there *appear* to be more factors.
2. The most interesting and powerful essentialist rally of which I am aware comes from an unpublished paper by Manolo Martínez, who argues that the circumstances of contingency would be *rare*. Though I disagree with that conclusion, it can be conceded for purposes here. Rare would be enough to belie a problem in principle with associating two jobs with "essences." A distinction between species essence and individual essence is wanted.
3. Species in effect serve as the standard measure. Again, consider the use of species as an initial marker, a rough index, of biodiversity for purposes of conservation. Of course, groups narrower than species merit recognition. The chihuahua merits preservation, by virtue of its unique heritage. Similar words hold for lineages wider than species, and categories wider than species, such as *genre*, are sometimes used as an index for biodiversity measurement (typically only in conjunction with *species*, though—for example, by California Academy of Sciences 2014); but species are the common coin used to translate and then get "the best information out as quickly as possible to all of the end users" like policy makers and consumers, who "really desperately need accurate information to make informed decisions" (CAS 2013: from 1.25). If there *were* a better measure for some or all purposes for which *species* serves, it would have to be implemented efficiently, in view of its own user-friendliness and in view of a clear path to its widespread appeal: we cannot wait for 200 years' improved understanding or adjustment to formulate or spread the idea. Entrenchment matters (as Ereshefsky observes, 2010a: 420–421).

REFERENCES

Andreasen, R. 2005. "The meaning of 'race': Folk conceptions and the new biology of race." *Journal of Philosophy* 102: 94–106.

Barker, G. & Kitcher, P. 2014. *Philosophy of Science* (Oxford University Press).

Brigandt, I. 2009. "Natural kinds in evolution and systematics: Metaphysical and epistemological considerations." *Acta Biotheoretica* 57: 77–97.

California Academy of Sciences. 2013. "189 new species." January 28. www.calacademy.org/explore-science/189-new-species

California Academy of Sciences. 2014. "Biodiversity and extinction, then and now." November 25. www.calacademy.org/explore- science/biodiversity-and-extinction-then-and-now

de Queiroz, K. 2011. "Branches in the lines of descent: Charles Darwin and the evolution of the species concept." *Biological Journal of the Linnean Society* 103: 19–35.

Devitt, M. 2008. "Resurrecting biological essentialism." *Philosophy of Science* 75: 344–382.

Dumsday, T. 2012. "A new argument for intrinsic biological essentialism." *Philosophical Quarterly* 62: 486–504.

Ellis, M. 2011. "The problem with the species problem." *History and Philosophy of the Life Sciences* 33: 343–364.

Ereshefsky, M. 2010a. "Darwin's solution to the species problem." *Synthese* 175: 405–425.

Ereshefsky, M. 2010b. "Species." In E. Zalta (ed.), *The Stanford Encyclopedia of Philosophy*. http://plato.stanford.edu/archives/spr2010/entries/species/

Ghiselin, M. 1987. "Species concepts, individuality, and objectivity." *Biology and Philosophy* 2: 127–143.

Gould, S. 1991. *Bully for Brontosaurus: Reflections in Natural History* (W.W. Norton).

Hardimon, M. 2013. "Race concepts in medicine." *Journal of Medicine and Philosophy*. doi:10.1093/jmp/jhs059.

Hey, J. 2001. *Genes, Categories, and Species: The Evolutionary and Cognitive Causes of the Species Problem* (Oxford University Press).

LaPorte, J. 2004. *Natural Kinds and Conceptual Change* (Cambridge University Press).

Locke, J. [1689] 1995. *An Essay Concerning Human Understanding* (Prometheus Books).

Okasha, S. 2002. "Darwinian metaphysics: Species and the question of essentialism." *Synthese* 131: 191–213.

Pedroso, M. 2012. "Essentialism, history, and biological taxa." *Studies in History and Philosophy of Science Part C* 43: 182–190.

Rieppel, O. 2010 "New essentialism in biology." *Philosophy of Science* 77: 662–673.

Rubin, M. 2008. "Is goodness a homeostatic property cluster?" *Ethics* 118: 496–528.

Velasco, J. 2013. "Phylogeny as population history." *Philosophy and Theory in Biology* 5. doi:10.3998/ptb.695 9004.0005.002.

14

What Is Human Nature (If It Is Anything at All)?

Louise Barrett

INTRODUCTION

Talk about human nature, particularly in an evolutionary framework, can get you into trouble with two sets of people: philosophers of biology and anthropologists. In both cases, there is an objection to the essentialism that seems inherent in such a concept. As David Hull (1986) and Michael Ghiselin (1997) famously pointed out, biological species are not natural kinds in the way of chemical elements; that is to say, they are not the sort of things that can possess an intrinsic essence. Hydrogen is hydrogen, for example, because it has the atomic number 1, an intrinsic feature shared by all hydrogen atoms that is always and forever the case (i.e., it possesses a particular micro-structural property that accounts for all of its macro-structural properties, no matter when or where you find it). Species are not like this: they come into being at a particular time and place, they come to an end when they go extinct, and they are made up of populations that can vary in their attributes—attributes that can themselves vary over time. As Hull (1986) puts it, species are individuals, not kinds (see also Sober 1980). It makes no sense to propose some intrinsic, hidden essence or "nature" that can explain why species are the way they are.[1] Although species are real entities, they are also simply analytical conveniences that reflect the ecological spatiotemporal scale at which humans view and understand the world around them.

This kind of objection to talk of "human nature" is often directed at efforts to demarcate the boundary between humans and other species, and identify those features that are supposedly unique to humans. The anti-essentialist argument is used to make the case that any attempt to find the defining "spark of humanity" is futile—there is no such spark, because species cannot be defined or distinguished in that way. Although these arguments obviously have relevance to any evolutionarily oriented approach, I am not concerned with the comparative argument specifically here. Instead, I want to focus on another objection from within evolutionary biology, one that hinges on the way in which the concept of "human nature" is pressed into service as the opposite of culture, creating a distorted view of both development and evolutionary processes; a view that remains incredibly resilient in the face of all attempts to eradicate it (see, e.g., the interviews with Kevin Laland and Patrick Bateson in Fuentes & Visala 2016). Specifically, I am interested in the extent to which recent attempts to define and describe human nature promote a dichotomous separation of nature from culture (an interesting question given that, as Lévi-Strauss pointed out, the separation of nature from culture is itself a cultural decision), and the tension between viewing humans as members of a single evolutionary species while simultaneously recognizing them as the product of highly particular historical processes.

This brings us to the objections from the social sciences. These rest in part on the way anthropologists of the 19th century used biological differences between human populations to argue for essential differences between peoples, and rank them according to an evolutionary scheme in which they became progressively more "civilized"; an interesting twist on the philosophical objection to essentialism (see, e.g., Marks 2009, 2013 for discussion). Although these ideas are now thoroughly discredited, there is continuing objection to the notion that it is possible to generalize across human populations, because this fails to recognize that they are highly specific products of time and space. There is no "human nature" that escapes the influence of culture, thus any attempt to naturalize humans is seen as incoherent and, what is more, continues to smack of the same kind of biologically rooted essentialism that led to the racist theories of earlier times (Bloch 2005, 2012; Ingold 2006, 2008; see Kronfeldner, Chapter 15 this volume). Even if these concerns can be put aside, there is still a worry that attempts at generalization will not only be inaccurate but also completely illusory, given that the attempt to construct a general view of human life will have been filtered through the culturally specific lens of the anthropologist undertaking the effort. Thus, over the course of the 20th century, there has been a turn away from theorizing about people in general, and anthropology has almost entirely focused on ethnography: its mandate is to carefully and sensitively document diversity and interpret how other people understand their societies and make meaning for themselves within them—to particularize, not generalize (see Bloch 2005 and Ingold 2008 for a discussion of this issue).

The same argument against generalization has been applied to our understanding of the past. In *The Great Cat Massacre*, Robert Darnton's cultural history of 18th-century France, he emphasizes the importance of receiving a good dose of culture shock, one that shakes us out of our false sense of familiarity with the past (Darnton 2009). We need this, he argues, because "nothing is easier than to slip into the comfortable assumption that Europeans thought and felt two centuries ago just as we do today—allowing for the wigs and wooden shoes" (2009: 4). We therefore need to capture the "otherness" of our recent ancestors' ways of thinking, just as ethnographers do in their studies of contemporary humans. As Darnton suggests, the minute you come across a proverb or joke that was once common wisdom, but which now seems utterly bizarre, you have the means to tug at the threads of another way of thinking, and potentially unspool a strange and intriguing world view.

So far, so good. As anyone who had to read Shakespeare at school knows, the jokes are not funny and, as Laura Bohannan details in her classic article "Shakespeare in the bush," the "universal themes" embodied by his plays proved not to be so when she tried to explain the plot of Hamlet to some Nigerian Tiv elders (Bohannan 1966). Instead, the elders questioned every cultural assumption contained within the plot, leaving Hamlet's motives much more difficult to explain. (Right off the bat, the elders considered Gertrude's marriage to Claudius as the only sensible response to the death of Hamlet's father.) Although Bohannan's piece is intended to demonstrate the dangers of assuming that others will think as we do, it also provides a beautiful illustration of Darnton's point: the Tiv elders were all too willing to "slip into the comfortable assumption" that the antics of Hamlet, Gertrude, and Claudius were explicable in terms of their own cultural traditions. It seems, then, that we continually run the risk of misrepresenting how other people think and feel, and there is no way to come up with a general understanding of what humans are like. If true, this creates a real problem for any evolutionary analysis of human behavior and psychology. If we cannot understand how our contemporaries think, or our ancestors from only 400 years ago, how can we possibly hope to understand how people thought 40,000 years ago, or 400,000?

But is it true? After all, people across the world continue to read and watch Shakespeare: they understand Hamlet's indecision, along with Othello's jealousy and Lady Macbeth's corrosive ambition. Although they questioned Hamlet's way of going about it, the Tiv elders nevertheless understood his desire to avenge his father's death. By the same token, although history can sometimes be opaque, we can and do make sense of past events on their own terms—they are not completely inexplicable to us, and historians are continually on their guard against Whiggish interpretations in which the "triumphs" of the present are pre-figured in the past. Anthropologists and their informants also manage to understand each other and communicate in their everyday interactions, even if their interpretations of precisely what goes on might differ. Surely, then, it is not too far-fetched to suppose that all humans have at least something in common, and that this can transcend both history and place?

HUMANS AS "INTERESTINGLY SIMILAR"

The notion that humans are, in fact, "interestingly similar" (as Edouard Machery puts it (2016: 63)) has driven most of the recent evolutionary theorizing on the nature of human nature. Machery (2008), for example, has developed what he calls a non-essentialist, "nomological," or descriptive concept of human nature. According to this account, human nature just is the set of species-typical attributes that humans tend to possess as a result of evolutionary processes. Thus, being bipedal is part of human nature, as is the capacity to speak, our tendency to startle when we hear an unexpected noise, along with our dependence on culture. Notably, this definition includes traits that we share with other species (i.e., it is not attempting to pick out human uniqueness), and it says nothing about the precise nature of these evolutionary processes; the traits that comprise human nature can be the products of natural selection, genetic drift, byproducts of adaptations, or the outcome of developmental constraints. The nomological account also recognizes that the traits that make up human nature can change over time. Machery therefore likens his concept to a "field guide" for humanity much like those used in ornithology (2008: 323), designed so that "we don't lose track of the shared aspect of human existence" (Machery 2016: 58). In this sense, the nomological concept maps onto views of some prominent social anthropologists, who have argued similarly that anthropology should be more than just ethnographic description and that some form of "acceptable generalization" is not only possible but necessary (Bloch 2005, 2012; Ingold 2008). In contrast, Tim Lewens has argued against Machery's view on several grounds, most notably because it explicitly excludes traits that are culturally produced or socially learned, thus reinforcing an unhelpful nature–nurture dichotomy, and seemingly denying that learning is itself the product of an evolutionary process (Lewens 2012). As we shall see below, this division between nature and culture is, however, often the precise reason why certain authors defend the human nature concept.

In contrast to this wholly descriptive approach, Richard Samuels (2012) takes Machery's "field guide" position one step further, defending what he calls a "causal essentialist" concept: in this view, human nature is the suite of mechanisms that underpin the species-typical suite of traits identified by the nomological concept. These mechanisms are themselves a species-typical suite of traits (e.g., language production systems) that can explain why similarities and differences are manifest at the surface level (e.g., differences in the languages that are actually spoken). As with Machery, the main aim here is to enable the similarities that exist across all humans to be considered within an evolutionary framework, by showing that these do not automatically fall foul of the "bad" biological essentialism identified by Hull (1986).

By far the most vehement and vociferous defenders of the concept of human nature, however, are members of the Santa Barbara School of Evolutionary Psychology (e.g., Tooby & Cosmides 1990a, 1990b, 2005; Buss 2005; Pinker 2003). (Indeed, the nomological and causal essentialist views within philosophy can be regarded as the "dialing down" of the Santa Barbara human nature concept; see also Samuels 1998, 2000.) The original impetus for an evolutionary concept of human nature was an immense frustration with the anthropological taboo on generalization, and a particular disdain for Clifford Geertz's (1973, 2000) approach to interpreting cultures as texts. John Tooby and Leda Cosmides (1992) therefore set about developing a manifesto for the study of our evolved human nature, as a way to tie the social sciences back to biology. More specifically, their aim was to integrate anthropology and sociology into psychology under the banner of evolutionary theory. To do so, and somewhat ironically perhaps, they imported wholesale Adolf Bastian's notion of the "psychic unity of mankind," one of the few organizing principles of anthropology (i.e., the idea that genetic differences cannot explain the manifest differences in behavior seen across different populations, and that all such differences are cultural in origin). From there, however, they went on to demolish what they referred to as the "Standard Social Science Model" (SSSM) in favor of their own evolutionary "Integrated Causal Model."

THE CUPCAKE MODEL OF MIND

The SSSM was presented as a naïve "Blank Slate" view. Tooby and Cosmides argued that the social sciences (mostly characterized by sociology and anthropology) regarded humans as infinitely malleable, and shaped entirely by culture through a small number of general-purpose learning mechanisms. As this move revealed human nature to have no content, it removed human nature as a legitimate and worthwhile area of study because there was simply nothing there *to* study: human nature was "an empty vessel, waiting to be filled by social processes" (Tooby & Cosmides 1992: 8). Anthropologists turned to the study of culture alone because of an adamant refusal to accept any biological influences on human behavior. Why study the blank page, as Tooby and Cosmides would have it, when you could study what was written on it? (See also Pinker 2003.)

Whether anyone, anywhere, ever subscribed to this grossly simplistic blank slate view is open to debate,[2] but it served its rhetorical purpose. Tooby and Cosmides (1992) were able to offer a radically different model of a universal human nature, constituted by a large number of highly specialized evolved (psychological) information-processing mechanisms, each of which represented a solution to the set of recurring problems faced by our ancestors. In many of their writings Tooby and Cosmides also mention the innate contents of our psychological mechanisms (Tooby & Cosmides 1990a; Tooby et al. 2005); that is, dedicated stores of knowledge built in by natural selection to help solve problems that would be impossible to achieve through individual learning—for example, the fitness costs of inbreeding with close relatives (see L. Barrett et al. 2014 for an argument against this position). Tooby and Cosmides's (1992) model was, then, an unapologetic argument for an evolved biological core of human traits, with cultural differences seen as the variable output generated in response to variable environmental inputs (which presumably themselves are often cultural, as well as simply climatic or environmental). This was brought out even more prominently in their notion of "evoked culture," with its use of a jukebox metaphor, where environmental inputs select the variant of our evolved psychology that will manifest in a particular setting, much as a single jukebox can play a variety of records.

In this and subsequent work, Tooby and Cosmides took great pains to explain why their view was not deterministic, as well as showing why it neither denied nor constrained the production of behavioral variability (Tooby et al. 2003; Tooby & Cosmides 2005). As they note, all traits result from the interaction of genetic and environmental inputs, and there is nothing in their formulation that denies this. In addition, they have argued that selection has acted on the entire developmental system, and not genes alone, and that each generation thus represents a process of "design reincarnation" whereby the phenotype is constructed through the interaction of genetic and environmental resources (see also H.C. Barrett 2006). While the incorporation of these elements ensures their argument is robust against simplistic accusations of genetic determinism, there remains a clear-cut divide between nature and culture: in their view, culture is not something that penetrates or alters our "human nature" in any fundamental way. Tooby and Cosmides are not alone in this, however, as Peter Richerson (forthcoming) identifies a similar tendency in all modern evolutionary theorizing about human nature; a stance he attributes to a desire to avoid acknowledging the inadequacies of standard evolutionary theory (i.e., the Modern Synthesis) in accounting for human behavior, and a resistance to the idea of an Extended Evolutionary Synthesis.

Another notable aspect of Tooby and Cosmides's formulation is that they seem to leave no room for human agency—cultural artifacts, rituals and beliefs are not the product of human ingenuity or foresight, but simply the manifestation of our evolved nature. Although Tooby and Cosmides concede that some knowledge is transmitted by social processes, they refer to this as "epidemiological culture," where certain ideas are more likely than others to "infect" us and get passed on because they accord with our evolved psychology, hence psychology shapes culture, but not vice versa (see Sperber 1996 for a more nuanced view relating to ideas of "cultural attraction," along with Morin 2016). More recently, however, Cosmides and Tooby have suggested that humans are also in possession of "improvisational intelligence" (i.e., the ability to come up with solutions to entirely unprecedented evolutionary novel problems), which evolved via the bundling together of a number of evolved functional specializations, and embedding them in a cognitive architecture that has a scope syntax, that is, "an elaborate set of computational adaptations for regulating the interaction of transient and contingent information sets within a multi-modular mind" (2002: 147). As much as culture is deemed "natural" by Tooby and Cosmides, then, both their original model and the more "improvisational" model of the mind resemble a cupcake, with our evolved biological nature as the sponge base onto which the variable icing of culture is plopped. This is why, for example, Steven Pinker, advocating for a universal human nature, is able to argue that "our brains are not wired to cope with anonymous crowds, schooling, written language, governments, police, courts, armies, modern medicine, formal social institutions, high technology and other newcomers to human experience" (2003: 42).

More recent work by H. Clark Barrett builds on Tooby and Cosmides's notion of an evolved human nature by incorporating aspects of developmental systems theory (Griffiths & Gray 1994). In this way, he attempts to deal with the cupcake objection head-on: "there is no evolved part of the mind and then the cultural part added on like icing on a cake; culture and experience percolate throughout. And conversely, evolved mechanisms have to be part of the explanation for mental functioning all the way up" (H.C. Barrett 2014: 261). Barrett then presents an argument in which environmental inputs (culture and experience) lead to functional specializations that are attuned to present circumstances, via the developmental process.[3] The opposition between nature and culture remains, however, because culture is still viewed simply as an input to which our evolved architecture responds during development, and an output generated by an evolved developmental process. Although Barrett argues convincingly that any evolved developmental system is shaped by and incorporates the influence of external environmental resources available to the organism, along with internal genetic and other cellular resources, there is nothing in his argument that incorporates cultural factors in a more thoroughgoing way, that is, as something that can shift or alter the structuring or the components of the developmental system itself. In his terms, "phenotype space" is the visible manifestation of the area of multidimensional "design space" that environmental and genetic resources tap into, but the design space itself remains unchanged.

Thus, as Barrett notes elsewhere (Frankenhuis et al. 2013), the incorporation of developmental systems theory into his theoretical framework is self-confessedly "soft," in that the developmental system remains internal to the organism. As such, it continues to promote a dichotomous account of development, where two classes of resources—genes and "all the rest"—interact to produce the adult phenotype. This is quite unlike the "hard" developmental systems theory (DST) formulation where genes are seen as one resource among many available to the developmental process, and not always as its central drivers (Griffiths & Gray 1994; Griffiths & Stotz 2000); that is, DST moves away from drawing a rigid contrast between genes and environment, instead emphasizing that genetic and environmental resources can often play virtually equivalent roles (a classic example being phenocopying, whereby similar changes in phenotype can be brought about by changes in the environment or due to mutation—for example, bithorax mutants in *Drosophila* can be induced by ether shocks, in a way that does not involve generating mutations in the genes). Thus, in "hard" DST, the developmental system is no longer internal, but comprises the sum total of resources that go into constructing the phenotype in each generation, which means it includes any and all cultural artifacts and practices, along with the social institutions that shape human life on earth. As a result, the organism plays a more active role, shaping its own environment in ways that can help guide its future evolution. This opens the way for human agency (i.e., the capacity to make choices, and engage in goal-directed activity) to enter into the process, rather than the environment simply evoking a particular kind of response.

Thus, according to this view, cultural artifacts and practices are constitutive of human nature—they literally make us what we are (see below)—and do not act simply as inputs or outputs. Another way to put this is that they serve to actively enlarge the design space, generating more than variations on a single theme, but entirely new modes of thought and behavior. Although Barrett (2014) argues explicitly against an essentialist view of human nature on the grounds proposed by Hull (1986), there remains an essentialist tinge to the idea that cultural (phenotypic) variation is the product of a universal, evolved "design space" in which there is no causal influence that can feedback from cultural practices and so alter the nature of the design space itself. As with Tooby and Cosmides, the view on offer seems to actively promote a particular kind of nature–culture dichotomy, even as it works to undermine it.

Moreover, the very abstract concept of "design space" used by Barrett seems to suffer from the some of the same conceptual problems that are often laid at the door of social anthropologists (see, e.g., Bloch 2005, 2012 for a discussion of this from the latter perspective). That is, it seems to generate a form of idealism that strips humans of the historical and cultural trappings that help make them what they are, in much the same way that structuralist theories focus exclusively on signification and symbolism, removing these away from the particular human activities with which they are associated (although see Bloch 2012 for an interesting discussion of Lévi-Strauss as an evolutionary anthropologist).

This is something that can be seen very clearly in the Evolutionary Psychological-notion of human nature (EPNHN) recently proposed by John Klasios (2016). This incorporates elements of Machery's (2008) and Samuels's (2012) concepts of human nature, with Richard Boyd's (1999) more general ideas concerning homeostatic property clusters,[4] and combines these with Barrett's (2014) developmental systems approach. Although Klasios makes all the right noises about cultural variability, phenotypic plasticity, and the ability to deal with novel environments, it is telling that he goes on to argue that the EPNHN construal will enable us to capture the "truly universal aspects" of our psychological adaptations because "it would peel away the historically-, ecologically-, and culturally-contingent aspects which feed into the open-ended parameters of the adaptation that ultimately generate the tokens and instead reveal the invariant aspects which actually constitute the function of the adaptation" (2016: 110). This view is problematic because it seems to suggest that we can identify a "pure" human nature (a form of Platonic ideal) that is separate from the contingent, messy world in which human beings actually live (see also Bloch 2005; Derksen 2007). Furthermore, it suggests that it is actually possible to disentangle culture from nature, a suggestion akin to an attempt "to uncook spaghetti" (Peter Richerson, pers. comm.); given that we have been cultural creatures since the emergence of our genus, and that cultural processes have been crucial to our evolutionary trajectory, we cannot now "uncook" the causally entangled genetic and cultural components of human spaghetti, and "somehow recover the separate straight bundles of genes and culture" (Richerson, pers. comm.).

GETTING CULTURE OUT OF THE HEAD

Other evolutionary views of human nature do not fall into the same trap because they incorporate culture more fully into their theoretical framework. Boyd and Richerson's "dual-inheritance" or gene-culture co-evolutionary theory (Boyd & Richerson 1985; Richerson & Boyd 2005), for example, along with subsequent work by Joseph Henrich (2015), states that behavior is a result of both genetic and cultural influences, and incorporates both genes and culture as distinct but intertwined inheritance systems. Dual inheritance theories also conceive of a universal set of psychological adaptations but, in this case, they comprise a small suite of specialized social learning mechanisms that make cultural evolution possible. Hence, these views do not argue for the same level of functional specialization as current thinking in evolutionary psychology.

According to Maarten Derksen (2007), however, the dual inheritance approach encounters a similar problem to classic evolutionary psychology. Derksen locates this in the way that culture is defined solely as social information (knowledge, beliefs, values) transmitted between individuals via teaching and imitation. As a result, although biology is influenced by culture and culture influences our biology, Derksen argues that it does so via a mind that acts simply as conduit between the two. Although the characteristics of this "channel" modify the mutual influence of genes and culture on each other, Derksen notes that this formulation grants humans only a very limited amount of individual agency. To this, I would add that perhaps it is not agency so much as a lack of the reflexivity that is typical of human engagement; that is, the way that our knowledge about ourselves serves to alter the manner in which we think, act, and construct our identities (Hacking 1995; Brinkmann 2004). It may also be the case that human agency is over-rated, and human choices are more constrained than we realize. In addition, formal modeling inevitably requires a number of simplifying assumptions in order to be tractable, and does not aim at capturing the full complexity of a given system. Derksen's criticism, although valid at one level, may also misunderstand the aim of gene-cultural co-evolutionary modeling, which is to appreciate how cultural processes give rise to new evolutionary dynamics. Finally, Derksen also seems to underestimate the extent to which Boyd and Richerson (1985, 2005) have modeled various bias forces (e.g., guided variation, context and content biases) in order to reflect human choice-making processes.

Having said this, there are some gene-culture co-evolutionists who sometimes seem to favor a position similar to the more hard-core evolutionary psychologists. Henrich (2015), for example, identifies a key set of evolved social learning mechanisms that generate the capacity for culture (and hence give rise to our "cultural nature"), which he links to other "innate" psychological tendencies. For example, among other things, Henrich suggests that "marriage norms found in most societies—for better or worse—operate to reinforce our otherwise flimsy pair-bonding instincts" (2015: 150), that "stripped of our social norms and beliefs, we aren't nearly as cooperative and cultural as we might seem" (154), and that "there is reason to suspect that we humans have an innate susceptibility to picking up meat aversions" (158). Such a view, perhaps inadvertently, suggests that it remains possible to divide up human tendencies into those that are biological (genetic) and those that are cultural, despite the overarching notion of genes and culture as deeply intertwined, and exerting complex forms of feedback on each other. As we have seen, for evolutionary psychologists, this is unproblematic. For social and cultural anthropologists, however, it is precisely this demarcation of genetic from cultural resources that continues to provoke their ire. Tim Ingold (2007), for example, considers all human behaviors to be as biological as each other—whether this be learning to walk or learning to play the cello—because both emerge as a result of a developmental process that takes place in a rich, supporting environment. He therefore argues against gene-culture co-evolutionary theory because, in his view, it continues to treat culture as some form of "add-on" to an evolved biological substrate. Ingold (2013) also disputes the idea that humans possess a general "capacity for culture" that somehow exists in advance of the diverse social practices that come to fill it. As he sees it, such theorizing is simply an attempt to prop up a fatally flawed neo-Darwinian approach that cannot account for the complexities of human lifeways. Ironically, given that he is firmly in the gene-culture co-evolution camp, Richerson (forthcoming) identifies the very same problem in many modern evolutionary views of human nature: specifically, his claim is that they contain a "Modern Synthesis fundamentalism" that insists on characterizing "human nature" as those abilities that are not penetrated or changed by culture, thus allowing the standard evolutionary model to persist unchallenged. (Richerson's brand of gene-culture co-evolution escapes this criticism precisely because it recognizes and models cultural evolutionary effects.) There is, then, more common ground between some gene-culture co-evolutionary theorists and social anthropologists than first appearances would suggest. This is tempered by the fact that, Ingold, no doubt, would continue to revile the recognition of culture as a separate evolving system, while Richerson would question the idea that one can meld genes with culture as seamlessly as Ingold suggests, given the greater conservatism of genes compared to the social and cultural practices. Richerson, and other gene-culture co-evolutionists would also maintain that psychological traits related to social learning and teaching are independent of the cultural practices they enable.

Another point of difference between Ingold (along with other social anthropologists) and gene-culture co-evolutionists is that, until very recently, the latter have worked with a purely psychological definition of culture (i.e., culture as information in the head) that excludes material artifacts and behavior from consideration (i.e., aspects of our embodied experience of the world). In this view, tools and other artifacts are considered to be the products of culture, but not potential producers, shapers, or even constituents of human cultural and psychological processes. That is, although cultural processes enable and sustain the manufacture of material artifacts, the artifacts themselves have not been considered "psychologically," that is, to partly constitute human cognitive systems (e.g., Malafouris 2013). There is recent evidence to suggest, however, that this view is being reconsidered. Henrich, for example, describes how the use of an abacus "provides a mental prosthesis that, by harnessing aspects of our visual memory, can deliver mental powers that seem almost unimaginable to those unfamiliar with this simple but elegant tool" (2015: 230). What remains unclear is the extent to which these "mental powers" are seen as the plastic, flexible manifestations of an evolved psychology attuned to pick up various kinds of cultural information and social practices, or are viewed as being transformative of human cognition itself (i.e., as generating entirely new kinds of minds). That is, to what extent might there be cumulative ratchet-like effects on human cognitive functioning, as a consequence of incorporating external tools and resources into our cognitive system (e.g., ideas relating to the "extended mind": Clark 2004, 2008; Menary 2010; see also Malafouris 2013) in the same way that we see cumulative ratchet effects on other forms of material culture?

Arguing along somewhat similar, although more classically psychological, lines, Cecilia Heyes (2012) has called into question the idea that humans have evolved highly specific social learning mechanisms as adaptations for functioning in cultural groups, suggesting instead that these are themselves the outcome of cultural learning processes. That is, she questions the idea that we possess biologically inherited "mills" (the processes that allow us to learn from others), which enable us to inherit culturally the "grist" of our lives (the knowledge and know-how needed to deal with the world). In contrast, she argues that both grist *and* mills are culturally inherited: our social learning mechanisms are cultural adaptations, not evolved functional specializations. Using the example of imitation, Heyes argues that a combination of perceptual biases and "asocial" general learning mechanisms can result in abilities equivalent to the specialized social-learning adaptations proposed by gene-culture co-evolution theorists, and argues that this provides an added advantage of greater built-in flexibility to respond to changing circumstances.

Similarly, Daniel Hutto's (2008) theory of narrative practice argues that Western folk psychology—the way we make sense of others by attributing private and invisible mental states—is a product of the stories we are read and taught as children: we talk about minds in a particular way as a result of this particular kind of social participation (an idea first expressed by Bruner 1990). In this account, the hidden, private, internal mind that we take as biologically given (and which is treated as a distinct modular adaptation by evolutionary psychologists) is a sociocultural construction generated by the social, public process of "minding" (see Heyes & Frith 2014 for a similar argument). For Hutto (2008), as for B.F. Skinner (1957, 1976) before him, "mind" is simply a way that we have learned to speak about our actions in the world as part of our cultural heritage. Heyes's and Hutto's ideas, although theoretically speculative at present, make a strong case for more serious consideration of how various sociocultural practices must be folded into, and considered as constitutive of, human cognition, and not merely causally related to it.

It is also important to note that these views do not deny a role for the evolved nature of the human brain. Instead, they emphasize mutual enhancement and integration: the nature of the human brain is instrumental in the process of creating minds (and thus we should not expect to find chimpanzees to be capable of developing fully human-like minds, even with appropriate social and cultural scaffolding), but minds cannot be reduced to brain function alone. In addition, other aspects of our embodiment may contribute crucially to our cognitive success, by affecting to our ability to construct and use material artifacts; our hands are obvious examples here, and have been considered as "organs of the mind" (Zdravko 2013).

COGNITIVELY INTEGRATED CULTURAL PRACTICES

The idea that human cognitive systems (or, if you prefer, minds) are not limited to brains alone but incorporate various environmental resources and tools—that is, the notion of "cognitive integration" (Menary 2010) or the "extended mind" (Clark 2008)—seems crucial to the generation of a satisfactory evolutionary account of human behavior, not least because there is abundant evidence to suggest that the invention of certain artifacts and practices has transformed our understanding of ourselves and the world around us. That is, we come to think both through and with objects, and we can document how human thought has changed in profound ways, across both evolutionary and historical time, through the invention and use of technologies of various kinds. Lambros Malafouris, for example, argues that human ideals of bilateral symmetry with respect to hand-axe manufacture may have been the product of stone hand-axe production, rather than a reflection of the prior intentions of the knapper about how an axe should look—that is, the stones themselves help shape knappers' choices by presenting suitable striking surfaces. Notions of intentionality thus cut across the standard categories of the internal mind and its external products. Instead, it is the "directed action of stone knapping . . . [that] brings forth the knapper's intention" (Malafouris 2010: 17; see also Malafouris 2008; Clark 2010).

In a similar but perhaps less radical vein, Kevin Laland et al. (2000) have made a convincing case that human evolutionary history should be viewed as a process of (ongoing) niche construction: the world to which humans have adapted over evolutionary time has been largely one of our own making. We are, as Kim Sterelny argues, "creatures of feedback" (2012: 75). Indeed, for Sterelny, "human nature" lies precisely in the uniqueness of the feedback mechanisms that connect, among other things, the cultural environments we construct and inhabit, human social learning, individual expertise, and human life history processes. As he notes: "Human brains are *developmentally* plastic, so transforming hominid developmental environments transformed hominid brains themselves. As hominids made their own worlds, they indirectly made themselves" (Sterelny 2003: 17). Here, then, human behavior and material culture act as selection pressures on human brains and bodies, and do not simply represent their proximate products, or the "novel" inputs to which our evolved human natures must respond.

Karola Stotz (2010) has recently combined these ideas, and incorporated them with developmental systems theory, to generate the concept of the "ontogenetic niche" (an idea first proposed by West & King 1987). This can be viewed as the suite of resources, both genetic and extra-genetic, that is vital to the reproduction of a life style. The use of the word "niche" brings in both the way of life of the organism and the environment in which the young are raised. The key claim here is that human nature does not lie in the possession of particular kinds of traits, but is instead represented by a particular kind of developmental process: young humans are born into a niche that has been "epistemically engineered" for them in various ways (i.e., there are various "non-neural structures

[that] are used to transform the shape of the problem-solving activity required of individual brains" (Wheeler & Clark 2008: 3565)). Past generations thus structure and scaffold the developmental context of those that succeed them, providing resources that are essential to the production of species-typical behavior. In addition, past generations provide ever more sophisticated forms of cognitive scaffolding that itself augments the scaffolding that previous generations bequeathed to them, and by this means they enhance what can be achieved by future generations (Griffiths & Gray 1994; Sterelny 2012). This process therefore gives rise to adults who possess commonalities with their parents (due to a similar developmental process), but who will also differ from them (often quite dramatically at times) due to changes in the suite of resources comprising its ontogenetic niche.

My colleagues and I have argued for something similar, using Derksen's (2005, 2007) notion of human nature as a process of "cultivation" over both evolutionary and historical time (L. Barrett et al. 2014). The ongoing construction of developmental niches, and the flexible response to them, also means we come to a different picture of the ultimate source and cause of human behavior as public shared phenomena, rather than simply as private knowledge and beliefs residing inside individual heads (L. Barrett et al. 2014); that is, cultural phenomena simultaneously exist as both shared social practices and as individual knowledge, and forcing a distinction between these two (as evolutionary psychological theorizing does, with its commitment to a strongly individualistic adaptationist stance based on the Modern Synthesis) is as false as the opposition between nature and culture (see also Boyd et al. 2011, for allied criticism of this strongly individual cognitivist stance). It also suggests that we need to take seriously Ian Hacking's (1995) "dynamic nominalism" or "looping effect"—the idea that our practices of labeling and naming interact with the things so named. More specifically, the idea is that, in describing ourselves psychologically, we are able to react to these self-descriptions in ways that generate or change the kinds of people that we are (Sugarman 2009); the developmental niche concept can no doubt be extended to include this process also.

Finally, it suggests that ideas of evolutionary "mismatches" of the kind identified by Pinker (2003) are misplaced. Such arguments imply that, having achieved perfectly optimal solutions to life's problems during our ancestral past, we have now fallen from this adaptive peak as a consequence of recent cultural evolutionary processes. Instead, as Boyd and Richerson (1985; Richerson & Boyd 2005) have detailed, a certain proportion of maladaptive behavior is the price we pay for learning socially, cumulatively, and culturally, and this is something that characterizes the entire history of our species; it is not a recent manifestation generated by rapid cultural change that leaves our "stone age minds" out of kilter with the modern world.

All of this means that, as Michael Wheeler and Andy Clark put it: "A child whose early experience is shaped by the special environments provided by books and software programs, and whose own emerging cognitive profile favors certain elements within that culturally enabled nexus over other elements, will end up with a cognitive system that is not just superficially, but profoundly different from that of a differently encultured child" (2008: 3572)—and this is especially true if we recognize that the cognitive system includes the extra-neural body as well as the brain, where aspects of nutrition and exercise can exert equally powerful effects on both. Put more broadly, this "neuroconstructivist" perspective (e.g., Sporns 2007) suggests that what characterizes human psychology is not some specific suite of traits acquired over the course of our evolution, but the ability to enter into "deep, complex and ultimately architecture-determining relationships with an open-ended variety of culturally transmitted practices, endowments and non-biological constructs, props and aids" (Wheeler & Clark 2008: 3572). Material culture can thus transform, enhance, and augment human cognitive functioning, and cultural practices are therefore both cause and consequence of our psychological and cultural variability (L. Barrett 2011; L. Barrett et al. 2014; Malafouris 2008, 2010, 2013). This results in a self-transforming, "snowballing/bootstrapping" extended cognitive architecture "whose constancy lies mainly in its continual openness to change" (Wheeler & Clark 2008: 3572). At present, some of the most prominent evolutionary theorizing about the human mind and nature leaves out the more material, embodied aspects of human cognitive life by treating culture as a mode of thought, rather than a means of active bodily engagement with the world. By contrast, a more relational, embodied view regards social and material practices to form a central element in understanding why humans think and act in the ways that we do; one that can help form a bridge between the social and natural sciences, without eradicating or denying the relevance of either one.

CONCLUSION

Human nature, if it is anything at all, is an ongoing and historically contingent process: we are constantly being transformed and adapted over the course of our individual development, as well as over historical and evolutionary time, by the inclusion of various psychological tools and material culture into our daily lives. As we take on the social practices that give rise to new psychological tools, so we take on the entire history of our culture. In this way, human lives become variable across space and time, and do not conform to any "universal" human nature. This process can explain why we are different psychological creatures to the peoples of 400, 40,000 and 400,000 years ago, while people 4000 years hence will be very different psychological creatures to us. Yet, this shared evolutionary and historical ancestry—this taking on of our cultural history—also links us together across time and space, allowing us to recognize and sustain some form of common humanity. Our lives are an entanglement of our biological bodies and brains with the cultural practices and contingent historical events to which they help give rise. The "universality" of human minds therefore inheres in the processes by which they are created and maintained: through social interaction, within a rich cultural milieu.

ACKNOWLEDGMENTS

Many thanks to Pete Richerson and Gert Stulp and for discussions on this and related issues, and for their very constructive comments and edits on the manuscript that helped improve the final product. Any remaining errors are, of course, my own. My work is supported by the NSERC Canada Research Chair and Discovery Grant programs.

NOTES

1. Although it should be noted that Hull did endorse the notion of some form of human nature, stating that "If by 'human nature' all one means is a trait which happens to be prevalent and important for the moment, then human nature surely exists" (Hull 1986: 9).
2. Tooby and Cosmides concede that even Geertz did not subscribe to the blank slate, although they do insist he was wedded to the idea of the mind as a "general purpose computer," which in their view is just as bad.
3. Notably, many aspects of Barrett's reconceptualization of classic evolutionary psychological theory bear a strong resemblance to Karmiloff-Smith's (1984) classic argument for progressive modularization across development, which seems to undercut the claim that evolutionary psychology promises a new and revolutionary approach to understanding the human mind.
4. The homeostatic property cluster view suggests that natural kinds are a consequence of shared causal mechanisms rather than shared essences. That is, a kind is composed of a cluster of properties shared by a group of entities, along with a causal mechanism that accounts for why these shared properties should co-occur. It should be apparent that Samuels's account of human nature is very close to this.

REFERENCES

Barrett, H.C. 2006. "Modularity and design reincarnation." In P. Carruthers, S. Laurence, & S. Stich (eds.), *The Innate Mind: Culture and Cognition* (Oxford University Press) 199–217.

Barrett, H.C. 2014. *The Shape of Thought: How Mental Adaptations Evolve* (Oxford University Press).

Barrett, L. 2011. *Beyond the Brain: How Body and Environment Shape Animal and Human Minds* (Princeton University Press).

Barrett, L., Pollet, T., & Stulp, G. 2014. "From computers to cultivation: Reconceptualizing evolutionary psychology." *Frontiers in Psychology* 5: doi: 10.3389/fpsyg.2014.00867.

Bloch, M. 2005. "Where did anthropology go? Or the need for human nature." In M. Bloch (ed.), *Essays on Cultural Transmission* (Berg) 1–20.

Bloch, M. 2012. *In and Out of Each Other's Bodies: Theory of Mind, Evolution, Truth, and the Nature of the Social* (Paradigm).

Bohannan, L. 1966. "Shakespeare in the bush." *Natural History* 75: 28–33.

Boyd, R. 1999. "Homeostasis, species, and higher taxa." In R. Wilson (ed.), *Species: New Interdisciplinary Essays* (MIT Press) 141–186.

Boyd, R. & Richerson, P. 1985. *Culture and the Evolutionary Process* (University of Chicago Press).

Boyd, R. & Richerson, P. 2005. *The Origin and Evolution of Cultures* (Oxford University Press).

Boyd, R., Richerson, P., & Henrich, J. 2011. "The cultural niche: Why social learning is essential for human adaptation." *Proceedings of the National Academy of Sciences* 108 (suppl. 2): 10918–10925.

Brinkmann, S. 2004. "Psychology as a moral science: Aspects of John Dewey's psychology." *History of the Human Sciences* 17: 1–28.

Bruner, J. 1990. *Acts of Meaning* (Harvard University Press).

Buss, D. 2005. *The Handbook of Evolutionary Psychology* (John Wiley & Sons).

Clark, A. 2004. *Natural-born Cyborgs: Minds, Technologies, and the Future of Human Intelligence* (Oxford University Press).

Clark, A. 2008. *Supersizing the Mind: Embodiment, Action, and Cognitive Extension* (Oxford University Press).

Clark, A. 2010. "Material surrogacy and the supernatural: Reflections on the role of artefacts in 'off-line' cognition." In L. Malafouris & C. Renfrew (eds.), *The Cognitive Life of Things: Recasting the Boundaries of the Mind* (David Brown) 23–28.

Cosmides, L. & Tooby, J. 2002. "Unraveling the enigma of human intelligence: Evolutionary psychology and the multimodular mind." In R. Sternberg & J. Kaufman (eds.), *The Evolution of Intelligence* (Erlbaum) 145–198.

Darnton, R. 2009. *The Great Cat Massacre: And Other Episodes in French Cultural History* (Basic Books).

Derksen, M. 2005. "Against integration: Why evolution cannot unify the social sciences." *Theory & Psychology* 15: 139–162.

Derksen, M. 2007. "Cultivating human nature." *New Ideas in Psychology* 25: 189–206.

Frankenhuis, W., Panchanathan, K., & Barrett, H.C. 2013. "Bridging developmental systems theory and evolutionary psychology using dynamic optimization." *Developmental Science* 16: 584–598.

Fuentes, A. & Visala, A. (eds.) 2016. *Conversations on Human Nature* (Left Coast Press).

Geertz, C. 1973. *The Interpretation of Cultures* (Basic Books).

Geertz, C. 2000. *Local Knowledge: Further Essays in Interpretive Anthropology* (Basic Books).

Ghiselin, M. 1997. *Metaphysics and the Origin of Species* (SUNY Press).

Griffiths, P. & Gray, R. 1994. "Developmental systems and evolutionary explanation." *The Journal of Philosophy* 91: 277–304.

Griffiths, P. & Stotz, K. 2000. "How the mind grows: A developmental perspective on the biology of cognition." *Synthese* 122: 29–51.

Hacking, I. 1995. "The looping effects of human kinds." In D. Sperber, D. Premack, & A. Premack (eds.), *Causal Cognition: A Multidisciplinary Debate* (Clarendon Press) 351–394.

Henrich, J. 2015. *The Secret of Our Success: How Learning from Others Drove Human Evolution, Domesticated our Species, and Made us Smart* (Princeton University Press).

Heyes, C. 2012. "Grist and mills: On the cultural origins of cultural learning." *Philosophical Transactions of the Royal Society, Series B: Biological Sciences* 367: 2181–2191.

Heyes, C. & Frith, C. 2014. "The cultural evolution of mind reading." *Science* 344: doi: 10.1126/science.1243091.

Hull, D. 1986. "On human nature." *PSA: Proceedings of the Biennial Meeting of the Philosophy of Science Association* 2: 3–13.

Hutto, D. 2008. *Folk Psychological Narratives* (MIT Press).

Ingold, T. 2006. "Against human nature." In N. Gontier, J.-P. van Bendegem, & D. Aerts (eds.). *Evolutionary Epistemology, Language and Culture* (Springer) 259–281.

Ingold, T. 2007. "The trouble with 'evolutionary biology.'" *Anthropology Today* 23: 13–17.

Ingold, T. 2008. "Anthropology is not ethnography." *Proceedings of the British Academy* 154: 69–92.

Ingold, T. 2013. "Prospect." In T. Ingold & G. Palsson (eds.), *Biosocial Becomings: Integrating Social and Biological Anthropology* (Oxford University Press) 1–21.

Karmiloff-Smith, A. 1984. "Children's problem solving." In M. Lamb, A. Brown, & B. Rogoff (eds.), *Advances in Developmental Psychology, Vol. 3.* (Erlbaum) 39–90.

Klasios, J. 2016. "Evolutionizing human nature." *New Ideas in Psychology* 40: 103–114.

Laland, K., Odling-Smee, J., & Feldman, M. 2000. "Niche construction, biological evolution, and cultural change." *Behavioral and Brain Sciences*, 23: 131–146.

Lewens, T. 2012. "Human nature: The very idea." *Philosophy and Technology* 25: 459–474.

Machery, E. 2008. "A plea for human nature." *Philosophical Psychology* 21: 321–329.

Machery, E. 2016. "Interview about human nature." In A. Fuentes & A. Visala (eds.), *Conversations on Human Nature* (Left Coast Press) 55–68.

Malafouris, L. 2008. "Beads for a plastic mind: The 'Blind Man's Stick' (BMS) hypothesis and the active nature of material culture." *Cambridge Archaeological Journal* 18: 401–414.

Malafouris, L. 2010. "Knapping intentions and the mark of the mental." In L. Malafouris & C. Renfrew (eds.), *The Cognitive Life of Things: Recasting the Boundaries of the Mind* (David Brown) 13–22.

Malafouris, L. 2013. *How Things Shape the Mind: A Theory of Material Engagement* (MIT Press).

Marks, J. 2009. *Why I Am Not a Scientist: Anthropology and Modern Knowledge* (University of California Press).

Marks, J. 2013. "The nature/culture of genetic facts." *Annual Review of Anthropology* 42: 247–267.

Menary, R. 2010. *The Extended Mind* (MIT Press).

Morin, O. 2016. *How Traditions Live and Die* (Oxford University Press).

Pinker, S. 2003. *The Blank Slate: The Modern Denial of Human Nature* (Penguin).

Richerson, P. Forthcoming. "The use and non-use of the human nature concept by evolutionary biologists." In E. Hanlon & T. Lewens (eds.), *Why We Disagree About Human Nature* (Oxford University Press).

Richerson, P. & Boyd, R. 2005. *Not by Genes Alone: How Culture Transformed Human Evolution* (University of Chicago Press).

Samuels, R. 1998. "Evolutionary psychology and the massive modularity hypothesis." *British Journal for the Philosophy of Science* 49: 575–602.

Samuels, R. 2000. "Massively modular minds: Evolutionary psychology and cognitive architecture." In P. Car-ruthers (ed.), *Evolution and the Human Mind: Modularity, Language and Meta-cognition* (Cambridge University Press) 13–46.

Samuels, R. 2012. "Science and human nature." *Royal Institute of Philosophy Supplement* 70: 1–28.

Skinner, B. F. 1957. *Verbal Behavior* (Appleton-Century-Crofts).

Skinner, B. F. 1976. *About Behaviorism* (Vintage).

Sober, E. 1980. "Evolution, population thinking, and essentialism." *Philosophy of Science* 47: 350–383.

Sperber, D. 1996. *Explaining Culture* (Blackwell).

Sporns, O. 2007. "What neuro-robotic models can teach us about neural and cognitive development." *Neuro-constructivism: Perspectives and Prospects* 2: 179–204.

Sterelny, K. 2003. *Thought in a Hostile World: The Evolution of Human Cognition* (Harvard University Press).

Sterelny, K. 2012. *The Evolved Apprentice* (MIT Press).

Stotz, K. 2010. "Human nature and cognitive–developmental niche construction." *Phenomenology and the Cognitive Sciences* 9: 483–501.

Sugarman, J. 2009. "Historical ontology and psychological description." *Journal of Theoretical and Philosophical Psychology* 29: 5–15.

Tooby, J. & Cosmides, L. 1990a. "On the universality of human nature and the uniqueness of the individual: The role of genetics and adaptation." *Journal of Personality* 58: 17–67.

Tooby, J. & Cosmides, L. 1990b. "The past explains the present: Emotional adaptations and the structure of ancestral environments." *Ethology and Sociobiology* 11: 375–424.

Tooby, J. & Cosmides, L. 1992. "The psychological foundations of culture." In J. Barkow, L. Cosmides, & J. Tooby (eds.), *The Adapted Mind: Evolutionary Psychology and the Generation of Culture* (Oxford University Press) 19–136.

Tooby, J. & Cosmides, L. 2005. "Conceptual foundations of evolutionary psychology." In D. Buss (ed.), *The Handbook of Evolutionary Psychology* (John Wiley & Sons) 5–67.

Tooby, J., Cosmides, L., & Barrett, H.C. 2003. "The second law of thermodynamics is the first law of psychology: Evolutionary developmental psychology and the theory of tandem, coordinated inheritances: Comment on Lickliter and Honeycutt (2003)." *Psychological Bulletin* 129: 858–865.

Tooby, J., Cosmides, L., & Barrett, H.C. 2005. "Resolving the debate on innate ideas." In P. Carruthers, S. Laurence, & S. Stich (eds.), *The Innate Mind: Structure and Content* (Oxford University Press) 305–337.

West, M. & King, A. 1987. "Settling nature and nurture into an ontogenetic niche." *Developmental Psychobiology* 20: 549–562.

Wheeler, M. & Clark, A. 2008. "Culture, embodiment and genes: Unravelling the triple helix." *Philosophical Transactions of the Royal Society, Series B: Biological Sciences* 363: 3563–3575.

Zdravko, R. (ed.). 2013. *The Hand, an Organ of the Mind: What the Manual Tells the Mental* (MIT Press).

15

The Right to Ignore: An Epistemic Defense of the Nature/Culture Divide

Maria Kronfeldner

THE EPISTEMIC FRAGMENTATION OF THE HUMAN

If we are lucky, we experience ourselves as a unity—as human beings. Yet, the moment in which the phenomenon of "being human" is transferred to science, this unity gets lost. In the disciplinary differentiated science of today, this unity has no home. As an *epistemic object* (i.e., an entity studied by sciences) human beings get lost. They literally *disappear* on their way to science. The phenomenon of "being human" becomes cut into pieces, apportioned, fragmented—into an evolved human nature, a culture, an immune system, a neuronal system, a mind, a society, etc. Through this *epistemic fragmentation* (as I call it) humans become epistemically "available" for science, but not *as* humans. In this sense, there is no unitary account of being human, no anthropology in the all-inclusive sense, no simple answer to what it means to be human in contemporary science. Evolutionary thinking, anthropology, sociology, psychology, history, philosophy, etc.— provide separate fragments of knowledge about humans, with often complex relations among the fragments.

This epistemic fragmentation of the human has often been judged negatively, first and foremost for existential and metaphysical reasons. What is lost is a unitary *Menschenbild,* a meaningful picture of (and vision for) humans. Such a unified picture of being human would provide an ultimate answer to who and how we are (and should be), a picture that would have to be synthesized out of the bits and pieces offered by the multitude of sciences, even if it would simultaneously transcend the bits and pieces by its existential or metaphysical orientation. Second, independent of the need that some might still feel for an existential or metaphysical vision for being human, the fragmentation of "being human" can be criticized on the basis of epistemic reasons, namely by stressing that it prevents us from gaining valuable scientific *knowledge* about humans (*Menschenkenntnis*). This epistemic issue is the subject of this chapter.[1]

Mary Midgley (1995: xxv–xxvi), for instance, in her introduction to a new edition of her *Beast and Man* (1978), deplores the "sharp division[s] between mind and body, between culture and nature, between thought and feeling." According to her, these conceptual splits are "the bad side of our inheritance from the Age of Reason" and the foundation for the epistemic fragmentation of the human. She writes:

> In modern times, science, because of its tremendous prestige, has been invoked to dramatize all these splits in a way that often has little to do with any real scientific work, but that seems to bring an unanswerable authority to the side that can exploit it. This . . . adds a damaging warfare between the "two cultures" to the general chaos, deepening the gaps already opened by specialization between different studies and generally fragmenting the intellectual scene in a way that wastes endless time and resources. (ibid.)

For Midgley, the gaps she mentions are unquestionably negative—because of the motivation for them and because of their consequences for producing knowledge about humans. Conceptual dichotomies such as the nature/culture divide are bad since they have to do with claiming authority (for detail, see Kronfeldner forthcoming) and because they lead to a waste of time and resources. In a similar vein, Clifford Geertz, in a famous paper "The impact of the concept of culture on the concept of man," criticized, in the name of a "synthetic view," what he calls "the 'stratigraphic' conception of the relations between biological, psychological, social, and cultural factors in human life." He writes:

> In this [stratigraphic] conception, man is a composite of "levels," each superimposed upon those beneath it and underpinning those above it. As one analyzes man, one peels off layer after layer, each such layer being complete and irreducible in itself, revealing another, quite different sort of layer underneath. (Geertz 1973: 37)

Like Midgley, Geertz mentions the quest for disciplinary autonomy and authority as reason for the attractiveness of the stratigraphic view. Both critics focus on the divide having a pragmatic function as a symbolic autonomy- and authority-securing device between scientific fields or disciplines, a function I elsewhere (Kronfeldner forthcoming) call *epistemic demarcation*, and both regard the consequences as detrimental.

Ontologically, the nature/culture divide can be justified, despite critique, be it from Midgley, Geertz or from developmental systems theory. It can be defended since there are differences among developmental resources in the world that allow us to successfully use the nature/culture divide to denote two different kinds of developmental resources traveling over time in distinct channels of inheritance. In the following, I thus assume that the nature/culture divide is ontologically meaningful.[2] What I aim to defend here is that the nature/culture divide can also be epistemically useful.[3] To do so, I will use the example of Alfred L. Kroeber, student of Franz Boas, who hardened the nature/culture divide to have the right to ignore one side of it, namely nature. According to him, nature and culture are separate evolutionary processes, and because of this, the cultural anthropologist can safely ignore nature. Kroeber had, I will claim, epistemically the *right to ignore* nature since to do so was epistemically fruitful.[4]

The argument is roughly the following: first and foremost, Midgley and Geertz forget that (whatever the motivations of those using it) the divide was fruitful in establishing and keeping an autonomous cultural anthropology alive, the very perspective from which they criticize the divide. Second, they seem to implicitly or explicitly not allow for a separationist stance—the right to ignore—as an epistemically fruitful research strategy, a heuristic. Interestingly, they share this with their most ardent contemporary critics, namely evolutionary psychologists, who challenge the autonomy of cultural anthropology (and other social sciences and humanities). Evolutionary psychologists do the latter by arguing for unity, that is, an integration of knowledge about humans, to overcome the epistemic fragmentation of being human. As part of this call for integration, they despise social scientists who (like Kroeber) take a separationist stance.

In the following section, I shall illustrate the evolutionary psychologists' argument from integration against what they call the "standard social science model," that is, any model that entails claims toward the autonomy of social sciences on the basis of the separation of nature from culture (and society). I will reply with an argument from *fruitful epistemic separation*.

THE ARGUMENT FROM INTEGRATION

Outlining why social scientists should pay attention to the insights of evolutionary psychology, Jerome Barkow, Leda Cosmides, and John Tooby write:

> Conceptual integration generates this powerful growth in knowledge because it allows investigators to use knowledge developed in other disciplines to solve problems in their own. . . . At present, crossing such boundaries is often met with xenophobia, packaged in the form of such familiar accusations as "intellectual imperialism" or "reductionism." But by calling for conceptual integration in the behavioral and social sciences we are neither calling for reductionism nor for the conquest and assimilation of one field by another. . . . Conceptual integration simply involves learning to accept with grace the irreplaceable intellectual gifts offered by other fields. To do this, one must accept the tenet of mutual consistency among disciplines, with its allied recognition that there are causal links between them. Compatibility is a misleadingly modest requirement, however for it is an absolute one. Consequently, accepting these gifts is not always easy, because other fields may indeed bring the unwelcome news that favored theories have problems that require reformulation.   (Barkow et al. 1992: 12–13)

Nobody involved in serious contemporary scholarly debates about evolution and culture asks for any kind of reductionism that would involve giving up the disciplinary structure of science and the pluralism that follows from that (compatible but incongruent perspectives), not even evolutionary psychologists, at least not Barkow et al. (1992), it seems. The disciplinary structure of science is a bulwark against any call for imperialist unification via ontological, theoretical, or methodological reduction of social sciences to biological ones.[5]

Barkow et al. do not ask for any of these kinds of reductions or for imperialism; they ask for *corrective integration*, which is a process (and research strategy) of creating or checking for external consistency, consistency between one's own theory and fields external to one's own. One should note that integration is less global than unity since the perspectives or fields which are integrated stay separate and are not reduced, but they are integrated, connected, and consequently constrain each other. The constraining or correction can have a direction. When we check whether a theory from one field of studies is consistent with a theory from another field, and then adapt the one according to its conflicts with the other, then we integrate the corrected theory *to* the correcting theory. Thus integration can have a direction of adaptation. In effect, Barkow et al. ask cultural anthropologists to make sure that what they claim is consistent with well-established knowledge from evolutionary theory, while considering their version of evolutionary psychology as providing the new "irreplaceable intellectual gifts" to be taken into account.

Interestingly, at least in the quote above, this corrective consistency-checking integration goes vice versa (i.e., the corrective direction goes both ways). Barkow et al. do not assume a corrective *asymmetry* between disciplines; they ask for *corrective symmetry*, for mutual integration.[6] Corrective *asymmetry* would mean that while the historian, sociologist or anthropologist will have to correct her theories when these conflict with evolutionary theory, the biologist does not have to correct anything just because a social scientist finds out something about this or that behavior that conflicts with evolutionary theory. And why is that, one might ask? "Because biology is more fundamental, more general," or so the argument might go. Claims for a corrective asymmetry probably rest on an implicit hierarchical ordering of scientific disciplines, from physics, to chemistry, to biological sciences, to psychology, to social sciences—an ordering that is often derived from a so-called layer-cake model of reality (often associated with Oppenheim & Putnam 1958). I am no fan of a corrective asymmetry between disciplines, but will not develop an argument against it here.[7] The focus shall rather be on the *value* of integration. My argument against the value of integration holds even if there were corrective asymmetries between disciplines.

The reason integration is regarded as good in the above plea for integration from Barkow et al. is that it is believed to be fruitful (i.e., generative): leading to new insights, theories, or even fields. Yet there seems to be a further assumption, even if it is only implicitly made in the quote. Barkow et al. seem to assume that integration and generativity (i.e., the production of new insights, new methods, or new theories representing whole new interdisciplinary fields) are so closely connected that *therefore* separation cannot be fruitful. This is what I label the *argument from integration*.

That they entertain this argument from integration (and even should do so for their own consistency) can be deduced from the context of the quotation: the explicit enemies of their text are people like Kroeber, involved in the development or defense of what Barkow et al. call the "standard social science view," a view that ignores *any* evolutionary explanations of our behavior, and thus exhibits what I call a separationist strategy.

Barkow et al. bring in the fruitfulness of integration to argue against Kroeber and scientists like him.

Made explicit, the argument from integration is this: anthropologists should not ignore human nature in their explanations since otherwise generativity is lost. Social scientists in general should always aim at integration: they should correct their views in the face of the knowledge in other fields because this secures the generativity of research (i.e., the production of novel, reliable, justified beliefs). But Barkow et al. can argue against Kroeber and the like in this way only if it is indeed the case that the latter's separationist stance has failed to be generative. And this is the assumption that I will criticize. A view that claims that only integration can be generative suffers from (what I call) a *synthesis bias*.[8] The argument from integration is biased regarding the historical cases since it disregards the actual historical fruitfulness of the separationist stance.

A similar synthesis bias might surface in the assumptions of the direct critics of evolutionary psychology. John Dupré (2010), for instance, allying himself with the developmental systems perspective, criticizes evolutionary psychologists for ignoring culture. Even though I agree with Dupré on most of his points against evolutionary psychologists, I shall develop a path between these polar positions. The argument is that it can be fruitful for both sides to ignore certain phenomena or causal factors and to thus utilize a nature/culture divide to establish a division of labor (and thus disciplinary primitives) in the explanation of what we are, think, and do.

The argument that I want to establish—the pragmatic–pluralist argument from the fruitfulness of epistemic demarcation—is that, in principle, separation (as a heuristic research strategy) can be as fruitful as integration; whether it is fruitful depends on context. Separating oneself in the sense of assuming disciplinary primitives—not in *willfully accepting known inconsistencies*, but in *not checking for consistency*—can have fruitful potential. If integration is valued because it encourages progress, then separation has to be taken as equally valuable if it encourages progress. To use an analogy: as evolutionary theory showed us that geographic isolation can be a creative factor in the evolution of species, this chapter aims to show that epistemic demarcation can be a creative factor in the evolution of disciplines and theories.

THE ARGUMENT FROM FRUITFUL SEPARATION

To argue against the synthesis bias, it is sufficient to establish that separation can be equally fruitful. Thus, one example of a separationist stance that has been fruitful is all that is needed. I will present Kroeber's above-mentioned case as such an example, that is, as a separationist stance that was epistemically fruitful.

The Case of Kroeber's Cultural Determinism

Kroeber was Franz Boas's first PhD in anthropology and the ninth in the whole US. He became famous at the beginning of the 20th century, when anthropology emancipated itself from a museum-based profession to an academic discipline. At that time, anthropologists were not alone in maturing as a science, a process which includes: asking for money, jobs, and institutional as well as intellectual power. Psychology and the expanding field of genetics were their most prominent competitors at that time. Kroeber was very engaged in the respective disciplinary "identity politics" and developed what has been called a radical "cultural determinism."[9] His case shows how—through the reconstitution of a field-defining phenomenon: culture—certain phenomena or causal factors can be made epistemically irrelevant.

Kroeber defended the identity and importance of cultural anthropology by using the biologist's own concept of heredity: he defined culture as heredity of another sort that is, at the same time, opposed to biological heredity. He moved from an analogy (culture as heredity) to a contrast: biological heredity (nature) versus cultural heredity (culture). Culture derives from previous culture, as a cell derives from previous cells, and (as we would add today) bits of DNA derive from previous bits of DNA. Since culture derives from previous culture it is autonomous from cells and biological heredity. This move helped him to establish not only the autonomy of cultural change from biological evolution, but also the identity and autonomy of those studying cultural change, which he termed (interchangeably) historians or anthropologists.

As Figure 15.1 nicely illustrates, there is not one evolution of organisms anymore, but three: the evolution of superorganic culture, of organic evolution, both contrasted with physical persistence (of stones, for instance, which do not reproduce and die, but persist and thus do not evolve). The evolutionary process is partitioned into three separate and autonomous processes. There are a couple of epistemic aspects of this separation of distinct evolutionary processes that need to be taken seriously:

- Kroeber partitioned one phenomenon (being human and the evolution of being human) into three distinct autonomous processes.
- He thereby established a new separate explanandum—namely, culture. Culture is not explaining life or the evolution of organisms; it is to be explained, that is, it is the new field-defining explanandum for the new specialist on stage, the cultural anthropologist.
- As part of this, nature becomes a disciplinary primitive: one can safely assume it and then ignore it since it does not make a difference for the historical change in culture that the cultural anthropologist aims to explain and which is depicted in Figure 15.1 with its characteristic take-off (change of it without a concomitant change in nature).

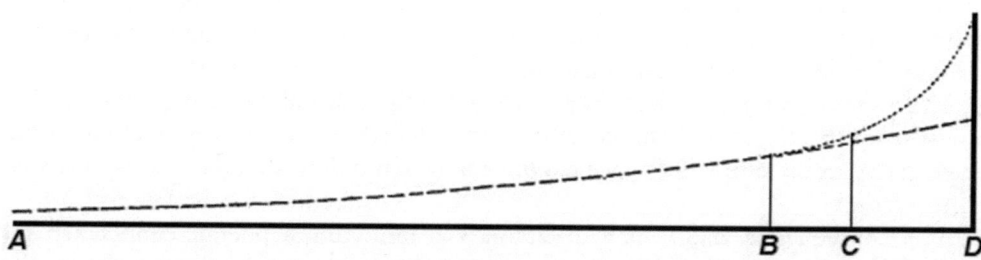

Figure 15.1 From Kroeber's famous paper "The superorganic" (1917: 211). It depicts three distinct and autonomously changing evolutionary processes: superorganic culture (dotted line) on the top, organic evolution (dashed line) and physical persistence (continuous line) beneath.

To think this way was far from trivial and quite fruitful in the fight against the scientific racism of the time. To understand why, we have to look at the importance and history of Lamarckian inheritance and the history of the concept of culture in more detail. Herbert Spencer (1893, 1894) famously argued for the necessity of Lamarckian inheritance to explain the evolution of musical sense and higher cognition generally. According to Spencer, there could not be a being like Mozart in Samoa (an example used by August Weismann (1892)) because people in Samoa cannot have the same musical sense as people of Mozart's "kind" and "time," simply because if they do not have the same kind of music, they do not have the same kinds of musical minds. Given Lamarckian inheritance, this was a valid inference. Spencer assumed that one can directly infer respective natural differences from cultural differences (and vice versa) since nature becomes culture, which becomes nature in turn, prohibiting the take off one sees in Kroeber's figure. In a Lamarckian picture, nature and culture are one indivisible system of inheritance and evolution.

On the basis of a Weismannian point of view, however, one could not infer racial differences from cultural differences since the two are independent, decoupled from the very first moment when the first animal managed to learn socially from another one, that is, from the very "birth" of culture (B in the figure). Since Lamarckism is wrong, as Weismann stressed, *there can be a Mozart in Samoa*, that is, it is an empirical possibility that there can be a human being in Samoa with the same cognitive capacities as the real historical Mozart. Kroeber used Weismann's claims to not only fight racism but also to establish the autonomy of cultural anthropology.[10]

Both Weismann and Kroeber believed that *only if Lamarckian inheritance is replaced with the concept of cultural inheritance* is one able to maintain (a) that cultural change is historically not correlated with biological change, that is, that one can change without the other and is autonomous in that sense. There is then also (b) no evidence to *not* believe in the psychic unity of humankind—if all one has is access to cultural differences (and this is all one had at that time since genetic factors were still hypothetical entities only). Therefore, from Weismann's and Kroeber's perspective, the unity of humankind could safely be assumed, that is, taken as a base line on top of which culture "takes off." If so, then (c) cultural change and differences are to be explained by cultural change and differences, rather than by a shared human nature.

It is important to acknowledge that with this decoupling Kroeber did not deny that there *is* a shared human nature as well as individual natures (i.e., differences in nature between individual people) that are important to explain each individual's behavior. He only denied that individual innate differences are important for what he wanted to explain—shared culture—since they average out at the cultural (i.e., group) level, and he also stressed that the shared human nature does not make a difference for cultural differences at the group level either since culture changes and differs on the basis of a shared human nature.

As a consequence, nature is a disciplinary primitivum, a phenomenon taken for granted, on top of which culture happens. And this holds despite the fact that at the individual level of development, nature and nurture certainly interact. But the individual level (explaining behavior of humans) is not the core explanandum anymore, and the co-evolution of nature and culture, of biologically and socially inherited developmental resources, was not yet an explanandum.

Kroeber epistemically reconstituted what "culture" means, he reconstituted the explanandum in the following sense.

1. "Culture" referred to traits of a group of people (playing the role of an explanandum) in the classic 19th-century evolutionist picture of anthropologists such as Edward B. Tylor, irrespective of how these traits are caused.
2. It predominantly referred to factors in development, playing a role as part of the explanans for explaining behavior in Boas's work, who reformed anthropology in many senses.
3. With Kroeber, "culture" became a system of inheritance, a new explanandum in its own right that is explained by previous culture, analogous to biological inheritance.
4. Finally, on the basis of this, culture recently became a factor in evolution (as part of the explanans in contemporary co-evolutionary theory) to explain—not behavior as such, but evolution of behavior.[11]

HOW TO EVALUATE HIS CASE

How shall we evaluate this case in terms of epistemic fruitfulness? I introduced epistemic fruitfulness as *generativity*, which is regarded by Barkow et al. (and many others) as an epistemic value, something a theory should exhibit. Generativity of a perspective (or approach, theory, model, etc.) consists in the production of new insights, new methods, or new inter-field theories representing whole new interdisciplinary fields. A new insight can be positive or negative, that is, producing new knowledge or establishing the dismissal of wrong beliefs.

- *A new insight*. Even though people defended (and still defend) scientific racism (let alone racism) on all kinds of grounds, I regard Kroeber's claim of the autonomy of culture as a historically important insight that helped to fight the false scientific racism of the early 20th century, which was supported by the belief in the inheritance of acquired characteristics. Thus, Kroeber developed an important and fruitful thesis that blocked one argument pattern for scientific racism of that time.
- *New field with new methods*. Kroeber's analogy between culture and nature as two systems of decoupled inheritance was not used for integration but for separation: of culture from nature, and of the alleged experts studying these, that is, cultural anthropologists on the one hand, and physical anthropologists plus geneticists on the other hand.[12] Kroeber, as mentioned, did not want to say that physical anthropologists or geneticists do not have a word to say on *humans*. He believed that science is fruitful only if each of these has a domain of its own. In the context of his time, I believe, he was right: it certainly was more fruitful at that time that each had a domain of his own in order to establish a field within science that has produced a tremendous amount of knowledge essential to understand what humans are, think, and do.
- *Precursor of new interfield*. Since Kroeber used the concept of distinct channels of inheritance, he can be considered as a precursor of contemporary *dual* or *multiple* inheritance theories studying the co-evolution of nature and culture, that is, the

evolutionary interaction of the two systems of inheritance.[13] These theories go beyond Kroeber since they indeed integrate in an interactive sense what became separated in Kroeber's hands. These theories are also in and of themselves quite productive, in at least three senses: in contrast to approaches such as evolutionary psychology, they can show in a statistical manner how biologically maladaptive behavior can evolve on the basis of specific cultural transmission settings; they can also show that different modes of learning (individual learning, prestige bias, conformist bias, success bias, etc.—all settings analogous to the biological mechanisms of heredity) lead to different macro-evolutionary patterns; finally, they can describe the evolutionary interaction of nature and culture, for example, how settings about legal inheritance and cultural learning influence the probability that a trait such as lactose tolerance (widespread, for instance, in Northern but not Southern Europe) can evolve.[14]

Kroeber ignored that nature and culture interact as factors in the evolution of humans since for him culture was first and foremost an explanandum. In contemporary co-evolutionary theory, culture takes on a different epistemic role: it is regarded as part of what explains evolution (a part of the explanans), rather than as a system of inheritance that is explained (explanandum). Kroeber was working at a different time and had to first separate what later became kinds of causes for a more general explanandum (co-evolution). Culture and nature thus have become—with but also beyond Kroeber—not just separate systems of inheritance but special kinds of causal factors in the evolution of human traits, be these traits physical or mental. In his own time, Kroeber regarded nature and culture not just as separate, but he ignored any possible interaction of them on a longer time scale, too focused on fighting the enemy of his time: scientific racism, which did not allow for culture to change independently of nature. That Kroeber's perspective is today too radical does not conflict with the claim that his separationist stance was quintessential for the historical development described in this chapter and justified given the generativity it exhibited.

SUMMARY OF THE CASE AND TWO ARGUMENTS ESTABLISHED

To summarize: Kroeber defended the place of cultural anthropology against the social and political hegemony of racist hereditarianism and the scientific force of the new genetics. He crossed the boundary between anthropology and biology and used Weismann's theory of heredity, at a time when both disciplines were expanding their scientific and institutional setting. The goal of the boundary crossing was, however, to make the "fence" between the two disciplines even higher. He did this in order to establish a clear specialization, a differentiation, that is, a clear division of labor—divide and conquer—between anthropology and biology, and between physical anthropology and cultural anthropology. When disciplines emerge, it is unlikely that their representatives are open-minded, for pragmatic reasons: they have to establish themselves first and secure a place in the midst of other disciplines. They have to appropriate phenomena.

The case shows that, as research strategies, integration and separation can both lead to fruitful scientific results. I named three such results for Kroeber's case of separating culture and nature as two different systems of inheritance. The separationist stance has been fruitful, since:

- it helped to block one detrimental argumentation pattern for racism in the early 20th century;
- it helped stabilize a young and at that time still fragile but productive sub-discipline, namely cultural anthropology;
- it was a historical precondition for contemporary co-evolutionary theories, productive approaches to specific problems regarding how culture and nature interact on the evolutionary level.

To conclude: there are then cases where it is more useful, in the service of scientific productivity, to batten down the hatches of one's scientific horizon—for a while at least. Sometimes it is fruitful to separate oneself: to ignore, for specific goals, that everything in reality hangs together in complex ways and that therefore nothing is autonomous.

The argument from fruitful separation is a general one (applying to the relationship between all kinds of phenomena and fields attached to phenomena), but it entails a more specific argument, specific to the relationship between nature and culture. The divide between nature and culture can be an epistemic mean for fruitful research in that it provides disciplinary primitives that allow for fruitful separation of kinds of explanations typical of a discipline or field. It is thus epistemically justified even though in the world everything hangs together and is on a par.

"IT TAKES TWO TO TANGO," AND INTEGRATION AS REGULATIVE IDEAL

Over the long run, it certainly takes two to tango, that is, separation and integration must be in a dialectic balance with each other in order to get ahead in a specific field, that is, to reach an "oscillatory progress" (Holton 2010: 249). Or, as Hans-Jörg Rheinberger (1997: 253) wrote: "Fragmentation, aiming at simplicity, finally creates complexity"—a manageable, palpable complexity instead of an undifferentiated one. Those who draw distinctions first and connect later, see more.

If separation and integration are in the long run both necessary for the generativity of science, they should have in principle equal value. But the philosophical literature by and large seems to still place a higher priority on integration (or "unification" as some still prefer to call it). It is a higher, if not *the* ultimate "regulative ideal" (Kitcher 1999) of science, a higher-level norm. As a regulative ideal, it is, according to Philip Kitcher (1999: 342), "good where we can get it, but not to be imposed willy-nilly." There are contexts where we should not apply it. Richard Burian (1993) defended a similar nuanced position, taking separation as transiently useful in certain contexts. He writes: "It is a matter of judgment when to take the norm seriously. . . . But this in no way undercuts the legitimacy of unification as a generally applicable higher-level norm" (1993: 314).

Kitcher and Burian are on the right track but miss the crucial point nonetheless. The context in which separation or integration can be useful includes a specific way of understanding the phenomenon at issue. Burian thus formulates the higher-level norm also as a conditional: "*when* work in different disciplines bears on a *given problem*, their practices, terminologies, and standards for evaluation of experimental evidence must be brought into accord with respect to the matter at hand for it to count as satisfactorily solved" (Burian 1993: 313; emphasis added). Well, problems are not simply "given." They are reconstituted, as in Kroeber's case, so that there is a functioning division of labor. Kroeber reconstituted culture in a way so that it allows the autonomy of cultural anthropology.

Burian in principle recognizes this option (and he seems to be the only one in the literature on the value of unity or integration): problems and phenomena can be reconstituted so that there is no need to integrate knowledge from diverse disciplines. "Ultimately," Burian writes, only one of the following options has to be fulfilled in an interdisciplinary encounter: "(1) achieve coherence between different scientific descriptions of the phenomena of concern and also between those descriptions and any theoretical explanations of those phenomena, or (2) transform the problem of concern so as to remove the obligation to take into account one or more of the competing descriptions or theories" (1993: 302–303).[15] Even though he thus admits that separation and integration are on a par, two distinct ultimate research strategies, he regards separation as a makeshift: a good strategy if things get too complex.[16] Unity stays the regulative ideal. This ignores that separation can be permanent without preventing local integration. Local integration can consist in the integration of knowledge for a concrete problem or stabilize itself as a new interfield (to solve a specific type of problem). To consider separation and integration as of equal value means allowing for a permanent separation without denying that at certain points there should be local integration.[17]

Stressing that separation is permanent does not imply that priority is given to separation. Giving priority to separation would involve claiming that only if there is no division-of-labor that allows one to reconstitute a phenomenon should one look for integration. No such claim has been defended here. The differentiation of disciplines historically correlates with a tremendous increase in knowledge. This could be taken as evidence that separation is more important than integration. Yet, there is evidence in the other direction too: as a regulative ideal separation (with integration as makeshift) would lead to an indefinite proliferation of problems and would lower the probability that errors are found via integration (thanks to its corrective force). Science, after all, is a systematic knowledge production endeavor and is robust thanks to its network-like structure. In a network, separation and integration are of equal value.

Kroeber (1948: 260) postulated a similar network structure for culture, and depicted it, as in Figure 15.2, as a tree—at that time the most conventional depiction of differentiating structures. He described it as the "tree of [the knowledge of good and evil—that is, of human] culture" and wrote of its form: "There is a constant branching-out, but the branches also grow together again, wholly or partially, all the time. Culture diverges, but it syncretizes and anastomoses too" (ibid.). Science too is a form of culture that evolves as a system of inheritance of knowledge with differentiation and local integration.

Figure 15.2 Kroeber's tree of knowledge (from Kroeber 1948: 260)

CONCLUSION

As long as separation is paired with later integration, it can result in fruitful scientific results since (if successful) it elucidates phenomena and is in that sense of epistemic value. The separationist strategy is a piecemeal heuristic but a fruitful one, as is the integrationist strategy. We can thus consider it as a good thing that there are different kinds of experts devoted to their "own" kind of causes, so that there is partial knowledge (from studying one factor in isolation, that is, while the others are ignored) that can later be integrated. In the long run, science needs both separation of disciplines like evolutionary biology, psychology, cultural anthropology, and philosophy, and it needs integration.

ACKNOWLEDGMENTS

I would like to thank the Department of Philosophy at Bielefeld University, the Center for Philosophy of Science at the University of Pittsburgh, and the Fishbein Center for the History of Science and Medicine at the University of Chicago for supporting my research during the time when the research that is at the foundation of this chapter was done. In particular, I would like to thank Hanno Birken-Bertsch, Martin Carrier, Heather Douglas, John Jackson, Richard Joyce, Maureen O'Malley, Sandra Mitchell, Alexander Reutlinger, Betty Smocovitis, Kyle Stanford, David Stump, and the audiences at the 2015 Meeting of the Nordic Network for Philosophy of Science and the 2015 Human nature course at the CEU for helpful discussions related to this chapter.

NOTES

1. I take the distinction between view/knowledge (Menschenbild/Menschenkenntnis) from Mühlmann ([1948] 1986: 140). See Birken-Bertsch (2013) for the issue about losing a unified picture of being human. This chapter is based on ideas first presented in Kronfeldner (2010, 2013).
2. A detailed defense of this has to wait for another occasion.
3. Some critics of the divide mention that pragmatically it can be useful; for example, Tim Lewens (2015: 91–92). Yet, even Lewens does not spell out the ontology that allows the pragmatic use in detail or describe the details of the pragmatic use.
4. I use the term "the right to ignore" in analogy to William James's (1897) appeal to a "right to believe."
5. See Mäki (2014) on what imperialism in these contexts can mean.
6. I owe the term "corrective symmetry" to Daniel Steel (2004), though he uses it in slightly but philosophically decisively different ways.
7. One would have to deal with reduction and the concept of a hierarchy of levels, which are intricate concepts that would lead us too far away. See Wimsatt (2007) and Brooks (2014) instead.
8. I owe the idea of comparing integration and separation as values to a short piece from Gerald Holton (1978), who shows that in the history of science *en gros*, synthesis has often been valued more than analysis, even though both should be regarded as being of equal value since sciences need both for progress. Oppenheim & Putnam (1958) are another example for synthesis bias. They write that the unity of science should be a working hypothesis since it "is, as has often been remarked, fruitful in the sense of stimulating many different kinds of scientific research. By way of contrast, belief in the irreducibility of various phenomena has yet to yield a single accepted scientific theory" (1958: 16). This claim was already unjustified in 1958.
9. His case has historically been analyzed in detail in Kronfeldner (2009) and Jackson (2010).
10. Although important, I have totally left out a discussion of how further political and social dimensions entered Kroeber's felt need for boundary work, that is, how industrialization, democracy, capitalism, imperialism, and the regulatory control of behavior involved in the "disciplining" of humans played a role. I have done so deliberately since it would by far exceed the space available here. Useful entry points for this are Ross (2003) or Patterson (2001). One thing is important to note, though: it has often been stressed that the Vienna Circle's rally for a unity of science had a political motivation, against the nationalism and racism of the early 20th-century Europe. Kroeber had the same political enemies but the exact opposite epistemic strategy to beat them. Thus, even if political contexts play a role, "we will not find . . . a single-valued, transhistorical function that plots assessment of unity onto a fixed political map," as Galison (Galison & Stump 1996: 8) already stated.
11. For details on this dialectic history of the concept of culture, see Kronfeldner (2009). For more on changing the explanandum from explaining individual development of a trait to explaining *differences* between individuals, see Kronfeldner (2015).
12. With psychologist and sociologist also studying something different. For the separation of sociology and anthropology see the paper that Kroeber wrote together with Talcott Parsons (1958).
13. See Boyd & Richerson (1985) as an instance of the first, and Jablonka & Lamb (2005) as an instance of the second. For comparison and general analysis, see Lewens (2015).
14. For more on the productivity of multiple inheritance theories, see Kronfeldner (2013).
15. Burian even gives a similar example: "One way of transforming a problem is by dividing it into strongly separate problems. For example, the rejection of Haeckel's biogenetic law facilitated just such a separation between the problem of determining the causes of evolution and the problem of establishing phylogenies" (1993: 303).
16. The same holds for Kitcher's "modest unificationism." Kitcher (1999) admits that there is no "fixed totality of questions," but does not consider the consequences of this for separation as equally valuable. Van der Steen (1993) also deplores a synthesis bias. I agree with Steen on the diagnoses, but not with the therapy.
17. Closest to my account is thus Mitchell (2003, 2009).

REFERENCES

Barkow, J., Cosmides, L., & Tooby, J. 1992. *The Adapted Mind: Evolutionary Psychology and the Generation of Culture* (Oxford University Press).

Birken-Bertsch, H. 2013. "Zur Kritik anthropologischer Wenden im Ausgang von Joachim Ritter." In A. Hügli, A. Horn, A. Kottmann, C. Steiner, S. Tietz, F. Wüstholz, & A. Wunder (eds.), *Die anthropologische Wende* (vol. 72) (Schwabe) 315–327.

Boyd, R. & Richerson, P. 1985. *Culture and the Evolutionary Process* (University of Chicago Press).

Brooks, D. 2014. *The Concept of Levels of Organization in Biology*. Dissertation. Bielefeld.

Burian, R. 1993. "Unification and coherence as methodological objectives in the biological sciences." *Biology and Philosophy* 8: 301–318.

Dupré, J. 2010. "Causality and human nature in the social sciences." *Kölner Zeitschrift für Soziologie, Sonderheft* 50: 507–525.

Galison, P. & Stump, D. (eds.). 1996. *The Disunity of Science: Boundaries, Contexts, and Power* (Stanford University Press).

Geertz, C. 1973. "The impact of the concept of culture on the concept of man." In C. Geertz, *The Interpretation of Cultures: Selected Essays* (Basic Books) 33–54.

Holton, G. 1978. "Analysis and synthesis as methodological themata." In G. Holton, *The Scientific Imagination: Case Studies* (Cambridge University Press) 111–151.

Holton, G. 2010. "On unity and disunity in the sciences: Variations of ancient themata." *Vienna Circle Institute Yearbook* 14: 245–262.

Jablonka, E. & Lamb, M. 2005. *Evolution in Four Dimensions: Genetic, Epigenetic, Behavioral and Symbolic Variation in the History of Life* (MIT Press).

Jackson, J. 2010. "Definitional argument in evolutionary psychology and cultural anthropology." *Science in Context* 23: 121–150.

James, W. 1897. *The Will to Believe, and Other Essays in Popular Philosophy* (Longmans, Green, and Co.).

Kitcher, P. 1999. "Unification as a regulative ideal." *Perspectives on Science* 7: 337–348.

Kroeber, A. 1917. "The superorganic." *American Anthropologist* 19: 163–213.

Kroeber, A. 1948. *Anthropology: Race, Language, Culture, Psychology, Prehistory*, 2nd ed. (Harcourt, Brace).

Kroeber, A. & Parsons, T. 1958. "The concept of culture and of social system." *American Sociological Review* 23: 582–583.

Kronfeldner, M. 2009. "If there is nothing beyond the organic . . .: Heredity and culture at the boundaries of anthropology in the work of Alfred L. Kroeber." *NTM: Journal of the History of Science, Technology and Medicine* 17: 107–133.

Kronfeldner, M. 2010. "Won't you please unite? Darwinism, cultural evolution and kinds of synthesis." In E. Barahona & H.-J. Rheinberger (eds.), *The Hereditary Hourglass: Genetics and Epigenetics, 1868–2000* (Max Planck Insititute for the History of Science) 111–125.

Kronfeldner, M. 2013. "Die epistemische Fragmentierung des Menschen: Wie der Mensch zwischen Natur und Kultur verschwindet." In A. Hügli, A. Horn, A. Kottmann, C. Steiner, S. Tietz, F. Wüstholz, & A. Wunder (eds.), *Die anthropologische Wende* (vol. 72) (Schwabe) 287–313.

Kronfeldner, M. 2015. "Reconstituting phenomena." In U. Mäki, I. Votsis, S. Ruphy, & G. Schurz (eds.), *Recent Developments in the Philosophy of Science: EPSA13 Helsinki* (Springer) 169–182.

Kronfeldner, M. Forthcoming. "Divide and conquer: The authority of nature and why we disagree about human nature." In T. Lewens & E. Hannon (eds.), *Why We Disagree About Human Nature* (Oxford University Press).

Lewens, T. 2015. *Cultural Evolution: Conceptual Challenges* (Oxford University Press).

Mäki, U. 2014. "Scientific imperialism: Difficulties in definition, identification, and assessment." *International Studies in the Philosophy of Science* 27: 325–339.

Midgley, M. 1995. *Beast and Man: The Roots of Human Nature* (Routledge).

Mitchell, S. 2003. *Biological Complexity and Integrative Pluralism* (Cambridge University Press).

Mitchell, S. 2009. *Unsimple Truths: Science, Complexity, and Policy* (University of Chicago Press).

Mühlmann, W. [1948] 1986. *Geschichte der Anthropologie* (Aula-Verlag).

Oppenheim, P. & Putnam, H. 1958. "Unity of science as a working hypothesis." In H. Feigl (ed.), *Concepts, Theories and the Mind-Body Problem* (University of Minnesota Press) 3–36.

Patterson, T. 2001. *A Social History of Anthropology in the United States* (Berg).

Rheinberger, H.-J. 1997. "Experimental complexity in biology: Some epistemological and historical remarks." *Philosophy of Science* (suppl. vol.) 64: 245–254.

Ross, D. 2003. "Changing contours of the social science disciplines." In T. Porter & D. Ross (eds.), *The Modern Social Sciences* (Cambridge University Press) 205–237.

Spencer, H. 1893. "The inadequacy of 'natural selection,' I and II." *Contemporary Review* 63: 153–166, 439–456.

Spencer, H. 1894. "Weismannism once more." *Contemporary Review* 63: 592–608.

Steel, D. 2004. "Can a reductionist be a pluralist?" *Biology and Philosophy* 19: 55–73.

Van Der Steen, W. 1993. "Towards disciplinary disintegration in biology." *Biology and Philosophy* 8: 259–275.

Weismann, A. 1892. "Gedanken über Musik bei Thieren und beim Menschen." In A. Weismann, *Aufsätze über Vererbung und verwandte biologische Fragen* (Gustav Fischer) 587–637.

Wimsatt, W. 2007. *Re-Engineering Philosophy for Limited Beings: Piecewise Approximations to Reality* (Harvard University Press).

IV

Evolution and Mind

16

Evolution and Mind: An Overview

Valerie Hardcastle

In the future I see open fields for far more important researches. Psychology will be securely based on the foundation . . . of the necessary acquirement of each mental power and capacity by gradation. Much light will be thrown on the origin of man and his history.

(Charles Darwin 1859)

Evolutionary psychology, in the broadest sense of the term, began with Charles Darwin's rather shocking claim in *The Descent of Man and Selection in Relation to Sex* (1871) that human minds are similar to the minds of other species. People differ from other animals only by degree and not by kind. That is, our abilities to reason, to remember, to speak, to learn, and to feel are the same fundamental abilities that many creatures have, though we and they perhaps have them to greater and lesser extents.

Acceptance of this view has allowed researchers to develop and use animal models to help us understand our own cognitive capacities as well as our psychopathologies. We do experiments on fruit flies, rats, pigeons, gerbils, guinea pigs, cats, dogs, monkeys, even yeast, because we believe that these organisms can serve as useful and informative substitutes for experiments on humans. And we do this because we believe that there are important structural and functional similarities among all these organisms, and that these similarities are there because we evolved from common ancestors. We all act in similar ways to solve similar problems driven by our similar environment. We all need to find food, shelter, and mates (except yeast). We need to protect and nurture our young. And we need to survive long enough to ensure their success in finding food, shelter, and their own mates, so the cycle can repeat.

In fact, despite humans' temptations to see themselves as the pinnacle of evolution, recent work in mapping diverse species' genomes has demonstrated that there are multiple, basic, developmental processes that insects, fish, and mammals all have in common (Robinson 2004; Papini 2008). Work in evolutionary biology has helped to underscore this very important point: we are not as special or as unique as we would like to think we are. In this overview chapter, I will give some brief examples to illustrate how applying evolutionary theory to the study of mind expands our scientific toolkit, as well as keeps us humble. Let us first look, though, at how evolutionary theory might aid philosophers trying to understand the metaphysics of mind.

EVOLUTION AND THE PHILOSOPHY OF MIND

By and large, there are two ways that contemporary analytic philosophers understand the mind. They are either type–type identity theorists or they are functionalists. Both approaches have challenges. Darwinian evolutionary theory opens a third path, and perhaps a new way to explain the mind that skirts the difficulties of the other views. In addition, Darwinian theory provides some guidance regarding how to investigate the mind profitably—and using methods that fall outside the scope of what identity theory or functionalism would sanction.

Type-identity theory presumes that each mental kind—beliefs, desires, emotions, memories, and so on—is reducible to or corresponds to a set of brain states. All mental states or processes are type-identical to underlying brain states or processes. Functionalists point out an obvious problem with this sort of physicalism: it is chauvinistic (Fodor 1975). If type–type identity is true, then things without brains, like computers or space aliens, could not possibly be minded. Of course, we do not know whether such things could have minds, but it seems problematic to reject the possibility out of hand before even understanding exactly how cognition works.

Functionalists argue that we should understand mental states in terms of inputs, internal functional or causal relationships, and outputs. The causal roles that beliefs, desires, and the like play in the overall cognitive economy define what they are. But of course, this view of mentality is too liberal. As Ned Block (1978) and John Searle (1980) argue, if functionalism is true, then we can envision all sorts of extreme examples in which the causal relations among mental states are instantiated, but we cannot intuitively agree that these are therefore minded events. For example, we could envision that Chinese citizens coincidentally used their landlines to call one another in the exact same patterns that we find with our mental states when see a red apple. But few would readily agree that the nation of China therefore has a mind.

Both type-identity theories and functionalism rely solely on our intuitions to generate support. It would be preferable if we could hang our metaphysics of mind on something more than personal reflection from an armchair. Evolutionary theory offers us a modicum of a different type of support: empirical data, methodological criteria, and biological theory could help inform our views about mind. Larry Shapiro (2001) argues that because both type–type identity theory and functionalism intuitively have basic problems, and because Darwinian approaches have the possibility of offering us varied types of support, we should take evolutionary theory seriously and think of the mind as a biological adaptation. Is he right?

We identify adaptations in terms of both structure and function. "Structures" are the physical substrates or features of an organism. To greatly oversimplify evolutionary theory, structures are what genes code for and they are what link to the environment. Genes, by and large, do not interface directly with the surrounding world. They code for proteins, which interact and combine in a variety of ways to produce larger features that then causally interact with the environment. The function of any structure is the effect the structure produces that accounts for its being there. From an evolutionary point of view, functions are the effects of structures that explain why they were selected.

To take a hackneyed example, hearts are large chambered muscles that do many things: they contract; they make noise; they consume energy. But among the myriad things they do is pump blood. That hearts pump blood explains why we have hearts. Animals that have hearts have a selective advantage over those who do not. The function of our hearts is to pump blood. Therefore, the heart is an adaptation, for it is a structure that was selected for one of the things it did (see Neander, Chapter 5 this volume).

With this tool in hand, we can now approach investigating the mind as an adaptation. We can see right away that both identity-theorists and functionalists have only half of the story. If the mind is an adaptation, then simply identifying the components of the mind with components of the brain, as identity theories do, is not enough. In order to be an adaptation, these structures also need to have particular mental functions that have been selected by evolution over time because they provided a reproductive advantage over conspecifics that lacked such components.

The astute reader might notice that assuming the mind is an adaptation suggests that virtually identical brains could be different types of adaptations, if they have different evolutionary histories. Never mind whether computers or aliens might have minds; it could be the case that creatures with extremely complex brains which support extremely intelligent behavior lack minds as well. This result seems even more counter-intuitive than a type-identity theory!

Let us think about this issue by comparing it to other cases in which the same structures have evolved via different evolutionary histories. For example, complex image-forming eyes have evolved independently somewhere between 50 and 100 times (Land & Nilsson 2002). Do the creatures that have these structures all see via eyes, or do only some of them see with eyes and others do other things? Part of the vagary in counting the number of independent adaptations for the eye stems from ambiguities in defining what counts as an eye, a fact which also makes answering the question regarding whether all eyed-creatures see difficult. Some of the genetic materials used to create light-sensitive organs are common to all evolved eyed organisms. Is being photo-receptive enough for counting as an eye? Once we move beyond simple reactivity to light, we find two distinct eye designs: microvilla for mollusks, annelid worms, and arthropods; and cilia for chordates and echinoderms (Land & Fernald 1992). Are both or either of these structures needed to be an eye? Or do eyes require the ability to sense shapes as well? Do they have to have transparent humors? Corneas? Lenses? How we answer these queries impacts how we individuate and count eyes.

I mention this challenge of vague or fuzzy definitional boundaries because the same fate is true of minds. How do we draw the boundaries around being minded, even assuming that the mind is an adaptation? Do minds have to have memory storage? Affect? Executive functioning? Language? Consciousness? What we include under the definition of mind will determine which structures and functions we track over evolutionary time. It will also determine how likely it is that different sorts of brained creatures evolved different types of mind or perhaps no mind at all. The possibility of a brained creature that lacks a mind will be a challenge for adaptationist perspectives on the mind only under certain definitions of mind. Similarly, it should be a problem for physicalists only under certain definitions as well.

Of course, right now we do not have a tight list of necessary and sufficient conditions for mindedness. At best, we perhaps have a list of components that comprise minds via family resemblances. Some minds might have affect; others, like Spock's (or aardvarks'), may not. What the adaptationist perspective suggests is that physicalist perspectives are half right. We can perhaps identify the component pieces of the mind with underlying structures of the brain. But this is not the full story for what counts as a mind from an adaptationist perspective. The component pieces have to have the right sort of functions as well—they have to confer a selective advantage and be heritable—in order to comprise mind.

Functionalists individuate minds on the basis of causal or functional role alone. They pay no attention to the underlying substrate or supporting structures. This minimalist approach gets them in trouble with our intuitions, because we can envision all sorts of things that might mimic the causal structure of mental processing but these things are not minded in any sense of the word. Mind as adaptation provides a response for functionalism's overly liberal metaphysics.

In order to be an adaptation, one must have the appropriate structure, which has evolved over time. This is why the activity of a group of Chinese citizens cannot be a mind. Not only do they not have the same structure as organismal minds, but they also do not have the correct function. That is, the activity of the Chinese citizens was not selected in virtue of its ability to promote survival and reproduction. Of course, this argument would also show that appropriately configured robots could not have minds either, for they have neither the appropriate structures nor a history driven by evolution. (I am leaving aside the possibility of using genetic algorithms with neural nets to evolve something at least resembling a mind.)

Lawrence Shapiro (2001) provides reasons why interacting Chinese citizens are likely not minded, but a robot might be. Even if eyes might have evolved multiple times over, there are still only a few ways to build a photosensitive organ (cf. Land 1991). The same biological solution to some adaptive problem appears over and over in evolutionary history. For example, there are only a few ways that animals can propel themselves forward in water. As Simon Conway-Morris explains, "It hardly matters if we choose to illustrate the method of swimming by reference to water beetles, palagic snails, squid, fish, newts, ichthyosaurs, snakes, lizards, turtles, dugongs, or whales; we shall find that the style in which the given animal moves through the water will fall into one of only a few basic categories" (1998: 204–205). The point is simply that "function constrains structure. It is simply false that any function can be multiply realized in numerous ways" (Shapiro 2001: 34).

It is highly likely that anything with a mind that resembles our minds will be structured very similarly to the way our minds are, for there are just not that many ways to build complex cognitive engines. Any true artificial intelligence is going to be strongly analogous to our intelligence. Similarly, it would be highly unlikely that something that lacks structures similar to our brains would have something that resembles our minds. So: Chinese citizens are out, but future robotic machines might be in.

However, there is a further issue to consider: adaptations have both structure and a historically defined function. If we build a robotic "brain" to resemble our organic brains, then it still would not have anything with an evolutionary function. The robot might have a brain-like structure, but the activities of that structure would, by presumption, not have been shaped by natural selection. Instead, we would just build the structures so that they behave in a particular way. For this reason, whatever it is that our mechanical counterparts have driving their cognition and behavior, it cannot be a mind, if minds are adaptations.

But again, this might also just be a question of how tightly we want to cleave to our definitions in understanding mindedness or cognition. I have a metal plate in my left wrist that takes the place of some bone I used to have there. The plate functions as part of my radius and ulna. But, of course, this plate has no evolutionary history. We would not mistake it for a biological adaptation, in the way my actual bones are. Nonetheless, we can also understand the plate as an artificial bone—it performs the same functions as my real bones. In the same way, we could understand an artificial intelligence as an artificial mind. It might have the same functions as organic minds do (because we designed it to have the same functions), only it did not evolve to have those characteristics. We would have created it deliberately to mimic our evolved functions.

Of course, if this argument is correct, then we may not have any good reason to eliminate coordinated Chinese behavior from being an artificial mind any more than we could rule out programmed robots as artificial minds. At the same time, since we do not know what actually goes into making a mind—biological or otherwise—it might be premature for us to try to definitively rule in or out various artificial structures as potentially being minded. Given our lack of basic knowledge, our intuitions here should count for little.

Though looking at the mind as an adaptation does not solve all the problems presented by type-identity theories or functionalism, it does help clarify the conceptual landscape. In particular, it helps us appreciate how many of our philosophical positions are driven by intuition, how undefined much of what we consider in philosophy of mind is, and how little we really know about what counts as having a mind in the first place. An additional advantage of adopting an evolutionary perspective on mind is that it also will help us fill in these lacunas by providing new ways for analyzing and studying the mind.

COMPARATIVE METHODOLOGIES

One advantage of adopting an evolutionary perspective on mind is that it helps define the appropriate methodologies for learning about minded creatures. An evolutionary perspective on mind starts with the assumption that minds, or rather brains, do not start out as blank slates. Nervous systems are shaped and organized to process incoming stimuli in ways that are useful to the particular species. If minds, or various mental capacities, are adaptations to certain environmental niches, then the structures underlying these functional capacities should vary in measurable ways across species with different functional requirements.

From this basic assumption, we can then move to two central ways to investigate mental phenomena based on Darwinian theory: phylogenetically and ecologically (Heyes & Huber 2000). At its most general, evolutionary theory uses historical patterns of similarity and difference to make arguments for how things have changed (or not) over time. Phylogenetic approaches highlight similarities through inheritance; they focus on commonalities in descent, which emphasize how we might be similar to our genetic neighbors. A phylogenetic approach might examine how certain qualities of mind have remained constant over evolutionary time or across species.

Ecological approaches focus on the phenotypes that different species have which allow them to survive in their unique environmental niches; it emphasizes the phenotypic outcomes of various selection pressures. An ecological approach might investigate how mental capacities for a particular group are attuned to the demands of its environment (which could include other conspecifics of that group) and how those capacities are different from other similar groups.

What is important about both approaches, however, is that they promote comparative studies. Evolutionary psychology at its best looks at how we are alike and how we differ—from one another, from conspecifics in different niches, and from our genetic relatives. And it uses these analyses to help us understand what is universal and what is particular about our minds and cognitive capacities. Are bee dances, bird songs, and human language reflective of some general communication capacity, or does our capacity for speech stand apart as something unique? Comparative studies can help sort out these types of questions (Sherry 2006). Let us look briefly at some specific examples.

Theoretically, the application of evolutionary theory to psychological issues must start with what is known as Morgan's Canon. British psychologist C. Lloyd Morgan wrote in 1894: "In no case may we interpret an action as the outcome of the exercise of a higher psychical faculty, if it can be interpreted as the outcome of the exercise of one which stands lower in the psychological scale" (1894: 53). Of course, the Canon has problems—it is not clear that eschewing explanations based on complex reasoning to solve problems is always justified (Shettleworth 2010)—but it still remains an important counter to our tendency to anthropomorphize intelligent behavior in other animals. Because we might reason our way through a problem using propositional logic does not mean that other creatures do the same thing. Moreover, simple associative learning is present in every creature in the animal kingdom; therefore, one should first assume that intelligent-seeming behavior is a product of basic conditioning before hypothesizing other mechanisms (Sober 2005).

Evolutionary approaches to studying animal cognition, like behavioral ethology, cognitive ethology, comparative psychology, or sociobiology (Krebs & Davis 1993; Real 1993; Wilson 1975), focus on the behavior of animals in the wild and ecologically meaningful cognitive processes. They aim to understand how animals decide where to forage for food, how they track their location across intricate terrain, how they learn to use tools, how they navigate social relationships in large groups, and so on. The ultimate goal is to use these studies to inform and expand our understanding of human cognition and behavior. To do this, we have to adopt two more theoretical axioms: the principle of allometry and the principle of proper mass. By and large, animals with bigger bodies also have bigger body parts, a principle named as "allometry." Consequently, we can plot the size of any structure against total body size, which should have a characteristic slope for that structure across species. We can use the regression line to discern any outliers, which might indicate a different or enhanced function for that structure.

Comparative neuroanatomists then turn to Harry Jerison's (1973) "principle of proper mass," which claims that the more important a function is for a creature, the greater the area the brain devotes to it. The superior colliculus, which processes visual information, is nine times larger in a diurnal squirrel than in a nocturnal rat, for example (Striedter 2005). Putting together Morgan's Canon, the notion of allometry, and the principle of proper mass, we now have a proper methodology for doing research into mental capacities from an evolutionary perspective: we should examine the relative size of brain structures in close and distant relatives to help us discern diverging or enhanced cognitive functionality of different brain areas.

For example, it turns out that the volume of the hippocampus, a structure dedicated to processing memory for spatial location, varies in relation to body weight and brain volume for different types of birds. Birds that store food for winter (and so have to remember lots of spatial locations) have larger hippocampi relative to birds that do not store food (Healy & Krebs 1993; Healy et al. 1994; Sherry et al. 1989). This suggests that the ability to store food in birds co-evolved with larger hippocampi. (Note that this correlation does not give us the direction of causality: maybe a relatively larger hippocampus allowed birds to start storing food, or maybe birds that started storing food put selection pressure on the size of the hippocampus.) Birds that store a lot of food also appear to have larger brains overall in comparison to their body size, which might reflect additional cognitive machinery they need to engage in food storing and food retrieving behaviors (Garamszegi & Eens 2004).

Similarly, birds that migrate have larger relative hippocampi than birds that stay put over their lifetimes (Cristol et al. 2003; Mettke-Hofmann & Gwinner 2003). We would expect that migratory birds have to remember more spatial information regarding where to acquire food and shelter, because, at the least, they need to know these things at either end of their voyage, while stationary birds only need to learn this information for one environment. Different ecological demands shape analogous structures differently across species, and we can use these selection pressures to help us understand what each of these structures do in our and others' cognitive economies.

Moving from birds to mammals, we can use similar approaches to understand brain differences among primates. It is hypothesized that primates that forage for fruit would need to remember more detailed spatial location information than primates who eat leaves (cf. Healy & Rowe 2007; van Schaik & Deaner 2003). Consequently, we would expect to find differences in relative hippocampi size among primates, just as we do with birds, based on their feeding patterns and preferences.

In addition, we can take the basic function we have assigned to the hippocampus to explore further differences in these neural components. We know, for example, that some mammals are monogamous and some are polygynous. In many monogamous species, males and females occupy the same territory, which suggests that they should have comparably sized hippocampi because (we assume) their spatial abilities should be the same, or close to it. In most polygynous species, females have relatively small territories where they rear the young, while males roam over much larger areas, seeking females with which to mate. This suggests that in polygynous mammals, males should have larger hippocampi relative to the females, because the males need to have greater spatial memory abilities than females. These hypotheses have been confirmed for several groups of voles and mice (Gaulin 1995; Jacobs 1995; Sherry 2006), and they are at least consistent with sex differences in brain structures and spatial cognition in people (Jones et al. 2003). (As a side note: female cowbirds who lay their eggs in other birds' nests have larger relative hippocampi than their male counterparts because they need to remember where all the potential host nests might be, so that they can quickly move in when the host parent is gone to deposit their eggs (Sherry et al. 1993). Other types of cowbirds that are not nesting parasites do not show these sex differences in size of hippocampus (Reboreda et al. 1996).)

The point in going through the details of this example is to suggest that our brain structures, along with their adaptive functions, have developed slowly over time in response to varied environmental pressures. Consequently, our minds reflect a multi-dimensional evolutionary history, which has resulted in mental capacities that strongly mirror the brain structures and cognitive capacities found in our biological relatives. In order to understand how spatial memory works in us, we need to understand how it works in other creatures and why, and how our underlying brain structures are similar to and different from our evolutionary neighbors.

Consider as a second example the evolution of the cerebellum. Often it is thought to be the structure tied to sensorimotor processing, like visually guided grasping, while the neocortex is presumed to underlie our more sophisticated information processing, like planning and executive control (cf. Lui et al. 2011; Rakic 2009). However, it has become clearer over the past decade or so that the cerebellum is intimately involved in our higher thought processes as well, including affect and empathy, associative learning, working memory, long-term memory, behavioral imitation, decision-making, and planning (Bel-lebaum & Daum 2011; Colibazzi et al. 2010; Ito 2008; Jackson et al. 2005; Leiner 2010; Rochefort et al. 2011; Steinlin 2008; Strick et al. 2009).

Robert Barton (2012) uses phylogenetic comparative analysis, a statistical modeling technique that combines data about the relatedness among species with data about their traits, to examine changes in the cerebellum and the neocortex in mammals over time. He compares the component volumes as well as the number of neurons with their behavioral correlates and demonstrates that the cerebellum and the neocortex evolved as a coupled unit, not only in primates (see also Barton & Harvey 2000; Herculano-Houzel 2010; Whiting & Barton 2003), but also across all mammal lineages.

This co-evolution lends support to the idea that the division between the "higher" cognition of the neocortex and the "lower" motor processing of the cerebellum is artificial at best. Human cognition is likely not due to adding supervisory modes of cognition on top of more primitive behavioral control mechanisms, but rather is "embodied" through and through. Thought and action perhaps are two sides of the same coin, developing and evolving together as a single piece (Anderson 2010; Barrett et al. 2012; Chiel & Beer 1997; Clark 1997; Damasio 1994; Wilson 2002).

The cerebellum might be fundamentally involved in modeling, predicting, organizing, and learning complex sequences of events and behaviors (cf. Barton 2012). These talents would support all sorts of diverse complex behavioral patterns—tool use, speech, working memory, and social organization. Human cognition might then turn out to be "the adaptation of sensory-motor brain mechanisms to serve new roles in reason and language, while [also] retaining their original function" (Gallese & Lakoff 2005: 456). It could be, for instance, that language is nothing more than a way for us to mark and maintain our places in a complex social hierarchy, or it could be an extension of our foraging behaviors and tool-use (cf. Barrett et al. 2012; Sterelny 2012; Sterelny, this volume).

Evolutionary perspectives on the mind open new ways of thinking about what it is to be human, what cognitive capacities might set us apart from our genetic neighbors (if any), and why our minds are structured the way they are. While not perfect, this presents a far better start to thinking about being minded than daydreaming in an armchair or assuming human exceptionalism.

REFERENCES

Anderson, M. 2010. "Neural reuse: A fundamental organizational principle of the brain." *Behavioral and Brain Science* 33: 245–313.

Barrett, L., Henzi, S., & Lusseau, D. 2012. "Taking sociality seriously: The structure of multi-dimensional social networks as a source of information for individuals." *Philosophical Transactions of the Royal Society B* 367: 2108–2118.

Barton, R. 2012. "Embodied cognitive evolution and the cerebellum." *Philosophical Transactions of the Royal Society B* 367: 2097–2107.

Barton, R. & Harvey, P. 2000. "Mosaic evolution of brain structure in mammals." *Nature* 405: 1055–1058.

Bellebaum, C. & Daum, I. 2011. "Mechanisms of cerebellar involvement in associative learning." *Cortex* 47: 128–136.

Block, N. 1978. "Troubles with functionalism." In C. Savage (ed.), *Perception and Cognition: Issues in the Foundations of Psychology* (University of Minnesota Press) 261–325.

Chiel, H. & Beer, R. 1997. "The brain has a body: Adaptive behavior emerges from interactions of nervous system, body and environment." *Trends in Neuroscience* 20: 553–557.

Clark, A. 1997. *Being There: Putting Brain, Body, and World Together Again* (MIT Press).

Colibazzi, T., Posner, J., Wang, Z., Gorman, D., Gerber, A., Yu, S., Zhu, H., Kangarlu, A., Duan, Y., Russell, J., & Peterson, B. 2010. "Neural systems subserving valence and arousal during the experience of induced emotions." *Emotion* 10: 377–389.

Conway-Morris, S. 1998. *The Crucible of Creation: The Burgess Shale and the Rise of Animals* (Oxford University Press).

Cristol, D., Reynolds, E., Leclerc, J., Donner, A., Farabaugh, C., & Ziegenfus, C. 2003. "Migratory dark-eyed juncos, *Junco hyemalis*, have better spatial memory and denser hippocampal neurons than nonmigratory conspecifics." *Animal Behavior* 66: 317–328.

Damasio, A. 1994. *Descartes' Error: Emotion, Reason, and the Human Brain* (Putnam).

Darwin, C. 1859. *The Origin of Species* (John Murray).

Darwin, C. 1871. *The Descent of Man and Selection in Relation to Sex* (John Murray).

Fodor, J. 1975. *The Language of Thought* (Thomas Cromwell).

Gallese, V. & Lakoff, G. 2005. "The brain's concepts: The role of the sensory-motor system in conceptual knowledge." *Cognitive Neuropsychology* 22: 455–479.

Garamszegi, L. & Eens, M. 2004. "The evolution of hippocampus volume and brain size in relation to food hoarding in birds." *Ecology Letters* 7: 1216–1224.

Gaulin, S. 1995. "Does evolutionary theory predict sex differences in the brain?" In M. Gazzaniga (ed.), *The Cognitive Neurosciences* (MIT Press) 1211–1225.

Healy, S. & Krebs, K. 1993. "Development of hippocampal specialization in a food-storing bird." *Behavioural Brain Research* 53: 127–131.

Healy, S. & Rowe, C. 2007. "A critique of comparative studies of brain size." *Proceedings of the Royal Society B* 274: 453–464.

Healy, S., Clayton, N., & Krebs, J. 1994. "Development of hippocampal specialization in two species of tit (*Parus spp.*)" *Behavioral Brain Research* 61: 23–28.

Herculano-Houzel, S. 2010. "Coordinated scaling of cortical and cerebellar numbers of neurons." *Frontiers in Neuroanatomy* 4: 12.

Heyes, C. & Huber, L. 2000. *The Evolution of Cognition* (MIT Press).

Ito, M. 2008. "Control of mental activities by internal models in the cerebellum." *Nature Review Neuroscience* 9: 304–313.

Jackson, P., Meltzoff, A., & Decety, J. 2005. "Neural circuits involved in imitation and perspective-taking." *NeuroImage* 31: 429–439.

Jacobs, L. 1995. "The ecology of spatial cognition." In E. Alleva, A. Fasolo, H.-P. Lipp, L. Nadel, & L. Ricceri (eds.), *Behavioral Brain Research in Naturalistic and Semi-Naturalistic Settings* (Kluwer) 301–322.

Jerison, H. 1973. *Evolution of the Brain and Intelligence* (Academic Press).

Jones, C., Braithwaite, V., & Healy, S. 2003. "The evolution of sex differences in spatial ability." *Behavioral Neuroscience* 117: 403–411.

Krebs, J. & Davies, N. 1993. *An Introduction to Behavioral Ecology* (Blackwell).

Land, M. 1991. "Optics of the animal kingdom." In J. Conley-Dillon & R. Gregory (eds.), *Evolution of the Eye and Visual System* (Macmillan) 118–135.

Land, M. & Fernald, R. 1992. "The evolution of eyes." *Annual Review of Neuroscience* 15: 1–29.

Land, M. & Nilsson, D.-E. 2002. *Animal Eyes* (Oxford University Press).

Leiner, H. 2010. "Solving the mystery of the human cerebellum." *Neuropsychological Review* 20: 229–235.

Lui, J., Hansen, D., & Kriegstein, A. 2011. "Development and evolution of the human neocortex." *Cell* 146: 18–36.

Mettke-Hofmann, C. & Gwinner, E. 2003. "Long-term memory for a life on the move." *Proceedings of the National Academy of Sciences (USA)* 100: 5863–5866.

Morgan, C. L. 1894. *An Introduction to Comparative Psychology* (Walter Scott).

Papini, M. 2008. *Comparative Psychology* (Psychology Press).

Rakic, P. 2009. "Evolution of the neocortex: A perspective from developmental biology." *Nature Reviews Neuroscience* 10: 724–735.

Real, L. 1993. "Toward a cognitive ecology." *Trends in Ecology and Evolution* 8: 413–417.

Reboreda, J., Clayton, N., & Kacelnik, A. 1996. "Species and sex differences in hippocampus size in parasitic and non-parasitic cowbirds." *Neuroreport* 7: 505–508.

Robinson, G. 2004. "Beyond nature and nurture." *Science* 304: 397–399.

Rochefort, C., Arabo, A., André, M., Poucet, B., Save, E., & Rondi-Reig, L. 2011. "Cerebellum shapes hippocampal spatial code." *Science* 334: 385–389.

Searle, J. 1980. "Minds, brains, and programs." *Behavioral and Brain Sciences* 3: 417–457.

Shapiro, L. 2001. "Mind as adaptation." In M. Walsh (ed.), *Naturalism, Evolution, and Mind* (Cambridge University Press) 23–41.

Sherry, D. 2006. "Neuroecology." *Annual Review of Psychology* 57: 167–197.

Sherry, D., Vaccarino A., Buckenham, K., & Herz, R. 1989. "The hippocampal complex of food-storing birds." *Brain, Behavior, and Evolution* 34: 308–317.

Sherry, D., Forbes, M., Khurgel, M., & Ivy, G. 1993. "Females have a larger hippocampus than males in the brood-parasitic brown-headed cowbird." *Proceedings of the National Academy of Sciences (USA)* 90: 7839–7843.

Shettleworth, S. 2010. *Cognition, Evolution, and Behavior*, 2nd ed. (Oxford University Press).

Sober, E. 2005. "Comparative psychology meets evolutionary biology: Morgan's canon and cladistic parsimony." In L. Daston & G. Mitman (eds.), *Thinking with Animals: New Perspectives on Anthropomorphism* (Columbia University Press) 85–99.

Steinlin, M. 2008. "Cerebellar disorders in childhood: Cognitive problems." *Cerebellum* 7: 607–610.

Sterelny, K. 2012. *The Evolved Apprentice: How Evolution Made Humans Unique* (MIT Press).

Strick, P., Dum, R., & Fiez, J. 2009. "Cerebellum and nonmotor function." *Annual Review Neuroscience* 32: 413–434.

Striedter, G. 2005. *Principles of Brain Evolution* (Sinauer Associates).

Van Schaik, C. & Deaner, R. 2003. "Life history and cognitive evolution in primates." In F. de Waal & P. Tyack (eds.), *Animal Social Complexity* (Harvard University Press) 5–25.

Whiting, B. & Barton, R. 2003. "The evolution of the cortico-cerebellar complex in primates: Anatomical connections predict patterns of correlated evolution." *Journal of Human Evolution* 44: 3–10.

Wilson, E. 1975. *Sociobiology* (Belknap Press).

Wilson, M. 2002. "Six views of embodied cognition." *Psychological Bulletin Review* 9: 625–636.

17

Routes to the Convergent Evolution of Cognition

Edward W. Legg, Ljerka Ostojić, and Nicola S. Clayton

Crows may be smarter than apes.[1]
Is Your Toddler as Smart as a Crow? No.[2]

These are just two examples of headlines of news articles discussing scientific reports of evidence for the convergent evolution between crows and primates, including human and non-human ones. However, evidence for convergent evolution of cognition is not just an interesting quirk. The process of convergence can provide a valuable insight into the evolution of cognition.

Here we demonstrate how possible cases of the convergent evolution of cognition can provide an insight into the nature of the evolution of cognition. First, we introduce the concept of convergent evolution. Second, we discuss how natural selection and other processes may act on cognition and the way in which these processes can lead to the convergent evolution of cognition. Third, we discuss candidates of convergence and the processes that can best explain these candidates. Finally, we reflect on what these proposed cases of convergence tell us about the nature of the evolution of cognition.

THE CONVERGENT EVOLUTION OF COGNITION

The definition of convergent evolution is a matter of contention. Typically, the term refers to the evolution of similar phenotypes in at least two distantly related species. For instance, the evolution of wings in both birds and bats is thought of as a classic example. Convergent evolution is contrasted with parallel evolution, which refers to the evolution of similar phenotypes in two related species and to cases where similar phenotypes are the result of similar developmental pathways. However, it is unclear at what point the relatedness between species is close enough for traits to be classified as parallel rather than convergent. This issue is further complicated by evidence that similar genes and developmental pathways can lead to similar phenotypic traits in distantly related species and that the converse pattern can emerge with closely related species exhibiting similar phenotypes through different genes and developmental pathways. These cases could,

237

arguably, fit the definition of both convergent evolution and parallel evolution. One solution to this contention has been to refer to all cases where the same trait has evolved independently as "convergent evolution" (Arendt & Reznick 2008). Thus, throughout this chapter we refer to convergent evolution *sensu* Arendt and Reznick (2008).

What Can the Convergent Evolution of Cognition Tell Us?

If a trait has evolved convergently, then the best explanation for that trait appears to be something other than the phylogenetic relationship between the two species with the trait. Consequently, non-phylogenetic similarities (e.g., ecology) between species can be used to produce hypotheses about why the two species possess similar traits. One of the benefits of convergent evolution of a phenotypic trait is that it helps answer the question "Why has this trait evolved?" without resorting to the "just so" stories criticized by Stephen J. Gould and Richard Lewontin (1979). If multiple instances of a trait serve the same function, then this makes it more likely that the current function of the trait reflects its historical adaptive value. For instance, the evolution of echolocation in bats and dolphins, both of which use echolocation for navigating their environments, is likely to have evolved because echolocation is a good solution for navigation when an environment contains limited visual information—the key similarity between the ecological niche of bats and cetaceans is a limit on visual information.

However, claims about the evolutionary origin of a trait must always be taken with caution, even in cases of convergent evolution (Losos 2011). First, the current utility of a trait need not be the reason why that cognitive ability evolved (Gould & Lewontin 1979). This caution may be particularly pertinent when considering cognitive processes because of their plasticity. For instance, the ability to process written words is likely to be a by-product of humans' ability to process visual features, and should not be viewed as the reason why humans can process visual features. Second, different species could exhibit two different traits that serve the same function. For instance, different jaw bone structures can produce the same biting force. Moreover, different cognitive abilities can serve the same function in different species. For example, desert ants return home using a dead reckoning system, whereby they integrate the motion vectors from the outward leg of the journey to produce a "homeward" vector (Müller & Wehner 1988). In contrast, bees rely on landmarks to relocate their hive (Collett & Graham 2004). Thus, there can be convergence in behavioral traits without there being convergence in cognitive traits.

The Multiple Realizability Problem

Cognition is not directly observable and must be inferred from the brain or behavior. Thus, similarities in brain or behavior are used to indicate that two species may possess similar cognitive abilities. However, as we have alluded to in the previous section, similar behaviors could be produced by different cognitive processes. In addition, these behaviors could also be produced by brains with a different structure (e.g., mammalian brains vs. avian brains). Consequently, there is no one-to-one mapping between brain, cognition, or behavior, and convergence on any level need not lead to convergence on another (Seed et al. 2009a).

Despite the lack of a one-to-one map between brain and cognition or between behavior and cognition, these two proxies can help to guide investigations into the convergent evolution of cognition. Species that exhibit similarities in their behavior or the structure of their brains can be seen as candidates for convergence of cognitive abilities.[3] To stand as cases of convergent evolution of cognition these candidates must be the result of at least two isolated examples of that particular cognitive process. In the next section we will discuss the theories about how these cognitive processes can be postulated and how they might occur.

APPROACHES TO STUDYING COGNITION

Historically there are two overarching approaches to the comparative study of cognition that differ in their expectations about what type of cognitive mechanisms different species should possess: the *general process* approach and the *ecological* approach (Papini 2002). The general process approach can trace its origins to Darwin, who suggested that there is no qualitative difference between the intelligence of man and other animals. He proposed that any differences were in degree rather than in kind. Thus, following Edward Thorndike (1911) the aim of comparative cognition research is to generate general rules that can account for the behavior of all species. In contrast, the *ecological* approach proposes that novel cognitive mechanisms may be selected for when they are beneficial for an individual. Thus, species in different ecological niches may possess different cognitive mechanisms that operate by different laws. This means that the ecological approach is more promiscuous than the general process view in making claims about the evolution of novel cognitive mechanisms.

The current approach within comparative cognition research is a hybrid between these two historical approaches. Specifically, the basic law of the general process account— associative learning—is used as a null hypothesis such that any claim for a specialized mechanism needs to be dissociated from associative learning. This means that convergent evolution of cognitive mechanisms would involve two species that possess the same cognitive mechanism and that this cognitive mechanism does not abide by the laws of associative learning.

However, other forms of convergence may occur that do not involve alterations to a cognitive mechanism itself. This can be illustrated by considering an analogous situation involving the various ways national leaders can be chosen. Different mechanisms allow national leaders to be selected—for example, democratic elections or a hereditary system. However, the outcome of the leader selection process can be influenced by changes that do not directly influence the mechanism. Universal suffrage, as opposed to male-only suffrage, does not alter the mechanisms used in an election but might lead to a different leader being selected. Thus, changes that occur upstream of a mechanism can influence the output of that mechanism.

In the case of cognitive mechanisms it is possible for a mechanism to remain intact but for alterations to the inputs of the mechanism to change. These altered inputs can be purely non-cognitive such as perceptual mechanisms, and could be the result of innate or environmental influences. For instance, animals are faster to associate tastes with sickness than pairings between sickness and an external stimulus such as a light, suggesting an innate bias that does not impact on the learning mechanism itself (Domjan 1980). Similarly, pigeons trained to discriminate on the basis of color between two lines that differ in both color and orientation, make more errors when learning a transfer task—involving different colors and orientations—when this task involves making an orientation-based discrimination rather than when the task involves making a color-based discrimination (Mackintosh & Little 1969). (This effect also occurs in the opposite direction when the initial training involves discriminating between lines on the basis of orientation.) This difference in performance is not the result of a change to how these discriminations are learnt but to a change in the feature that subjects attend to. Consequently, the outcome of a cognitive process may diverge or converge in two distinct ways: by the assembly of a novel cognitive mechanism or by alterations to the inputs to a mechanism.

The Four Routes Approach

Natural selection has undoubtedly shaped cognition but it would be incorrect to single it out as the only factor explaining the convergence or divergence of cognition. Where developmental processes are not constrained, it is possible for outcomes that are not solely influenced by natural selection to occur.

The need to distinguish the role of development in cognition from the role of natural selection is highlighted in an ongoing debate about whether there is an evolved mechanism involved in the recognition of faces. Evidence suggests that face processing can be localized to a specific brain region, and there are hallmarks of face processing such as inverted faces being much harder to discriminate than other inverted objects, indicating that face recognition involves a different mechanism to other forms of feature recognition (Kanwisher & Yovel 2006; Kanwisher et al. 1997). This has led some to propose that face processing is the result of an innate mechanism that has been selected for. However, an alternative view is that face processing could be the result of a general system that has been tuned, during development, to detect faces. Evidence for the latter hypothesis shows that people with expertise judging cars, dogs, or birds find it harder to process inverted images of the stimuli within their area of expertise than images of stimuli outside their area of expertise. In addition, it appears that the same brain region is involved in processing objects of expertise and faces (Gauthier et al. 2000; Diamond & Carey 1986). Thus, a face-processing mechanism may develop during an individual's own lifetime because of the importance of discriminating faces during development rather than through natural selection. There remains a debate about which of these hypotheses best explains the face-processing mechanism found in adult humans, but this debate highlights the importance of distinguishing the evolutionary and developmental influences on cognitive mechanisms.

As discussed previously, the evolution of cognition may occur for two reasons: through the assembly of a novel cognitive mechanism or through altering inputs to a mechanism. Developmental factors influencing cognition may also operate through either of these processes. Bringing all of these factors together, Cecilia Heyes (2003) has argued that it is advantageous to classify cases of the evolution of cognition within a framework highlighting the roles of phylogeny vs. development and the construction of mechanisms vs. alterations to inputs. Thus, Heyes has proposed that there are four routes to cognitive evolution, which she labels "phylogenetic construction" (natural selection leading to the assembly of a cognitive mechanism), "developmental construction" (developmental processes leading to the assembly of a cognitive mechanism), "phylogenetic inflection" (natural selection leading to biases in the inputs to a mechanism), and "developmental inflection" (developmental inputs leading to biases in the inputs to a mechanism).

The importance of Heyes's "four routes" framework is that it highlights the distinct routes by which cognition may be adapted. Consequently, this framework allows the identification of the basis of putative cases of convergent evolution of cognition, and provides a method to classify the nature of the convergence. Furthermore, the framework allows us to identify areas where more research is required to establish how convergence has come about.

In the next section we turn to discussing three examples where convergent evolution of cognition is thought to have occurred. For each putative case of convergence we will first describe the evidence that allows for the claims of convergence to be made. We will highlight evidence for similarities in the behaviors, brains, and cognitive abilities that may be best explained by ecological factors rather than relatedness. Second, we will turn to fitting the evidence into the "four routes" model so we can consider the causes of these similarities.

EVIDENCE FOR CONVERGENT EVOLUTION OF COGNITION

Spatial Memory

Spatial memory is by far the most well-studied example of the convergent evolution of a cognitive process. Different species vary in how much they need to rely on remembering different locations. Where survival and/or reproductive success depends on remembering locations, it is possible that there has been pressure to develop a specialized system for storing and processing spatial information. Examples include animals with large home ranges, brood parasites who must remember which nests they have laid eggs in, and food storers (cachers) who need to remember where they stored food (Sherry et al. 1992; Reboreda et al. 1996).

We will focus on the best-researched example—namely, food storing birds—although similar arguments could be made about the role of spatial memory in other cases. Importantly, in the case of caching there are two families of birds where the cognitive abilities of both caching and non-caching species have been extensively studied: paridae (e.g., blue tits and great tits) and corvidae (e.g., crows and jackdaws). Thus, it is possible to investigate whether ecological or phylogenetic factors might be the best explanations for variation in spatial memory.

Spatial information is stored in the hippocampus (O'Keefe & Nadel 1978). Thus, it is possible to test whether variations in a species' reliance on spatial memory influence the volume or the structure of the hippocampus. In the case of caching, there is good evidence for a link between dependence on caching and hippocampal volume.[4] In the parids and corvids, species that are heavily reliant on caching have larger hippocampi than species that are less reliant on caching (Healy & Krebs 1992; Hampton et al. 1995; Krebs et al. 1989). There is also some evidence that the hippocampal volume of a single species varies in relation to the amount they have to rely on caching—in the northern hemisphere, cachers from more northerly areas face a harsher climate, which makes them more reliant on their caches as a food source. In line with this hypothesis, the hippocampi of black-capped chickadees that live in Alaska are larger than the hippocampi of individuals living in Colorado (Pravosudov & Clayton 2002). Thus, hippocampus size is thought to be positively correlated with ecological pressures that promote caching.

In addition to the neurophysiological data there have been numerous studies investigating whether reliance on caching influences a species' performance on spatial memory tasks. According to the adaptive specialization hypothesis, species that are reliant on caching should outperform non-caching species on spatial memory tasks but not on tasks testing other forms of memory. A number of studies claim to support this hypothesis in both parids and corvids (Kamil et al. 1994; Hilton & Krebs 1990; Balda & Kamil 1988). Among the strongest of these is a study on four species of corvid that vary in the amount they cache. In this study the subjects were presented with both a spatial memory task and a color memory task on a touch screen monitor. Both tasks involved learning a "non-matching to sample" rule which requires subjects to press a feature that is different to one which has previously been displayed (Olson et al. 1995). In the spatial memory task stimuli could appear in two locations. Within each trial a stimulus would appear in one location and then the display would show the same stimulus in two locations and subjects were required to press the location that did not match the previously presented location. In the color memory task one of two colors was displayed first followed by a display of the two colors. The subjects had to press the color that did not match the previously presented color. All four species showed similar rates of learning in the color memory task but in the spatial memory task the Clark's nutcracker, the species most reliant on caching, showed a much faster rate of learning. This result led the authors to conclude that "these data indicate that evolution has acted differently on [the] processing of spatial and nonspatial information among these corvids" (Olson et al. 1995: 181).

Which Route Led to Convergence?

Is the enhanced spatial memory of cachers the result of the construction of a novel memory system or the result of biases in inputs? To date there is little clear evidence that the spatial memory system of cachers is qualitatively distinct from the spatial memory system of non-cachers. Alterations in the rate of learning and in the size of the hippocampus could be indicative of a novel spatial memory system but they need not be. A number of researchers have expounded an explanation of the differences found in the hippocampal and behavioral data between cachers and non-cachers as being linked to perceptual biases.

In line with the perceptual bias hypothesis, there is evidence that cachers prioritize spatial information over other forms of information. This evidence comes from tasks where spatial or visual information provides reliable cues about where to search for food. In these tasks subjects watch as food is hidden in one visually distinct object in one location and that object is moved to another location and replaced by a second visually distinct object. Here, cachers search first in the location indicated by spatial information but non-cachers use the visual and spatial information equally (Clayton & Krebs 1994).

Thus, differences between cachers' and non-cachers' performance on spatial memory tasks may simply be because it is "easier" for cachers to pick out the spatial information from the environment. (This attentional bias could be learnt (see Mackintosh & Little 1969) or innate.)

Moreover, Heyes (2003) has implied that there is a possibility that caching itself may cause the differences seen between cachers and non-cachers because making caches in distinct spatial locations increases the amount of spatial information that cachers perceive. Coupled with a bias to perceive spatial information this would mean that cachers are likely to perceive substantially more spatial information than non-cachers. Evidence that increased exposure to spatial information can influence the size of the hippocampus comes from humans. London taxi drivers have been found to have a larger hippocampus than other people, and this increase in hippocampal size develops during their training (Maguire et al. 2000; Woollett & Maguire 2011). If the size of a bird's hippocampus is related to levels of exposure to spatial information then differences between the size of the hippocampus in cachers and non-cachers could be the result of experience rather than the result of a novel cognitive mechanism for processing spatial memory.

Tool Use and Causal Cognition

One of the defining features of humans was previously thought to be our ability to manufacture and use tools (Oakley 1944). This attitude had to be re-assessed with the discovery of chimpanzee tool use, and this re-assessment has continued with the discovery of tool use in a wide variety of species from birds to cephalopods (Finn et al. 2009; Hunt 1996). Examples include New Caledonian crows inserting tools into crevices to retrieve grubs, and chimpanzees using sticks to fish for ants and using leaves or moss to drink water (McGrew 1974; Hunt 1996; Hobaiter et al. 2014).

Although tool use is no longer considered to be a uniquely human ability, it is still considered to reflect intelligent behavior. Sue Parker and Kathleen Gibson (1977) propose that for omnivores, extractive foraging (i.e., foraging that requires acquiring food from within an inedible container such as the hard shell of nuts) could require sophisticated cognition when it is not restricted to a specific context. Although there is no evidence that extractive foraging is related to brain size (Dunbar 1995), tool-using birds and primates do have a larger relative brain size (Reader et al. 2011; Lefebvre et al. 2004). Thus, it is seems reasonable to investigate whether or not tool use involves a specialized form of cognition.

If tool use has led to an adaptive specialization in cognition, then the specialization is likely to be in causal reasoning (the ability to reason about physical and causal events). Theoretically, evidence for an adaptive specialization in causal cognition requires demonstrating, first, that tool-users' physical reasoning abilities are different from those predicted by associative learning and, second, that there are differences between tool-users' and non-tool-users' use of causal reasoning. In reality, single studies do not always test both criteria, and comparisons need to be made using studies that employ similar methods.

Attempts to distinguish associative processes from a specialized causal reasoning mechanism have not always been successful. One of the most commonly used ways of dissociating associative learning and physical/causal reasoning involves testing whether a tool-using animal can learn about the functional properties of objects and transfer this knowledge to novel situations. If an animal is learning via associative learning, then during these transfer tasks it should rely on perceptual rather than functional similarity of objects. An early manifestation of this approach used a trap-tube—a cylindrical tube with a trap in the bottom. Researchers thought animals that learnt associatively could be distinguished from those with enhanced physical reasoning based on the animals' performance on a transfer task. In this transfer task the trap was present but was no longer functional because the tube had been rotated so that the "trap" was in the ceiling of the cylinder. If the animals had learnt to avoid the trap associatively, then they were expected to have learnt that the trap led to food loss and would continue to avoid the non-functional ceiling trap because of this association. In line with the latter associative learning-based hypothesis, tool-using capuchin monkeys continue to avoid the trap in the inverted tube whereas tool-using birds (namely, woodpecker finches) act in line with the former physical reasoning-based hypothesis, and ignore the trap (Tebbich & Bshary 2004). However, what the inverted trap transfer task actually tests is unclear because humans, who can verbalize that they understand that the trap in the ceiling is not functional, continue to avoid the inverted trap. This suggests that the avoidance of the inverted trap need not demonstrate a lack of physical cognition in humans or non-human animals (Silva et al. 2005).

Variants of the trap tube task that systematically manipulated the perceptual cues and functionality of the "trap" in a series of transfer tasks found that one chimpanzee (tool-user) could correctly solve all the transfers (Seed et al. 2009b). As for non-tool-using species, the rook demonstrated the equivalent level of performance (Tebbich et al. 2007; Seed et al. 2006). These results suggest that the same physical understanding may be present in tool-using and non-tool-using species that are distantly related.

Similar comparisons have been made between closely related species that differ in their use of tools. In one series of experiments, a comparison was made between tool-using woodpecker finches and non-tool-using small tree finches as well as between New Caledonian crows (tool-users) and carrion crows (non-tool-users) (Teschke et al. 2013). These experiments tested whether the birds would choose a functional hook to pull food toward them rather than a non-functional hook. (One hook is functional because it is placed around food and one hook is non-functional because it is placed next to food.) No difference between the performance of tool-using and non-tool-using finches was found, but tool-using corvids learnt faster than non-tool-using corvids. The task itself is very similar to the New Caledonian crows' natural tool use (which involves retrieving grubs from crevices) which may have made the task less demanding for them. Thus, this task may have been solved by New Caledonian crows because their learning was facilitated by a propensity to pull sticks toward them.

Further studies suggest that there may be little difference in what tool-users and non-tool-users understand about physics. Some of the best evidence against the "associative learning only" view comes from a study on Eurasian jays (a non-tool-using species) in a task reminiscent of the Aesop's fable about the crow and the pitcher (Cheke et al. 2011). The jays learnt to drop stones into tubes, and in a choice task rapidly learnt to drop stones into water rather than woodchips to retrieve a floating worm. In the critical transfer task, the jays were presented with an apparatus with three tubes. The central tube contained a floating worm and was too narrow for the jays to drop stones into. The two outer tubes were large enough for the jays to drop stones into and one of these was connected to the central tube but the jay did not know about this connection. If jays were solely reliant on the feedback to their actions, then they should be able to learn to drop a stone into the tube that made the worm come closer. However, if jays used their causal knowledge, which in this case was that there was no reason for dropping stones into the outer tubes to raise the central water level, then they would struggle to learn where to drop the stones. This is precisely what happened: contrary to their performance on other transfer tasks, the jays did not learn which of the two outer tubes raised the water level. This finding provides evidence that associative learning by itself is not enough to account for the jays' actions. Follow-up experiments on New Caledonian crows (tool-users) suggest their performance in the transfer tasks is similar to the jays' performance, indicating that similar cognitive abilities could underlie the physical cognition of both species (Jelbert et al. 2014).

Which Route Led to Convergence?

It is not clear that there has been convergent evolution of causal reasoning mechanisms between tool-using species. Instead, the majority of evidence indicates that similar causal reasoning mechanisms account for the behavior of both tool-using and non-tool-using species. It remains an open question whether the causal reasoning abilities of some animals can be dissociated from associative learning. Evidence from Eurasian jays and New Caledonian crows suggests this might be the case; if so, then further research will be needed to establish why these species might possess this type of cognitive mechanism.

Even if specialized cognitive mechanisms are not involved in tool use, we can still look at the phylogenetic and developmental influences on this behavior. The role of developmental influences on the nature of tool use has been highlighted by Ellen Meulman and colleagues (2013) who provide evidence that many tool-users acquire competency in the use of tools only toward the end of their developmental period. Social influences during development are thought to have a profound impact on tool-use with the tool preferences and activities of parents in chimpanzees, dolphins, and New Caledonian crows influencing infants' acquisition of tool use and the infants' preferences for tool types (Mann et al. 2008; Kenward et al. 2006). Moreover, in species that manufacture their tools, the ability to handle and use successful tools that have been discarded by adults may facilitate infants' mastery of tool use and manufacture. Whether these extended developmental periods are adaptations to tool use or, more likely, tool use is acquired because of relatively unconstrained learning during these periods is currently unclear.

In terms of phylogenetic influences on tool-using behavior, it is unclear whether there have been any adaptations that specifically facilitate tool use. In part this is because the ability to use tools may be a by-product of other traits (Taylor & Gray 2014). Possible examples include New Caledonian crows, which have straight beaks and highly binocular vision that allows tools to be manipulated more readily and for the tool tip to be seen (Troscianko et al. 2012). There may also be important constraints on whether tool use is possible based on the speed at which brains process and integrate information.

For instance, it has been suggested that the reason New Caledonian crows learn to pull strings, and goldfinches and siskins do not, is because corvids are much faster at integrating perceptual information (the reward coming closer) with motor information (pulling and then stepping on a string (Taylor et al. 2012)). Of course, neither of these traits needs to have been selected for their role in tool use, and so further investigations are necessary.

In summary, it appears that tool use may not have resulted in the evolution of special cognitive mechanisms. However, both developmental and phylogenetic influences may have led to the convergent evolution of tool-using behavior but the precise role of these influences requires more substantive investigation.

Mental Time Travel

Another aspect of cognition thought to set humans apart from other animals is our ability to mentally travel in time (Suddendorf & Corballis 1997). The concept of mental time travel has its origins in the distinction between episodic memory and semantic memory made by Endel Tulving (1972). Tulving initially classified episodic memory as involving what happened (the event), where it happened, and when it happened. However, the definition of episodic memory has since been re-formulated to refer to the re-experiencing of a past event (Tulving 2005). In contrast, semantic memory involves the recollection of information and does not require re-experiencing an event. Mistakes are much more common in episodic memories than semantic memories and this raises the question of why humans possess an episodic memory system. One of the most prevalent explanations of why there are two systems is the suggestion that semantic memory may function solely for the recollection of information and episodic memory may function to construct possible future scenarios from past experiences (Schacter et al. 2007).

Questioning the evolution of these two systems is difficult when in humans the episodic memory system is defined in subjective and phenomenological terms. Rather than asking whether animals consciously re-experience past episodes or pre-experience constructed episodes, investigators have had to develop behavioral tests based on Tulving's initial definition of episodic memory—namely, an integrated memory of what, where, and when (Clayton et al. 2009).

In the case of caching species, it may be beneficial to anticipate future needs and future states, which may have led to a strong selection pressure on episodic memory (Griffiths et al. 1999). In line with this hypothesis, a number of caching species appear to satisfy the basic what–where–when criteria of episodic memory. The first demonstration of this type of memory came from western scrub jays that were found to alter whether they retrieve preferred but perishable items from a caching tray or less preferred non-perishable items, depending on the time delay since they had cached (Clayton & Dickinson 1998). (The jays had learnt that they could retrieve from the trays four hours or 124 hours after caching.) Further results have indicated that providing the jays with additional information in the interval between caching and recovery, such as food degrading more quickly than expected, leads to the jays updating their expectation about the rate of decay such that they avoid the items they have learnt perish very quickly when their original caching tray is returned (Clayton et al. 2003). These abilities do not appear to be limited to western scrub jays because experiments on rats, magpies, and black-capped chickadees have revealed that they too encode what–where–when information (Zinkivskay et al. 2009; Babb & Crystal 2005; Feeney et al. 2009).

The future planning component of episodic memory has also been tested in a large number of studies that seek to demonstrate that animals can disengage from their own state to plan for a future state. For instance, squirrel monkeys who have learnt that selecting one thirst-inducing date will lead to water being received soon, and selecting four dates will lead to a delay before they receive water, end up reversing their initial preference for the four dates (Naqshbandi & Roberts 2006). In addition, Eurasian jays and western scrub jays alter where and what they cache based on what their future state will be when recovering their caches (Cheke & Clayton 2012; Raby et al. 2007; Correia et al. 2007).

To date, evidence for episodic-like memory and future planning has been found in a variety of species including pigeons and rats. Before claims of convergence can be made there needs to be some evidence that other species lack this ability. However, one reason to think that this type of memory may have evolved in a convergent manner is that one species of invertebrate, the cuttlefish, has been shown to learn a what–where–when rule (Jozet-Alves et al. 2013).

Which Route Led to Convergence?

Are behavioral indicators of mental time travel evidence for a specialized cognitive mechanism? This question comes in two forms. The first is whether what–where–when memory in non-human animals could indicate that animals have phenomenological states (i.e., that they re-experience a past event). This first question remains unanswered. The second, and most important question in terms of the convergent evolution of mental time travel, is whether evidence for what–where–when memory in non-humans is a different kind of memory.

To answer the second question requires testing whether the mechanism that allows an animal to act on what–where–when information is different from other forms of memory. For instance, Thomas Suddendorf and Janie Busby (2003) have proposed that an individual could act on what–where–when information without having an episodic-like memory system. For instance, exposure to what–where–when information could influence an individual's behavior without that individual's having a memory of "what–where–when" information. They claim that "event A might produce a cognitive change B that affects behavior C at a later point in time, but this need not imply that B carries any information about A itself—the mediator B might be causal rather than informational" (2003: 392). If this were the case, then the behavior at the latter time point would be highly inflexible because it is simply based on the presence of B (which contains no information) and the (potentially) learnt response C. An animal using this strategy would not be able to update its response C if new information came to light about what it cached at time A. However, as we have already discussed, western scrub jays do make these kind of updates and are capable of using information about changes to decay rates or a cached item's value to alter their behavior when they are allowed to recover their caches (Clayton et al. 2003). This suggests that the jays' behavior may well be based on a different kind of memory system. Whether other species exhibit this level of flexibility remains to be seen.

CONCLUSION

We have discussed three very different examples where convergence of cognition may have occurred. In the case of spatial memory, species that cache tend to have an enlarged hippocampus and perform better on spatial memory tasks than non-cachers. In contrast, in the case of tool use there is limited evidence that tool-using individuals have enhanced physical reasoning skills compared to non-tool-users. Finally, in the case of mental time travel it is currently unclear why species possess this ability, and further research is required to establish the reasons why they are capable of remembering an integrated representation of what–where–when.

The causes of these cases of convergence also differ from each other. In the case of spatial memory, there is a debate about whether the cognitive adaptation of cachers is due to differences in the cognitive mechanism or in perceptual inputs. In the case of tool use, it appears that the ecological pressures involved in tool use have not constructed a novel cognitive mechanism. Instead they might influence the developmental period or physiological features of tool users. Finally, in the case of mental time travel, a novel memory system might have been constructed. These examples show that the evolution of cognition does not occur in the same manner for all traits, and to understand how these traits evolved we need to consider the interactions between development and phylogeny as well as the role of non-cognitive mechanisms on cognitive processes via the information they feed to a cognitive mechanism.

These nuances in the way in which cognition has evolved can be easily overlooked if cognitive mechanisms are viewed as *either* specialized mechanisms *or* domain-general mechanisms. When cases of convergent evolution are considered within a framework that takes into account the role of development and upstream non-cognitive mechanisms, it is possible to generate hypotheses about the way in which cognition has evolved.

NOTES

1. www.telegraph.co.uk/news/science/science-news/3351960/Crows-may-be-smarter-than-apes.html.
2. http://time.com/42068/crows-intelligence-animals/
3. A drawback of this approach it that one could fail to spot species who have similar cognitive abilities that manifest themselves in different behaviors.
4. There has been some controversy over the evidence linking levels of caching and the size of the hippocampus on the basis that a meta-analysis of previous studies was unable to find a correlation between these variables (Brodin & Lundborg 2003). However, follow-up studies demonstrate that this was due to a difference between the sizes of hippocampi in North American and European species, and the correlation is found when the analysis is conducted for each continent separately (Lucas et al. 2004). It remains to be seen why there is a difference in the size of hippocampi between the two continents.

REFERENCES

Arendt, J. & Reznick, D. 2008. "Convergence and parallelism reconsidered: What have we learned about the genetics of adaptation?" *Trends in Ecology and Evolution* 23: 26–32.

Babb, S. & Crystal, J. 2005. "Discrimination of what, when, and where: Implications for episodic-like memory in rats." *Learning and Motivation* 36: 177–189.

Balda, R. & Kamil, A. 1988. "The spatial memory of Clark's nutcrackers (*Nucifraga columbiana*) in an analogue of the radial arm maze." *Animal Learning & Behavior* 16: 116–122.

Brodin, A. & Lundborg, K. 2003. "Is hippocampal volume affected by specialization for food hoarding in birds?" *Proceedings of the Royal Society B: Biological Sciences* 270: 1555–1563.

Cheke, L. & Clayton, N. 2012. "Eurasian jays (*Garrulus glandarius*) overcome their current desires to anticipate two distinct future needs and plan for them appropriately." *Biology Letters* 8: 171–175.

Cheke, L., Bird, C., & Clayton, N. 2011. "Tool-use and instrumental learning in the Eurasian jay (*Garrulus glandarius*)." *Animal Cognition* 14: 441–455.

Clayton, N. & Dickinson, A. 1998. "Episodic-like memory during cache recovery by scrub jays." *Nature* 395: 272–274.

Clayton, N. & Krebs, J. 1994. "Memory for spatial and object-specific cues in food-storing and non-storing birds." *Journal of Comparative Physiology A* 174: 371–379.

Clayton, N., Russell, J., & Dickinson, A. 2009. "Are animals stuck in time or are they chronesthetic creatures?" *Topics in Cognitive Science* 1: 59–71.

Clayton, N., Yu, K., & Dickinson, A. 2003. "Interacting cache memories: Evidence for flexible memory use by Western Scrub-Jays (*Aphelocoma californica*)." *Journal of Experimental Psychology: Animal Behavior Processes* 29: 14–22.

Collett, T. & Graham, P. 2004. "Animal navigation: Path integration, visual landmarks and cognitive maps." *Current Biology* 14: R475–477.

Correia, S., Dickinson, A., & Clayton, N. 2007. "Western scrub-jays anticipate future needs independently of their current motivational state." *Current Biology* 17: 856–861.

Diamond, R. & Carey, S. 1986. "Why faces are and are not special: An effect of expertise." *Journal of Experimental Psychology: General* 115: 107–117.

Domjan, M. 1980. "Ingestional aversion learning: Unique and general processes." *Advances in the Study of Behavior* 11: 275–336.

Dunbar, R. 1995. "Neocortex size and group size in primates: A test of the hypothesis." *Journal of Human Evolution* 28: 287–296.

Feeney, M., Roberts, W., & Sherry, D. 2009. "Memory for what, where, and when in the black-capped chickadee (*Poecile atricapillus*)." *Animal Cognition* 12: 767–777.

Finn, J., Tregenza, T., & Norman, M. 2009. "Defensive tool use in a coconut-carrying octopus." *Current Biology* 19: R1069–1070.

Gauthier, I., Skudlarski, P., Gore, J., & Anderson, A. 2000. "Expertise for cars and birds recruits brain areas involved in face recognition." *Nature Neuroscience* 3: 191–197.

Gould, S. & Lewontin, R. 1979. "The spandrels of San Marco and the Panglossian paradigm: A critique of the adaptationist program." *Proceedings of the Royal Society B: Biological Sciences* 205: 581–598.

Griffiths, D., Dickinson, A., & Clayton, N. 1999. "Episodic memory: What can animals remember about their past?" *Trends in Cognitive Sciences* 3: 74–80.

Hampton, R., Sherry, D., Shettleworth, S., Khurgel, M., & Ivy, G. 1995. "Hippocampal volume and food-storing behavior are related in parids." *Brain, Behavior and Evolution* 45: 54–61.

Healy, S. & Krebs, J. 1992. "Food storing and the hippocampus in corvids: Amount and volume are correlated." *Proceeding of the Royal Society B: Biological Sciences* 248: 241–245.

Heyes, C. 2003. "Four routes of cognitive evolution." *Psychological Review* 110: 713–727.

Hilton, S. & Krebs, J. 1990. "Spatial memory of four species of *parus*: Performance in an open-field analogue of a radial maze." *Quarterly Journal of Experimental Psychology Section B* 42: 37–41.

Hobaiter, C., Poisot, T., Zuberbühler, K., Hoppitt, W., & Gruber, T. 2014. "Social network analysis shows direct evidence for social transmission of tool use in wild chimpanzees." *PLoS Biology* 12: e1001960. doi. org/10.1371/journal.pbio.1001960.

Hunt, G. 1996. "Manufacture and use of hook-tools by New Caledonian crows." *Nature* 379: 249–251.

Jelbert, S., Taylor, A., Cheke, L., Clayton N., & Gray R. 2014. "Using the Aesop's fable paradigm to investigate causal understanding of water displacement by New Caledonian crows." *PLoS ONE* 9: e92895. doi:10.1371/journal.pone.0092895.

Jozet-Alves, C., Bertin, M., & Clayton, N. 2013. "Evidence of episodic-like memory in cuttlefish." *Current Biology* 23: R1033–1035.

Kamil, A., Balda, R., & Olson, D. 1994. "Performance of four seed-caching corvid species in the radial-arm maze analog." *Journal of Comparative Psychology* 108: 385–393.

Kanwisher, N. & Yovel, G. 2006. "The fusiform face area: A cortical region specialized for the perception of faces." *Philosophical Transactions of the Royal Society B: Biological Sciences* 361: 2109–2128.

Kanwisher, N., McDermott, J., & Chun, M. 1997. "The fusiform face area: A module in human extrastriate cortex specialized for face perception." *Journal of Neuroscience* 17: 4302–4311.

Kenward, B., Rutz, C., Weir, A., & Kacelnik, A. 2006. "Development of tool use in New Caledonian crows: Inherited action patterns and social influences." *Animal Behaviour* 72: 1329–1343.

Krebs, J., Sherry, D., Healy, S., Perry, V., & Vaccarino, A. 1989. "Hippocampal specialization of food-storing birds." *Proceedings of the National Academy of Sciences USA* 86: 1388–1392.

Lefebvre, L., Reader, S., & Sol, D. 2004. "Brains, innovations and evolution in birds and primates." *Brain, Behavior and Evolution* 63: 233–246.

Losos, J. 2011. "Convergence, adaptation, and constraint." *Evolution* 65: 1827–1840.

Lucas, J., Brodin, A., de Kort, S., & Clayton, N. 2004. "Does hippocampal size correlate with the degree of caching specialization?" *Proceedings of the Royal Society B: Biological Sciences* 271: 2423–2429.

Mackintosh, N. & Little, L. 1969. "Intradimensional and extradimensional shift learning by pigeons." *Psychonomic Science* 14: 5–6.

Maguire, E., Gadian, D., Johnsrude, I., Good, C., Ashburner, J., Frackowiak, R., & Frith, C. 2000. "Navigation-related structural change in the hippocampi of taxi drivers." *Proceedings of the National Academy of Sciences USA* 97: 4398–4403.

Mann, J., Sargeant, B., Watson-Capps, J., Gibson, Q., Heithaus, M., Connor, R., & Patterson, E. 2008. "Why do dolphins carry sponges?" *PloS ONE* 3: e3868. doi:10.1371/journal.pone.0003868.

McGrew, W. 1974. "Tool use by wild chimpanzees in feeding upon driver ants." *Journal of Human Evolution* 3: 501–508.

Meulman, E., Seed, A., & Mann, J. 2013. "If at first you don't succeed . . . Studies of ontogeny shed light on the cognitive demands of habitual tool use." *Philosophical Transactions of the Royal Society B: Biological Sciences* 368: 20130050. doi:10.1098/rstb.2013.0050.

Müller, M. & Wehner, R. 1988. "Path integration in desert ants, *Cataglyphis fortis.*" *Proceedings of the National Academy of Sciences USA* 85: 5287–5290.

Naqshbandi, M. & Roberts, W. 2006. "Anticipation of future events in squirrel monkeys (*Saimiri sciureus*) and rats (*Rattus norvegicus*): Tests of the Bischof-Kohler hypothesis." *Journal of Comparative Psychology* 120: 345–357.

O'Keefe, J. & Nadel, L. 1978. *The Hippocampus as a Cognitive Map* (Oxford University Press).

Oakley, K. 1944. "Man the tool-maker." *Proceedings of the Geologists' Association* 55: 115–118.

Olson, D., Kamil, A., Balda, R., & Nims, P. 1995. "Performance of four seed-caching corvid species in operant tests of nonspatial and spatial memory." *Journal of Comparative Psychology* 109: 173–181.

Papini, M. 2002. "Pattern and process in the evolution of learning." *Psychological Review* 109: 186–201.

Parker, S. & Gibson, K. 1977. "Object manipulation, tool use and sensorimotor intelligence as feeding adaptations in cebus monkeys and great apes." *Journal of Human Evolution* 6: 623–641.

Pravosudov, V. & Clayton, N. 2002. "A test of the adaptive specialization hypothesis: Population differences in caching, memory, and the hippocampus in black-capped chickadees (*Poecile atricapilla*)." *Behavioral Neuroscience* 116: 515–522.

Raby, C., Alexis, D., Dickenson, A., & Clayton, N. 2007. "Planning for the future by western scrub-jays." *Nature* 445: 919–921.

Reader, S., Hager, Y., & Laland, K. 2011. "The evolution of primate general and cultural intelligence." *Philosophical Transactions of the Royal Society B: Biological Sciences* 366: 1017–1027.

Reboreda, J., Clayton, N., & Kacelnik, A. 1996. "Species and sex differences in hippocampus size in parasitic and non-parasitic cowbirds." *Neuroreport* 7: 505–508.

Schacter, D., Addis, D., & Buckner, R. 2007. "Remembering the past to imagine the future: The prospective brain." *Nature Reviews: Neuroscience* 8: 657–661.

Seed, A., Emery, N., & Clayton, N. 2009a. "Intelligence in corvids and apes: A case of convergent evolution?" *Ethology* 115: 401–420.

Seed, A., Call, J., Emery, N., & Clayton, N. 2009b. "Chimpanzees solve the trap problem when the confound of tool-use is removed." *Journal of Experimental Psychology: Animal Behavior Processes* 35: 23–34.

Seed, A., Tebbich, S., Emery, N., & Clayton, N. 2006. "Investigating physical cognition in rooks, *Corvus frugilegus.*" *Current Biology* 16: 697–701.

Sherry, D., Jacobs, L., & Gaulin, S. 1992. "Spatial memory and adaptive specialization of the hippocampus." *Trends in Neurosciences* 15: 298–303.

Silva, F., Page, D., & Silva, K. 2005. "Methodological-conceptual problems in the study of chimpanzees' folk physics: How studies with adult humans can help." *Animal Learning and Behavior* 33: 47–58.

Suddendorf, T. & Busby, J. 2003. "Mental time travel in animals?" *Trends in Cognitive Sciences* 123: 391–396.

Suddendorf, T. & Corballis, M. 1997. "Mental time travel and the evolution of the human mind." *Genetic Social and General Psychology Monographs* 123: 133–167.

Taylor, A. & Gray, R. 2014. "Is there a link between the crafting of tools and the evolution of cognition?" *Wiley Interdisciplinary Reviews: Cognitive Science* 5: 693–703.

Taylor, A., Knaebe, B., & Gray, R. 2012. "An end to insight? New Caledonian crows can spontaneously solve problems without planning their actions." *Proceedings of the Royal Society B: Biological Sciences* 279: 4977–4981.

Tebbich, S. & Bshary, R. 2004. "Cognitive abilities related to tool use in the woodpecker finch, *Cactospiza pallida*." *Animal Behaviour* 67: 689–697.

Tebbich, S., Seed, A., Emery, N., & Clayton, N. 2007. "Non-tool-using rooks, *Corvus frugilegus*, solve the trap-tube problem." *Animal Cognition* 10: 225–231.

Teschke, I., Wascher, C., Scriba, M., von Bayern, A., Huml, V., Siemers, B., & Tebbich, S. 2013. "Did tool-use evolve with enhanced physical cognitive abilities?" *Philosophical Transactions of the Royal Society B: Biological Sciences* 368: 20120418. doi:10.1098/rstb.2012.0418.

Thorndike, E. 1911. *Animal Intelligence: Experimental Studies* (Macmillan).

Troscianko, J., von Bayern, A., Chappell, J., Rutz, C., & Martin, G. 2012. "Extreme binocular vision and a straight bill facilitate tool use in New Caledonian crows." *Nature Communications* 3: 1110. doi:10.1038/ncomms2111.

Tulving, E. 1972. "Episodic and semantic memory." In E. Tulving & W. Donaldson (eds.), *Organization of Memory* (Academic Press) 381–403.

Tulving, E. 2005. "Episodic memory and autonoesis: Uniquely human." In H. Terrance & J. Metcalf (eds.), *The Missing Link in Cognition: Origins of Self-reflective Consciousness* (Oxford University Press) 3–56.

Woollett, K. & Maguire, E. 2011. "Acquiring 'the Knowledge' of London's layout drives structural brain changes." *Current Biology* 21: 2109–2114.

Zinkivskay, A., Nazir, F., & Smulders, T. 2009. "What-where-when memory in magpies (*Pica pica*)." *Animal Cognition* 12: 119–125.

18

Is Consciousness an Adaptation?

Kari L. Theurer and Thomas W. Polger

INTRODUCTION

Philosophers who inquire about whether consciousness is an adaptation often have a slightly different question in their sights. The question they want to answer is what consciousness does, and they reason that if we can explain why consciousness came to be—or why human beings (and our ancestors) came to be conscious—then we can hope to explain what consciousness does—or what consciousness does in us, at any rate.[1] In particular, the thought goes, if consciousness was selected for by natural selection—if consciousness is an adaptation—then there must be something that consciousness "did" to increase the fitness of its bearers. And that, we can infer, is what consciousness does. So answering the question of whether consciousness is an adaptation is instrumental to figuring out what consciousness does, or does in creatures like us.

Our goal in this chapter is to expose the flaws in this line of reasoning and to outline and illustrate an alternative approach to the evolution of consciousness. Those who adopt what we think is a flawed line of reasoning typically presume that answering the question of whether consciousness is an adaptation—how and why consciousness came to be, as a product of natural selection—is a way of answering the question of what consciousness does. We begin by offering some reasons for thinking that these questions are not the same, and that answering one does not inevitably imply an answer to the other. We then argue that those who reason this way are tempted by a confused picture of evolution on which an adequate account of the evolution of consciousness must specify for what function consciousness is necessary—that is, why we had to be conscious—and that this misunderstanding depends on a suppressed premise which we think is entirely implausible and not even remotely tempting for ordinary traits.

We recommend an alternative strategy. Rather than attempting to explain why consciousness as a whole evolved, we propose a piecemeal approach to the study of consciousness that begins with small examples of consciousness, examines how and why they evolved, and then builds back out to the evolution of consciousness itself. We then apply this strategy to one small part of a paradigmatic example of consciousness: pain. We explain how nociceptors, specialized receptors involved in the pain sensory system, seem to have evolved; and we discuss what sorts of evidence are required for showing that nociception is an adaptation. In closing, we return to the question of how and why consciousness itself evolved.

THE QUESTION OF ADAPTATION

There are several problems with using the question of whether consciousness is an adaptation as an indirect approach to the question of what consciousness does. First, there are ways of answering the question about the etiology of a trait that shed no light at all on what it does. For example, one could observe that the increased representation of the trait in a population is greater than can be accounted for by random drift and from that infer that the trait is an adaptation—was selected for by natural selection—but without knowing what the trait does or for what virtues it was selected. Of course, it is widely thought that we cannot "observe" the presence or absence of consciousness in members of a population over evolutionary time, so this sort of approach to answering the question of whether consciousness is an adaptation may not be germane. But the present point nevertheless serves to highlight that the question of whether consciousness is an adaptation is not the very *same* question as the question of what consciousness does.

Second, it may be that consciousness is an adaptation but that what it "did" in the past that explains its initial selection is not what it does (or fails to do) now that explains its maintenance in our population. What consciousness is "for" in the sense of what it did (often enough) to increase the fitness of its historical bearers need not be what it does most often or at all in its current bearers. This is because it may be co-opted to do something new, as in the widely discussed example of insect wing buds that likely originated for thermoregulation and only later came to be used for flight (Kingsolver & Koehl 1985). But it could also be that consciousness is vestigial, a trait that once did something but does not now. Or its maintenance in the population could be due to neutral selective pressures, or to having been driven to fixation in the population at an earlier time.

A third possibility is that consciousness does something but that what it does is not what it is "for" doing, either because it was selected "for" something else or because consciousness itself was never selected for at all. The latter case is the widely discussed possibility that consciousness is a "spandrel," the byproduct of selection in action but not itself favored by natural selection (Gould & Lewontin 1979; Flanagan 1995, 2000).

So answering the question of whether consciousness is an adaptation is not guaranteed to help answer the question of what consciousness does, or vice versa.

That being said, there is still a possible path from evidence about the evolution of consciousness to conclusions about what consciousness does. If consciousness was adapted or maintained by selective pressures, and if we can find some evidence concerning how consciousness enhanced the fitness of its bearers, then we may hope to draw some conclusions about what consciousness used to do. And then we may inquire about whether that is also what consciousness currently does for creatures like us.

This line of reasoning is fraught with risks, but not entirely hopeless. We are not convinced that the kinds of evidence that are called for are currently available. For example, there are theoretical frameworks for examining the evolution of consciousness by linking the presence of consciousness to the presence of larger cortical mass (as discussed in Bering & Bjorklund 2007). But we fear that this hypothesis over-values cortex as opposed to other ways of arranging neural tissues; for example, nuclei in avian brains (Jarvis et al. 2005). And, even if validated, the hypothesis is one that would allow us to track the increased representation of consciousness as a phenotype without learning what consciousness itself does, or whether it was itself the object of selection pressures. Rather than taking the macroscopic approach of looking for evidence of the evolution of consciousness taken whole, we prefer to take a micro-based approach and look for evidence of the adaptation of some very small and very old examples of consciousness that most likely evolved long before mammals came onto the scene. We know that this raises certain doubts about whether we are discussing consciousness or only its precursors, and we shall return to those concerns in the end. But first we need to examine one more philosophical pitfall.

WHAT CONSCIOUSNESS DOES, NOT WHAT CONSCIOUSNESS MUST DO

Why is the flawed line of reasoning so tempting? We think the answer is that when philosophers think about the evolution of consciousness, they are often tempted by a confused picture of natural selection and its products. The confused picture has its origins in the old worry that consciousness might turn out to be entirely epiphenomenal, that is, that it might have absolutely no causal effects at all. If there is nothing that consciousness does and no difference that being conscious makes, then consciousness cannot be selected for—natural selection cannot sort out differences that make no difference. But, the questionable line of reasoning continues, whenever we think about candidates for what consciousness might do—what difference it might make—we always seem to be able to imagine or conceive of a non-conscious creature that does those same activities or has those same abilities. So it is hard to see how consciousness could have evolved if we cannot even conceive of what benefits it confers that could not also be accumulated by some non-conscious "zombie" counterpart (cf. Moody 1994; Bringsjord & Noel 2002). According to this line of reasoning, an adequate account of the evolution of consciousness will have to tell us what consciousness does that could not be done otherwise—for what it is that consciousness is necessary.[2]

Consider, for example, the following *Argument from Functionalist Intuitions* discussed by Zack Robinson and colleagues in an article entitled, "Is consciousness a spandrel?" (2015: 380):

1. Functionalist intuition: Any biological function f performed by conscious organism O^* could have been performed by a non-conscious, functionally equivalent organism O.

2. Properties that make no difference to biological function cannot be selected for.
3. Therefore, BAV: consciousness is either a spandrel or an evolutionary accident.

BAV is the *By-product or Accident View*, according to which consciousness is not an adaptation. Robinson et al. formulate this argument in order to show that BAV is implied by some familiar versions of functionalism, and therefore should not be shunned by philosophers. But if some versions of functionalism imply BAV, this argument does not show it. The functionalist intuition cited in the first premise is one source of the epiphenomenalist worry we entertained above. And we have granted that selection can act only on traits that make a difference. But there is a suppressed premise that is entirely implausible:

Replaceable is Epiphenomenal:
For any trait T*, if T* could possibly be replaced by a functionally equivalent trait T, then T* makes no difference.

According to *Replaceable is Epiphenomenal*, the only traits that can evolve are those that make a difference *that could not be made in any other way*. For if the purported effects of T* could also be produced by T, then we conclude that T* does not have those effects after all.

Fortunately for us—that is, both for all living organisms and for our present philosophical quandary—*Replaceable is Epiphenomenal* is false. It is not the case that the only traits that can evolve are those that produce effects that could not be produced in any other way. *Replaceable is Epiphenomenal* mistakes a comparative distinction for an absolute one. It is true that if T* and T have all and only the same effects, then if traits T* and T are present in the same evolutionary environment, then they will be equally fit and there would be no selective differentiation. But this is only relevant if the selective environment contains both T* and T. The fact that we can imagine a "functionally equivalent" T for any T* does not show that T* could not have evolved, for T* may have evolved in an environment that did not, as a matter of fact, contain any "functionally equivalent" T. In particular, we should not suppose that T* evolved from T. Which competing organisms and traits were present in the evolutionary environment wherein T* evolved can be established by finding evidence about those environments, but not by imagining what they could have been. The possibility of some "functionally equivalent" T is not a defeater for the hypothesis that T* is an adaptation.

Notice that the invalid line of reasoning is not even tempting for ordinary traits, that is, for traits other than consciousness. The fact that I can imagine birds flying by means of propellers rather than wings (or by insect-like wings rather than bird-like wings) is obviously irrelevant to the hypothesis that bird wings are adaptations for flight. This is because we know that birds have wings, and we are inquiring about how bird wings, as a matter of historical etiology, came to be. It seems to us that the *Replaceable is Epiphenomenal* trap tempts philosophers only because the epiphenomenalist worry about consciousness raises the specter that we do not even know whether we ourselves are conscious. If so then, in contrast to the example of wings, we cannot be confident that the trait is present at all. So there is a temptation to think that we need an answer to the question of what consciousness does that simultaneously accounts for its evolution and rebuts the epiphenomenalist by establishing that consciousness is in some sense necessary. But this is too much to ask from the evolutionary explanation. The theory of evolution by natural selection is useful precisely for explaining how traits that are not necessary come to be. Natural selection is an engine of contingency. If some trait were necessary—if that even makes sense, and we are not certain—then we do not need to explain its presence by citing natural selection.[3] What is necessary exists because it could not fail to do so, not because it increased the fitness of its bearers. And even if some trait T* were necessary for a particular capacity C, that would not explain the presence of T* unless C were also necessary. But we are doubtful that there are any such capacities or traits, or that the selectionist framework is the right way to think about such putative traits. To entertain one slightly plausible example: everything in an evolutionary environment has mass, so having mass is not, as such, fitness enhancing. Of course having mass might be a necessary condition on having any traits that are relevant to fitness, and it is plain that variances in mass can be relevant to fitness. But just the necessary "trait" as we are supposing it to be—having mass at all—is not a candidate for selection precisely because it is necessary.[4]

At this point it will be useful to make explicit some of the claims about natural selection and selectionist explanations to which we have already appealed. Natural selection is a process that causes the frequency of creatures bearing some particular trait to increase over generational time relative to the total population in its environment over time. The conditions under which this occurs require that the population include the trait in question as well as others—that is, that there is variation in the population—and that the trait in question be heritable. This is why we emphasized that traits that are necessary—insofar as we understand that idea at all—are not explained by natural selection. For we imagine that if a trait is necessary, then there is no variation in the population.

Furthermore, in order for selection to occur, the trait has to increase the fitness of its bearers relative to the fitness of the other members of the evolutionary environment, and there has to be some non-accidental connection between the increased fitness of the bearers of the trait and their increased frequency in the population (Lewontin 1970; Brandon 1990). If the trait is neutral with respect to fitness, then there would be no reason to expect that its frequency would increase in the population—it would not matter whether a creature has the trait or not.

Notice that answering the question of whether trait T* evolved by natural selection requires information both about T* and about the other traits that were present in the actual evolutionary environment. This is why we have emphasized that T* evolves in an actual population with actual and historically contingent environmental companions. T* does not have to be fitness enhancing compared to an imagined population of creatures that can do everything that creatures with T* can do, but without T*. It is enough that T* evolves in a population that does not have whatever fitness enhancement is in fact provided by T*, however accidental.

So the right question to ask about the evolution of consciousness is: What fitness advantage was actually provided by consciousness to its bearers in some actual evolutionary environment that explains why that trait—being conscious—became prevalent in some actual populations? For this question, our ability to imagine creatures who do the things that we do but who are not conscious is irrelevant. And that is fortunate, because it is hard enough to find actual evidence for evolution even when there is no doubt about the distribution of a trait in current or ancestral environments, much less if we are also worried about imaginary populations.

TOWARD AN UNDERSTANDING OF THE EVOLUTION OF CONSCIOUSNESS

Now that we have a better grip on the questions that we should ask and the evidence we should seek, how shall we proceed?

One approach would be to formulate a theory of consciousness, or at least a model of the correlates of consciousness, that would allow us to seek evidence for its presence and frequency in ancestral populations. The theory that consciousness is energy-hungry and requires big brains and bodies with certain metabolic capacities is one such model. This will lead us directly into questions about whether we have the formulated the correct theory of consciousness or whether we have identified the correct evidential markers. The task is not hopeless, but it is fraught with difficulties, not least of which the danger that we will end up begging the questions about what consciousness does that we hoped to investigate by inquiring about the evolutionary origins of consciousness.

An alternative approach is to forgo the grand theorizing about consciousness, or even formulation of a total evidential model, and instead focus individually on the markers of consciousness as we find it now. Compare the question of the evolution of, say, birds. When and how did the trait *being a bird* (i.e., *birdness*) evolve? Of course, just as with the complex trait of consciousness, we do not expect to find that *birdness* is the kind of trait (if there are any) that is turned on or off by a particular gene on a particular chromosome. Rather, *birdness* involves the coordinated presence of some array of traits. We could pick out some as the markers of *birdness*, and inquire as to when they were first co-present in some kind of creature, and what their co-presence in that kind of creature allowed them to do that other kinds of creatures could not—creatures that did not have all of those traits together, or perhaps that had none, as the evolutionary environment might contingently have been. If there is broad agreement about the array of features that are constitutive or indicative of *birdness*, then the inquiry will be straightforward—which is not to say that it is an easy question to answer, but only that the problems are the usual problems of doing science, especially historical science. Alternatively, we can delay the question of *birdness* itself, and instead inquire about the particular traits that might (or might not) be constitutive or indicative of *birdness*. This will be a promising approach if there is consensus about some but not all the features associated with *birdness*. And it offers a way of moving forward, however provisionally, even if there is no consensus about the traits associated with *birdness*. After all, we would have to look for that kind of evidence even if we had a theory or analysis of *birdness*.

This second approach is what we recommend for the investigation of the evolution of consciousness. Lacking a general theory or consensus about the correlates of *consciousness* per se, we can inquire about the evolution of some of the aspects of consciousness about which there are higher degrees of consensus. Now some will object that by taking this approach we are changing the subject, we are substituting an inquiry into the "easy" problems of consciousness for what seemed to be an inquiry about the "hard" problem of consciousness (Chalmers 1995, 1996). But this distinction between "hard" and "easy" requires us to think about consciousness in comparison to an imaginary population that includes zombie creatures who do everything that their conscious counterparts can do but without being conscious. And we have already pointed out that that is no way to think about evolution. Zombie "birds" that fly without having wings are relevant to thinking about the evolution of birds only if we have reason—evidence—to think that birds evolved in an evolutionary environment that was also inhabited by zombie "birds." And we have no reason to think so. Likewise, we currently have no reason to think that consciousness, or its components, evolved in an environment (or series of environments) inhabited by non-conscious "functional" duplicates. The default assumption, for which we now have good evidence, is that birds evolved in environments that were not also populated by zombie "birds." So too we suppose that consciousness (and its constituents) evolved in environments that were not also inhabited by zombie creatures that somehow pre-evolved the capacities enabled by consciousness in us but without being conscious.

We propose that the question of the evolution of consciousness should be approached by looking at the evolutionary origins of the components of consciousness as we experience it, and even of aspects of those components—pieces of the pieces, as it were. This seems to us to be an approach that is tractable at present, that does not await consensus on a grand theory of consciousness, and that can indeed contribute to that theorizing (as we hoped it might) rather than begging any questions. We will illustrate the piecemeal approach by considering the evolution of nociceptors, specialized receptors involved in the pain sensory system.

THE EVOLUTION OF PAIN: LESSONS FROM NOCICEPTORS

We begin our piecemeal investigation into the evolution of consciousness with a paradigmatic example of a conscious phenomenon: pain. Pain is perhaps a fundamental piece of the puzzle of consciousness that must be properly placed if we are to determine whether and why consciousness itself evolved. And yet when we begin our piecemeal investigation into the evolution of pain, a problem analogous to the one we have described with investigating consciousness wholesale immediately emerges, this time at a different level of analysis. The evolution of pain, much like the evolution of consciousness, is not so easily investigated because pain, like consciousness as a whole, itself involves a complex constellation of psychological, neurophysiological, and behavioral capacities that are not readily teased apart. Pain cannot be reduced to a single trait or simple cluster of traits about whose evolution we could hope to inquire. That is, pain, like consciousness itself, is more like *birdness* than it is like any simple trait or cluster of traits—such as industrial melanism in moths (Bowater 1914; Kettlewell 1955; Cook & Saccheri 2013)—that lends itself to easy empirical investigation. In the section "What Consciousness Does, Not What Consciousness Must Do," we suggested that if there is any hope for an evolutionary account of consciousness at all, then it is likely to come into focus only at the end of the long and arduous process of stitching together individual evolutionary explanations of bits of consciousness, and bits of those bits, in a patchwork fashion. Only once some of these smaller bits are firmly in place can we begin to assemble a clearer picture of how something as complex as pain, or attention, or color vision evolved. And only when those larger pieces begin to fall into place can we assemble a clearer picture of how conscious-ness as a whole evolved.[5]

Which pieces of the puzzle of the evolution of pain can we successfully investigate now? A natural place to begin illustrating our strategy is with an inquiry into the evo-lution of nociception. The ability to detect and respond to change in the environment, especially to the presence of noxious stimuli, is crucial for an organism's survival. Nociception, the process by which organisms detect actual or threatened damage to non-neural tissue, is an important part of the pain sensory system. Nociceptors are high-threshold receptors found on specialized neurons in the peripheral nervous sys-tem; they are responsible for detecting potentially or actually damaging stimuli. When activated in response to a potentially or actually noxious stimulus, nociceptive neurons transduce and encode that stimulus and send the appropriate signals to the rest of the nervous system. It is the nociceptors located on neurons in your fingertips that sense the heat of a hot stove, sending the appropriate signals to the central nervous system, causing you to pull your hand away from the stove. This process in humans is, of course, experienced as a distinct sensation: pain.

Nociceptors come in numerous forms corresponding to a variety of nociceptive capacities, including the detection of noxious mechanical, thermal, and chemical stimuli. The variety and sophistication of nociceptors present increases roughly with the overall complexity of the organism, but their basic molecular and electrophysiological proper-ties, especially at the level of ion channels, are strikingly similar across a wide array of animal species.

Nociceptors protect the organism via their responsiveness to potentially harmful stimuli in the environment. But their extraordinary plasticity serves to protect the organism from additional harm in another way: after injury, sensitization often occurs, during which the organism's nervous system protects the injured body part through increased nociceptor sensitivity and/or plasticity of nociceptor-related neural circuits. A hand that has touched a hot stove remains sensitive, thus prompting us to protect it from any further harm. The result is that the organism tends to the injured body part, thus preventing any further damage to the tissue (Lewin & Moshourab 2004; Woolf & Ma 2007).

So how did nociception, and nociceptive sensitization, evolve? The available data from comparative neurobiology suggests that nociceptive capacities evolved very early in phylogenetic history. Some bacteria, such as *Escherichia coli*, have primitive nocicep-tive capacities in the form of chemo-sensitive ion channels. However, this alone does not amount to full-blown nociception because bacteria lack the specialized neuronal tissue necessary for detecting noxious stimuli. In order to determine when genuine nociception began to evolve, we need to look at the development of the nervous system. Nervous tissue began to evolve around the time that animals with specialized tissues of any sort began to evolve, beginning with Eumetazoa. Beginning with Cnidaria, the ability to sense noxious mechanical stimuli evolved, along with simple nerve nets. But in Cnidaria there are no defined nociceptors. The evolution of genuine nociception seems to have kicked into high gear along with the development of a single major integra-tive area of the nervous system, for which bilateral symmetry cleared a path (Smith & Lewin 2009). So it is with the evolution of bilateral symmetry that we see the first nociceptors. Specialized neurons responding to noxious mechanical stimuli have been identified in molluscs and annelids. With the evolution of mammals, more complicated nociceptors emerged, which are capable of detecting a wide range of mechanical, thermal, and chemical stimuli (Figure 18.1).

The ubiquity of nociception in the animal kingdom is strongly suggestive of two things. First, it is clear that nociception evolved very early in the history of animals. This

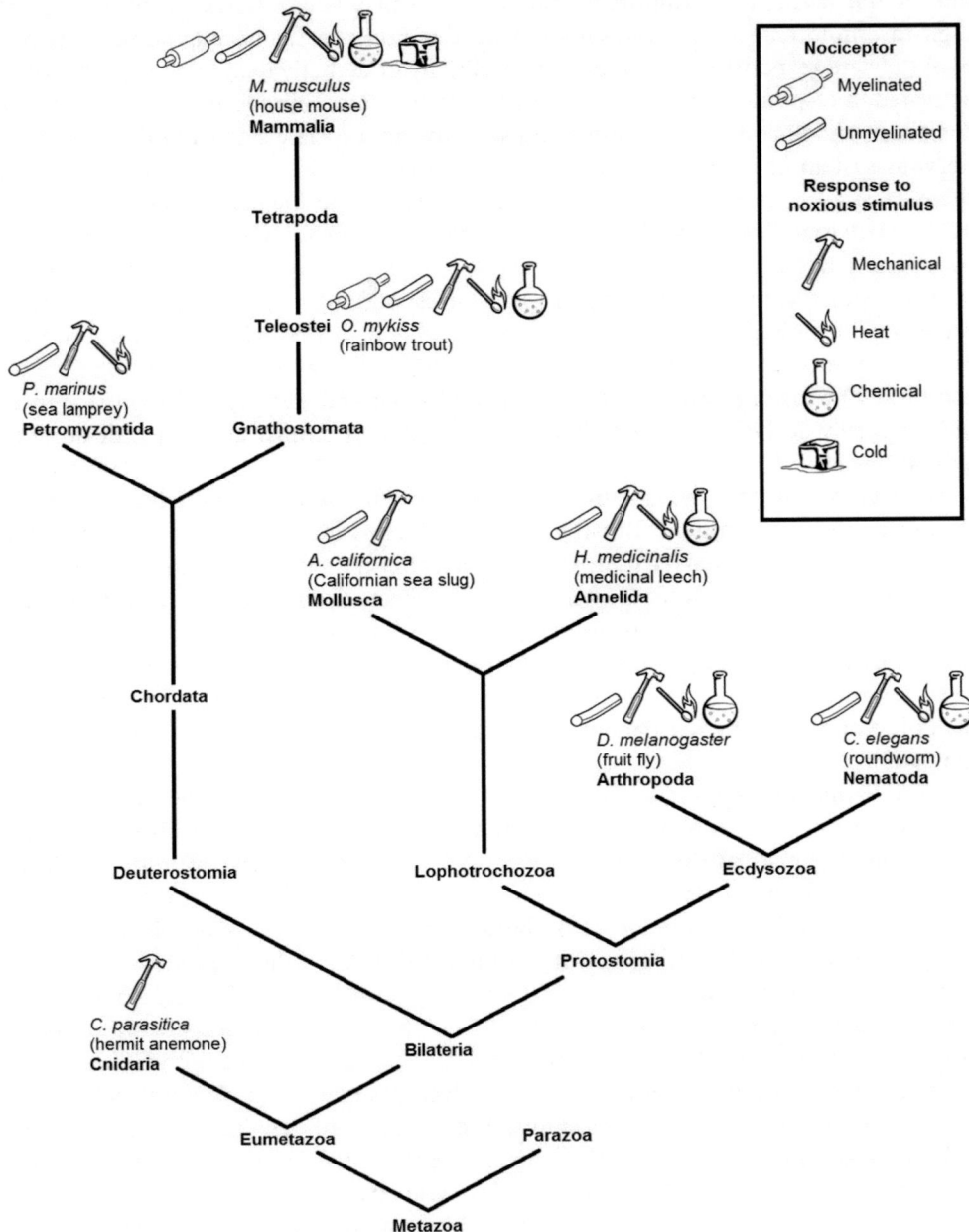

Figure 18.1 The evolution of nociceptors (from Smith & Lewin 2009)

is reflected in the similarities in the very basic properties of nociceptors themselves: while distinct sets of nociceptors, along with their respective capacities, are differentially distributed throughout the animal kingdom, the electrophysiological properties are strikingly similar, particularly at the level of the ion channels themselves. This is strong evidence that some of the basic components of nociceptors evolved only once and very early in phylogenetic time (Barr et al. 2008; Elwood & Appel 2009; Smith & Lewin 2009; Sneddon 2004; Sneddon et al. 2003). This kind of story is familiar from the evolution of parts of the visual system: while photosensitive eyes evolved many times, some of their basic components—like opsins, photosensitive pigments involved in the visual system—appear to have evolved from a common ancestor very early on in evolutionary time (Frings 2009; Shichida & Matsuyama 2009). Indeed, the ability to detect life-threatening stimuli is one of the most highly conserved traits in the animal kingdom (Smith & Lewin 2009; Neely et al. 2012), and the presence of mechano-sensitive channels has been demonstrated in bacteria, including *Escherischa coli* (Sukharev et al. 1994; Levina et al. 1999). In the case of both nociceptors and opsins, the diversity of macro-structures throughout the animal kingdom likely owes to the diversification of function over time in response to a variety of selection pressures—that is, to the processes of adaptation or exaptation. Moreover, the ubiquity of nociception is consistent with its conferring a strong selective advantage. Nociceptive sensitization also occurs in a wide variety of animals, which supports the claim that it, too, is the result of strong selective pressures in its favor (Walters 1994).

FROM HOW NOCICEPTION EVOLVED TO WHAT NOCICEPTION DOES

None of the comparative data or natural history we have just discussed is enough to determine decisively that nociception is in fact an adaptation. Nor does it allow us to take the additional step of inferring the function of nociception—that is, what nociception *does*. It is certainly possible for a gene or a trait to be driven to fixation in the absence of natural selection. Drift, exaptation, sexual selection, or the so-called "spandrel" effect can have this result (Gould & Lewontin 1979; Gould & Vrba 1982; Kimura 1968). Similarly, the ubiquity of pain or consciousness in a phylogenetic group would not provide incontrovertible evidence that either is an adaptation. So what sorts of evidence are required for showing that a trait is an adaptation?

We have already discussed in the section "The Question of Adaptation" some of what is required. We are now in a position to return to this point, applying these criteria specifically to the case of nociception.

Elisabeth Lloyd (2005) argues that several lines of evidence are required in demonstrating that a trait is an adaptation. Here we investigate each in turn, focusing primarily on the lines of evidence on which we have already mentioned and on which we believe selectionist explanations of consciousness must crucially depend.

First, and most crucially, the trait must have a differential effect on fitness. That is, the trait must make a difference to the expected reproductive success of the organism. In the absence of a connection between the trait and fitness, we cannot conclude that the trait is an adaptation. Evidence that the trait has a differential effect on fitness can take multiple forms, but it must be shown that the trait enhances the performance of the organism in some way that is relevant to reproductive success. And as we have emphasized, the relevant comparison is to the performance of actual evolutionary companions, not to imagined "zombie" competitors.

By now, we hope that the potential survival advantage afforded organisms with nociceptors is clear. Here we can distinguish between the survival benefit conferred by two distinct nociceptive capacities that we have already discussed. Nociception is advantageous for an organism's survival. If some organisms have this capacity and others do not, then those that can detect and avoid harmful or life-threatening stimuli are more likely to survive long enough to reproduce. Having a built-in mechanism that alerts you to stay away from things that might kill you increases the chances that you might find something to mate with and leave some offspring before you die. This much is intuitively plausible but it is not mere armchair theorizing: extensive experimental and clinical evidence suggests that immediate nociception during injury confers selective advantages that are far more specific than the "detect and protect" behavior that might seem obviously advantageous from the armchair. For example, when exposed to noxious stimuli, *Drosophila* (fruit fly) larvae produce stereotyped defensive behavior distinct from their characteristic locomotive behavior. Some evidence suggests that this unique nociceptive behavior evolved in response to selective pressure imposed by parasitoid wasps (Hwang et al. 2007). In addition to the survival benefits conferred by nociception, nociceptive sensitization after injury confers an additional fitness advantage by ensuring that the organism will continue to protect the injured tissue from further harm (Walters 1994). Minor injury increases the risk of predation, but nociceptive sensitization after injury appears to enhance responsiveness to threat, thereby increasing fitness of injured animals by decreasing the risk of predation (Crook et al. 2014).

One key element in this line of evidence is that it must be demonstrated that the trait is the result of selection pressures that *actually occurred* in the ancestral environment. This is more difficult than it may seem. Because our aim is to show that a given trait is an adaptation, we cannot reason from the assertion that the trait is an adaptation to the fact that the ancestral environment must have contained the relevant selection pressures. Thus the threat of circular reasoning looms and must be carefully avoided. Evidence that the environment actually contained the posited selection pressures must be confirmed independently of the trait's status as an adaptation. And indeed, we have good evidence that the environments in which nociception and nociceptive sensitization evolved contained heat, acid, cold, noxious mechanical stimuli, and predators of various sorts (Walters 1994).

Second, the trait in question must have a heritable basis. Without a heritable basis (including, but perhaps not restricted to, a genetic basis), the trait cannot be passed down from generation to generation, which is required for selection. And indeed the genetic underpinnings of nociception have been well documented (Clapham et al. 2001; Neely et al. 2012; Khuong & Neely 2013).

Third, there must be variation in the environment, that is, the trait in question must be one among many that could have been selected *for*. Although nociception is an ancient trait in animals, it is not universal to all multicellular organisms. We have good reason to think that nociception evolved in an environment that contained living things that did not have nociceptive capacities, so we also have good reason to think that there was variation in the selective environment and that the direction of change was from beings that lack nociception into those that have the capacity.

Selectionist explanations are made stronger by two further lines of evidence. Ideally, we would identify a mechanism linking the trait to increased fitness. A mere correlation between the trait and differential survival is not enough; a causal link between survival and possession of the trait is required. In the case of industrial melanism in moths, it was not enough merely to demonstrate the differential rates of survival between dark and light colored moths. Bernard Kettlewell's (1955) experiments showed *why* the dark-colored moths had a better shot at survival, nearly driving the dark-colored phenotype to fixation: in a polluted environment, light-colored moths are more easily visible to predators.

Finally, the hypothesis that the trait is an adaptation should be independently confirmed by manipulating either the environment or the trait itself, either artificially in the laboratory or by observing natural variation in the trait or the environment in the wild.

What do we make of all of this? The evidence strongly suggests that nociception is an adaptation. The amount of evidence and the detail in which we have discussed it may seem like overkill to some readers; they may think, "*Of course* nociception is an adaptation." But our goal here is not to *convince* anyone that it is—that should be clear enough. Our motivation for discussing the evidence in detail is to highlight what it actually takes to *demonstrate* that a trait is an adaptation. One cannot merely infer this from the fact that the trait is ubiquitous in a population, nor can one simply engage in armchair theorizing about the link between function and adaptation—those are "just-so stories" (Gould & Lewontin 1979). If we were to reason in this manner, we would run the risk of engaging in the kind of fantastical reasoning about *what consciousness does* that we have so far implored philosophers of mind to avoid.

IS CONSCIOUSNESS AN ADAPTATION?

As things stand, there is very little evidence meeting our demands that bears on whether consciousness as a whole is an adaptation. We suspect that this is because consciousness, like *birdness,* is not the kind of capacity that is amenable to this sort of explanation. But eventually, we may be able to construct a patchwork explanation of consciousness as we fill in the gaps about how various pieces of consciousness evolved—pain, color vision, hunger, conscious recollection, and the like.

We took some steps in that direction with our example of nociception. We aimed to meet the evidential standards for showing that a trait is an adaptation, which are, rightfully, quite high. We detailed the sorts of evidence required for showing that nociception (or any other piece of consciousness) is an adaptation. Armchair theorizing about the function of any trait is not enough, even when a trait's function seems intuitively obvious.

Our goal was to make some progress on the question of what consciousness does by showing what a piece of one bit of it does. Of course, nociceptor activation is not pain. Neuroscientists, evolutionary biologists, and clinicians are careful to maintain this distinction. And nociceptive pain is not the only way of being in pain. Because nociception is not pain, and because there is more than one way of being in pain, some philosophers may complain that we have only made some strides toward addressing one of the "easy" problems of consciousness, leaving the infamous "hard" problem untouched. As evidence, philosophers tempted by this line of reasoning might notice that they can imagine organisms who could do everything we can do—including avoid injury—without pain. Or that such organisms might have nociception but not pain. Or that alien or robotic organisms might manage to avoid harmful stimuli without either pain or nociception.

One can certainly imagine that organisms could respond to harmful stimuli and tissue damage via nociception but without pain and its accompanying and unpleasant affective states. And in fact we need not appeal to our imaginations or to philosophical thought experiments to see this. Many organisms are likely capable of nociception but not pain, because they have nociceptive capacities but lack the more sophisticated cognitive and neural machinery needed for the full-blooded mammalian pain with which we are all too intimately familiar.

Yet we should remind ourselves, first, that if our argument is persuasive, then we should admit that we are unlikely to make any headway on the problem of pain without first addressing a key step in its evolution. Understanding pain is, in turn, a key step in understanding consciousness. We have not explained consciousness, nor have we explained pain; but one cannot get to either of those destinations without making the requisite stops along the way. Understanding nociception is one of those stops. Similarly, chromosomal crossover during meiosis was a key step in the evolution of sexual reproduction. While sexual reproduction and chromosomal crossover are certainly not the same, one could not hope to inquire about the evolution of the former without understanding why and how the latter increases fitness.

Second, as we have emphasized, *of course* we can imagine other ways that the problem of injury avoidance could have been solved. In fact, there *are* other ways. For this is how natural selection works. The fact that we can imagine injury avoidance without pain or even without nociception does not entail that pain and nociception have no functional role whatsoever. This, we have claimed, is akin to claiming that because birds could have evolved with propellers rather than wings, or because different *kinds* of wings evolved independently in birds and insects (or because organisms found more than one way to reproduce, or because some animals navigate the environment by seeing whereas others navigate using echolocation and so on) that wings (or sexual reproduction, or vision) could not be adaptations at all. Yet that conclusion is absurd.

Nociception may occur in animals that lack the full-blooded subjective experience of pain that we humans find so compelling. But it does not follow that nociception or pain have no causal powers, or that they are not adaptations in us. That would follow only if we adopted the suppressed premise that we have warned against, namely, that a capacity that could have been performed in another way has no causal powers at all—*Replaceable is Epiphenomenal*. But there is no good reason to accept this suppressed premise, and plenty of reason to reject it.

CONCLUSION

In this short chapter we have sketched out, however provisionally, a viable strategy for answering questions about the evolution of consciousness, and particularly about whether consciousness is an adaptation. Our approach aims to avoid certain errors in reasoning about consciousness that have all-too-often tempted philosophers—particularly the assumption that *replaceable is epiphenomenal*. Our approach is piecemeal; it engages questions about consciousness from the "bottom up" by looking at particular examples of conscious experience, or even their precursors, with the hope of eventually assembling a fuller picture of consciousness as a whole (or such bigger bits of it that there are, as the case may be) and its arrival on the evolutionary scene. We thus chart a path that can be followed in the here and now, without waiting on a grand consensus about the nature of consciousness from philosophers or neuroscientists. This has the dual advantages of giving us a practical way forward in the face of widespread disagreement, and of avoiding the accusation that it begs questions about the nature of consciousness. We may begin our inquiries with the various and sundry pieces of consciousness around us whether or not they turn out to form a unified phenomenon of consciousness. Our experience is that many philosophers find this sort of piecemeal, particularist, and incremental approach unsatisfying; some are convinced that our kind of approach is inherently unsuited for answering "fundamental" questions about the world, such as those about metaphysics in general and about the metaphysics of consciousness in particular. But the piecemeal, particularist, and incremental approach is entirely typical of scientific inquiry in general and of evolutionary biology in particular—not just because they are as good as we can do in the here and now, but because they are the best than can be done for the study of evolution. The strategy is not merely heuristic, it is recommended. We should not expect that explanations of the evolution of consciousness will look different from explanations of the evolution of any other evolved phenomenon. What is good for the quail is good for the qualia.

NOTES

1. We assume that it is uncontroversial that normal adult human beings are conscious. We shall further assume that at least most mammals are conscious—that is, that they are the bearers of mental states or processes for which there is "something it is like" to undergo those states or processes. Later in the chapter, and more controversially, we shall assume that at least some varieties of non-mammalian animals are subjects of conscious mental states, but that assumption plays no part in the distinctions we are highlighting in this opening section.

2. We have previously examined this line of reasoning in Polger & Flanagan 2002 and Polger 2007. But as we shall demonstrate, it continues to tempt philosophers.

3. And in fact we do not. Features of phylogenetic groups that are universal, or very nearly so, are not always explained by natural selection. Some are explained by developmental constraints or are embedded in the basic body plan, or *bauplan*, of the organism. Bilateral symmetry in Bilateralia, for example, is maintained not because of direct selection pressure but because at this point in evolutionary history bilateral symmetry is locked into the body plan of all of the lineages that have it. Such features are indeed relatively impervious to manipulation in both laboratory and natural settings.

4. We are ignoring abstracta in this example, and supposing that all existents are concrete existents in space and time. Or, at least, everything in the selective environment has mass, that is, mass/energy.

5. For present purposes we assume that consciousness is a complex trait but we admit that we harbor some doubts about that assumption. It may be that the syndrome of processes that we recognize as conscious are insufficiently unified to form a genuine kind. In that case there is nothing to the phenomenon beyond the various pieces, no big picture formed by the puzzle. But the assumption of unity is harmless for us because the piecemeal approach we recommend is the right one either way.

REFERENCES

Barr, S., Laming, P., Dick, J., & Elwood, R. 2008. "Nociception or pain in a decapod crustacean?" *Animal Behavior* 75: 745–751.

Bering, J. & Bjorklund, D. 2007. "The serpent's gift: Evolutionary psychology and consciousness." In P. Zelazo, M. Moscovitch, & E. Thompson (eds.), *The Cambridge Handbook of Consciousness* (Cambridge University Press) 597–630.

Bowater, W. 1914. "Heredity of melanism in Lepidoptera." *Journal of Genetics* 3: 299–315.

Brandon, R. 1990. *Adaptation and Environment* (Princeton University Press).

Bringsjord, S. & Noel, R. 2002. "Why did evolution engineer consciousness?" In J. Fetzer (ed.), *Consciousness Evolving* (John Benjamins) 111–138.

Chalmers, D. 1995. "Facing up to the problem of consciousness." *Journal of Consciousness Studies* 2: 200–219.

Chalmers, D. 1996. *The Conscious Mind: In Search of a Fundamental Theory* (Oxford University Press).

Clapham, D., Runnels, L., & Strübing, C. 2001. "The TRP ion channel family." *Nature Neuroscience* 2: 387–396.

Cook, L. & Saccheri, I. 2013. "The peppered moth and industrial melanism: Evolution of a natural selection case study." *Journal of Heredity* 110: 207–212.

Crook, R., Dickson, K., Hanlon, R., & Walters, E. 2014. "Nociceptive sensitization reduces predation risk." *Current Biology* 24: 1121–1125.

Elwood, R. & Appel, M. 2009. "Pain experience in hermit crabs?" *Animal Behavior* 77: 1243–1246.

Flanagan, O. 1995. "Deconstructing dreams: The spandrels of sleep." *Journal of Philosophy* 5–27.

Flanagan, O. 2000. *Dreaming Souls* (Oxford University Press).

Frings, S. 2009. "Primary processes in sensory cells: Current advances." *Journal of Comparative Physiology A: Neuroethology, Sensory, Neural, and Behavioral Physiology* 195: 1–19.

Gould, S. & Lewontin, R. 1979. "The spandrels of San Marco and the Panglossian paradigm: A critique of the adaptationist programme." *Proceedings of the Royal Society, Series B: Biological Sciences* 205: 581–598.

Gould, S. & Vrba, E. 1982. "Exaptation—a missing term in the science of form." *Paleobiology* 8: 4–15.

Hwang, R., Zhong, L., Xu, Y., Johnson, T., Zhang, F., Deisseroth, K., & Tracey, W. 2007. "Nociceptive neurons protect Drosophila larvae from parasitoid wasps." *Current Biology* 17: 2105–2116.

Jarvis, E., et al. (The Avian Brain Nomenclature Consortium). 2005. "Avian brains and a new understanding of vertebrate brain evolution." *Nature Reviews Neuroscience* 6: 151–159.

Kettlewell, H. 1955. "Selection experiments on industrial melanism in the Lepidoptera." *Heredity* 9: 323–342.

Khuong, T. & Neely, G. 2013. "Conserved systems and functional genomic assessment of nociception." *The FEBS Journal* 280: 5298–5306.

Kimura, M. 1968. "Evolutionary rate at the molecular level." *Nature* 217: 624–626.

Kingsolver, B. & Koehl, M. 1985. "Aerodynamics, thermoregulation, and the evolution of insect wings: Differential scaling and evolutionary change." *Evolution* 39: 488–504.

Levina, N., Totemeyer, S., Stokes, N., Louis, P., Jones, M., & Booth, I. 1999. "Protection of Escherichia coli cells against extreme turgor by activation of MscS and MscL mechanosensitive channels: Identification of genes required for MscS activity." *The EMBO Journal* 18: 1730–1737.

Lewin, G. & Moshourab, R. 2004. "Mechanosensation and pain." *Journal of Neurobiology* 61: 30–44.

Lewontin, R. 1970. "The units of selection." *Annual Review of Ecology and Systematics* 1: 1–18.

Lloyd, E. 2005. *The Case of the Female Orgasm: Bias in the Science of Evolution* (Harvard University Press).

Moody, T. 1994. "Conversations with zombies." *Journal of Consciousness Studies* 1: 196–200.

Neely, G., Rao, S., Costigan, M., Mair, N., Racz, I., Milinkeviciute, G., et al. 2012. "Construction of a global pain systems network highlights phospholipid signaling as a regulator of heat nociception." *PLoS Genetics* 8: e1003071. doi:10.1371/journal.pgen.1003071.

Polger, T. 2007. "Rethinking the evolution of consciousness." In M. Velmans & S. Schneider (eds.), *The Blackwell Companion to Consciousness* (Blackwell) 72–86.

Polger, T. & Flanagan, O. 2002. "Consciousness, adaptation and epiphenomenalism." In J. Fetzer (ed.) *Consciousness Evolving* (John Benjamins) 21–42.

Robinson, Z., Maley, C., & Piccinini, G. 2015. "Is consciousness a spandrel?" *Journal of the American Philosophical Association* 1: 365–383.

Shichida, Y. & Matsuyama, T. 2009. "Evolution of opsins and phototransduction." *Philosophical Transactions of the Royal Society, Series B: Biological Sciences* 364: 2881–2895.

Smith, E. & Lewin, G. 2009. "Nociceptors: A phylogenetic view." *Journal of Comparative Physiology A* 195: 1089–1106.

Sneddon, L. 2004. "Evolution of nociception in vertebrates: Comparative analysis of lower vertebrates." *Brain Research Reviews* 46: 123–130.

Sneddon, L., Braithwaite, V., & Gentle, M. 2003. "Do fishes have nociceptors? Evidence for the evolution of a vertebrate sensory system." *Philosophical Transactions of the Royal Society, Series B: Biological Sciences* 270: 1115–1121.

Sukharev, S., Blount, P., Martinac, B., Blattner, F., & Kung, C. 1994. "A large-conductance mechanosensitive channel in E. coli encoded by mscL alone." *Nature* 368: 265–268.

Walters, E. 1994. "Injury-related behavior and neuronal plasticity: An evolutionary perspective on sensitization, hyperalgesia, and analgesia." *International Review of Neurobiology* 36: 325–427.

Woolf, C. & Ma, Q. 2007. "Nociceptors—noxious stimulus detectors." *Neuron* 55: 353–364.

19

Plasticity and Modularity

Edouard Machery

Since the 1970s, theories of cognitive architecture have vacillated between two opposites: to caricature somewhat, hypotheses of non-modular cognitive architecture dominated the 1970s and 1980s; in contrast, the massive modularity hypothesis put forward by evolutionary psychologists played an important role in cognitive theorizing in the 1990s and early 2000s, while the recent rise of predictive coding represents a shift toward less modular theories of cognitive architecture. The goal of this chapter is to discuss the prospects of modular theories of cognitive architecture in light of what is perhaps the strongest objection that has been raised against them: that such theories cannot account for the plasticity that is characteristic of humans' and other animals' cognition. I will argue that while plasticity raises a challenge for the strongest modular theories, it also reveals that much of human cognition involves an interplay between modular systems and non-modular cognitive control: in effect, there can be no plasticity without modularity.

Here is how I will proceed. In the first section, I introduce modular theories of cognitive architecture, and I discuss several influential evolutionary arguments for and against the claim that evolved cognitive architectures should be expected to be modular. In the next section, I distinguish different types of plasticity and present the challenge that plasticity raises for modular architectures. The following section examines three responses to this challenge. The final section argues that, while plasticity may challenge some forms of modularity, it also requires a substantial amount of modularity.

MODULARITY

Concepts of Modularity

What are modular theories of cognitive architecture committed to? Answering this question is difficult because the notion of module has been explicated in very different ways (for an overview, see Machery 2011a). It is common to start with Jerry Fodor's influential notion of modules (Fodor 1983), according to which modules are cognitive processes that tend to share the following nine properties:

1. A module has a specific type of input.
2. It produces shallow or non-conceptual outputs.
3. It is fast.
4. It is automatic.
5. It is cognitively opaque (i.e., other systems have no access to the representations produced within a module).
6. It is informationally encapsulated (i.e., it does not have access to some information accessible to other cognitive processes).
7. It is realized in a discrete brain area.
8. It is innate.
9. It has specific breakdowns.

Let's call "Fodor-module" any cognitive process that has most of the properties on this list. While none of them is necessary for a process to be a Fodor-module, Fodor often highlights informational encapsulation as the most important property: "A module sans phrase is an informationally encapsulated cognitive mechanism, and is presumed innate barring explicit notice to the contrary" (Fodor 2000: 58). Fodor takes the senses to be paradigmatic of Fodor-modules (but see Prinz 2005).

The notion of a Fodor-module is clearly inadequate to characterize most cognitive processes that make up the human mind: many are acquired, not innate; they are not localized—most processes are realized by networks of brain areas; many do not have characteristic breakdowns, etc. In response to this observation, philosophers and cognitive scientists have proposed various weakenings of the notion of module.

Peter Carruthers (2005, 2006) distinguishes two notions of informational encapsulation: wide-scope encapsulation and narrow-scope encapsulation. A cognitive process is wide-scope encapsulated or "frugal" if and only if there is some information that it does not have access to; it is narrow-scope encapsulated if it does not have access to some information (always the same body of information). Carruthers proposes that a process is a module if it is widely encapsulated: at any given time, there is some information that this process does not have access to. Let us call "Carruthers-module" any cognitive process that is widely encapsulated.

The notion of a Carruthers-module is undoubtedly weaker than the notion of a Fodor-module, and, in contrast to the former, it applies to many cognitive processes. However, the weakening proposed by Carruthers trivializes the notion of module since for all cognitive processes it is always the case that some information is not accessible to them. Only processes that would execute a complete search through the information available to a system would not be modular on this account, and it is extremely unlikely that any actual cognitive process is built this way. Searching takes time, and after some point the marginal benefits one could derive by finding some useful information are smaller than the marginal costs incurred by searching further through memory.

According to Steven Pinker (2005), a module is merely a cognitive process that has a specific function: Pinker refers to "modules in a much weaker sense of domain-specific functional organization" (2005: 16). He draws an analogy between the cognitive processes that make up human cognition and the components of audiosystems: the latter are made of parts, each of which is dedicated to a given function. Let us call this notion of module a "Pinker-module." Pinker-modules suffer from the same problem as Carruthers-modules (Fodor 2005): because cognitive processes are individuated functionally, it would seem that every cognitive process counts as a module under this characterization.

On an earlier occasion (Machery 2007), I attempted to address the shortcomings of Pinker's characterization. I call the cognitive processes that fit the improved characterization "Darwinian modules":

> What characterizes Darwinian modules is that they are designed to fulfill a specific function. That is, first, Darwinian modules are adaptations—the products of evolution by natural selection. Second, they fulfill a specific function: they evolved to underlie a specific cognitive competence. Third, being committed to some form of adaptationism, evolutionary psychologists often assume that modules are well designed for fulfilling their function.
> (Machery 2007: 826; see also Samuels 1998)

On this view, some modules will have many of the properties singled out by Fodor, including informational encapsulation, if these properties were adaptive. Other modules will have few of these properties if it was not adaptive to be a Fodor-module: they could be informationally open (vs. encapsulated), they could deliver conceptual outputs instead of shallow ones, they may be acquired largely by learning, and so on.

The Massive Modularity Hypothesis

To my knowledge, Dan Sperber (1994: 42) was the first to use the expression "massive modularity," but the massive modularity hypothesis had already been described and defended by evolutionary psychologists (Tooby & Cosmides 1992; Cosmides & Tooby 1994; for discussion, see Samuels 1998, 2000, 2005; Barrett & Kurzban 2006; Machery 2011b; Machery & Barrett 2006). The massive modularity hypothesis asserts that all or most cognitive processes are modules. As John Tooby and Leda Cosmides put it, "Our cognitive architecture resembles a confederation of hundreds or thousands of functionally dedicated computers (often called 'modules')" (1995: xiii).

Naturally, the exact content of the massive modularity hypothesis depends on which notion of module is used to unpack it. We can thus distinguish at least three massive modularity hypotheses (for additional distinctions, see Samuels 2000):

Fodorian Massive Modularity Hypothesis
All cognitive processes are Fodor-modules.

"Informational-Encapsulation" Massive Modularity Hypothesis
All cognitive processes are narrow-scope informationally encapsulated.

Evolutionary Massive Modularity Hypothesis (Machery 2007)
The mind is likely to consist of many systems, each having been designed by natural selection to fulfill a specific function.

Massive modularity hypotheses of this ilk are said to be "strong" because they assert that *all* processes are modules (in some sense of modules). By contrast, weak versions of these three hypotheses hypothesize that the mind is *mostly* made of modules (Samuels 2000).

Evolutionary Arguments For and Against the Massive Modularity Hypothesis

Evolutionary psychologists have put forward several arguments for the massive modularity hypothesis (Samuels 2000, 2005; Carruthers 2005, 2006). In a nutshell, "the optimality argument," as Richard Samuels (2000) calls it, argues for the massive modularity hypothesis on the grounds that, first, a collection of cognitive processes, each dedicated to solve a single adaptive challenge during the evolution of our lineage, would do better than a single cognitive process dedicated to solving all these challenges, and that, second, our ancestors were confronted with many different kinds of adaptive challenges. The optimality argument says nothing about the properties characteristic of Fodor-modules, including encapsulation; rather, it supports a weak or a strong version of the Darwinian massive modularity hypothesis. The so-called "solvability argument" (Samuels 2000) notes that the information needed to solve many problems cannot be learned, and is specific to these problems. Again, this argument supports either a weak or a strong version of the Darwinian massive modularity hypothesis. A third argument, which we may call "the mosaicity argument," goes further. The argument notes that evolution by natural selection requires the semi-independence or mosaicity of cognitive processes: when a system is made up of semi-independent components, natural selection can tinker with each of them without threatening the functioning of the other components and thus without needing to tinker with them. This argument suggests that the many psychological adaptations that make up the human mind have sparse communication channels: that is, cognitive processes receive information and send information to few other processes, a form of informational encapsulation.

Evolutionary arguments have also been put forward against the massive modularity hypothesis. In an important series of articles, Barbara Finlay (e.g., Finlay & Darlington 1995) has presented evidence that among mammals the volume of the large components of the brain (e.g., neocortex, thalamus) is allometrically related to the volume of the brain. That is, their volume varies as a function of the overall volume of the brain, exactly as across mammals the volume of the heart varies as a function of body size. Steven Quartz (1999, 2002) has appealed to this fact to challenge the idea that cognitive processes are modules. As he puts it (2002: 189):

> [D]espite a 10,000-fold range in neocortex size across mammals, the relative size of many brain structures is highly correlated. I review evidence indicating that heterochronic changes in the duration of neurogenesis result in the coordinated pattern of brain size across a variety of mammalian species. These results suggest that neural systems covary highly with one another as a consequence of the restricted range of permissible alterations that evolutionary psychology can act upon. This makes the massive modularity hypothesis of narrow evolutionary psychology untenable.

The gist of the argument goes as follows: versions of the massive modularity hypothesis that assume that cognitive processes are adaptations—particularly the Darwinian

Massive Modularity Hypothesis—must hypothesize that the neural systems that realize the modules evolved in a mosaic manner (see "the mosaicity argument" above): each module must have been under selective pressure independently of the other modules, at least to a large extent. However, Finlay and colleagues' work suggests that brain evolution was not mosaic.

Quartz's evolutionary argument against the massive modularity is much too quick (Machery 2007). First, while the volume of the large divisions of the brain is indeed related to the volume of the whole brain across mammals, there are also divergences from the expected regression line for many large brain structures in many lineages, suggesting that selection acted on such brain structures mosaically. Second, there is more to brain evolution than the evolution of volume. Evidence suggests that many other types of changes took place during the evolution of the primate brain.

THE MASSIVE MODULARITY HYPOTHESIS AND
THE CHALLENGE FROM PLASTICITY

Plasticity

"Plasticity" refers to the variation in the phenotype of organisms depending on environmental conditions. The norm of reaction of a species describes the plasticity of an organism with respect to an environmental factor. Plasticity fails when the phenotype of an organism is insensitive to environmental variation, for instance when it is canalized with respect to some varying environmental parameter (on the need to relativize canalization to particular environmental parameters, see Griffiths & Machery 2008); the phenotype is then "rigid." Plasticity may or may not be adaptive; in some circumstances, lack of plasticity (perhaps due to canalization) may be adaptive.

Plasticity takes various forms. In particular, plasticity is *developmental* when conspecifics develop different phenotypes in different environments. For instance, marsh plants (*Sagittaria sagittifolia*) look very different depending on whether they are partially submerged, completely terrestrial, or completely submerged. Some developmental variation is continuous—plants, for instance, can be more or less tall depending on their environment—while others are discontinuous—as is the case of the different phenotypes of the marsh plants. Learning can be thought of as an instance of developmental plasticity, one that involves rational changes in an organism's phenotype. Other forms of developmental plasticity are not naturally characterized as rational, although they can often be explained evolutionarily.

Behavioral and Psychological Plasticity

Different types of plasticity may give rise to distinct challenges to the massive modularity hypothesis (Carruthers 2006). First, we can distinguish *behavioral* from *psychological plasticity*. While human behavior is sometimes rigid—think, for example, of withdrawal behavior caused by anything looking like a snake or a spider—the diversity of human beings' behavior is remarkable. This type of plasticity includes the behavioral diversity found across cultures. Furthermore, human beings are often able to behave in inventive ways when they encounter novel circumstances; in particular, they often find new solutions to new problems. While behavioral plasticity is fascinating, it plausibly results from psychological plasticity: behavior is diverse because thought is diverse; behavior is inventive because thought is inventive. I now turn to psychological plasticity.

Psychological plasticity takes many forms (Carruthers 2006). We can distinguish context-sensitivity and "isotropy" (Fodor 1983). *Context-sensitivity* is the fact that information is processed differently depending on the context. *Isotropy* (which H. Clark Barrett (2005) calls "globality") is the fact that people are able to bring together all the thoughts they possess. Isotropy includes *conceptual isotropy*—the capacity to combine any two concepts—and *inferential plasticity*—the capacity to draw conclusions on the basis of any premise or evidence, a capacity deployed when one brings unexpected pieces of evidence to support or undermine a claim. Fodor refers to the latter as "the comparatively domain-free inferential capacities which apparently mediate the higher flights of cognition" (1983: 43). Human beings' thought is characterized by a remarkable degree of conceptual and inferential isotropy.

Carruthers (2006: 218) holds that context-sensitivity does not pose any problem for modular architectures: "I don't believe that the context-sensitive form of flexibility raises any particular problem for a massively modular conception of the human mind." Carruthers proposes, first, that an organism could collect information about the context and uses that information as input to the relevant module; second, that different organisms could focus on different aspects of the context. This proposal fails because, as Carruthers's proposal makes clear, the challenge that context sensitivity raises for modular architectures is tightly connected to the challenge that isotropy raises: a module processes information in a context-sensitive manner only if it is able to take as inputs the information that is constitutive of the relevant context, which is just another form of isotropy. Thus, Carruthers's proposed solution to the problem raised by context-sensitivity assumes that the isotropy challenge can be solved.

Behavioral and psychological plasticity contrasts with behavioral and psychological rigidity. An organism's behavioral repertoire is rigid to the extent that its behavior remains the same across circumstances. An organism's psychological repertoire is rigid to the extent that it processes information in a context-insensitive manner or to the extent that it is unable to bring together some concepts or to draw some inferences on the basis of some evidence (a form of encapsulation). Research on spatial orientation suggests that young children and adults who are prevented from engaging in inner speech are unable to bring together the color of a beacon with the shape of a room to orient themselves (e.g., Hermer & Spelke 1994, 1996; Hermer-Vazquez et al. 1999). Perceptual illusions also show that the perceptual systems are to some extent rigid.

The Challenge of Plasticity

Plasticity is often taken to raise a challenge to the massive modularity hypothesis. Sperber (1994: 40–41) writes that "the modularity of thought thesis seems to deny . . . plasticity," while Carruthers (2006: 224) asks whether "massive modularity theorists [should] expect that there will be limitations on the flexibility with which concepts (the components of thought contents) can be combined?" and responds that "the answer to this question is 'Yes.'" Barrett begins his article defending the massive modularity hypothesis with the following observation (2005: 259): "Currently, there is widespread skepticism that higher cognitive processes, given their apparent flexibility and globality, could be carried out by specialized computational devices, or modules."

Whether or not plasticity actually raises a challenge for the massive modularity hypothesis depends on which version of the hypothesis one considers (see Table 19.1 for a summary). First, plasticity does not challenge weak versions of this hypothesis, all of which are compatible with the existence of non-modular cognitive processes that could explain human beings' psychological plasticity. To see this, let us distinguish two versions of psychological plasticity. The *radical* version of psychological plasticity asserts that, perhaps outside perception and action, cognitive processes are all context-sensitive and isotropic. The *moderate* version concedes that some cognitive processes may well fail to be context-sensitive and isotropic, but adds that at least *some* processes are not so limited. For instance, on the former view, whenever one assesses a potential mate, the process underlying mate choice has always access to *all* possible information about him or her instead of relying on a limited set of cues (perhaps cues that have been indicative of high heritable fitness during evolution); on the latter view, the mate-choice process may well be limited in considering particular kinds of cues (as evolutionary psychologists have proposed), but we are also able to make decisions about whom to mate with by using a non-modular process able to access any type of input. The radical version of psychological plasticity would threaten even weak versions of the massive modularity hypothesis since on this view *no* process is restricted in the type of information it can process, which is inconsistent with the weak versions of Fodorian and "informational-encapsulation" massive modularity hypotheses. However, this version of psychological plasticity is unlikely to be correct, since evidence suggests that some cognitive processes have access to only a limited amount of information.

Let us now turn to the strong versions of the massive modularity hypothesis. Psychological plasticity challenges the strong versions of Fodorian and "informational-encapsulation" massive modularity hypotheses because both assert that cognitive processes are informationally encapsulated and thus seem unable to account for human beings' apparent capacities for entertaining thoughts with novel content or for bringing unexpected pieces of evidence to bear on inferences. Psychological plasticity also challenges the strong version of the Darwinian modularity because for cognitive processes the quasi-independence apparently required for the evolution of a system of adaptations amounts to a more or less extensive informational encapsulation.

Proponents of a strong version of the massive modularity hypothesis could of course respond by denying that human beings are psychologically plastic to a degree that is

Table 19.1 Summary (X = incompatible; □ = compatible)

	Strong modularity hypotheses			Weak modularity hypotheses		
	Fodorian	Informational-encapsulation	Evolutionary	Fodorian	Informational-encapsulation	Evolutionary
Radical plasticity	X	X	X	X	X	X
Moderate plasticity	X	X	X	□	□	□

incompatible with massive modularity, but instead they have typically attempted to show how a massively modular architecture is compatible with psychological plasticity. The next section reviews three influential attempts.

THREE RESPONSES TO THE CHALLENGE OF PLASTICITY

Sperber's Relevance

Sperber (2001) focuses on how sensitive, non-monotonous inferences could take place in a modular architecture. We often draw inferences that we would not draw if additional information or premises were added. We infer from the wetness of the street that it has rained, but if we remembered that it is the first day of the month and that street cleaning occurs on that day, we would not draw this inference. (Notice here again how context sensitivity and isotropy are tightly connected, a point made earlier against Carruthers's take on context sensitivity.) Sperber's solution is to appeal to a non-computational mechanism, salience or, in the more elaborate relevance-theoretic version, relevance (2001: 54):

> Does there have to be a higher-order computational process that triggers my remembering the day and suspending the default inference? Of course not. The allocation of resources among mental devices can be done in a variety of non computational ways without compromising the computational character of the devices. Saliency is an obvious possible factor. For instance the premise [the street is washed on the first day of the month] may be more salient when both the information that it is the first day of the month, and that the street is wet are activated. A device that accepts this salient premise as input is thereby more likely to receive sufficient processing resources.

The thought is that we draw contextually appropriate non-monotonous inferences because different pieces of information happen to be salient in different contexts, triggering different modular processes.

Sperber is right that critics of modular architectures that appeal to psychological plasticity have not addressed the possibility that a non-computational factor such as salience (or for that matter any other) could explain context-sensitivity, but, as he himself acknowledges, this is the wrong standard with which to assess his proposal: what is needed is a (by hypothesis non-computational) mechanism by which salience gets to be attributed in a context-appropriate manner. More important, Sperber's proposal seems to be more a restatement of the problem raised by psychological plasticity than the sketch of a possible solution; in particular, to say that inference is isotropic just is to say that, when one assesses whether to endorse a conclusion, any premise can be salient. Appealing to salience does not explain this problem, because the attribution of salience is what needs to be explained.

Barrett's Enzymatic Architecture

Barrett develops an influential analogy between enzymes and modules to argue that "specialized, modular processing can occur in an open system" (2005: 259): his proposed cognitive architecture is "an enzyme-like system in which modules all have access to a central blackboard or bulletin board, and in which there are no rigid routing procedures between modules" (2005: 261).[1] Barrett contrasts his position with Fodor's along two main dimensions. First, as most do (see above), he liberalizes the notion of module, embracing a notion that combines an evolutionary functional-specificity (as proposed by the evolutionary notion of module) with some form of domain specificity (see also Barrett 2015; Barrett & Kurzban 2006). Second, he proposes a particular solution to "the routing problem"—how information is routed from process to process inside the mind: all modules have access to all the outputs of perceptual processes and of other modules; these outputs are broadly broadcast on a bulletin board that is accessed by all (or nearly all) modules (an architecture known as a "blackboard architecture"). Similarly, enzymes have "access" to, or can encounter, any molecule. The routing problem is solved, Barrett thinks, because each module has input criteria that determine which representations it happens to take as inputs; other representations are simply ignored. This procedure alleviates the need for both rigid information channels (Fodor's (1983) solution) or a "meta-module" functioning to route information between modules, since it combines "access generality" with "processing specificity" (2005: 270).

The enzymatic model of modular architecture is meant to solve the problems raised by plasticity for modular architectures. First, this model claims to explain how thought could be global by appealing to access generality. Each module has access to all the representations broadcast on the bulletin board. Second, this model allows for contextual modulation of the processing carried by each module. Because the outputs of the processes are displayed on the common bulletin board, each module can be influenced by the outputs of the other modules, which would not be possible in a hierarchical system, where processes at one level forward their inputs to the next level.

There are, however, reasons to doubt that the enzymatic model of cognitive architecture satisfyingly addresses the problems raised by plasticity. First, it is unclear whether it satisfyingly solves the routing problem. Enzymes encounter the molecules they help catalyze by chance. A high concentration of enzymes of a particular type and of the relevant molecules ensures that the required encounters take place. A chance-plus-high-concentration strategy for matching inputs and processes is inadequate for matching input representations and cognitive processes. While there are numerous enzymes of a given type, there is only a single token process of a given kind; for example, there is only one cheater-detection module (if there is one at all, as Cosmides (1989) has proposed). As a result, the chancy nature of the matching between processes and inputs cannot be compensated by a high concentration of processes of a given type and of inputs. Chance matching between input representations and cognitive process would make the processing of inputs by processes uncertain, but the often subtle timing of processing excludes a form of matching that makes it uncertain whether processing occurs at the right time.

Second, Barrett's enzymatic model of cognitive architecture does not really show how a modular system could be "global," as Barrett calls it (or, as we called it, "isotropic"). It is true that in his model every process has access to every representation (provided that the latter is displayed on the common bulletin board), but not every process can be triggered by every representation; rather, each process has particular input conditions that exclude most representations. Thus, each process can be influenced only by very particular types of information, and it is then dubious that in an enzymatic architecture an inference can really be influenced by any information whatsoever.

Carruthers's Architecture: Plasticity through Inner Speech

Carruthers (2006) embraces a particular cognitive architecture, which in his view explains how a massively modular mind could be psychologically flexible. Of central importance is the fact that perceptual information is broadcast to all modules—a point reminiscent of Barrett's enzymatic model, except for the fact that Carruthers restricts the global broadcast of information to perceptual representations—and that inner speech involves the production of perceptual information (the representations of the articulation and of the predicted sounds). Conceptual modules feed into the module involved in the production of sentences. Representations of the articulatory structure of sentences are then produced. In inner speech, these representations feed into the auditory modules, resulting in efferent copies of these sentences (i.e., auditory representations of what would have been heard if these sentences had been uttered). These efferent copies, which are broadcast broadly in virtue of their perceptual nature, feed into the modules underlying language comprehension. As Carruthers puts it (2006: 232):

> Speech, of course, is a form of action. So one would expect that the precursor capacity for action rehearsal would carry over into this new domain. In which case there should be a capacity for creative generation of speech schemata (i.e. representations of possible utterances in a code appropriate for receipt by the motor systems), and a capacity to map those representations into a sensory modality (normally hearing), so that they can be globally broadcast and received as input by the conceptual modules.

According to Carruthers, this architecture solves the problem of psychological flexibility because the modules underlying speech can combine concepts coming from different modules and because articulatory and auditory representations with all types of content can be broadcast. He writes (2006: 236):

> [T]he language system may have the power to combine some of these into a single utterance, whose content will therefore be different from the content of any single thought currently being entertained. And one might expect that when an utterance is rehearsed (even one that is a direct encoding of an existing thought) a whole new set of inferences and emotional reactions might result once that thought has been globally broadcast and received as input by the full range of conceptual modules.

Unfortunately, Carruthers's solution to the challenge from plasticity fails (Machery 2008). Remember that the conceptual resources of each module are supposed to be limited. That is, each module can form the representations that are relevant for its function—and only those. For instance, our folk biology can form representations about the essence of biological kinds, about those biological processes (birth, death, disease, etc.) that explain changes in the biological properties of animals and plants, about the hierarchical organization of biological kinds, and so on, but, it cannot form representations about the intentional states of agents; it does not have the conceptual resources to do so. How, then, can the representation of the content of the sentence "My brother is a cheater" be processed by the desire-generating module underlying punishment? This module does not have the conceptual resources needed to form representations about kinship—it just has the conceptual resources needed to form representations involved in making decisions about punishment. More generally, how can the representation of the content expressed by the sentence "My brother is a cheater" be processed by any module? By hypothesis, this thought combines concepts that could not be combined, were it not for the capacity of the language-production module to produce the representation of the sentence "My brother is a cheater" out of the representations of the sentences "John is my brother" and "John is a cheater." Producing and globally broadcasting representations of sentences that combine words expressing concepts that belong to different modules are of little use in a modular structure, because modules can process only specific kinds of thoughts.

The psychological forms of plasticity examined in this chapter raise a challenge to strong versions of the three massive modularity hypotheses introduced in the first section. Massive modularity theorists have responded by weakening the notion of module and by presenting sophisticated architectures meant to show the compatibility of plasticity—including context-sensitivity and isotropy—and modularity, but the three accounts we have examined in this section are unsatisfying. Their failure suggests that a purely modular architecture is unable to account for the psychological plasticity characteristic of human thinking.

NO PLASTICITY WITHOUT MODULARITY

The previous section highlighted the challenge raised by psychological plasticity for modularity, and reached a negative verdict: as far as we know, a truly massively modular architecture leaves no room for human psychological plasticity. It would, however, be erroneous to infer from this the verdict that human cognition is not modular to a substantial extent. In fact, while psychological plasticity probably shows that a mind cannot be purely modular, it *also* shows that it must be modular to a large extent. That is, psychological plasticity shows the need for a subtle combination of interactive modular and non-modular processes. This, I argue, is the lesson to be drawn from recent research on expertise.

A common but outdated approach to expertise suggests that expertise results from an automatization of psychological processes: the expert (e.g., the world-class tennis player) can respond quickly to his or her environment (a tennis ball arriving on her left) because she does not have to think about what to do; rather, the detection of a given cue automatically results in the contextually appropriate behavior (e.g., Dreyfus & Dreyfus 1986). On this approach, expertise results in the modularization of thinking, including motor processes and decision heuristics, that is, our competencies depend on automated processes that respond quickly, although efficiently and adequately, to a limited set of cues.

In an important series of papers, Wayne Christensen and colleagues have shown that this approach is only partially correct (Christensen et al. 2015, 2016). Expertise does result in the automatization of some processes, but it also results in an increased role for a non-modular process of cognitive control. The expert not only reacts to stimuli in an automatic manner but she also constantly monitors her environment in order to respond to unexpected and novel circumstances. Automated processes and top-down non-modular control are thus tightly integrated according to Christensen and colleagues. They describe their theory, called "Mesh," as follows (2016: 43):

> Mesh sees a broadly hierarchical division of control responsibilities, with cognitive control usually focused on strategic aspects of performance and automatic processes more concerned with implementation. . . . Mesh proposes that controlled and automatic processes are closely integrated in skilled action, and that cognitive control directly influences motor execution in many cases.

It is only in easy circumstances, according to Mesh, that cognitive control becomes unnecessary, and that expert performance appears automatized. As soon as difficulty increases, the expert pays extreme attention to her environment to direct decisions and modulate the processing done by automatized processes.

How does the psychology of expertise bear on the relation between plasticity and modularity? First, expert behavior is plastic par excellence: what characterizes the expert, as Christensen and colleagues rightly insist, is her capacity to behave appropriately in novel circumstances. For example, the expert chess player knows how to deal with routine situations and with extraordinary ones. The psychology of expertise thus gives us a better understanding of the psychological underpinning of plasticity. In addition, while the psychology of expertise tends to focus on rare forms of expertise—competitive chess playing, the psychology of athletes, etc.—all of us have acquired many forms of expertise—ranging from reading, to walking, to biking, to driving, etc. The psychology of expertise thus casts light on mundane forms of plasticity.

What we learn from the psychology of expertise is that plasticity requires both modularization of component processes and top-down, non-modularized (non-automatic, broad range of inputs) control of these modularized components, and that the two interact constantly. So, while plasticity highlights the limitations of purely modular architecture, it also reveals the importance of modularization in enabling plasticity.

CONCLUSION

The plasticity of human behavior and of its underlying psychology has often been thought to be a serious challenge to modular architectures of cognition. While weakly modular architectures are immune to this challenge, strongly modular architectures do not seem able to deal with psychological plasticity, particularly with various forms of isotropy. On the other hand, psychological plasticity requires a great deal of modularization tightly integrated with non-modular processes. In this sense, there is no plasticity without modularization.

NOTE

1. The analogy was first proposed by Sperber (1994).

REFERENCES

Barrett, H.C. 2005. "Enzymatic computation and cognitive modularity." *Mind & Language* 20: 259–287.

Barrett, H.C. 2015. *The Shape of Thoughts: How Mental Adaptations Evolve* (Oxford University Press).

Barrett, H.C. & Kurzban, R. 2006. "Modularity in cognition: Framing the debate." *Psychological Review* 113: 628–647.

Carruthers, P. 2005. "The case for massively modular models of mind." In R. Stainton (ed.), *Contemporary Debates in Cognitive Science* (Wiley-Blackwell) 3–21.

Carruthers, P. 2006. *The Architecture of the Mind: Massive Modularity and the Flexibility of Thought* (Oxford University Press).

Christensen, W., Sutton, J., & McIlwain, D. 2015. "Putting pressure on theories of choking: Towards an expanded perspective on breakdown in skilled performance." *Phenomenology and the Cognitive Sciences* 14: 253–293.

Christensen, W., Sutton, J., & McIlwain, D. 2016. "Cognition in skilled action: Meshed control and the varieties of skill experience." *Mind & Language* 31: 37–66.

Cosmides, L. 1989. "The logic of social exchange: Has natural selection shaped how humans reason? Studies with the Wason selection task." *Cognition* 31: 187–278.

Cosmides, L. & Tooby, J. 1994. "Origins of domain specificity: The evolution of functional organization." In L. Hirschfeld & S. Gelman (eds.), *Mapping the Mind* (Cambridge University Press) 85–116.

Dreyfus, H. & Dreyfus, S. 1986. *Mind Over Machine: The Power of Human Intuition and Expertise in the Era of the Computer* (Free Press).

Finlay, B. & Darlington, R. 1995. "Linked regularities in the development and evolution of mammalian brains." *Science* 268: 1578–1584.

Fodor, J. 1983. *The Modularity of Mind* (MIT Press).

Fodor, J. 2000. *The Mind Does Not Work That Way* (MIT Press).

Fodor, J. 2005. Reply to Steven Pinker: "So how does the mind work?" *Mind & Language* 20: 25–32.

Griffiths, P. & Machery, E. 2008. "Innateness, canalization, and 'biologicizing the mind.'" *Philosophical Psychology* 21: 397–414.

Hermer, L. & Spelke, E. 1994. "A geometric process for spatial reorientation in young children." *Nature* 70: 57–59.

Hermer, L. & Spelke, E. 1996. "Modularity and development: The case of spatial reorientation." *Cognition* 61: 195–232.

Hermer-Vazquez, L., Spelke, E., & Katsnelson, A. 1999. "Sources of flexibility in human cognition: Dual-task studies of space and language." *Cognitive Psychology* 39: 3–36.

Machery, E. 2007. "Massive modularity and brain evolution." *Philosophy of Science* 74: 825–838.

Machery, E. 2008. "Massive modularity and the flexibility of human cognition." *Mind & Language* 23: 263–272.

Machery, E. 2011a. "Modularity." In D. Pritchard (ed.), *Oxford Bibliographies Online: Philosophy* (Oxford University Press). doi:10.1093/obo/9780195396577-015.

Machery, E. 2011b. "Developmental disorders and cognitive architecture." In A. De Block & P. Adriaens (eds.), *Maladapting Minds: Philosophy, Psychiatry, and Evolutionary Theory* (Oxford University Press) 91–116.

Machery, E. & Barrett, H.C. 2006. "Debunking *Adapting Minds*." *Philosophy of Science* 73: 232–246.

Pinker, S. 2005. "So how does the mind work?" *Mind & Language* 20: 1–24.

Prinz, J. 2005. "Is the mind really modular?" In R. Stainton (ed.), *Contemporary Debates in Cognitive Science* (Blackwell) 22–36.

Quartz, S. 1999. "The constructivist brain." *Trends in Cognitive Sciences* 3: 48–57.

Quartz, S. 2002. "Toward a developmental evolutionary psychology: Genes, development, and the evolution of the human cognitive architecture." In S. Scher & F. Rauscher (eds.), *Evolutionary Psychology: Alternative Approaches* (Kluwer) 185–210.

Samuels, R. 1998. "Evolutionary psychology and the massive modularity hypothesis." *British Journal for the Philosophy of Science* 49: 575–602.

Samuels, R. 2000. "Massively modular minds: Evolutionary psychology and cognitive architecture." In P. Carruthers & A. Chamberlain (eds.), *Evolution and the Human Mind: Modularity, Language and Meta-Cognition* (Cambridge University Press) 13–46.

Samuels, R. 2005. "The complexity of cognition: Tractability arguments for massive modularity." In P. Carruthers, S. Laurence, & S. Stich (eds.), *The Innate Mind: Structure and Contents* (Oxford University Press) 107–121.

Sperber, D. 1994. "The modularity of thought and the epidemiology of representations." In L. Hirschfeld & S. Gelman (eds.), *Mapping the Mind* (Cambridge University Press) 39–67.

Sperber, D. 2001. "In defense of massive modularity." In E. Dupoux (ed.), *Language, Brain and Cognitive Development: Essays in Honor of Jacques Mehler* (MIT Press) 47–57.

Tooby, J. & Cosmides, L. 1992. "The psychological foundations of culture." In J. Barkow, L. Cosmides, & J. Tooby (eds.), *The Adapted Mind* (Oxford University Press) 19–136.

Tooby, J. & Cosmides, L. 1995. "Foreword." In S. Baron-Cohen (ed.), *Mindblindness: An Essay on Autism and Theory of Mind* (MIT Press) xi–xviii.

20

The Prospects for Teleosemantics: Can Biological Functions Fix Mental Content?

Justine Kingsbury

Teleosemantic theories are naturalistic theories of mental content distinctive for their appeal to biological functions. Ruth Millikan's *Language, Thought and Other Biological Categories* (1984) and David Papineau's *Reality and Representation* (1987) presented teleosemantic theories that were the subject of much critical discussion in the 1980s and 1990s. Less attention has been paid to teleosemantics since, not because it has been shown to face insurmountable obstacles, or because some competing naturalistic theory of content has triumphed, but rather because naturalizing content is no longer center stage in the philosophy of mind.[1] However, teleosemantics remains one of the leading contenders in the field of naturalistic theories of content.

NATURALIZING CONTENT

What is it for a mental state to be about, say, the whiteness of snow, rather than the price of fish or the raininess of the weather here and now? This is the question that theorists of content seek to answer.

There are good reasons to want an answer to the question that is in some sense naturalistic. Naturalism goes along with seeing the philosophy of mind as an empirical rather than purely conceptual enterprise—an enterprise that is not different in kind from natural science. Mental states cause behavior, and what they are about makes a difference to what behavior they cause; on the face of it, intentionality (or "aboutness") is both real and causally efficacious. And as Jerry Fodor puts it: "if aboutness is real, it must really be something else."

> I suppose that sooner or later the physicists will complete the catalogue . . . of the ultimate and irreducible properties of things. When they do, the likes of *spin*, *charm* and *charge* will perhaps appear on the list. But *aboutness* surely won't;

intentionality simply doesn't go that deep . . . If the semantic and the intentional are real properties of things, it must be in virtue of their identity with (or maybe their supervenience on?) properties that are themselves *neither* intentional *nor* semantic. (Fodor 1987: 97)

There is little agreement on what exactly the naturalism requirement amounts to, but at a minimum it involves giving an account of mental content that does not make reference to anything intentional or semantic. It must not appeal to truth, or satisfaction, or reference; it must not appeal to the content or meaning of other kinds of entities (such as linguistic entities). The naturalizer's task is to characterize in non-intentional terms the relationship between a mental state and the state of affairs it is about.

An obvious first thought is that the relationship is causal. In its simplest form, the causal theory of content says that a mental state is about whatever it is caused by. No one believes the simple causal theory, but a brief consideration of its problems illustrates the challenges faced by naturalistic theories of content more generally and shows why teleosemantics looks like a promising response to them. Mental states are caused by all kinds of things. My first tokening of the belief that snow is white might have been caused directly by perceiving the whiteness of snow, but subsequent tokenings of that belief may be many causal steps away from that experience. When I token that belief nowadays (suppose a child asks me about snow, and in the process of thinking about what to say to her about snow, I think "Well, it's white"), there are many causes more proximal than my having perceived the whiteness of snow. And of course even my first tokening of it might have been several causal steps distant—if, for example, I found out that snow was white not by seeing that it was but by asking someone or by looking at the snowy landscape pictured on a Christmas card. How do we pick out the whiteness of snow as the content-relevant cause? Even supposing there is a principled way to do so, the situation for the simple causal theory appears hopeless when you consider *false* beliefs. If Mark falsely believes that it is now raining, the current raininess of the weather cannot have caused his belief, because it *isn't* raining. No false belief can be caused by the state of affairs it is about, because what it means to say is false is precisely that that state of affairs does not obtain (see Stegmann, Chapter 6 this volume).

Simple *informational* theories of content, according to which a mental state is about the state of affairs it carries information about, face the same problem. Being able to account for misrepresentation is a central requirement of a theory of mental content, and most naturalistic theories of content have trouble with it.[2] However, teleosemantics is an exception.

THE APPEAL OF TELEOSEMANTICS

Teleosemantic theories account for content in terms of the proper functions of mental states or of the mechanisms that generate those states. The proper function of something is what that thing is in some sense *supposed to do,* as livers are supposed to dispose of toxins and corkscrews are supposed to remove corks from bottles. Something may have a proper function that it does not perform: perhaps it is defective or malfunctioning; perhaps current conditions are not ones in which the function can be

performed. This feature of proper functions makes them a promising basis for a naturalistic account of content, since it suggests an answer to the question of how mental
states can misrepresent.

Appealing to proper functions in order to naturalize content will work only if proper
functions themselves can be naturalistically cashed out. The theory of proper function
accepted by most teleosemanticists is an etiological account: it satisfies this requirement.

THE ETIOLOGICAL ACCOUNT OF FUNCTIONS

The etiological account of function is most clearly illustrated by looking at the functions of biological items whose characteristics can be explained with reference to natural
selection. My heart, for example, does a number of things, including occupying space
in my chest cavity, circulating my blood, and making noises that can be detected with
a stethoscope. Which (if any) of these is its biological function? It so happens that the
noises my heart makes as it pumps blood are useful for diagnostic purposes, but making
those noises is surely not what the heart is *for*. It is for pumping blood: that is its biological function.

According to the etiological account of functions, what makes this so is that pumping
blood is what the hearts of my ancestors did that contributed to my ancestors' fitness. It
is because my ancestors' hearts were efficient blood pumps that those ancestors survived
and reproduced, and that explains (in part) why my heart is the way it is.

More formally, Karen Neander defines proper function in biology as follows:

> It is the/a proper function of an item (X) of an organism (O) to do that which
> items of X's type did to contribute to the inclusive fitness of O's ancestors, and
> which caused the genotype, of which X is the phenotypic expression, to be
> selected by natural selection.
>   (Neander 1991: 174; see also Neander, Chapter 5 this volume)

The etiological account of proper function can be extended and generalized to cover the
functions of things that are not straightforwardly the products of natural selection. The
extension is that if a mechanism (X) has the proper function of producing a particular
thing (Y) in order to bring about a certain further effect (E), then the (derived) proper
function of Y is to bring about E (Millikan 1989b: 288–289). For example, if a chameleon's
pigment-changing mechanism has the proper function of making the chameleon's skin
match its background in order to conceal it from predators, then the chameleon's skin
pattern has the proper function of concealing it from predators. The *generalization* is
that selection processes other than natural selection can generate proper functions; for
example, some learning processes, such as operant conditioning, may do so (Papineau
1987: 65–67; Kingsbury 2008).

TYPES OF TELEOSEMANTICS

Some teleosemantic theories account only for the content of very basic representations;
others are intended as fully fledged theories of content that cover all mental states including beliefs and desires. The theories put forward by Neander (1995) and Mohan Matthen

(1988) are of the former kind, although Neander hopes that her theory will provide a foundation on which a (not necessarily teleosemantic) theory of content for beliefs and desires can be built. The theories put forward by David Papineau (1987, 1993) and Millikan (1984, 1986, 1989a) are of the latter kind; Papineau and Millikan think teleosemantics can account for the content of structured, complex representations like beliefs and desires.

A further distinction is between *producer-oriented* and *consumer-oriented* teleosemantic theories—producers being mechanisms that generate representations and consumers being mechanisms that use them. I will illustrate these different kinds of teleosemantic theory using a standard case from the literature on teleosemantics: the frog that snaps at passing flies. The example also serves to illustrate some issues that arise for teleosemantic accounts of the content of basic representations. In particular, the most common criticism of teleosemantics is that it does not deliver determinate content attributions. My discussion of the frog-snap case will suggest that this is true of teleosemantic accounts of the content of basic representations, but that it is not a problem for such accounts. I will then go on to discuss teleosemantic accounts of more complex representations: in this context, indeterminacy *would* be a problem, but I will suggest that Millikan's teleosemantics in particular has the resources to avoid it.

BASIC REPRESENTATIONAL CAPACITIES: THE FROG'S FLY-SNAP MECHANISM

The Northern leopard frog (*Rana pipiens*) snaps its tongue out at small dark things moving past it and ingests them. Presumably it has some representational state R that is caused by small dark moving things (via the effect they have on its retina) that in turn causes the frog to snap at them. In the frog's natural environment, and presumably in the environment in which the frog evolved, almost all of the small dark moving things are flies. However, the frog is not discriminating about which small dark moving things it snaps at; if you toss ball bearings past it, it responds to them exactly as it does to flies.

R is a basic representation in the sense that it is a single on/off response to a particular stimulus; it is not part of any larger system of representation, it has no component parts that themselves have content, and it does not play a role in inferences. It is both indicative (it represents a state of affairs in the outside world) and imperative (it directly causes behavior).

What is the function of the frog's state R? What R does is to cause the frog to snap at and ingest certain objects in its environment, and presumably this is what it was selected for. These objects are usually flies; they are also small black moving things, and frog food. When the objects snapped at are frog food, ingesting them nourishes the frog; when they are not, snapping at them does the frog no good. So it looks as though R has the function of causing the frog to snap at food.

However, that is not R's only function (or at least it is not the only sensible way to *describe* R's function). Most if not all of the food snapped at by the frog in the environment in which it evolved was in the form of flies. So causing the frog to snap at *flies* is also something that R does that has led to the survival and multiplication of frogs (or, "snapping at flies" is a different way of describing the *same* thing that R does that has led to the survival and multiplication of frogs), and thus it is a function of R. For that matter, whenever R helped the frog to get nourished by making it snap at flies, we could also say it did so by making it snap at small dark moving things, so that too looks like a function of R. The following diagram (from Neander 1995) shows what R has historically done for the frog:

contributed to gene replication
by
helping to feed the frog
by
helping the frog to catch flies (prey? food?)
by
[causing it to snap at] small dark moving things.[3]

There is a multiplicity of ways of describing the function of R, and that is potentially a problem for the teleosemanticist if the aim is to use R's proper function to fix the content of R.

What is R's content? There are a number of possible answers. You might think, with Neander, that it is <small dark moving thing>; after all, the frog's mechanism does not distinguish between objects more finely than by identifying them as small dark moving things or *not* small dark moving things. In that case, the frog is not making a mistake when it snaps at a ball-bearing, even though ingesting ball-bearings does it no good. R is *supposed* to detect small dark moving things, and when a frog snaps at a ball-bearing, the problem is not that there is something wrong with R, but that the connection in the external world between small-dark-movingness and nutritiousness has broken down. Alternatively, you might think that the content is <fly>, or, as Millikan suggests (Millikan 1991: 163), that it is <food>, because it is only in the presence of food that R has contributed to the survival and multiplication of frogs—in which case the frog misrepresents when it snaps at a ball-bearing.

Neander's teleosemantics is *producer*-oriented: the discriminatory capacities of the mechanism that produces R are limited to distinguishing between small dark moving things (which trigger the snap response) and everything else, and that is a reason to think the content of R is <small dark moving thing>. Millikan's teleosemantics is *consumer*-oriented: the mechanisms that *use* R perform their proper functions when R is produced in response to frog food (regardless of whether or not the frog food is small, dark and moving), and that is a reason to think that the content of R is <food>.

An imaginary case due to Paul Pietroski (1992) vividly illustrates the way in which what triggers a representation can come apart from what that representation is used for. The kimus climb their local mountain every morning in response to the redness of the sunrise behind it, thus avoiding their predators the snorfs, who pass by the foot of the mountain each morning. The evolutionary history of this kimu response is that past kimus that did not respond this way got eaten by the snorfs before they had a chance to breed. The response is a one-off, basic response: kimus do not have an integrated system of representations of colors or representations of types of predator of which the representation underlying the mountain-climbing response is a part.

What is the function of the kimu mechanism, and what is the content of the representation that is triggered by the red sunrise and causes the kimu to move toward it? Are the kimus representing redness, or predator-freeness, when they respond to a perceptual stimulus every morning by climbing the mountain?

A *consumer-oriented* teleosemantics will say that what is represented is the direction of predator-freeness. It is because the mechanism causes the kimus to move to an area where there are no predators that the mechanism has spread through the population. That it is by responding to red light that the mechanism achieves this is neither here nor there. However, on the face of it, <no predators that way> seems an unlikely content attribution, in the same way as it might seem odd to attribute a <food> representation to the frog. The frog cannot tell food from non-food, as evidenced by its snapping at ball-bearings, and the current generation of kimus have never seen a predator (since they are always up the mountain when they pass by), so it seems a stretch to attribute to them a <no predators that way> representation. The *producer-oriented* teleosemanticist thinks so: in their view, it is what the representation has the function of being produced in response to that determines the content of the representation—in this case, red light.

The claim that the kimu cannot have an <avoid predators> representation because it cannot distinguish predators from anything else, and the claim that the frog cannot have a <food> representation if it cannot distinguish food from ball-bearings, have considerable intuitive force. However, this is undermined if you really manage to think of these representations as not being part of any wider system of representation. The kimu's representation does nothing except cause it to climb a hill and thus avoid being eaten—it is unrelated to any (other) representations of predators or to any (other) representations of color. As Graham MacDonald and David Papineau put it, "if these states never do anything except trigger simple avoidance behavior, it seems natural enough to read them as representing the danger they are designed to avoid" (Macdonald & Papineau 2006: 9). Likewise, *mutatis mutandis*, with the frog.

Different teleosemantic theories give different answers to the question "What is the content of R?" and it is not clear which of them is the right answer. (Consequently, the frog case is best used as an illustration of the differences between teleosemantic theories rather than as an argument for one or other of them.) Furthermore, given the situation represented in Neander's diagram above, even if we had reason to prefer consumer-oriented teleosemantics, we would still have a range of content-attributions available to us (<food> and <fly>, at least).

At this point it may be useful to step back a little. For what purpose do we need to assign determinate content to basic representations? The reason we think we need to for representations that form part of complex representational systems is in order to account for phenomena like inference. It may be that the content of the frog's state R just is indeterminate: for some purposes (such as predicting frog behavior) it makes sense to attribute the content <small dark moving thing>, but for others (such as explaining how the frog gets nourished) it makes sense to attribute the content <food>. (This amounts to being a *pluralist* about content.) Alternatively, we could pick one or the other: if the content is <small dark moving thing>, then the explanation of how the frog gets nourished is that it snaps at small dark moving things in an environment in which enough of the small dark moving things are food, and if the content is <food>, then the explanation of the frog's behavior goes via small dark movingness being the characteristic of the available food that the frog's mechanism responds to.[4] Nothing seems to turn on this. It is unclear that leaving the content of basic representational states such as R indeterminate is a fault in a theory of content.

TELEOSEMANTICS FOR COMPLEX REPRESENTATIONAL SYSTEMS

However, indeterminacy would be a fault in a teleosemantic account of content for states that form part of complex representational systems. I turn now to a more detailed consideration of Millikan's teleosemantics, which is the most fully worked-out teleosemantic account of the content of beliefs and desires.

If a desire has a proper function, plausibly its function is to bring about a particular state of affairs. The *content* of a desire, according to the teleosemanticist, is the state of affairs the desire has the function of bringing about (Millikan 1984; Papineau 1993).

It is less obvious how to characterize the proper function of a belief. It is tempting to suppose that "the function of any given belief type is to be present when a certain condition obtains" (Papineau 1987: 64). This would give us a tidy teleosemantic account of belief-content: the content of a belief is the state of affairs that belief has the function of being co-present with. However, the proper function of a thing is always to *do* something—to bring something about. Being co-present with some state of affairs is not the right kind of thing to be a proper function.[5]

What, then, are the proper functions of beliefs? Belief, as Frank Ramsey (1931) has it, is a map by which we steer. It is only because I have true beliefs about the locations of the pieces of furniture in a room relative to me that I can walk across the room without bumping into anything. And it is only because I have an accurate picture of how things are in the world that I succeed in acting in ways that change the world in the ways I want it changed. This is what beliefs do for us, and it seems unlikely to be accidental that they do. Presumably the reason we have cognitive mechanisms that generate beliefs and desires that work in this way is that having them increased the fitness of our ancestors. Millikan proposes that one of the proper functions of a belief is "to participate in inferences in such a manner as to help produce fulfillment of desires" and another is "to participate in inferences to yield other beliefs" (Millikan 1986: 67).

How, though, do such general proper functions generate an account of the content of particular beliefs? On Millikan's view, it is the conditions under which a particular belief performs those general belief-functions that determine the content of the belief. Consider my belief that there is coffee in the mug in front of me (call this belief "C"). When C is functioning properly, it participates in processes like this one: I want coffee, and because I have C, I drink the contents of the mug, and my desire for coffee is satisfied. This last stage, the satisfaction of my desire, will not (usually) happen unless there is coffee in the mug. There being coffee in the mug is the condition in the world that enables C to perform its functions, and so it is the content of C. Or consider my belief that it is going to rain today (call it "W" for *wet*). I have the desire to stay dry, and the belief that the best way to stay dry when it rains (if it is not windy, which I believe it is not going to be) is to use an umbrella—so I take my umbrella when I leave for work. I have other desires that are relevant to my picking up the umbrella on this occasion, including the desire not to cart an umbrella around unnecessarily. I also have other non-umbrella-related desires that W may help me to satisfy, including the desire to get my laundry dry and the desire not to water the garden unnecessarily. W plays a role in various inference processes that in turn generate behavior (taking the umbrella, putting the laundry in the dryer rather than on the washing line, refraining from watering the garden). What enables all of these W-involving processes to contribute to the satisfaction of my desires is that it actually is going to rain today—and so <rain today> is the content of W.

Two issues arise immediately. One is that this is not yet looking like a naturalistic account of belief-content, since desire-satisfaction plays an important role in it, and desire-satisfaction is itself an intentional notion. This is not an objection provided the teleosemantic account of content for desires succeeds, since that account is intended precisely to be a naturalistic account of satisfaction conditions.[6] The other is that the truth of a belief is neither necessary nor sufficient for the satisfaction of associated desires— sometimes things go wrong, and sometimes they go right by accident. A true belief can act in concert with a desire and yet fail to satisfy that desire, if other relevant beliefs are false. If the coffee mug is stuck to the table with superglue and I do not know that it is, my attempts to drink from it are more likely to result in the breaking of the mug than in the satisfaction of my desire for coffee. And a belief can help to satisfy the desires it is acting in concert with even if it is false—if it does not rain after all, but there is a burst water-main in the car park at work, the umbrella may help me stay dry.

On Millikan's view, the truth condition of a belief is neither necessary nor sufficient for that belief performing its proper functions; rather, it is a *Normal condition* for the performance of those functions. A *Normal explanation* is an explanation of how, historically, a type of thing has performed its proper function, and Normal conditions figure in Normal explanations. For any proper function and type of object that has that proper function, there will be more proximate and less proximate Normal explanations. Where R is a type of object and F is the function of objects of that type:

> [the most proximate Normal] explanation is the *least detailed* explanation possible that starts by noting some features of the structure of members of R, adds some conditions in which R has historically been when it actually performed F—these conditions being uniform over as large a number of historical cases as possible—adds natural laws, and deduces, i.e., shows in detail without gaps, how the setup leads to the performance of F.   (Millikan 1984: 33)

The *Normal conditions* for the performance of a mechanism's proper function are the conditions that must be mentioned in the most proximate Normal explanation of the proper functioning of that mechanism.

So, if a belief has the function of participating in processes that lead to the satisfaction of desires, the way to find the Normal conditions for the performance of that function is to consider the cases in which the belief has succeeded in performing that function, give an explanation of how it has performed it that covers as many of those cases as possible, and then see what conditions are mentioned in that explanation. These conditions will include the believer's having desires, having other beliefs, and having certain inferential abilities. They will also include certain conditions in the world outside the believer's head, and *these* are the conditions that are relevant to the determination of content. It is a Normal condition for the proper functioning of my belief that it is going to rain that it actually is going to rain: on those occasions when the belief has contributed to desire-satisfaction, that is the external condition that has (not always, but often) been in place. There may be occasions when the belief helps to satisfy my desires by accident even though it is false; they will not, however, be covered by a Normal explanation. There may be occasions when the belief does not help me to satisfy my desires even though it is true (because other Normal conditions are not satisfied—for example, my legs fail, or other beliefs that are acting in concert with it are false); these cases are not covered by a Normal explanation either, since a Normal explanation is an explanation of the cases when the belief *does* perform its function. Thus a Normal condition for the performance of a function is neither a necessary nor a sufficient condition for the performance of that function. Note that Millikan's "Normal" (with a capital "N") does not mean statistically normal: some things have proper functions that they hardly ever perform, and in such cases the Normal conditions for the performance of the function are ones that are hardly ever in place.

The story so far makes it sound as though there are lineages of beliefs of the same type (e.g., beliefs that snow is white), with some sort of selection process determining which beliefs proliferate—it seems to assume that beliefs have direct proper functions. The full story is more complicated. There are two ways, for Millikan, in which things can acquire functions: one is by being the direct products of selection processes like natural selection, and the other is by being the products of those products. So if belief-forming mechanisms have direct proper functions, the beliefs that they produce can have proper functions in the second way: *derived* proper functions.

Say that my belief that it is going to rain has a function which it performs only when it actually *is* going to rain, and that it has that function because the mechanism that produces it has the function of producing beliefs that bear a relationship to the world that will enable them to participate in processes which lead to the satisfaction of my desires. There still remains the question: Why does the mechanism have the function of producing precisely *that* belief in *that* circumstance? For some beliefs, we could answer this question by talking about the conditions under which that particular belief has in the past contributed to the satisfaction of desires. But there are some beliefs—for example, beliefs that have never been tokened before, and beliefs that have never succeeded in contributing to the satisfaction of desires—that would come out as contentless on this approach.

Millikan's answer appeals to the complex and structured nature of systems of representation like human belief/desire systems. My belief that it is going to rain today is not like the basic mental state that causes the frog to snap at flies or the one that causes the kimus to climb the mountain every morning. Rather, it is part of a system of mental states that have complex structural relationships to each other, and that system maps onto the (also complex and structured) external world in a way that determines the content of particular mental states.

In Millikan's view, our evolutionary history picks out a particular mapping from beliefs onto states of affairs.[7] It is the mapping in accordance with which beliefs must map onto the world if the consumers of the belief are to perform their proper functions in accordance with a Normal explanation. Beliefs (like my belief that it is going to rain, here, today) do not map piecemeal onto states of affairs (like its actually raining here today). Beliefs are systematically related to each another, and the states of affairs onto which they map are systematically related to each other in a way that mirrors the relations between the beliefs.

This story about mapping provides a way of accounting for the fact that novel beliefs and ineffectual beliefs (and novel or ineffectual desires) have content even though there is no lineage of states of their type that have performed proper functions: they have content in virtue of their position in a network of states. But the mapping account does something more general: it delivers determinacy of content for mental states that are part of complex systems of representation.

When a fly crosses *my* visual field, I have a wide range of possible behavioral responses (I might swat it, or open the window to let it out, or put away the food that otherwise it might land on), and there are a wide range of mental states I might token in response to seeing it (as well as "That's a fly," I might think "Flies are annoying" and "This is the sort of weather that brings out the flies" and "I hope my sheep aren't being bothered by flies"). Sometimes (e.g., if I have just put eye-drops in) I might instead think "What is that moving black thing?" But the "fly" thoughts and the "moving black thing" thoughts have different content from each other in virtue of their different relations—not causal relations, but structural relations—to other mental states.

CONCLUSION

I have focused on indeterminacy as a potential problem for teleosemantics for two reasons. Doing so provides a useful way of illustrating the differences between different teleosemantic theories, but also indeterminacy is generally regarded as the problem that will sink teleosemantics if anything will. I have suggested that states of simple representational systems (such as the frog's fly-snap mechanism) may not in fact have determinate content, and I concluded that the failure of teleosemantics to assign determinate content to such states is a feature, not a bug. A theory of content for states of more complex representational systems *does* need to assign determinate content: but the complexity of the systems provides the teleosemanticist with the means to do so.

NOTES

1. See, for example, Mendelovici & Bourget 2014: 327; Schroeder 2007. As Schroeder puts it, "[teleosemantics] was trampled into sociological oblivion by the horde of philosophers rushing to study consciousness. Since about 1995, teleosemantics has received the attention accorded to the merely effable by people rediscovering ineffable mysteries."

2. More sophisticated informational theories that try to get around these problems include Dretske 1981, 1986 and Fodor 1990.

3. Neander 1995: 125. Neander says "*detecting* small dark moving things" rather than "*causing it to snap at* small dark moving things," but this might be taken to beg the content question.

4. Neander disagrees (Neander 2006): she thinks that content-attributions such as <fly> or <food> cannot serve the purposes of information-processing explanations of the frog's capacities.

5. Millikan 1990: 154–155; Godfrey-Smith 1989: 542: Papineau acknowledges this in Papineau 1993: 59, note 3. But see Neander 2013, 23–24 for an alternative view.

6. See Kingsbury 2006: 34–36 for a defense of the teleosemantic account of desires against the suggestion that it is not a naturalistic account. Note that *inference* also plays a role in Millikan's account of belief-content, and so a naturalistic account of inference is also necessary to the overall naturalizing project.

7. This is oversimplified in that it suggests that we all share the same language of thought. Millikan (forthcoming) emphasizes that my <cat> representation need have nothing in common with your <cat> representation other than being about cats: thus a more precise statement of the evolutionary story would be that we each have a mechanism (the same one in all of us) whose function is to design a consistent mental representation system (not the same one in each of us).

REFERENCES

Dretske, F. 1981. *Knowledge and the Flow of Information* (MIT Press).

Dretske, F. 1986. "Misrepresentation." In R. Bogdan (ed.), *Belief: Form, Content and Function* (Oxford University Press) 17–36.

Fodor, J. 1987. *Psychosemantics* (MIT Press).

Fodor, J. 1990. "A theory of content II: The theory." In J. Fodor, *A Theory of Content and Other Essays* (MIT Press) 89–136.

Godfrey-Smith, P. 1989. "Misinformation." *Canadian Journal of Philosophy* 19: 533–550.

Kingsbury, J. 2006. "A proper understanding of Millikan." *Acta Analytica* 21: 23–40.

Kingsbury, J. 2008. "Learning and selection." *Biology and Philosophy* 23: 493–507.

Macdonald, G. & Papineau, D. 2006. "Introduction: Prospects and problems for teleosemantics." In G. Macdonald & D. Papineau (eds.), *Teleosemantics* (Oxford: University Press) 1–22.

Matthen, M. 1988. "Biological functions and perceptual content." *Journal of Philosophy* 85: 5–27.

Mendelovici, A. & Bourget, D. 2014. "Naturalizing intentionality: Tracking theories versus phenomenal intentionality theories." *Philosophy Compass* 9: 325–337.

Millikan, R. 1984. *Language, Thought and Other Biological Categories* (MIT Press).

Millikan, R. 1986. "Thoughts without laws: Cognitive science with content" *Philosophical Review* 95: 47–80.

Millikan, R. 1989a. "Biosemantics." *Journal of Philosophy* 86: 281–297.

Millikan, R. 1989b. "In defense of proper functions." *Philosophy of Science* 56: 288–302.

Millikan, R. 1990. "Compare and contrast Dretske, Fodor, and Millikan on teleosemantics." *Philosophical Topics* 18: 151–161.

Millikan, R. 1991. "Speaking up for Darwin." In B. Loewer & G. Rey (eds.), *Meaning in Mind: Fodor and His Critics* (MIT Press) 151–165.

Millikan, R. forthcoming. *Beyond Concepts: Unicepts, Language and Natural Information* (Oxford University Press).

Neander, K. 1991. "Functions as selected effects: The conceptual analyst's defense." *Philosophy of Science* 58: 168–184.

Neander, K. 1995. "Misrepresenting and malfunctioning." *Philosophical Studies* 79: 109–141.

Neander, K. 2006. "Content for cognitive science." In G. Macdonald & D. Papineau (eds.), *Teleosemantics* (Oxford University Press) 167–194.

Neander, K. 2013. "Toward an informational teleosemantics." In D. Ryder, J. Kingsbury, & K. Williford (eds.), *Millikan and Her Critics* (Wiley-Blackwell) 21–41.

Papineau, D. 1987. *Reality and Representation* (Blackwell).

Papineau, D. 1993. *Philosophical Naturalism* (Blackwell).

Pietroski, P. 1992. "Intentionality and teleological error." *Pacific Philosophical Quarterly* 73: 267–282.

Ramsey, F. 1931. "General propositions and causality." In F. Ramsey, *The Foundations of Mathematics* (Kegan Paul) 237–255.

Schroeder, T. 2007. "Review of *Teleosemantics: New Philosophical Essays.*" *Notre Dame Philosophical Reviews*. http://ndpr.nd.edu/news/25277-teleosemantics-new-philosophical-essays

V

Evolution and Ethics

21

Evolution and Ethics: An Overview

Catherine Wilson

INTRODUCTION

Ethics in its broader sense is the study of how best to live. In the narrower sense, identified with "morality," it is the study of right and wrong, obligations, prohibitions, and permissions, as these pertain to the actions of individuals and groups that bring harms and benefits to others.

A long tradition in Western philosophy teaches that moral awareness and agency distinguish human beings from non-human animals. The former are regarded as endowed with free will, or at least a deliberative faculty, and foresight, along with language and rationality. They are seen as subject not only to civil laws which they understand to some extent and can represent to themselves, but also to moral laws commanded by a divinity or issuing from some analogous authority, perhaps even from oneself. The position that human moral agency is a form of reason-responsiveness unavailable to other life forms continues to be defended (Katz 2000: 1–162; Korsgaard 1996).

Evolutionary ethics takes a different point of departure. It treats morality as a set of dispositions and behaviors that represent transformations of the "prosocial" or "proto-moral" dispositions and behaviors of extinct human ancestors. These dispositions and behaviors are theorized as adaptive, as having contributed to the chances of leaving a lineage, by the animals that possessed them. Insofar as human reasoning and emotionality depend on cortical mechanisms, they, like morphology and metabolism, can be supposed to have been shaped by evolutionary forces.

As human anatomy resembles that of the apes who are our closest living evolutionary relatives, whilst being distinctive with respect to posture, brain size, dentition, hairiness, and many other features, so, the evolutionary ethicist maintains, human social behavior will exhibit analogies to, but also marked differences from, theirs. Empathy, aggression, pacification, friendship, sexuality, and altruism have been studied in chimpanzees and bonobos, our closest living relatives, as well as in gorillas and baboons. The great apes and some monkeys appear to have the representational abilities to understand what other animals experience and know (Povinelli 1996; Russon et al. 1996). They form friendships and alliances, and display loyalty, reciprocity, and revenge (Smuts 1985; Silk 1992). They show distress at the deaths of offspring and comrades (Barley 2010).

Humans are less emotional and irritable than apes and their behavior is noteworthy for its impulse control. Human representational abilities extend to the remote past, the distant future, and to merely possible states of affairs. We do not only behave according to discernible patterns, but also represent, conform to, and in some cases revere norms, including those of grammar and decorum. Language and literacy permit the consolidation and transmission of knowledge alongside social and technological development. These features give human morality its breadth, diversity, and cultural importance. They do not, however, imply the irrelevance of biology for normative and metaethical moral theory.

Eighteenth-century theorists of the moral sentiments such as David Hume and Adam Smith as well as ancient and early modern materialists rejected divine command theories of ethics and discussed the social function of morality, but it was Darwin's *Descent of Man* of 1871 that set the present course of investigation. Darwin found precursors of the moral and the aesthetic sense in birds and mammals. Although he regarded conscience as a particular human endowment, he speculated that "any animal whatever, endowed with well-marked social instincts, the parental and filial affections being here included, would inevitably acquire a moral sense or conscience, as soon as its intellectual powers had become as well, or nearly as well developed, as in man" (Darwin 1879: 120–121). He believed further that "habits" acquired in an individual's lifetime could become cemented into "instincts" (Darwin 1859: 208–218).

The notion of hereditary habit has since been supplanted by the notion of "niche construction" (Odling-Smee et al. 2003), and the neo-Darwinian framework assumes the existence of multiple units of heredity (the "genes") that direct growth and functioning in their "vehicle" (the "phenotype") so as to get themselves into the next generation (Dawkins 1976). Although the "gene's eye" view raises numerous problems that Darwin himself did not foresee, including difficulties associated with the definition of the gene (Godfrey-Smith 2009), the number and type of independent mechanisms driving evolution, and the levels at which it can occur (Lloyd 2012, Chapter 2 this volume), it has provided fruitful new models for understanding organisms and their interactions.

On the descriptive side, evolutionary ethics comprises a number of interrelated empirical and computational studies. First, there are field investigations of the behavior of other species that bear a resemblance to human moral agency. Second, there are investigations of practices and attitudes with respect to aggression, sexual behavior, cooperation, dominance and submission, and punishment in humans, and the presentation of adaptive hypotheses to explain them. The few remaining hunter-gatherer societies—although they are not Palaeolithic relics and although they show wide variation in their attitudes and practices from place to place—are favored targets of study, especially when they do not engage in extensive contact with outsiders or employ metal technology. Third, investigators employ simulation studies to show how behavioral strategies, such as vengefulness, promiscuity, or cooperative behavior, might fare in a population over many generations when they encounter the same or opposing strategies. Evolutionary stable strategies are those that, once established, cannot be invaded by alternative strategies (Axelrod & Hamilton 1981; Maynard Smith & Price 1973; Skyrms 1996).

On the prescriptive side, some philosophers maintain that what is known about heritable species-specific traits and dispositions ought to enter into our assessment of current social practices and institutions. This position has generated extended controversy. It first appeared in the form of Social Darwinism associated with the eugenics movement, imperialism, and genocide in the late 19th and early 20th centuries (Crook 1994, 2007); then, after the Second World War, it reappeared in a series of sensational popular books focused on territoriality and implying masculine supremacy (Ardrey 1961, 1966; Tiger & Fox 1971), reinforcing the distaste amongst the educated public and professional ethicists for this line of enquiry. Since the 1990s, the field has largely succeeded in shaking off this dismal legacy and presenting a pacific, emancipatory, and egalitarian face. However, it remains rife with oversimplification and hasty inference. In the popular literature, male–female relations and interactions have largely dominated the agenda (Ridley 1994; Buss 2003). Discussion of these issues is reserved for the section "Metaethics and Prescriptivity" below following a survey of descriptive approaches.

DESCRIPTIVE EVOLUTIONARY ETHICS

Altruism and Cooperation

Altruism in evolutionary theory is defined as any disposition to behavior that has costs in viability or fecundity to the donor whilst benefiting the recipient; cooperation implies the acceptance of short-term certain losses for longer-term larger but riskier gains. As Darwin observed, group-living animals undergo risks and deprivations to nourish and protect their conspecifics, citing many examples of warning, grooming, feeding, and rescuing which he supposed conducive to group survival (Darwin 1879: 123–136). Where apes share food in response to begging, groom one another, and tend one another's injuries (Köhler 1927: 308–310), humans offer food and perform numerous tasks cooperatively, including hunting, cooking, building, and child-minding. They care for the old when they cease to be economically productive and for their children—even when these children, by reason of some misfortune, are exceedingly unlikely to have children of their own. Some dive into icy rivers or dash into burning buildings to save the lives of strangers.

Darwin saw no conflict between his hypothesis that individual organisms are in competition with their conspecifics to survive and reproduce and his observations of helping behavior. Their hereditary moral habits, he supposed, gave some tribes a survival advantage over others, a proposal revived in recent theories of group selection (Darwin 1879: 110; Sober & Wilson 1998). The neo-Darwinian paradigm of the 20th century, by contrast, demanded explanations that could demonstrate the competitive advantage to the individual organism (or, more precisely, its genes) of altruistic behavior. Much altruistic behavior, it was argued, is directed at kin, and genes that direct the animal to incur some risk or sacrifice to benefit close relatives who share that gene can be selected for (Hamilton 1964). Altruistic behavior directed at non-kin could then be explained as undertaken in the expectation of reciprocity (Trivers 1971; Alexander 1987). If I feed you when food is scarce and I have some, I can expect that you will feed me when food is scarce and you have some. Bats and other mammals do this, without language or ratiocination

(Wilkinson 1988). Emotion-based mechanisms that punish selfishness with exaggerated retaliation might be expected to evolve alongside altruism and cooperation.

However, there is a great deal that such benefit-to-self-or-kin explanations leave mysterious. Why do we feel duty bound to look after our aging parents when they are not going to produce more copies of our genes? Why do we feel impelled to follow our leaders into war when we are likely going to get killed and produce no offspring at all? Why do we lavish our love and attention to men or women who will never give us any more children or even help us with the ones we already have? One possibility is that our most useful underlying altruistic dispositions can be hijacked. The behavior of organism A may be controlled in part by organism B for the latter's advantage (Dawkins 1982: 209–249). When a parent rushes to the assistance of their own crying child this is probably kin selection at work; when a parent rushes to the aid of a stranger's child (or writes a check to a welfare organization) this is probably manipulation that works via the arousal of anxiety, a "save the endangered child response," that is deeply and strongly wired in because the child is likely to be one's own. A complex approach to the problem of non-reciprocal altruism has been offered by John Tooby and Leda Cosmides (1996), who contrast the "selectionist" approach of artificial simulation exercises in favor of an "adaptationist" effort to identify the necessary psychological mechanisms and behavioral contingencies that underlie empirically observable behavior. They suggest that costless, unintentional benefits emitted by an individual (I allow you to follow me home, or toss my half-eaten apple over my shoulder where you hungrily pick it up) might be met with social rewards consisting of displays of warmth, loyalty, or gratitude on the part of the recipient, facilitating the transition to the emission of more costly intentional benefits. The "Baldwin Effect," whereby organisms are selected for latent morphological or behavioral traits initially exhibited only under stressful or unusual circumstances, may prove helpful in understanding how this might occur (Ananth 2005).

The performance of altruistic actions will in any case be facilitated by a psychological reward system that is not only gratified by a good outcome but also gratified by the agent's role in bringing about the good outcome and dismayed by its failure. This explains the popularity of the Ciceronian notion that "virtue is its own reward" and the focus in modern moral theory on agency and responsibility (Williams 1981). The heroic actions of sappers and firefighters are likely supported by a complex mixture of benefit-to-self-or-kin and manipulation-by-other mechanisms relating to the social rewards obtainable and the intrinsic satisfactions obtained from moral heroism. Feelings of kinship can be exaggerated by social learning. A charismatic leader who inspires me to donate resources and effort to his cause at my own risk is a classic manipulator.

Coercion and Control

In any population, individuals of the same species vary according to age, size, strength, cunning, irritability, risk-friendliness, and along many other physical and personality dimensions. In many mammalian species, these differences translate into dominance hierarchies or multiple sets of dominance hierarchies in one or both sexes that determine privileges with respect to feeding, mating, grooming, and choice of location (Noë et al. 1980). Dominant males in harem-forming species collect the majority of females for insemination and thus sire the most offspring, and high-ranking females in some species appear to suppress ovulation or access to resources in lower ranking females through behavioral or chemical means (Stockley et al. 2013).

Although the dominance enjoyed by a more aggressive, cunning, or hormonally well supplied animal can thus confer benefits in the form of prolonged survival and more and better reproductive opportunities, dominance behavior is obviously relational and not a heritable trait as such. Logic demands that we understand hierarchies as arising from the adaptive tendency of smaller, weaker, less aggressive members of a species to give way, for the sake of their own well-being, to larger and more aggressive animals, acting in their own interest (Rowell 1974). Dominance can perpetuate itself within a family lineage only to the extent that, within a given environment, forceful characteristics are significantly heritable.

Facile claims to the effect that human females prefer to mate with "alpha males" for evolutionary reasons should be regarded skeptically on both logical and empirical grounds. First, the term "mating" covers a range of human behaviors from lifelong marriage with a brood of co-parented children to fleeting, inconsequential encounters. At different stages of the life cycle, and under different socio-economic arrangements, women's preferences will vary, and people's actual situations do not in any case reliably reflect their preferences as stated in surveys. Women in one relatively careful study, however, did not prefer as "romantic partners" males described as aggressive or dominant, though they valued confidence and assertion along with "easygoingness" and sensitivity (Burger & Cosby 1999).

Most anthropologists believe that human beings evolved away from a putative common ancestor of chimpanzees, bonobos (pygmy chimpanzees), and humans, with selection for the inhibition of aggression, the reduction of the size of canine teeth, and reduction in sexual dimorphism (Shine 1990). Michael Chance has argued that the reduction in our species of a preoccupation with rank order has freed attention and energy for other purposes such as tool use (Chance 1978: 144). The bonobo (*Pan paniscus*), with its more fluid social structure, hypersexuality, and female dominance over males, has been appealed to as a better model for human society than the common chimpanzee (*Pan troglodytes*) (De Waal & Lanting 1997; though cf. Stanford 1998). Christopher Boehm ascribes a preference for equality to male groups of hunter-gatherers (Boehm 2000: 85). He notes, however, that this implies a readiness on the part of groups to exile or assassinate intensely disliked individuals, and that sexual jealousy accounts for a high homicide rate in these societies.

A disposition to obey authority—a trait Jonathan Haidt and Craig Joseph (2004) argue to be one of five fundamental to human psychology—implies that I may accept another individual's wanting me to do something as a reason for my doing it, even when it is by no means to my benefit (see also Geiger 1993). The disposition to submit and obey, and the psychological tendency to follow, and even to revere and protect violent, charismatic leaders can be supposed to have evolutionary roots. However, it assumes the economically productive but morally destructive forms it does in human societies only through late appearing cultural mechanisms. Under conditions of large population, technological development, ownership of the means of production, money, and writing, human groups spontaneously take on a despotic form, with kleptocratic, legislating governments, administrative bureaucracies, military and priestly castes, occupational hierarchies, and slavery. All ancient urban societies imposed control over the movements of elite females, and permitted the sexual use of male and female slaves. These institutions and practices raised new moral questions that could not have been asked in the environment of evolutionary adaptiveness (EEA).

Aggression and Pacification

Animals do not usually engage in lethal combat with members of their own species. Primates can, however, be remarkably hard on one another. In baboon and chimpanzee societies, males may attack, injure, and sometimes kill females, and infanticidal behavior is widespread, occurring in langurs, baboons, gorillas, and chimpanzees (Hrdy 1977, 1979). It is an effective reproductive strategy in males to kill the offspring of females to induce ovulation and receptivity when the probability of paternity is low. Martin Daly and Margo Wilson (1994) have argued that the remarkably prevalent aggression against infants and small children by stepfathers is the human version of this behavior. The evolution of "social monogamy"—implying association and interaction amongst a male–female pair and, shared parenthood—has been proposed as the basis of reduced infanticide in those primates that practice it (Opie 2013).

When it comes to the treatment of unfamiliar conspecifics, primate behavior depends on the species and the circumstances of the encounter. As Edvard Westermarck pointed out, in tribal societies and in the ancient world, the stranger was regarded as someone to whom the concepts of the sanctity of life and property did not apply, or did not apply as strictly as the prohibitions against harming fellow citizens (Westermarck 1932: 199–200). Cannibalism is observed in our own species and in chimpanzees, and Richard Wrangham and Dale Peterson created consternation in their 1997 book *Demonic Males*, which described chimpanzee male raiding parties that cooperated in order to kill, savagely and excitedly, red colobus monkey. Although the monkeys were of an unrelated species and constituted a food source, the authors suggested that this behavior was a precursor to human warfare. While the extent of Palaeolithic warfare is much debated, with some writers arguing for genocidal behavior (Keeley 1996), recent analyses have argued that conflict avoidance through migration of expanding populations was the more typical pattern (Kelly 2005). The manufacture of iron tools and weapons, the invention of the bow, and the use of horses, fortified dwelling places, and wheeled vehicles, together with high population densities, wealth accumulation, and scarcity of resources, contributed to the development and spread of militarism.

Frans de Waal has argued that bonobos—although they hunt small prey, do not engage in infanticide or raiding parties, and are relatively tolerant of immigrants and one another—employ sex to ease social tensions. He cites instances of peace-making efforts amongst both forms of chimpanzee and reconciliation amongst individuals who have been involved in some altercation (de Waal 1990). In any case, there is little point in arguing over whether humans are "innately aggressive" or are made aggressive by "society." Along with the capacity to be roused to homicidal rage and to enjoy participation in coolly coordinated military exercises, most humans are averse to killing those of whom they are not afraid and with whom they have no quarrel. The interest of groups to which they belong may diverge in this regard from their personal interests. Particular societies, from gangs to nations, strive to inform their members, by direct instruction and by the presentation of salient models, as to obligatory, permissible, and prohibited forms and targets of lethal aggression.

Reproduction and the Sexual Division of Labor

Of particular interest for evolutionary ethics are patterns of behavior related to reproductive strategies and parental behavior. Because this is an intriguing and popular subject, much oversimplification has infected the discussion. It is commonly observed, for example, that male fertility is "in principle" much higher than female fertility across a whole range of sexually reproducing taxa. Female fertility is accordingly regarded by biologists as a universal "scarce resource." From this it is held to follow that males and females have diverse and sometimes conflicting reproductive strategies: that females are choosy and coy, males bold and undiscriminating; that males provide "resources" in the form of subsistence to females to acquire "mating opportunities," and that human females choose, once for all, "good providers" or males with "the best genes" as lifelong mates, while human males make more "mating effort" and choose their many temporary mates on the more superficial criteria of face and figure. With reproductive strategies supposedly imprinted into the respective psychological templates of human beings, these generalities are held to explain many social observations and to have prescriptive implications. They can, however, be criticized as irrelevant, contradictory, or misleading, especially when the term "mating" is used sometimes to mean "marrying" and sometimes to mean "having sex with."

Chimpanzees have not one but three main patterns of sexual association: consortships, in which a female and a male sequester themselves from the rest of the group and remain together as a sexually exclusive pair for as long as a month; possessive relationships, in which a dominant male tries to monopolize a female or several females in estrus; and opportunistic mating, in which several males take turns mating with a single female in estrus. There are parallels to all these forms in human relationships, but an exceptional feature of our species is that human life reflects "the direct, deliberate, and conscious intervention of parents and other close kin on the sexual lives of their descendants" (Walker et al. 2011).

Arranged marriage is arguably a basic pattern of human pair-bond formation, dating back at least 50,000 years. In those hunter-gatherer societies that have been studied, such marriages produce somewhere between 70 and 99 percent of a woman's offspring, depending on the degree of social policing and internalization of norms (Anderson 2006). The preferences of parents may accommodate or coincide with personal attraction between the pair, but factors such as status, kinship relations, and intertribal alliances, as well as parental estimates of factors conducive to the production of a healthy lineage, are apt to enter into it. This influence may render much discussion of mate-preferences in evolutionary psychology, such as the male preference for a particular waist-to-hip ratio (Singh 2002), somewhat beside the point, along with the theory that good moral character and "creative intelligence" were targets of sexual selection in the EEA (Miller 2000). Extra-pair mating is, however, more likely to reflect the personal choices of the participants, and if it produces many offspring, it is clearly of evolutionary significance.

Humans are one of a small number of species in which males engage in paternal care (Geary 2005). Male chimpanzees do not know who their offspring are, and it does not concern them; human beings, by contrast, attach importance to social fatherhood even in conditions where biological fatherhood is not understood, or where it is less significant than the paternal or avuncular role played a man who has a relationship with the child's mother.

Reproductive skew in humans—the greater variance in the number of offspring born to males than to females—is also unusually small in humans relative to other mammals, making over-general points about the massive number of sperm theoretically available largely irrelevant to the human case. The average number of living offspring—four to five—outside of societies practicing contraception, abortion, and infanticide is obviously the same in both sexes. Although men can father children in advanced old age while new motherhood after age fifty is highly unusual, few males in fact become fathers after the age of fifty in hunter-gatherer societies (Buller 2005: 220). Both sexes have incentives to be choosy and coy, as well as indiscriminate and sexually aggressive, especially when economic dependencies enter the picture. Pursuit can be costly and choosiness a waste of time. "Promiscuity" can enhance fecundity or be a disliked trait and a basis for rejection. The notions that women do not suffer from sexual jealousy and have no biological motives to adultery can only be understood as some people's wishful thinking (Harris 2004).

The observation that both sexes are specialized for reproduction but in different ways raises questions as to the natural and the morally defensible division of labor. Foraging patterns and predation practices do not differ in most mammals, but they are subtly different in chimpanzees and bonobos, and most societies assign special roles and occupations to men and women respectively. The portrayal in countless museum dioramas of a nuclear cave-family consisting of a couple of children and a maternal parent waiting at the campfire and a male parent returning from a successful all-day hunt with the family's nourishment is nevertheless a fantasy. Sarah Hrdy (2009) has provided strong evidence that human were originally "cooperative breeders" with "allomothers" including boys and girls, aunts, and grandparents sharing childcare, permitting and encouraging the human pattern of long dependency, long life, and fairly rapid reproduction. In hunter-gatherer societies, women with or without their children in tow cover large territories, fish, hunt small game, and in the southern latitudes provide most of the calories (Lee & DeVore 1968; Estioko-Griffin & Griffin 1981)

Conscience and the Moral Sentiments

The "moral sentiments" described by Adam Smith in the 18th century (Smith [1759] 2009—including sympathy, concern for the welfare of kin, the diminishing intensity of moral concern with social distance, and regard for the "merit" and "propriety" of actions—accord well with the sort of psychological platform needed to support group living in the human mode. Morality is frequently opposed to immediate self-interest, and its psychological platform involves an understanding of and concern for the Other. It is strikingly manifest in situations in which an agent could perform an action in their own interest but to the disadvantage of another without experiencing retaliation, either because the action is carried out in secret or because the other is powerless to retaliate. Moral systems are in the main prohibitive; their "duties" are concerned with the avoidance of practices of deception, abandonment, disobedience, non-fulfillment of contracts and promises (Wilson 2004).

Both Smith and Darwin laid importance on "conscience" as a distinctive human mental feature. Smith pointed out that guilt and shame in the perpetrator tend to follow deceptive and unkind or unjust action, even when human punishment is not anticipated. Although there is the expected spectrum of guilt-proneness in any population, people who harm others and do not experience remorse are characterized as psychopaths and regarded as dangerous and as needing to be immobilized. Humans are disposed to believe that supernatural forces can perceive and evaluate all their motivations and actions and will punish them for infractions of moral rules and the breaking of other restrictive taboos by visiting them with death or illness. Religiosity in this sense has been proposed as an adaptation underwritten by cortical mechanisms that make human morality possible and co-existence productive (Atran & Norenzayan 2004). That selfish behavior, especially female sexual indulgence and unsanctioned male violence, entails punishment or death brought about by inexorable cosmic forces is a view expressed and inculcated in myths and stories not only in traditional societies but also in sophisticated fiction.

METAETHICS AND PRESCRIPTIVITY

All cultures demand restrictions on killing, bullying, some forms of sexual behavior, and deception, and operate with notions of fairness, purity, and propriety. Darwin, like many of his predecessors, was nevertheless impressed by the distance between the morality of "savages" and that of his own highly proper Victorian culture which was at the same time influenced by Benthamite ideals of the general welfare and wealthy enough to realize them. He noted and approved the existence of hospitals and orphanages that extended care and the preservation of life to the sick, disabled, insane, and abandoned, whose fate would likely have been otherwise in prehistoric times. He commented on the tendency of artificial helps to perpetuate socially undesirable physical and psychological traits, but argued that such institutions probably added little to the genetic burden since their inmates did not produce large families. In any case, he said, we could not check our sympathy "even at the urging of hard reason, without deterioration in the noblest part of our nature" (Darwin 1879: 134).

Darwin's worries on this score raise first the question of the relationship between human nature (to the extent that we can sensibly employ this term) and the evaluation of attitudes and practices; and second the question of the epistemological status of moral claims. For a number of writers, evolutionary ethics holds out the old promise of taking us back to a life according to nature, offering formulas for the construction of a social world which "matches our tendencies" and erases the distortions of civilization (Hobel et al. 1976; see also Dennett 1996: 268). Steven Pinker says that the new sciences of human nature, which include evolutionary psychology, "can help lead the way to a realistic, biologically informed humanism . . . They promise a naturalness in human relationships, encouraging us to treat people in terms of how they do feel rather than how some theory says they ought to feel" (Pinker 2002: xi).

Most moral theorists and some biologists have been skeptical of such claims (Kitcher 1985; Lewontin et al. 1984). Among the venerable 19th-century opponents of the natural life were the philosopher J.S. Mill and the Darwinian comparative anatomist Thomas Henry Huxley. Mill claimed that the term "nature" was "one of the most copious sources of false taste, false philosophy, false morality, and even bad law" (Mill [1874] 1904: 7). Huxley agreed with Darwin regarding the existence of an evolutionary basis for morality but declared in his Romanes Lectures that the struggle to exist at the expense of others was frankly opposed to morality (Huxley [1893] 2004: 27).

As noted earlier, appeals to evolutionary theory were associated with racist and eugenicist programs that took progress and improvement as their goals, also with a low estimation of the non-reproductive capabilities of women. The various "races" were seen as engaged in competition in a struggle for existence, with the imperial powers able to speed nature along in this regard. Demands for women's emancipation and access to the visible and lucrative positions were argued to be incompatible with conditions conducive to optimal reproduction. If the morally good was to coincide with the set of behaviors regarding kin, treatment of animals, sex, aggression, cooperation, and hierarchy that appears to be species-typical in the ways suggested above, and so to characterize human nature, we would have a prescriptive morality that endorsed the following:

1. Homosexuality is permissible.
2. Be kind to your close relatives.
3. Don't deceive your fellows for your own benefit.
4. Help your friends but not your enemies.
5. Be omnivorous.
6. Infanticide is permissible.
7. Polygynous and polyandrous relationships are permissible.
8. Marriages ought normally to be arranged by one's elders, but adultery is permissible if detection can be avoided.
9. Males and females should contribute differently to subsistence.
10. Respect for the lives of strangers belonging to other groups is optional.
11. Assassinate anyone in your group who is behaving too obnoxiously or endangering others.
12. Rape is a permissible human mating strategy.

Tested against the sensibilities of academy-trained moral theorists in the West, we would probably find general agreement with 1, 2, and 3; some degree of support for 4 and 5, but also resistance from peacemakers and vegetarians; serious reservations and demands for further qualifications about 6, 7, 8, and 9; and outright rejection of 10, 11, and 12. Few would sign up to this code as a "realistic, biologically informed humanism." A feature of moral theory as it has developed since the 18th century is its evolution toward generality and abstraction. Rights and obligations as enshrined in written laws have become less conditional on social class, sex, and race, and moral theory has introduced universalization and equalization motifs such as Kant's universality test and Bentham's greatest good for the greatest number criterion.

Hume's observation that "ought" cannot logically be derived from "is" (Hume [1740] 1978: 302) is apt to be triumphantly presented in any critical discussion of evolution and ethics. The fact that some trait or tendency has survived attempts by nature to eliminate it from the gene pool does not imply, it is frequently argued, that a modern human society would not be better off repressing its expression. But contrary to this assumption, reasonable arguments for why we ought to do or forbear from doing particular things are normally inferences from "is"s. We believe inebriated pilots cause plane crashes, and that therefore pilots ought not to be allowed to drink before flying planes. We think abused children suffer pain and confusion with lifelong effects, and ought therefore to be removed from abusive homes. The invisible premises needed to make these arguments into what look more like "logical" deductions are statements of value, which are themselves disguised "ought"s and "ought not"s (e.g., "Plane crashes, pain, and confusion are bad; they ought to be prevented") that are taken to be minimally controversial. The resulting arguments are not conclusive insofar as they invite objections and qualifications that appeal to competing values or additional facts.

On the minimally controversial assumption that a morally better world is one with less exploitation and parasitism (as opposed to productive symbiosis), less human-induced suffering and deprivation, and more opportunity for people to express and enjoy the fruits of their own capabilities, evolutionary ethics has a good deal to offer. It is not, as is sometimes claimed, essentialist: the observation that individuals within a population vary with respect to their physical and psychological traits and employ a multiplicity of context-sensitive strategies to live and reproduce is fundamental to biological thinking. The evolutionary perspective quickly reveals the absurdity of the intellectual tradition that teaches that women are the bodies of the world, seeing to the continuation of the species, while men are its minds, seeing to its cultural development. Everyone's "purpose"—what our minds and bodies have been shaped to do—is the same: to get our genes into future generations. But everybody can use them for other purposes, and the diminution of rationality that biological purpose implies affects both sexes equally.

Where exploitation and deprivation are concerned, a long and distinguished literary-anthropological tradition, memorably represented by Jean-Jacques Rousseau ([1754] 1990), depicts the trade-offs between economic and technological development and individual happiness. More recently, the anthropologist Marshall Sahlins describes the !Kung of East Africa, who had modest wants and technologies to address them that were simple but adequate, in a famous off the cuff remark, as "the original affluent society" (Sahlins 1968). He contrasts the classical economists' view of the human being as a creature of almost unlimited wants and limited means to achieve them. Meaningful work on behalf of oneself, kin, and friends, that involved personal control over foraging, manufacture, and decoration, were prominent features of the EEA that have been lost in the modern organization of labor. Moral progress would restore to our species as much as possible of the autonomy its members formerly enjoyed in this regard.

A fuller and more accurate understanding of the range and variety of male and female reproductive strategies and the psychology underlying them, might encourage a critical attitude toward punitive divorce laws and the double standard. Recognition of the economic role of females in prehistoric times and the importance of "alloparenting" offers a way of meeting arguments that women are not psychologically suited to working outside the home and that their full-time presence is critical for the intellectual and emotional development of their children. Further, we may learn to see abortion as a vastly preferable alternative to infanticide in coping with the age-old problem of unwanted offspring. We can understand that rape, though not amongst the most common male reproductive strategies, and likely much rarer in an era without automobiles, weapons, and apartments with soundproof walls, will occur where the incentives are powerful and the disincentives weak, because there is sufficient human dimorphism in size and strength to make it possible, and because female choice works powerfully enough to reject courtship overtures.

The evolutionary perspective can explain the robustness of manipulation and parasitism in the natural world and sensitize us to its effects in the social world. We can understand how cooperation and competition can operate simultaneously in dyads and in larger groups and cease to be surprised when this happens. We can become more cynical about the powers of charismatic leaders and less credulous about the excellence of the socially dominant. We can learn from a consideration of the behavior of our non-human ancestors that the liability to destructive homicidal rages on the part of some members of our species is not a set of baffling singularities resulting in unforeseeable tragedies, but to be expected in populations whose members vary in their coefficient of irritability. Because some persons lack the inhibitory mechanisms that function in most people's frontal lobes, we should design our laws so that such people cannot get their hands on lethal weapons unavailable to them in the EEA.

Moral realists (Brink 1989; Shafer-Landau 2003; Copp 2008) maintain that the facts about what to think, how to behave, and how to react morally, exist independently of all existing beliefs and practices. We can come to know some moral truths, other realists argue, by perceiving the supervenience of some moral qualities, including wrongness and permissibility, on some states of affairs, a position misleadingly characterized as "moral naturalism." To students of evolutionary ethics, these positions are extremely doubtful. Critics of realism argue that our intuitions are too tainted by our evolutionary heritage to deliver reliable moral conclusions (Street 2006), or alternatively that the existence of such properties is a beneficial cognitive illusion (Ruse & Wilson 1985) or fictional (Joyce 2001). For other philosophers who adopt an anthropologically sensitive model of moral knowledge, moral claims can neither be demonstrated nor confirmed by testing them against nature, but only against moral intuitions of a particular society or against "extra-moral" criteria of theoretical goodness (Prinz 2008: 288–308).

REFERENCES

Alexander, R. 1987. *The Biology of Moral Systems* (Aldine de Gruyter).
Ananth, M. 2005. "Psychological altruism vs. biological altruism: Narrowing the gap with the Baldwin Effect." *Acta Biotheoretica* 53: 217–239.
Anderson, K. 2006. "How well does paternity confidence match actual paternity?" *Current Anthropology* 47: 513–520.
Ardrey, R. 1961. *African Genesis* (Atheneum).
Ardrey, R. 1966. *The Territorial Imperative* (Atheneum).
Atran, S. & Norenzayan, A. 2004. "Religion's evolutionary landscape: Counterintuition, commitment, compassion, communion." *Brain and Behavioral Sciences* 27: 1–57.
Axelrod, R. & Hamilton, W. 1981. "The evolution of cooperation." *Science* 211: 1390–1396.
Barley, S. 2010. "How chimps mourn their dead." *New Scientist* 206: 9.
Boehm, C. 2000. "Conflict and the evolution of social control." In L. Katz (ed.), *Evolutionary Origins of Morality* (Imprint Academic) 79–102.
Brink, D. 1989. *Moral Realism and the Foundations of Ethics* (Cambridge University Press).
Buller, D. 2005. *Adapting Minds: Evolutionary Psychology and the Persistent Quest for Human Nature* (MIT Press).
Burger, J. & Cosby, M. 1999. "Do women prefer dominant men? The case of the missing control condition." *Journal of Research in Personality* 33: 358–368.

Buss, D. 2003. *The Evolution of Desire* (Basic Books).

Chance, M. 1978. "Sex differences in the structure of attention." In L. Tiger & H. Fowler (eds.), *Female Hierarchies* (Beresford Book Service) 135–162.

Copp, D. 2008. "Darwinian skepticism about moral realism." *Philosophical Issues* 18: 186–206.

Crook, P. 1994. *Darwinism, War and History: The Debate over the Biology of War from the "Origin of Species" to the First World War* (Cambridge University Press).

Crook, P. 2007. *Darwin's Coat-Tails: Essays on Social Darwinism* (Peter Lang).

Daly, M. & Wilson, M. 1994. "Some differential attributes of lethal assaults on small children by stepfathers versus genetic fathers." *Ethology and Sociobiology* 15: 207–217.

Darwin, C. 1859. *The Origin of Species* (Murray).

Darwin, C. [1879] 2004. *The Descent of Man and Selection in Relation to Sex* (Penguin).

Dawkins, R. 1976. *The Selfish Gene* (Granada).

Dawkins, R. 1982. *The Extended Phenotype* (Oxford University Press).

De Waal, F. 1990. *Peacemaking among Primates* (Harvard University Press).

De Waal, F. & Lanting, F. 1997. *Bonobo: The Forgotten Ape* (University of California Press).

Dennett, D. 1996. *Darwin's Dangerous Idea* (Touchstone).

Estioko-Griffin, A. & Griffin, P. 1981. "Woman the hunter, the Agta." In F. Dahlberg (ed.), *Woman the Gatherer* (Yale University Press) 121–154.

Geary, D. 2005. "Evolution of paternal investment." In D. Buss (ed.), *The Evolutionary Psychology Handbook* (John Wiley & Sons) 483–505.

Geiger, G. 1993. "Evolutionary anthropology and the non-cognitive foundation of moral validity." *Biology and Philosophy* 8: 133–151.

Godfrey-Smith, P. 2009. *Darwinian Populations and Natural Selection* (Oxford University Press).

Haidt, J. & Joseph, C. 2004. "Intuitive ethics: How innately prepared intuitions generate culturally variable virtues." *Daedalus* 133: 55–66.

Hamilton, W. 1964. "The genetical evolution of social behaviour I and II." *Journal of Theoretical Biology* 7: 1–16, 17–52.

Harris, C. 2004. "The evolution of jealousy." *American Scientist* 92: 62–71.

Hobel, P., Trivers, R., DeVore, I., & Wilson, E.O. 1976. *Sociobiology: Doing What Comes Naturally* (Document Associates Inc.).

Hrdy, S. 1977. "Infanticide as a primate reproductive strategy." *American Scientist* 65: 40–49.

Hrdy, S. 1979. "Infanticide among animals: A review, classification, and examination of the implications for the reproductive strategies of females." *Ethology and Sociobiology* 1: 13–40.

Hrdy, S. 2009. *Mothers and Others: The Evolutionary Origins of Mutual Understanding* (Harvard University Press).

Hume, D. [1740] 1978. *A Treatise of Human Nature*. L. Selby-Bigge (ed.) (Clarendon).

Huxley, T. [1893] 2004. *Evolution and Ethics* (Prometheus).

Joyce, R. 2001. *The Myth of Morality* (Cambridge University Press).

Katz, L. 2000. *Evolutionary Origins of Morality: Cross-Disciplinary Perspectives* (Imprint Academic).

Keeley, L. 1996. *War Before Civilization: The Myth of the Peaceful Savage* (Princeton University Press).

Kelly, R. 2005. "The evolution of lethal inter-group violence." *Proceedings of the National Academy of Sciences* 102: 24–32.

Kitcher, P. 1985. *Vaulting Ambition: Sociobiology and the Quest for Human Nature* (MIT Press).

Köhler, W. 1927. *The Mentality of Apes* (Routledge & Kegan Paul).

Korsgaard, C. 1996. *The Sources of Normativity* (Cambridge University Press).

Lee, R. & DeVore, I. (eds.). 1968. *Man the Hunter* (Aldine).

Lewontin, R., Rose, S., & Kamin, L. 1984. *Not in our Genes: Biology, Ideology and Human Nature* (Penguin).

Lloyd, E. 2012. "Units and levels of selection." In E. Zalta (ed.), *The Stanford Encyclopedia of Philosophy*. http://plato.stanford.edu/archives/win2012/entries/selection-units/

Maynard Smith, J. & Price, G. 1973. "The logic of animal conflict." *Nature* 246: 15–18.

Mill, J.S. [1874] 1904. *Nature, the Utility of Religion and Theism* (Watts & Co.).

Miller, G. 2000. *The Mating Mind* (Random House).

Noë, R., de Waal, F., & van Hooff, J. 1980. "Types of dominance in a chimpanzee colony." *Folia Primatologica* 34: 90–110.

Odling-Smee, J., Laland, K., & Feldman, M. 2003. *Niche-Construction: The Neglected Process in Evolution* (Princeton University Press).

Opie, C. 2013. "Male infanticide leads to social monogamy in primates." *Proceedings of the National Academy of Sciences* 110: 1332.

Pinker, S. 2002. *The Blank Slate: The Modern Denial of Human Nature* (Viking).

Povinelli, D. 1996. "Chimpanzee theory of mind? The long road to strong inference." In P. Carruthers & P. Smith (eds.), *Theories of Theories of Mind* (Cambridge University Press) 293–329.

Prinz, J. 2008. *The Emotional Construction of Morals* (Clarendon).

Ridley, M. 1994. *The Red Queen: Sex and the Evolution of Human Nature* (Penguin).

Rousseau, J.-J. [1754] 1990. *Discourse on the Origins of Inequality*. In R. Masters & C. Kelly (eds.), *Collected Writings of Rousseau*, 6 vols. (University Press of New England).

Rowell, T. 1974. "The concept of social dominance." *Behavioral Biology* 11: 131–154.

Ruse, M., & Wilson, E. 1985. "The evolution of ethics." *New Scientist* 1478: 50–52.

Russon, A., Bard, K., & Parker, S. (eds.). 1996. *Reaching into Thought: The Minds of the Great Apes* (Cambridge University Press).

Sahlins, M. 1968. "Notes on the original affluent society." In R. Lee & I. DeVore (eds.), *Man the Hunter* (Aldine) 85–89.

Shafer-Landau, R. 2003. *Moral Realism: A Defense* (Clarendon).

Shine, R. 1990. "Ecological causes for the evolution of sexual dimorphism: A review of the evidence." *Quarterly Review of Biology* 64: 419–461.

Silk, K. 1992. "The patterning of intervention among male bonnet macaques, reciprocity, revenge and loyalty." *Current Anthropology* 33: 318–25.

Singh, D. 2002. "Female mate value at a glance: Relationship of waist-to-hip ratio to health, fecundity and attractiveness." *Neuroendocrinological Letters* 23: 81–91.

Skyrms, B. 1996. *The Evolution of the Social Contract* (Cambridge University Press).

Smith, A. [1759] 2009. *The Theory of Moral Sentiments* (Penguin).

Smuts, B. 1985. *Sex and Friendship in Baboons* (Aldine).

Sober, E. & Wilson, D.S. 1998. *Unto Others: The Evolution and Psychology of Unselfish Behavior* (Harvard University Press).

Stanford C. 1998. "The social behavior of chimpanzees and bonobos: Empirical evidence and shifting assumptions." *Current Anthropology* 39: 399–420.

Stockley, P., Bottell, L., & Hurst, J. 2013. "Wake up and smell the conflict: Odour signals in female competition." *Philosophical Transactions of the Royal Society, Series B: Biological Sciences* 368: 1631.

Street, S. 2006. "A Darwinian dilemma for realist theories of value." *Philosophical Studies* 127: 109–166.

Tiger, L. & Fox, R. 1971. *The Imperial Animal* (Holt Rheinhart).

Tooby, J. & Cosmides, L. 1996. "Friendship and the Banker's Paradox: Other pathways to the evolution of adaptations for altruism." *Proceedings of the British Academy* 88: 119–143.

Trivers, R. 1971. "The evolution of reciprocal altruism." *Quarterly Review of Biology* 46: 35–57.

Walker, R., Hill, K., Flinn, M., & Ellsworth, R. 2011. "Evolutionary history of hunter-gatherer marriage practices." *PLoS ONE* 6: e19066. doi:10.1371/journal.pone.0019066.

Westermarck, E. 1932. *Ethical Relativism* (Kegan Paul, Trench, and Trubner).

Wilkinson, G. 1988. "Reciprocal altruism in bats and other mammals." *Ethology and Sociobiology* 9: 85–100.

Williams, B. 1981. *Moral Luck* (Cambridge University Press).

Wilson, C. 2004. *Moral Animals: Ideals and Constraints in Moral Theory* (Clarendon).

Wrangham, R. & Peterson, D. 1996. *Demonic Males: Apes and the Origin of Human Violence* (Houghton Mifflin).

22

The Evolution of Moral Intuitions and Their Feeling of Rightness

Christine Clavien and Chloë FitzGerald

INTRODUCTION

Humans have many kinds of responses that may be called "intuitions." On a common understanding, an intuition is akin to a gut feeling, hunch or instinct, encountered frequently in both moral and non-moral contexts in everyday life. According to this picture, a moral intuition could be a response to a real-life situation in which one is directly involved, such as deciding when to break it to one's mother that she has a terminal illness. However, the kind of moral intuitions most investigated and examined by moral philosophers and psychologists consist in compelling responses to hypothetical scenarios—for example, whether a doctor should sacrifice a healthy patient in order to save the lives of five other patients in an urgent organ donation (Foot 1967)—or to abstract moral principles—for example, the principle "It is wrong to lie." Finally, philosophers are also interested in metaethical intuitions, which are concerned with the nature of morality. For instance, one commonly held intuition is that there are universally valid moral truths.

Moral intuitions are puzzling phenomena because they are very resistant even when we face good reasons for questioning them or discounting their force. We are typically convinced that the moral norms and judgments we intuitively endorse are right and should be universally held. However, anthropological and worldwide survey studies indicate that the moral norms intuitively endorsed by people can vary widely from one culture to another and even within a culture (Haidt & Graham 2007; Sachdeva et al. 2011). We routinely experience this diversity in moral opinions and observe the etiology of others' moral intuitions without this leading us to doubt our own deeply felt moral intuitions. Let us illustrate.

Imagine that Aaron and Sabrina are engaged in a heated debate about immigration. Sabrina has a strong intuition that it is wrong to allow the country to be "swamped" by immigrants, while Aaron has a similarly strong intuition that it is wrong to ignore the suffering of fellow human beings. Each is aware that they cannot agree because they do not share each other's intuitions. Sabrina perceives that Aaron's liberal views are the result of his upbringing in a left-wing family that have inculcated him with "bleeding heart" values. She discounts his intuitions because she considers that their basis in his upbringing renders them arbitrary. However, Sabrina knows that her views stem from her own upbringing in a conservative household where patriotic values were prized. This knowledge does little to make her doubt her own moral intuitions because she simply feels fortunate to have grown up with the "right" values. This example illustrates how moral intuitions can remain compelling even when we reflect on information that leads us to question their objectivity. This aspect of moral intuition makes them attractive as an object of study.

The philosophical and psychological literature contains a variety of accounts of moral intuitions. However, a fully fledged explanation of the phenomenological, evolutionary, and mechanistic aspects of moral intuitions is currently lacking. In particular, existing accounts provide only cursory descriptions of the compelling psychological experience of having a moral intuition, which distinguishes it from the experience involved in making a similar non-intuitive evaluation. Moreover, we still lack a general explanation of *why* humans have intuitions in given circumstances, but not in others; is there a general pattern or a mechanism that could explain or predict the occurrence of moral intuitions? We answer this question with a novel evolutionary account of the capacity to experience moral intuitions.

Our aim is to shed light on what moral intuitions are and why they are remarkably resistant in the face of evidence of their interpersonal variability. We provide a novel descriptive picture of moral intuitions within an evolutionary framework that avoids the major pitfalls encountered by other accounts of intuition. The next section is an overview of how moral intuitions are understood in the literature in psychology and philosophy. We explain why we find something lacking in these views, particularly when it comes to explaining the compelling phenomenology of intuitions. The following section is an account of intuitions as a two-component phenomenon, comprising an evaluative mental state and a feeling of rightness. The penultimate section provides an evolutionary and mechanistic account of the feeling of rightness and introduces a necessary eliciting condition for intuitions: the existence of conflicting evaluations in the subject's mind. The final section explains how this feeling of rightness applies to moral intuitions, including metaethical intuitions.

INTUITIONS AS USUALLY UNDERSTOOD IN THE LITERATURE

Moral intuitions are a popular topic, especially in philosophy and psychology. There is a wide variety of views and we can only provide a brief overview. Two common features of moral intuitions highlighted in the literature are that they are compelling from a first-person perspective and that they are not the result of conscious inference on the part of the intuiter. Many regard just this latter feature as necessary and sufficient (e.g., Tersman 2007). When a more detailed characterization of intuitions is required, however, divergences arise. As we will see in this section, there are three broad categories of accounts: for some, intuitions are a particular form of judgment or belief; for others, intuitions are emotional or affective responses; and more recently, intuitions have been characterized as seemings, that is, states that present the way things seem to us.

Normative ethicists tend to think of moral intuitions as a kind of judgment, belief, conviction, or proposition (Cohen 2008; DePaul 2006; Kamm 2006; McMahan 2013; Rawls 1971). Most ethicists do not give a particularly detailed descriptive account of intuition because they are more interested in the practice of normative ethics than in moral psychology. The moral judgments that are thought deserving of the label "moral intuition" are characterized in different ways; for example, as strong (Sinnott-Armstrong 2008), self-evident (Audi 2009; Shafer-Landau 2005), or as the result of long and considered reflection (Kamm 2006). On these views, having the moral intuition that abortion is wrong involves having a strong, self-evident, or considered judgment that abortion is wrong.

One major difficulty with these accounts is that the way intuitions are distinguished from other judgments tends to be underdeveloped: characterizing intuitions as judgments that are "strong," "self-evident," or "considered" does not capture the phenomenology of intuitions. To illustrate, a Canadian, accustomed to the moral norm of leaving substantial tips in restaurants, is likely to feel intuitively that she must leave a tip while visiting Spain, even while aware that tipping is not an obligatory part of the local custom. In this case, she has the intuition that she should leave a tip, while simultaneously judging that there is no moral obligation to tip. In the reverse case, a Spaniard may have no moral intuition that she should leave a tip when in Canada, yet firmly judge that she should leave a tip because she knows that this is a local moral norm. Thus, a person may make a judgment while finding the opposite judgment intuitive. Furthermore, two different people may make the same judgment, one compelled by intuition, and the other deliberately going against her intuition. The phenomenological differences between the Canadian's judgment that she should tip and her judgment that she should not cannot be convincingly equated with the strength or self-evidence of these judgments. Moreover, this example reveals how a person's considered judgment can in fact be different from her intuition.

The phenomenology of moral intuitions has led some scholars to identify them, or a particular subclass of them, with emotions or affective responses (Greene 2007; Kauppinen 2013; Roeser 2011; Singer 2005). For instance, seeing a child being beaten by bullies is likely to make one feel indignant or compassionate, or imagining betraying one's closest friend is likely to make one feel ashamed or disgusted. According to this kind of account, moral intuitions spring from these emotional experiences. For example, when the Canadian feels strongly that she should leave a tip, this springs from an emotional response, in contrast to her judgment that she is not required to tip.

The difficulty with this account, however, is that not all cases that are typically cited as moral intuitions seem to have an emotional component or to involve emotional responses. For instance, reflecting on the principle "It is wrong to steal for fun" may not arouse any emotion, even while it strikes one as intuitive. One can have a strong moral intuition about how to fairly distribute a cake among children at a party without feeling particularly emotional about the matter. It is possible to avoid this problem by developing a more nuanced view of the link between emotions and intuitions, arguing that intuitions involve dispositions to feel emotions or judgments of the appropriateness of feeling certain emotions (Kauppinen 2013; Roeser 2011). Or one can argue in the opposite direction that moral intuitions that do not spring from emotional responses are superior to those that do (Greene 2007; Singer 2005).

Nevertheless, these views struggle to explain what makes the phenomenology of intuitions distinctive in those cases where emotions are not directly involved. A view of moral intuitions as *feelings* rather than emotions could also circumvent this difficulty, while retaining the advantage of explaining the phenomenology of intuitions via their affectivity (Haidt & Bjorklund 2008). Feelings are one possible element of fully blown emotions, but they are also involved in non-emotional mental and physical states. However, "feeling" is a vague term that is used to describe a wide range of affective mental and bodily phenomena; for example, "feeling of tiredness," "feeling of uncertainty," "feeling of hunger," "feeling of dread," "feeling of exclusion," "feeling of entitlement," "feeling of familiarity." More detail is needed to individuate the kind of feeling that is involved in moral intuitions, a point we will return to in the next section.

A third kind of account classes moral intuitions as seemings, which are neither judgments nor emotions, but a particular type of state that represents how the world appears to us (Bedke 2008; Huemer 2008; Tucker 2013). A classic example of a seeming in the non-moral domain involves perceiving a stick as bent when it is half-submerged in water; we have the perceptual seeming that it is bent even while we judge that it is straight. If moral intuitions are seemings, it is easy to see how one can have a moral intuition that conflicts with one's actual judgment. For instance, the Canadian in Spain has the seeming that she should tip, but she judges that she does not need to tip. Her seeming is not an emotional response; it simply presents the appearance to her that tipping is the morally appropriate action. In a way that is somewhat similar to the perceptual seeming of the stick, her intuition about tipping can be remarkably resistant despite her considered judgment to the contrary.

This view has the advantage of identifying a specific phenomenology that differentiates intuitions from judgments. Matthew Bedke argues that the seemings involved in moral intuitions are constituted by feelings such as felt appropriateness or veridicality (Bedke 2008: 261–263). Unfortunately, Bedke does not specify the nature of the feeling that constitutes a moral seeming any further. This is a general problem in the literature on moral intuitions: the notion of "feeling" is sometimes found relevant for defining intuitions but remains underspecified. No information is given about precisely what kind of feeling is involved, how it arises, and why it appears in certain situations but not in others. In this article, we provide such a contribution.

A DEFINITION OF MORAL INTUITIONS AND A PHENOMENOLOGICAL ACCOUNT OF THE FEELING OF RIGHTNESS

We argue that moral intuitions are a two-component phenomenon involving a *challenged evaluation* and a *feeling of rightness* (FOR). Thus when the Canadian has the moral intuition that she should tip the waiter in Spain, her intuition is composed of both (i) an evaluation that it is right to tip the waiter, which is challenged by the evaluation that she is not morally required to tip, and (ii) a FOR that reinforces her first evaluation. It is not necessary for our view to take a stance on the controversial matter of which kinds of mental states involve evaluations. "Evaluation" here is a placeholder for a judgment, belief, emotion, decision, or any other judgment-like state that the reader considers involves evaluation. However, as will be explained in the next section, for an evaluation to feel intuitive, it must be challenged by a contradictory evaluation. For now, let us concentrate on the second and most original individuation criterion for moral intuitions: the FOR. This section provides a phenomenological description of this feeling.

The FOR belongs to a particular class of feelings that is not usually mentioned in the literature on moral intuition. Accounts that liken moral intuition to emotion or feeling are correct to recognize an affective component to the experience of finding something intuitive. This affective component is what produces the particular phenomenology and makes a moral intuition *feel* different from a judgment. The Canadian who has the intuition to leave a tip in Spain, is likely to say: "I know that I don't have to leave a tip here, but I *feel* that I should nonetheless." However, we argue that such a feeling episode is not necessarily linked to bodily or mental states involved in emotions. It pertains to a class of feelings that psychologists call "metacognitive feelings," which are feelings *about our own thought processes and cognitive capacities*.

There are various forms of metacognitive feeling. In the non-moral domain, one much-discussed feeling is the tip-of-the-tongue experience, a vivid experience of being very close to accessing something from memory, such as someone's name, but not being quite able to do so (Schwartz & Metcalfe 2011). A different form of metacognitive feeling which has been identified by Valerie Thompson (2009) is the "feeling of rightness" or FOR, which is about the correctness of a belief or evaluation that arises quickly and fluently in one's mind. It is a vivid experience of having the right answer which feels so compelling that it tends to shut down or bias further reflection, even if the first answer is not necessarily correct. As an example, Thompson provides a simple mathematical problem from an IQ test:

> If it takes 5 machines 5 minutes to make 5 widgets, how long would it take 100 machines to make 100 widgets? (Thompson 2009: 172)

Most of us have the intuition that the answer is 100 minutes and do not bother resorting to effortful calculation because the answer is accompanied by an FOR. However, the right answer is 5 minutes.

Thompson does not discuss the FOR in moral cases, but we think that this psychological phenomenon is present whenever moral intuitions are felt. When we experience a moral intuition, the FOR is involved in a similar way as in the example above. It makes us feel that we have the right answer and renders us less likely to challenge it. For instance, if the Canadian in our tipping example forces herself to reason that Spanish moral norms do not require her to tip, the FOR renders this reasoning unconvincing and easy to ignore or to question. The Canadian may end up not leaving a tip, but not without a mental struggle to repress her intuition. In addition, she is likely to feel guilty about not leaving a tip, even while she judges that she behaved correctly according to local norms.

Our descriptive account of moral intuition encompasses elements from all three kinds of account discussed in the preceding section. First, it recognizes that moral intuitions involve moral evaluations, often in the form of judgments, an aspect highlighted by the judgment accounts. Second, it includes the FOR to account for the particular phenomenology of intuitions. With the FOR, we acknowledge the affective component of moral intuitions emphasized by emotion accounts, without assimilating moral intuitions to emotional reactions. Indeed, the FOR is distinct from emotional feelings. In highly emotional situations—for example, when we condemn the Nazis' crimes against humanity—it may be difficult to discern the presence of the FOR because it is phenomenologically overshadowed by emotional feelings. The FOR is easiest to identify when it occurs in non-emotional contexts; for example, when we evaluate the relevance of an abstract moral principle. Third, our account is sympathetic to the idea that moral intuitions are seemings, distinct from emotions or judgments. We go one step further by elaborating on the composition of seemings as a two-component phenomenon: a challenged evaluation and the FOR. In the next section, we will add an evolutionary and mechanistic explanation to this description. This will allow us to characterize the FOR more closely and to explain why all intuitive evaluations are *challenged*.

AN EVOLUTIONARY UNDERSTANDING OF THE FEELING OF RIGHTNESS

Human cognition is a complex phenomenon composed of numerous task-specific systems and sub-systems organized in a web of interactions that are largely unknown. Some simple systems have been studied and their biological function singled out. For instance, the food aversion system serves to avoid the consumption of toxic food (Bernstein 1999), or the mimicry of others' movements system serves to facilitate joint action with other individuals (Knoblich et al. 2011). Mechanistically, these evolved systems process a given type of stimuli (e.g., unpleasant experiences with a food item) and are made of simple and distinct task-specific operations (e.g., felt aversion toward foods that have been associated with an unpleasant experience). These task-specific operations are "information-frugal" in the sense that they are activated by very specific cues. They are also "processing frugal" in the sense that they use simple algorithms or rules tailored specifically to the task demands, and during the execution of these operations a small subset of the total available information in the mind is recruited (Carruthers 2007). Here we argue that the FOR is the expression of a simple task-specific system of this sort, whose biological function and mechanistic functioning can be singled out.

The FOR plays a role at the interface between Type 1 and Type 2 thinking. According to dual process theories (Evans & Stanovich 2013), human cognitive abilities can be divided into two kinds of thinking that can be roughly characterized as follows: Type 1 thinking involves fast, unconscious, autonomous processing, and is typically induced by the activity of relatively independent and primitive systems dedicated to particular decision problems. It does not make demands on working memory and is thus relatively fluent and effortless. Type 2 thinking, however, involves slow and conscious processing that gives rise to reasoning able to manipulate abstract concepts and rules. This type of thinking needs working memory resources, making it effortful and expensive. As Thompson (2009) originally argued, part of the function of the FOR is to constrain Type 2 thinking. The FOR favors Type 1 responses, which come easily and fluently to people's minds, and thus discourages or biases Type 2 thinking. Here we build on this idea and embed it in an evolutionary and mechanistic explanation.

On our view, the evolution of the capacity for Type 2 thinking created a problem for early humans. Abstract thinking and reasoning allowed humans to communicate with each other, to cooperate in large groups and to develop sophisticated tools and techniques (Barrett et al. 2010). However, there is another side to the coin. As Gert Gigerenzer and others have highlighted, it is sometimes more adaptive to make a decision based on one's "gut feelings," or Type 1 processes, rather than based on careful deliberation (Gigerenzer et al. 1999). Humans often need to make quick decisions about complex problems based on incomplete knowledge. In these circumstances, they may sometimes be better off abandoning rational weighting of all relevant cues, relying instead on evolved task-specific systems that react to the most important information and ignore the rest. Daniel Dennett makes a similar point when he compares basic moral principles as useful "conversation-stoppers" which alleviate the need for "endlessly philosophizing, endlessly calling us back to first principles and demanding a justification" (1995: 506). Thus, with the development of Type 2 thinking, a need to prevent the excessive interference of rational deliberation in the generally adaptive operation of evolved Type 1 task-solving systems emerged.

We hypothesize that this need emerging from the conflict of influence between Type 1 and 2 processing led to the co-evolution of a further system: the feeling of rightness *system* (hereafter FORs). The evolutionary function of the FORs is to facilitate humans' decision making. The FORs helps people to select optimally among competing evaluative responses that are simultaneously present in their minds. The FOR makes the responses produced by evolved task-specific systems more appealing. It thereby facilitates decision-making and biases it in favor of systems that have been selected for solving the task at hand, or a similar task. At the mechanistic level, as with any evolved mental system, the FORs responds to certain stimuli and produces distinct task-specific operations: the stimuli processed are conflicting evaluative responses; the operation performed is to give priority to the response that arises most quickly and fluently in the subject's mind because it is a reliable indicator for the activity of an adaptive task-solving system. Phenomenologically, the FORs expresses itself as an FOR, which pushes us to think that the rapidly produced evaluation is correct.

Let us illustrate this phenomenon with the case of limerence, a term given to the heady experience of being-in-love (Tennov 1979). People in limerence experience intense feelings of love toward their partner, sexual attraction, obsessive thinking about the loved one, and a tendency to idealize and see only their good traits. They are likely to judge that their partner is the best possible match for them and that their love will last forever, despite their awareness that few couples remain happy in monogamous relationships for their whole lives, and despite previous disappointments. Limerence contributes to the forming of long-term pair bonds between two partners, an important condition for rearing human children given the length of gestation and the vulnerability of the human neonate (Fletcher et al. 2015). If a lover pauses and engages in a rational evaluation of her partner's traits, she may be forced to recognize that there are many potential partners out there who have similar traits, possibly in even greater measure. Why should she not question her judgment that her partner is the best possible match? In the face of conflicting evaluation produced by the primitive limerence system and Type 2 reflective thoughts of this kind, the FORs is activated and renders the evaluation that is produced more quickly and fluently more compelling. This explains why people in the grip of limerence, including philosophers, scientists, and relationship counselors, tend to be intuitively convinced that their present lover is the best possible match for them, even when their rational mind tells them otherwise.

The FORs may reinforce evaluations that are adaptive but incorrect. Limerence is one example: the likelihood that a person has fallen in love with her best possible match is very slim, but believing this is an important bonding factor. A mother's distorted view of her child is another example. From an evolutionary point of view, mother–infant bonding is important because it secures maternal protection and caregiving, thereby ensuring children's survival, brain maturation, and social-emotional growth. Bonding in the mother expresses itself as a set of inclinations including seeking proximity and physical contact with her child, worry about her child's safety and welfare, and extensive mental focus on her child (Feldman et al. 2007). Bonding also includes a mother's inclination to think that her child is the most beautiful baby ever born (Feldman et al. 1999). Such a belief is both adaptive and objectively false in the vast majority of cases. In such situations, the FORs compels a mother to keep believing that her baby is the most beautiful, even when her analytic reasoning questions the reasonableness of this belief.

Once the FORs has evolved, it can also be recruited in support of rationally or culturally acquired operations. For instance, young children in the process of learning basic mathematics may not initially find it compelling that $2 + 2 = 4$. But after having become familiar with the calculation and understood the underlying logic of it, usually by physically adding objects together until the pattern "clicks," this type of calculation is processed in a Type 1 manner; whenever one asks them the sum of $2 + 2$, the answer comes quickly and fluently to their mind. At this point, if somebody tried to convince them that $2 + 2 = 3$, the FORs would be elicited, making them feel that 4 is the right answer. The same phenomenon can be observed with the example of the Canadian who feels intuitively that she should leave substantial tips even while visiting countries where this is not the custom. These examples illustrate a well-known phenomenon in evolutionary biology (Futuyma 2013): a system that has been selected for one task, here to facilitate optimal decision-making by preserving the adaptive responses of evolved task-solving systems, may be later recruited for other tasks, here to select among competing learned responses.

The FORs may also support incorrect and non-adaptive responses. In the previous section, we mentioned the example of the incorrect answer to the mathematical widget question prompting a strong FOR in most people. These sorts of errors may occasionally induce non-adaptive decisions in humans. However, occasional failures in generating an adaptive answer is the fate of evolved traits. Indeed, for a trait to evolve it is sufficient that it produces adaptive responses in the most fitness-relevant contexts (Nettle 2006).

Our explanation of moral intuitions is more complete than alternative existing accounts. We make an important distinction between the phenomenological aspect (the FOR) and the mechanistic aspect (FORs) of intuitions. This distinction enables us to single out the eliciting factor for intuitions: they are generated by conflicting evaluations. Thus, any evaluation that feels intuitive has been *challenged* by a conflicting evaluation while a Type 2 process took place. Thanks to this account of intuitions, it is possible to explain why a quickly and fluently produced Type 1 evaluation sometimes feels intuitive and sometimes not. Imagine a situation (inspired from Greenspan 1988: ch. 1) where a householder judges a salesman to be trustworthy on the basis of the evidence available to her (Type 2 process), but stops short of signing a contract because she senses that the salesman is not to be trusted (Type 1 process). In such a situation, the householder experiences her suspicions as "intuitive" only because she is also tempted to evaluate the salesman as trustworthy; her suspicions are challenged by her reasoning that he has the correct identification documents and that her neighbor has recommended him. If all relevant information had concurred in confirming the impression of untrustworthiness, the householder would have quickly dismissed the salesman as untrustworthy, sent him away, and experienced no intuition about him. Similarly, we can routinely and quickly make the calculation 2 + 2 = 4 without experiencing it as intuitive. The characteristic phenomenology of intuition occurs only when the answer is challenged, for example, by someone claiming that 4 is not the right answer.

Finally, our account highlights that the human faculty to experience intuitions is of general scope. The FORs is a blind task-specific system which may apply to non-moral domains, such as conflicting aesthetic and mathematical evaluations, in addition to moral domains, such as conflicting evaluations about how to behave in a social context. From an empirical point of view, there is nothing special about moral intuitions compared to other types of intuition, aside from their subject matter. However, since this chapter focuses on moral intuitions, our next section is devoted to how the FOR applies to the moral domain.

THE FEELING OF RIGHTNESS APPLIED TO MORAL INTUITIONS

In the previous two sections, we argued that moral intuitions are composed of a challenged moral evaluation reinforced by a feeling of rightness (FOR). This feeling is produced by a psychological system (the FORs) whose eliciting factors are conflicting evaluations provided by different task-solving systems. The FORs prioritizes the response that comes quickly and fluently to the subject's mind. In this section, we explain how the FORs functions in the case of moral intuitions, why moral intuitions are a frequent occurrence, and speculate on why humans are biased in favor of metaethical internalism and against moral relativism.

Much moral evaluation involves Type 1 thinking, which is influenced by genetically and culturally shaped systems, yet varies greatly from one person to another. There is empirical evidence that heritable systems are involved in determining which domains of interaction are considered suitable for moral evaluation; for example, domains involving matters related to harming, fairness, loyalty, authority, and spiritual purity (Graham et al. 2011). To some extent, heritable systems even seem to influence the content of moral evaluations; for example, in the direction of incest avoidance (Lieberman et al. 2003), or in the condemnation of violations of social exchange rules (Barrett et al. 2010). However, despite their partial genetic heritability and the automaticity of their operations, the psychological systems involved in human moral evaluations are developed and shaped during human ontogenesis in various cultural environments (for a review, see Sunstein 2005). Consequently, significant variation can be observed among the Type 1 moral evaluations that are produced around the world and even between people from the same culture (Haidt & Graham 2007; Sachdeva et al. 2011).

In moral contexts, Type 2 thinking typically leads us to challenge our Type 1 evaluations and thus to experience intuitions. Type 2 processing can take more information into account than Type 1 processing, rendering people aware of challenges to their evaluations, which is a trigger for intuitive experiences. For instance, to refer to our example from the first section, when she is confronted by Aaron's criticism of her anti-immigrant stance, Sabrina may become aware that her views seem harsh and unkind from the immigrants' point of view; her FORs will be elicited, and since this system favors evaluations that are processed quickly and fluently, Sabrina will experience her Type 1 anti-immigration view as intuitive. At the end of the discussion, two scenarios are possible. Either Sabrina is thoroughly convinced that she is right, or she may rationally agree that Aaron is right, while still finding it intuitive to condemn immigration. In the latter case, her considered judgment conflicts with her intuition and she may have difficulties standing by her considered judgment in future circumstances unless she reinforces it in other ways. We can thus see how moral reflection and debate triggers intuitions. Since in contemporary pluralistic societies we often experience differing moral points of view, this explains why moral intuitions are so frequently experienced.

The FORs can favor learned and rational evaluations. As a result of life experiences, social pressure, or extensive private moral reasoning, people often come to modify their Type 1 evaluations. For instance, if Sabrina falls in love with an immigrant, she is likely to become able to understand his perspective and thus to sympathize with all immigrants, which may, over time, modify her Type 1 moral evaluations. Following her discussion with Aaron, she could also train herself to evaluate immigration positively, just as a convert to a particular brand of consequentialism may train herself to give equal moral weight to certain non-human animals and to mentally impaired humans, in spite of her initial intuitions. Once these evaluative habits are firmly established as Type 1 processes, they are favored by the FORs whenever conflicting evaluations are revealed by a slower Type 2 form of thinking. This illustrates the fact that the FORs is a simple task-specific system that does not discriminate between evaluations according to how they have been produced, that is, genetically, rationally, under the influence of life experiences. The FORs blindly tracks the rapidity and fluency with which evaluations are processed.

To conclude, we will briefly elaborate in a more speculative manner on how the FORs appears to bias our metaethical beliefs. Moral disagreement is common, thus the FORs is regularly activated and moral intuitions pervade moral discourse. Since the experience of a FOR tends to make people believe strongly in the correctness of their evaluation, frequent experience of the FOR in moral contexts renders it more likely that people believe that there are universal moral truths, which is a metaethical stance. Indeed, empirical data attest that when ordinary people make moral evaluations within their own culture, that is, on matters they are closely concerned with, they tend to be very optimistic about the universal validity of their evaluations (Sarkissian et al. 2011). Moreover, as Joyce has pointed out, it may be practically useful to believe in universal moral validity because this belief silences selfish calculations and bolsters self-control against practical irrationality (Joyce 2005). However, as mentioned above, there is also significant variation among Type 1 evaluations and people regularly experience moral disagreements. This fact can tempt people to adopt a more relativist view of morality. Thus, facing the contradicting evaluations that moral beliefs are universally valid and relative, the FORs may be activated and is likely to render the universalist view intuitive because it is processed more fluently. Note, however, that this analysis is a description of what may happen in people's minds; it provides no direct argument in favor of moral relativism as theory.

Similarly, the FORs may bias people's metaethical beliefs about moral motivation. As demonstrated in empirical literature, people are usually very optimistic about the motivational power of moral evaluations (Björnsson et al. 2015), an attitude conducive to the view called "motivational internalism." This attitude is likely to result from Type 1 thinking. However, empirical evidence also shows that in the absence of additional external motivating factors, people are only mildly motivated to comply with the moral norms they openly advocate. Whenever their own interests are at stake, they tend to prefer strategic over moral behavior (Aarøe & Petersen 2013; Batson 2008; Dhami, 2003; Talwar et al. 2004). They try to appear moral to others, while, if possible, avoiding the cost of being moral, a tendency that Daniel Batson calls "moral hypocrisy" (2008). These failures to follow moral norms are routinely experienced in our everyday life. Awareness of these facts while engaging in Type 2 reasoning may lead to disturbing conclusions: people, including oneself, may not be genuinely motivated to comply with the norms they openly hold! These contradictory Type 1 and 2 thoughts prompt the activation of the FORs which favors the Type 1 internalist belief.

People are not typically aware of the psychological mechanisms involved in their thoughts, such as whether their evaluation has been processed in a Type 1 or Type 2 manner, or whether it is reinforced by the FORs. Moreover, by restricting and biasing rational and objective assessment of quickly and fluently produced responses, the FORs enables people to resolve or overlook their mental inconsistencies with minimal discomfort. There is empirical evidence that people happily confabulate and produce post-hoc rationalizations for their moral intuitions (Haidt et al. 1993). It is thus possible that the FORs is one of the crucial mechanisms underlying the behavior reported by cognitive dissonance theorists (Festinger 1957).

CONCLUSION

We claim that there are two components to moral intuitions: a challenged evaluation and a metacognitive feeling of rightness (FOR). The emergence of the psychological mechanism responsible for the FOR can be explained by its clear evolutionary function in core cases: to facilitate optimal decision-making by preserving the adaptive responses of evolved task-solving systems.

Our account helps to bridge the gap between different conceptions of moral intuition in psychology and philosophy. On the one hand, it recognizes the role of moral judgments, the most common form of moral evaluation, and of Type 2 thinking. On the other hand, it emphasizes the affective component of intuition: the FOR. Moreover, our account builds on the useful conception of moral intuitions as seemings and provides additional evolutionary and psychological information about the workings of this mental phenomenon. In particular, it highlights that the experience of intuition arises during Type 2 mental activity and only against a background of contradictory evaluations produced by different task-solving systems.

An interesting consequence of our account is that there is not much that is special about moral intuitions compared to other types of intuition. All are produced by the same blind task-specific system, the FORs. Intuitions qualify as "moral" when the FORs applies to conflicting moral evaluations.

Finally, our account explains why intuitions are a frequent occurrence in the moral domain, why they remain so compelling and resistant in the face of reasons to discount them, and why humans are biased toward the ideas of universal moral truth and intrinsic motivation to follow moral norms.

ACKNOWLEDGMENTS

We thank Bernard Baertschi, Richard Dub, Benoît Dubreuil, Richard Joyce, and Fabrice Teroni for their useful comments, and the audience at *The Moral Domain: Conceptual Issues in Moral Psychology* conference in Vilnius, Lithuania.

REFERENCES

Aarøe, L. & Petersen, M. 2013. "Hunger games: Fluctuations in blood glucose levels influence support for social welfare." *Psychological Science* 24: 2550–2556.

Audi, R. 2009. *The Good in the Right: A Theory of Intuition and Intrinsic Value* (Princeton University Press).

Barrett, H.C., Cosmides, L., & Tooby, J. 2010. "Coevolution of cooperation, causal cognition and mindreading." *Communicative & Integrative Biology* 3: 522–524.

Batson, C. 2008. "Moral masquerades: Experimental exploration of the nature of moral motivation." *Phenomenology and the Cognitive Sciences* 7: 51–66.

Bedke, M. 2008. "Ethical intuitions: What they are, what they are not, and how they justify." *American Philosophical Quarterly* 45: 253–269.

Bernstein, I. 1999. "Taste aversion learning: A contemporary perspective." *Nutrition* 15: 229–234.

Björnsson, G., Eriksson, J., Strandberg, C., Olinder, R., & Björklund, F. 2015. "Motivational internalism and folk intuitions." *Philosophical Psychology* 28: 715–734.

Carruthers, P. 2007. "Simple heuristics meet massive modularity." In P. Carruthers, S. Laurence, & S. Stich (eds.), *The Innate Mind: Culture and Cognition* (Oxford University Press) 181–198.

Cohen, G. 2008. *Rescuing Justice and Equality* (Harvard University Press).

Dennett, D. 1995. *Darwin's Dangerous Idea. Evolution and the Meanings of Life* (Penguin).

DePaul, M. 2006. "Intuitions in moral inquiry." In D. Copp (ed.), *The Oxford Handbook of Ethical Theory* (Oxford University Press) 595–623.

Dhami, M. 2003. "Psychological models of professional decision making." *Psychological Science* 14: 175–180.

Evans, J. & Stanovich, K. 2013. "Dual-process theories of higher cognition advancing the debate." *Perspectives on Psychological Science* 8: 223–241.

Feldman, R., Weller, A., Leckman, J., Kuint, J., & Eidelman, A. 1999. "The nature of the mother's tie to her infant: Maternal bonding under conditions of proximity, separation, and potential loss." *Journal of Child Psychology and Psychiatry and Allied Disciplines* 40: 929–939.

Feldman, R., Weller, A., Zagoory-Sharon, O., & Levine, A. 2007. "Evidence for a neuroendocrinological foundation of human affiliation: Plasma oxytocin levels across pregnancy and the postpartum period predict mother-infant bonding." *Psychological Science* 18: 965–970.

Festinger, L. 1957. *A Theory of Cognitive Dissonance* (Row & Peterson).

Fletcher, G., Simpson, J., Campbell, L., & Overall, N. 2015. "Pair-bonding, romantic love, and evolution: The curious case of *Homo sapiens*." *Perspectives on Psychological Science* 10: 20–36.

Foot, P. 1967. "The problem of abortion and the doctrine of double effect." *Oxford Review* 5: 5–15.

Futuyma, D. 2013. *Evolution*, 3rd ed. (Sinauer Associates, Inc.).

Gigerenzer, G., Todd, P., & ABC Research Group. 1999. *Simple Heuristics that Make Us Smart* (Oxford University Press).

Graham, J., Nosek, B., Haidt, J., Iyer, R., Koleva, S., & Ditto, P. 2011. "Mapping the moral domain." *Journal of Personality and Social Psychology* 101: 366–385.

Greene, J. 2007. "The secret joke of Kant's soul." In W. Sinnott-Armstrong (ed.), *Moral Psychology: The Neuroscience of Morality: Emotion, Brain Disorders, and Development* (MIT Press) 359–372.

Greenspan, P. 1988. *Emotions and Reasons: An Inquiry into Emotional Justification* (Routledge).

Haidt, J. & Bjorklund, F. 2008. "Social intuitionists answer six questions about moral psychology." In W. Sinnott-Armstrong (ed.), *Moral Psychology: The Cognitive Science of Morality: Intuition and Diversity* (MIT Press) 181–217.

Haidt, J. & Graham, J. 2007. "When morality opposes justice: Conservatives have moral intuitions that liberals may not recognize." *Social Justice Research* 20: 98–116.

Haidt, J., Koller, S., & Dias, M. 1993. "Affect, culture, and morality, or is it wrong to eat your dog?" *Journal of Personality and Social Psychology* 65: 613–628.

Huemer, M. 2008. *Ethical Intuitionism* (Palgrave Macmillan).

Joyce, R. 2005. "Moral fictionalism." In M. Kalderon (ed.), *Fictionalism in Metaphysics* (Oxford University Press) 287–313.

Kamm, F. 2006. *Intricate Ethics: Rights, Responsibilities, and Permissible Harm* (Oxford University Press).

Kauppinen, A. 2013. "A Humean theory of moral intuition." *Canadian Journal of Philosophy* 43: 360–381.

Knoblich, G., Butterfill, S., & Sebanz, N. 2011. "Psychological research on joint action: Theory and data." In B. Ross (ed.), *The Psychology of Learning and Motivation* (Elsevier) 59–101.

Lieberman, D., Tooby, J., & Cosmides, L. 2003. "Does morality have a biological basis? An empirical test of the factors governing moral sentiments relating to incest." *Proceedings of the Royal Society, Series B: Biological Sciences* 270: 819–826.

McMahan, J. 2013. "Moral intuition." In H. LaFollette & I. Persson (eds.), *The Blackwell Guide to Ethical Theory*, 2nd ed. (Wiley-Blackwell) 103–120.

Nettle, D. 2006. "The evolution of personality variation in humans and other animals." *American Psychologist* 61: 622–631.

Rawls, J. 1971. *A Theory of Justice* (Harvard University Press).

Roeser, S. 2011. *Moral Emotions and Intuitions* (Palgrave Macmillan).

Sachdeva, S., Singh, P., & Medin, D. 2011. "Culture and the quest for universal principles in moral reasoning." *International Journal of Psychology* 46: 161–176.

Sarkissian, H., Park, J., Tien, D., Wright, J., & Knobe, J. 2011. "Folk moral relativism." *Mind & Language* 26: 482–505.

Schwartz, B. & Metcalfe, J. 2011. "Tip-of-the-tongue (TOT) states: Retrieval, behavior, and experience." *Memory & Cognition* 39: 737–749.

Shafer-Landau, R. 2005. *Moral Realism: A Defense* (Clarendon Press).

Singer, P. 2005. "Ethics and intuitions." *Journal of Ethics* 9: 331–352.

Sinnott-Armstrong, W. 2008. "Framing moral intuitions." In W. Sinnott-Armstrong (ed.), *Moral Psychology: The Cognitive Science of Morality: Intuition and Diversity* (MIT Press) 47–76.

Sunstein, C. 2005. "Moral heuristics." *Behavioral and Brain Sciences* 28: 531–542.

Talwar, V., Lee, K., Bala, N., & Lindsay, R. 2004. "Children's lie-telling to conceal a parent's transgression: Legal implications." *Law and Human Behavior* 28: 411–435.

Tennov, D. 1979. *Love and Limerence: The Experience of Being in Love* (Scarborough House).

Tersman F. 2007. "Fresh air?" In T. Rønnow-Rasmussen (ed.), *Hommage à Wlodek: 60 Philosophical Papers Dedicated to Wlodek Rabinowicz.* www.fil.lu.se/hommageawlodek

Thompson, V. 2009. "Dual process theories: A metacognitive perspective." In J. Evans & K. Frankish (eds.), *Two Minds: Dual Processes and Beyond* (Oxford University Press) 171–195.

Tucker, C. 2013. "Seemings and justification: An introduction." In C. Tucker (ed.), *Seemings and Justification: New Essays on Dogmatism and Phenomenal Conservatism* (Oxford University Press) 1–29.

23

Are We Losing It? Darwin's Moral Sense and the Importance of Early Experience

Darcia Narvaez

THE MORAL SENSE

Charles Darwin was concerned to counter the now-pervasive views initially promulgated by Herbert Spencer—that self-interest was primary and morality did not emerge from the tree of life. He spent considerable effort in distinguishing between behaviors that might help an individual better survive in the short term and the characteristics that would help a group survive over generations. Along with natural selection, he emphasized moral evolution among humans through group selection as one of progressive increase (Darwin [1871] 1981; see also Loye 2000). As part of these efforts, Darwin proposed that humans have a "moral sense" that contributes to their evolution, beyond the role of natural selection (Gruber 1974). According to Darwin, humanity's moral sense arose from the sexual, parental, and social instincts that evolved in mammals generally, but especially in humans, giving rise to the golden rule. Accordingly, "moral behavior was embodied in the nature of the species, and not imposed on the natural world as something foreign to it. . . . Humans are not sacrificing their natures when they act morally; they are responding to them" (Schwartz 2009: 11). Darwin even toyed with the idea that the moral sense was the main propellant of human evolution (Loye 2000).

Here are quotes with slight paraphrase (modernized language) of Darwin describing the components of the moral sense:

> In the first place, the social instincts lead an animal to take *pleasure* in the society of its fellows, to feel a certain amount of *sympathy* for them, and to perform various services for them. . . . Secondly, as soon as the mental faculties had become highly developed, *images of all past actions and motives* would be incessantly passing through the brain of each individual. Out of a *comparison of past and present*, the feeling of dissatisfaction, or even misery, which invariably results from any unsatisfied instinct, would arise. Third, after the power of *language* had

been acquired, and the wishes of the community could be expressed, the *common opinion of how each member ought to act for the public good* would naturally become the guide to action . . . Lastly, *habit* in the individual could ultimately play a very important part in guiding the conduct of each member, for the social instinct together with sympathy, is, like any other instinct, greatly strengthened by habit, and so consequently would be *obedient to the wishes and judgment of the community*.   (Loye 2000: 128–129 (emphasis added))

Rejecting the liberal notion of rational self-interest as a prime motivator of human beings, Darwin viewed morality to be born of social instincts, instincts that make morality satisfying in and of itself. In other words, "the consummate moral individual is not one who conquers his/her base inclinations at the moment of choice, but rather one who does not experience them. . . . Morality is judged by the heart, and not only by deed" (Schwartz 2009: 12–13). Rationality instead plays a role in negotiating among conflicting interests. Interestingly, while Darwin contrasted the selfish rivalry among his male compatriots with the morality of women in his society (women are less selfish, more "tender," more intuitive, with greater perceptual capabilities), he noted that the members of "lower races" and those in "lower states of civilization" demonstrated the characteristics of "civilized women." As described below, early experience may have a lot to do with which type of morality—more selfish or more tender—that one develops.

Components of the moral sense—social pleasure, empathy, social concern, and habit control—are reportedly apparent all over the world among (male and female) small-band hunter-gatherer communities (immediate-return societies; hereafter called "SBHG"), the type of society in which 99 percent of human genus history is presumed to have been spent. Although hunter-gatherer societies vary in many ways and emerged independently around the world, they share striking commonalities (for reviews, see Fry 2006; Ingold 1999, 2011; Martin 1999). Illustrations related to the components of Darwin's moral sense are mentioned here and contrasts are drawn with trends in US culture (see Narvaez 2014 for fuller details and references).

Regarding desire and capacity for *social pleasure*, hunter-gatherers greatly enjoy one another's company; individuals rarely spend any time alone and cannot conceive that anyone would want to be alone, despite the efforts of visiting anthropologists to strive for alone time. In the USA of 2011, over 50 percent of adults are single (compared with 22 percent in the 1950s), with single-adult households the most common and growing type of household (Klinenberg 2012). Although many US adults feel fine spending many hours alone each day, at the same time isolation and loneliness have been increasing (Cacioppo & Patrick 2008).

Regarding moral capacities such as *empathy*, anthropologists note remarkable SBHG empathic care of young children (Konner 2005), kindness in needed situations (Everett 2009), inclusionary communal care (Ingold 1999), and cognitive empathy (perspective-taking) (e.g., Wolff 2001). In the USA, college student empathy has significantly decreased in the last decade (Konrath et al. 2011), with parallel increases in narcissism (Twenge & Campbell 2009) and avoidant attachment. (Both are signs of poor emotion system development and characterized by low emotional connection; Konrath et al. 2014.)

Regarding concern for *socially responsible* behavior (conscience), hunter-gatherers are concerned with certain values-in-action, specifically egalitarianism, generosity, and sharing; they repel coercion and hierarchy, and expel cheaters. In the USA, an increasing number of families exhibit anti-social behavior (Mooney & Young 2006; H. Walker 1993); cheating to get ahead is widespread in all walks of life, and many institutions and policies are designed to promote sociopaths (Callahan 2004; Derber 2013).

Regarding developing appropriate habits, *self-control* is a vital value among SBHG and its lack is grounds for expulsion (Everett 2009). In the USA, a greater and increasing number of young children arrive in kindergarten with behavior dysregulation (e.g., Gilliam 2005; Powell et al. 2003) and two-thirds of American teens report having experienced an explosive outburst (intermittent explosive disorder) in the previous year (McLaughlin et al. 2012).

Noting the variable societal differences, one might wonder whether there is a larger post-natal component to the moral sense than Darwin anticipated. Intensive parenting emerged with social mammals more than 30 million years ago and intensified further as hominins became more and more helpless and needy at birth to accommodate bipedalism and other factors (Konner 2010; Trevathan 2011). In comparison to the development of other primates at birth, humans are born 9–18 months early in terms of mobility and bone development. Full-term human babies (40–42 weeks gestation) have 25 percent of adult brain volume at birth, resulting in the fact that early life caregiving co-constructs multiple brain and body systems, in a type of "exterogestation," or external womb (Montagu 1986). Like social mammals, the human child is a dynamic system whose initial and early experiences with caregivers shape biosocial systems and establish trajectories for multiple aspects of the brain and body (Dettmer et al. 2014; Meaney 2001). Epigenetic effects are ongoing, with a constant interaction among maturational state, prior experience, and environmental influence (Murgatroyd & Spengler 2011).

What specific intensive caregiving practices evolved for humans? Anthropologists have documented the caregiving commonalities among hunter-gatherers (Hewlett & Lamb 2005; Konner 2005, 2010). These practices comprise what my colleagues and I call "the Evolved Developmental Niche" (EDN), slightly altered from the early care that characterizes other catarrhine mammals. In what follows I shall discuss them briefly, mentioning a few of their benefits. (For reviews, see Narvaez 2014; Narvaez et al. 2013b, 2014, 2016a.) Of course, for caregivers there can be costs to these practices, though adults in SBHG demonstrate pleasure in providing them (Hewlett & Lamb 2005).

Natural childbirth: this means vaginal birth, no separation of mother and newborn, and no induced pain. For example, mother–child separation after birth affects mother–child bonding, breastfeeding success, and child self-regulation (Ball & Russell 2013; Bystrova et al. 2009).

Breastfeeding: infants nurse at will and frequently (two to three times/hour initially). Nursing lasts for two to five years, with an average weaning age of age four. Breast milk provides thousands of ingredients to build a healthy brain and body (Goldman et al. 1990; M. Walker 1993). For example, breast milk provides "antimicrobial, prebiotic, and likely probiotic factors that function dually to promote the growth of beneficial gut bacteria and inhibit pathogenic bacteria from establishing and replicating in the infant" (Martin & Sela 2013: 239). In contrast, formula feeding introduces pathogenic bacteria that are related to various diseases (Rubaltelli et al. 1998).

Affectionate touch: in the first years of life, children are carried skin-to-skin, held, or kept near others constantly. Affectionate touch keeps the infant calm and growing, fostering, for example, important brain/body systems such as oxytocin function (Caldji et al. 1998; Champagne et al. 2006; Kramer et al. 2003).

Responsivity: caregivers respond promptly to the needs of babies, resulting in little distress. Responsiveness properly sets up multiple systems, including the vagus nerve which is critical for well-functioning digestion, cardiac, respiratory, stress, immune, and emotion systems (Calkins & Hill 2007; Donzella et al. 2000; Haley & Stansbury 2003; Propper et al. 2008; Siniatchkin et al. 2003; Stam et al. 1997).

Play: children of all ages enjoy free play in the natural world with multiage playmates. Free play, especially rough-and-tumble play, fosters better brain development and maturation, mental health and social skills (Panksepp 2007; Pellis & Pellis 2009; Spinka et al. 2001).

Alloparents and positive social support: mother–child dyads experience high social embeddedness, with young children frequently cared for by community members, keeping positive emotions active and needs met (e.g., Morelli et al. 2014). Maternal social support is linked to greater maternal responsiveness (Hrdy 2009).

These EDN components might represent a necessary context for optimal development. Early experience influences the design of physiological and psychosocial systems such as the stress response, endocrine system, and emotion systems (Narvaez et al. 2013b; Schore 1996). Deficiencies in these systems can impair sociality. For example, proper function of the vagus nerve, which underlies capacities for social intimacy, relies on caregiver touch and positive responsiveness in early life (Porges 2011). Poor vagal tone and poor capacities for controlling anxiety lead to greater stress response in social situations, whereas a good vagal tone is linked to greater compassion (Keltner 2009). The stress response draws energy for survival fight-or-flight, impairing executive controls in the prefrontal cortex, and higher order thinking, undermining social and moral imagination (Arnsten 2009; Sapolsky 2004).

Could the evolved developmental niche play a significant role in the development of the moral sense? A large number of studies have examined caregiver supportive responsiveness in early life, showing that it is linked to multiple positive child outcomes. For example, a "mutually-responsive orientation" in the mother–child relationship builds secure attachment and leads over the years to greater empathy, self-regulation, conscience, openness, and social competence (Kochanska 2002; Eisenberg 2000). But how about the other aforementioned parenting practices that form part of our mammalian heritage?

My colleagues and I have been examining whether the EDN influences well-being and morality. In several studies we have measured parenting behaviors and attitudes about the EDN and their relation to early signs of moral development in young children (3 to 5 year olds). In one study with an existing longitudinal dataset (Narvaez et al. 2013a), we examined responsivity, positive touch, breastfeeding initiation, and maternal social support. After controlling for the effects of responsivity, education, and income, there were several significant results. For example, we found that those who were breastfed at all were less aggressive at age two. Those mothers who provided more positive touch had children with greater intelligence and social engagement at age three. Mothers who reported greater social support had children who demonstrated less aggression and more social competence at 24 months, and greater cooperation at 18 and 30 months. These findings suggest that EDN-consistent care may have long-term effects in early life. We also developed a survey of attitudes and behaviors consistent with the evolved developmental niche (Family Life Attitude and Behavior Measure). Again, we controlled for the effects of education, income and responsivity. In a Chinese sample (n=383; Narvaez et al. 2013c), we found that each parenting practice was related to one or more child moral outcomes (self-regulation, empathy, conscience). Across these and other studies, we are finding consistent patterns of links between EDN-consistent care and child well-being and morality.

How does early experience play a role in how an individual's moral orientations are shaped? When early care does not match the EDN, it presents a toxic environment for the development of prosocial morality. Toxic stress keeps active the stress response, which is a different brain state from one of social pleasure (MacLean 1990). A global brain state shifts perception, information processing, and affordances (action possibilities). Immersed in a stress-inducing environment, the child will adapt to that environment and learn to be sensitive to threat. When a brain state guides behavior and is acted upon socially, it represents an ethic (Narvaez 2008, 2012, 2014). Those from stressful early environments are more likely to develop dispositions for self-protective ethics (unless other significant support was experienced later). When early care matches the EDN, the capacities underlying the relationally-attuned engagement ethic and communal imagination ethic develop as a matter of course. In this way, adult moral orientations are shaped by early experience.

We have tested whether there are long-term effects of the EDN on adult moral orientations. In one study (Narvaez et al. 2016b), over six hundred adults were asked about their childhood experiences. We found positive and negative pathways to moral orientation. The positive pathway moved from EDN-consistent childhood to secure attachment, then to mental health, then to perspective-taking capacities, and then to an engagement ethic. The negative pathways led from lower EDN-consistent childhood to lack of secure attachment, to psychopathology, then either to personal distress and a withdrawal (wallflower) ethic or to lack of perspective taking and an oppositional (bunker) ethic. Thus, the evidence from young children and from adults suggests that the EDN matters not only for human physiological and mental health, but also for moral outcomes. Denying children their evolved needs may undermine not only the nature of their sociality but alienate them from themselves. The message received is that one's impulses are wrong and untrustworthy. This may be transferred to all of the natural world (human and non-human), building in a basic alienation to life.

Although we have not studied this empirically directly, it appears that the punishment and coercion that characterizes most childhoods in the West, but not hunter-gatherer childhoods, may lead to the preconventional moral thinking that Lawrence Kohlberg found in children (stage 1: obey to avoid punishment; stage 2: instrumental hedonism or one-time bargaining) as well as deception. These stages of thinking are not apparent in hunter-gatherers or in groups where children are not coerced or punished but where generous sharing is the group norm. Children from societies where the EDN is provided appear to display stage 3 (interpersonal concordance) from a young age (Bolin 2010). In fact, babies from (natural) birth display expectation and readiness for the social life (Trevarthen 2005). The story of socio-moral development then should be flipped. It is not how a selfish being becomes social but how a deeply social being becomes selfish. Thus, even Western-derived theory may be misleading us on what is truly part of evolved human nature.

SPECIES-ATYPICAL OUTCOMES?

There are several common reactions when one points out species-atypical parenting in societies like the USA. One reaction is to argue that humans have continued to evolve and so it cannot be expected that today's humans would need the same things as our ancestors (and, besides, who really knows how our ancestors behaved?). The response of course is to describe how humans are still in the line of 30-million-year-old social mammals; humans still emerge from the womb at a remarkably early point in development, with many systems shaped by the post-natal environment and with the longest maturation schedule of any animal. Hence, post-natal life is critical for shaping the physiology that underlies health and well-being. Moreover, we can document the social mammalian parenting practices and their cultural correlations among hunter-gatherers (99 percent of human genus history). And, as briefly noted above, there is an increasing amount of neurobiological research demonstrating the importance of the evolved caregiving practices.

Another counterargument notes that the brain is plastic and malleable throughout life, so early life cannot matter that much. Yes, but plasticity decreases with age and it is still unclear how much one can alter some epigenetic effects that occur during early sensitive periods. For example, the glucocorticoid receptor protein well studied in rats by Michael Meaney and colleagues (2010) has a critical window for shaping its expression (first ten days of life). If a rat pup has a low-nurturing mother during that time, the gene is never properly expressed, leaving the individual anxious in new situations throughout life. Certainly, it is possible to control the subsequent anxiety with drug intervention later, as Meaney and colleagues have shown, but the gene is one of hundreds of genes affected by maternal care. Similar methylation patterns have been identified in human brains.

Another response is to argue that parenting varies because parents prepare their children for a particular culture (Harkness & Super 1995). Of course, even among SBHG there is cultural variability in child-raising, but the variability is relatively narrow and fits within humanity's social mammalian heritage and the EDN. In contrast, nations like the USA deny young children most of the EDN components, especially since the mid-20th century when mother–child alienating hospital births and practices became predominant, along with infant formula, prescriptions to let babies cry, and the isolation of babies and mothers from the community. The shift away from the EDN has continued with increased physical isolation (in carriers, strollers, car seats) and placement of young children in daycare centers where touch and free play are increasingly absent. The data are suggesting that EDN-inconsistent care may not be adaptive in any long-term sense.[1] When adults do not provide young children with what they evolved to need for healthy brain and body development, thriving is undermined, which, along with reproduction and survival, is critical for multigenerational adaptation. Though of course there are likely multiple causes of adult ill health beyond early life experience, the links between early life experience and well-being bear closer examination due to the deterioration of US children's and adult's health over the last half of the 20th century (e.g., National Research Council 2013; OECD 2009).

CHALLENGING IMPLICATIONS

There are several potential implications of the epigenetic nature of moral development. First, we must understand that brains and bodies raised outside the EDN are unlikely to exhibit full human capacities and instead are species atypical. It is then unwise and inaccurate to draw any conclusions about evolved human characteristics without contextualizing an individual's development niche. Early toxic stress undermines optimal development and there is no evidence that optimality can be recovered (Shonkoff et al. 2012). Therefore, we should not be generalizing about human morality or human nature from research on participants raised outside the EDN. In comparison with those raised with a full EDN, they are likely to have impaired sociality and be more self-centered. As noted above, the USA seems to display these species-atypical outcomes.

Second, culture matters. As Douglas Fry (2006) notes, theorists apply their own cultural lenses to the data they select to interpret. It will "make sense" to adults raised outside the EDN, who likely display stress reactivity and self-protective ethics, to describe universal human nature as selfish. But, as Marshall Sahlins has pointed out:

> For the greater part of humanity, self interest as we know it is unnatural in the normative sense; it is considered madness, witchcraft or some such grounds for ostracism, execution or at least therapy. Rather than expressing a pre-social human nature, such avarice is generally taken for a loss of humanity.
>
> (Sahlins 2008: 51)

In other words, ascribing selfishness to human nature may be a sign of a particular cultural niche—in terms of both perceptions but also outcomes that are brought about by the attitudes and behaviors that correspond to those cultural beliefs—and not a matter of human nature, per se. The shaping of worldview may start very early in life and have much to do with parent attitudes and childrearing behaviors (Tomkins 1965).

CONCLUSION

Margaret Mead's (1939) studies among South Pacific groups were guided by questions regarding human nature, its limits and potentialities, with the aim to "present evidence that human character is built upon a biological base which is capable of enormous diversification in terms of social standards" (Mead 1939: x). Neurobiological studies are proving her intuitions to be correct. From her studies she concluded that "human nature is flexible, but it is also elastic—it will tend to return to the form that was impressed upon it in earliest years" (xiii). Again, her conclusions are being supported by studies of attachment, sociality, and well-being in adulthood.

Childrearing baselines have shifted away from our mammalian heritage, at least in the USA. Undermining species-typical development will influence capacities of all sorts, including the moral sense. Dismissing inherited mammalian needs in early life may actually create the *wrong kind* of moral sense: one that is self-protective, mindless, and ultimately destructive of the species and life on the planet (Naess & Rothenberg 1989). With climate and ecological devastation close by, it seems that we are now staring our self-destruction in the face.

NOTE

1. Of course we must be careful with the term "adaptive." In developmental psychology, sometimes it is misused such as labeling early reproduction an evolved phenotype. Those who like to argue that reaching reproduction is a sufficient sign for fitness adaptation are too easy on themselves. To determine whether something is adaptive in the fitness sense can only be known after several generations and only in comparison to rivals (Lewontin 2010).

REFERENCES

Arnsten, A. 2009. "Stress signalling pathways that impair prefrontal cortex structure and function." *Nature Reviews Neuroscience* 10: 410–422.

Ball, H. & Russell, C. 2013. "Night-time nurturing: An evolutionary perspective on breastfeeding and sleep." In D. Narvaez, J. Panksepp, A. Schore, & T. Gleason (eds.), *Evolution, Early Experience and Human Development: From Research to Practice and Policy* (Oxford University Press) 241–261.

Bolin, I. 2010. "Chillihuani's culture of respect and the circle of courage." *Reclaiming Children and Youth Worldwide* 18: 12–17.

Bystrova, K., Ivanova, V., Edhborg, M., Matthiesen, A., Ransjö-Arvidson, A., Mukhamedrakhimov, R., Uvnäs-Moberg, K., & Widström, A. 2009. "Early contact versus separation: Effects on mother-infant interaction one year later." *Birth* 36: 97–109.

Cacioppo, J. & Patrick, W. 2008. *Loneliness: Human Nature and the Need for Social Connection* (W.W. Norton & Co.).

Caldji, C., Tannenbaum, B., Sharma, S., Francis, D., Plotsky, P., & Meany, M. 1998. "Maternal care during infancy regulates the development of neural systems mediating the expression of fearfulness in the rat." *Proceedings of the National Academy of Sciences* 95: 5335–5340.

Calkins, S. & Hill, A. 2007. "Caregiver influences on emerging emotion regulation: Biological and environmental transactions in early development." In J. Gross (ed.), *Handbook of Emotion Regulation* (Guilford Press) 229–248.

Callahan, D. 2004. *The Cheating Culture: Why More Americans are Doing Wrong to Get Ahead* (Harcourt Harvest).

Champagne, F., Weaver, I., Diorio, J., Dymov, S., Szyf, M., & Meaney, M. 2006. "Maternal care associated with methylation of the estrogen receptor-alpha1b promoter and estrogen receptor-alpha expression in the medial preoptic area of female offspring." *Endocrinology* 147: 2909–2915.

Darwin, C. [1871] 1981. *The Descent of Man* (Princeton University Press).

Derber, C. 2013. *Sociopathic Society: A People's Sociology of the United States* (Paradigm Publishers).

Dettmer, A., Suomi, S., & Hinde, K. 2014. "Nonhuman primate models of mental health: Early life experiences affect developmental trajectories." In D. Narvaez, K. Valentino, A. Fuentes, J. McKenna, & P. Gray (eds.), *Ancestral Landscapes in Human Evolution: Culture, Childrearing and Social Wellbeing* (Oxford University Press) 42–58.

Donzella, B., Gunnar, M., Krueger, W., & Alwin, J. 2000. "Cortisol and vagal tone responses to competitive challenge in preschoolers: Associations with temperament." *Development Psychobiology* 37: 209–220.

Eisenberg, N. 2000. "Emotion, regulation, and moral development." *Annual Review of Psychology* 51: 665–697.

Everett, D. 2009. *Don't Sleep, There are Snakes: Life and Language in the Amazonian Jungle* (Vintage).

Fry, D. 2006. *The Human Potential for Peace: An Anthropological Challenge to Assumptions about War and Violence* (Oxford University Press).

Gilliam, W. 2005. *Prekindergarteners Left Behind: Expulsion Rates in State Prekindergarten Systems* (Yale University Child Study Center).

Goldman, A., Goldblum, R., & Hanson, L. 1990. "Anti-inflammatory systems in human milk." *Advances in Experimental Medicine and Biology* 262: 69–76.

Gruber, H. 1974. *Darwin on Man: A Psychological Study of Scientific Creativity* (University of Chicago Press).

Haley, D. & Stansbury, K. 2003. "Infant stress and parent responsiveness: Regulation of physiology and behavior during still-face and reunion." *Child Development* 74: 1534–1546.

Harkness, S. & Super, C. 1995. *Parents' Cultural Belief Systems: Their Origins, Expression, and Consequences* (Guilford).

Hewlett, B. & Lamb, M. (eds.). 2005. *Hunter-gatherer Childhoods: Evolutionary, Developmental and Cultural Perspectives* (Transaction).

Hrdy, S. 2009. *Mothers and Others: The Evolutionary Origins of Mutual Understanding* (Belknap Press).

Ingold, T. 1999. "On the social relations of the hunter-gatherer band." In R. Lee & R. Daly (eds.), *The Cambridge Encyclopedia of Hunters and Gatherers* (Cambridge University Press) 399–410.

Ingold, T. 2011. *The Perception of the Environment: Essays on Livelihood, Dwelling and Skill* (Routledge).

Keltner, D. 2009. *Born to be Good: The Science of a Meaningful Life* (Norton).

Klinenberg, E. 2012. *Going Solo: The Extraordinary Rise and Surprising Appeal of Living Alone* (Penguin).

Kochanska, G. 2002. "Mutually responsive orientation between mothers and their young children: A context for the early development of conscience." *Current Directions in Psychological Science* 11: 191–195.

Konner, M. 2005. "Hunter-gatherer infancy and childhood: The !Kung and others." In B. Hewlett & M. Lamb (eds.), *Hunter-gatherer Childhoods: Evolutionary, Developmental and Cultural Perspectives* (Transaction) 19–64.

Konner, M. 2010. *The Evolution of Childhood* (Belknap Press).

Konrath, S., O'Brien, E., & Hsing, C. 2011. "Changes in dispositional empathy over time in college students: A meta-analysis." *Personality and Social Psychology Review* 15: 180–198.

Konrath, S., Chopik, W., Hsing, C., & O'Brien, E. 2014. "Changes in adult attachment styles in American college students over time: A meta-analysis." *Personality and Social Psychology Review* 18: 326–348.

Kramer, K., Cushing, B., & Carter, C. 2003. "Developmental effects of oxytocin on stress response: Single versus repeated exposure." *Physiology & Behavior* 79: 775–782.

Lewontin, R. 2010. "Reply to comment on 'Not So Natural Selection.'" *New York Review of Books* September 30.

Loye, D. 2000. *Darwin's Lost Theory of Love* (Writer's Press).

MacLean, P. 1990. *The Triune Brain in Evolution: Role in Paleocerebral Functions* (Plenum Press).

Martin, C. 1999. *The Way of the Human Being* (Yale University Press).

Martin, M. & Sela, D. 2013. "Infant gut microbiota: Developmental influences and health outcomes." In K. Clancy, K. Hinde, & J. Rutherford (eds.), *Building Babies: Primate Development in Proximate and Ultimate Perspective* (Springer) 233–258.

McLaughlin, K., Green, J., Hwang, I., Sampson, N., Zaslavsky, A., & Kessler, R. 2012. "Intermittent explosive disorder in the national comorbidity survey replication adolescent supplement." *Archives of General Psychiatry* 69: 1131–1139.

Mead, M. 1939. *From the South Seas: Studies of Adolescence and Sex in Primitive Societies* (William Morrow & Co.).

Meaney, M. 2001. "Maternal care, gene expression, and the transmission of individual differences in stress reactivity across generations." *Annual Review of Neuroscience* 24: 1161–1192.

Meaney, M. 2010. "Epigenetics and the biological definition of gene x environment interactions." *Child Development* 81: 41–79.

Montagu, A. 1986. *Touching*, 3rd ed. (Harper & Row).

Mooney, J. & Young, J. 2006. "The decline in crime and the rise of anti-social behaviour." *Probation Journal* 53: 397–407.

Morelli, G., Ivey Henry, P., & Foerster, S. 2014. "Relationships and resource uncertainty: Cooperative development of Efe hunter-gatherer infants and toddlers." In D. Narvaez, K. Valentino, A. Fuentes, J. McKenna, & P. Gray (eds.), *Ancestral Landscapes in Human Evolution: Culture, Childrearing and Social Wellbeing* (Oxford University Press) 69–103.

Murgatroyd, C. & Spengler, D. 2011. "Epigenetics of early child development." *Frontiers in Psychiatry* 16: 1–15.

Naess, A. & Rothenberg, D. 1989. *Ecology, Community and Lifestyle* (Cambridge University Press).

Narvaez, D. 2008. "Triune ethics: The neurobiological roots of our multiple moralities." *New Ideas in Psychology* 26: 95–119.

Narvaez, D. 2012. "Moral neuroeducation from early life through the lifespan." *Neuroethics* 5: 145–157.

Narvaez, D. 2014. *The Neurobiology and Development of Human Morality: Evolution, Culture and Wisdom* (W.W. Norton).

Narvaez, D., Gleason, T., Wang, L., Brooks, J., Lefever, J., Cheng, A., & Centers for the Prevention of Child Neglect. 2013a. "The evolved development niche: Longitudinal effects of caregiving practices on early childhood psychosocial development." *Early Childhood Research Quarterly* 28: 759–773.

Narvaez, D., Panksepp, J., Schore, A., & Gleason, T. (eds.). 2013b. *Evolution, Early Experience and Human Development: From Research to Practice and Policy* (Oxford University Press).

Narvaez, D., Wang, L., Gleason, T., Cheng, Y., Lefever, J., & Deng, L. 2013c. "The evolved developmental niche and child sociomoral outcomes in Chinese 3-year-olds." *European Journal of Developmental Psychology* 10: 106–127.

Narvaez, D., Valentino, K., Fuentes, A., McKenna, J., & Gray, P. (eds.). 2014. *Ancestral Landscapes in Human Evolution: Culture, Childrearing and Social Wellbeing* (Oxford University Press).

Narvaez, D., Braungart-Rieker, J., Miller-Graff, L., Gettler, L., & Hastings, P. (eds.). 2016a. *Contexts for Young Child Flourishing: Evolution, Family and Society* (Oxford University Press).

Narvaez, D., Wang, L., & Cheng, A. 2016b. "Evolved developmental niche history: Relation to adult psychopathology and morality." *Applied Developmental Science*. doi.org/10.1080/10888691.2015.1128835.

National Research Council. 2013. *U.S. Health in International Perspective: Shorter Lives, Poorer Health* (The National Academies Press).

Organization for Economic Cooperation and Development (OECD). 2009. *Doing Better For Children* (OECD Publishing).

Panksepp, J. 2007. "Can PLAY diminish ADHD and facilitate the construction of the social brain?" *Journal of the Canadian Academy of Child and Adolescent Psychiatry* 10: 57–66.

Pellis, S. & Pellis, V. 2009. *The Playful Brain: Venturing to the Limits of Neuroscience* (Oneworld).

Porges, S. 2011. *Polyvagal Theory* (W.W. Norton).

Powell, D., Fixen, D., & Dunlop, G. 2003. *Pathways to Service Utilization: A Synthesis of Evidence Relevant to Young Children with Challenging Behavior*. University of South Florida: Center for Evidence-based Practice: Young Children with Challenging Behaviors.

Propper, C., Moore, G., Mills-Koonce, W., Halpern, C., Hill-Soderlund, A., Calkins, S., Carbone, M., & Cox, M. 2008. "Gene-environment contributions to the development of infant vagal reactivity: The interaction of dopamine and maternal sensitivity." *Child Development* 79: 1377–1394.

Rubaltelli, F., Biadaioli, R., Pecile, P., & Nicoletti, P. 1998. "Intestinal flora in breast and bottle-fed infants." *Journal Perinatal Medicine* 26: 186–191.

Sahlins, M. 2008. *The Western Illusion of Human Nature* (Prickly Paradigm Press).

Sapolsky, R. 2004. *Why Zebras Don't Get Ulcers* (Holt).

Schore, A. 1996. "The experience-dependent maturation of a regulatory system in the orbital prefrontal cortex and the origin of developmental psychopathology." *Developmental Psychopathology* 8: 59–87.

Schwartz, E. 2009. *Human Nature, Ecological Thought and Education after Darwin* (SUNY Press).

Shonkoff, J., Garner, A., The Committee on Psychosocial Aspects of Child and Family Health, Committee on Early Childhood, Adoption, and Dependent Care, and Section on Developmental and Behavioral Pediatrics, Siegel, B., Dobbins, M., Earls, M., McGuinn, L., Pascoe, J., & Wood, D. 2012. "The lifelong effects of early childhood adversity and toxic stress." *Pediatrics* 129: doi:10.1542/peds.2011–2663.

Siniatchkin, M., Kirsch, E., Arslan, S., Stegemann, S., Gerber, W., & Stephani, U. 2003. "Migraine and asthma in childhood: Evidence for specific asymmetric parent-child interactions in migraine and asthma families." *Cephalalgia* 23: 790–802.

Spinka, M., Newberry, R., & Bekoff, M. 2001. "Mammalian play: Training for the unexpected." *Quarterly Review of Biology* 76: 141–168.

Stam, R., Akkermans, L., & Wiegant, V. 1997. "Trauma and the gut: Interactions between stressful experience and intestinal function." *Gut* 40: 704–709.

Tomkins, S. 1965. "Affect and the psychology of knowledge." In S. Tomkins & C. Izard (eds.), *Affect, Cognition, and Personality* (Springer) 72–97.

Trevarthen, C. 2005. "Stepping away from the mirror: Pride and shame in adventures of companionship." In C. Carter, L. Ahnert, K. Grossmann, S. Hrdy, M. Lamb, S. Porges, & N. Sachser (eds.), *Attachment and Bonding: A New Synthesis* (MIT Press) 55–84.

Trevathan, W. 2011. *Human Birth: An Evolutionary Perspective* (Aldine de Gruyter).

Twenge, J. & Campbell, R. 2009. *The Narcissism Epidemic: Living in the Age of Entitlement* (Free Press).

U.S. Department of Health and Human Services. 2011. *The Surgeon General's Call to Action to Support Breast-feeding* (U.S. Department of Health and Human Services, Office of the Surgeon General).

Walker, H. 1993. "Antisocial behavior in school." *Journal of Emotional and Behavioral Problems* 2: 20–24.

Walker, M. 1993. "A fresh look at the risks of artificial infant feeding." *Journal of Human Lactation* 9: 97–107.

Watson, D., O'Hara, M., Simms, L., Kotov, R., Chmielewski, M., McDade-Montez, E., Gamez, W., & Stuart, S. 2007. "Development and validation of the Inventory of Depression and Anxiety Symptoms (IDAS)." *Psychological Assessment* 19: 253–268.

Wolff, R. 2001. *Original Wisdom* (Inner Traditions).

24

The Evolution of Morality and the Prospects for Moral Realism

Ben Fraser

MORAL REALISM

Defining "realism" is difficult. Here I take realism about morality to have two aspects. The first is epistemic: moral realists are not skeptics. They think we can and sometimes do have true moral beliefs. The second is metaphysical. Moral realists think moral facts are somehow objective, independent of our moral opinions, attitudes, concepts, and so on. The notion of "mind independence" is tricky in social domains. Our attitudes matter causally in the social world: whether an action is objectively harmful can depend partly on agents' responses to the action. But if harms are objective (in the realist's sense), then whether an agent is harmed is not constituted by their, or anyone's, opinions on that matter. In meta-ethics, this idea has been captured by Russ Shafer-Landau's phrase: moral facts are "stance independent" (Shafer-Landau 2003). I will later suggest that moral facts are facts about social interactions that support stable cooperation. If this is true, then moral realism, as I will understand it, is the view that moral facts depend on facts about what we are like and how we live together but are not constituted by individual or collective opinions on what those facts are. The chapter from here on explores the relationship between moral realism, so understood, and our best hypotheses about the evolution of moral cognition.

Much empirical work has been done on the evolution of morality. (For recent work, see Richerson & Boyd 2001; Joyce 2006; Boehm 2012; Richerson & Henrich 2012; Chudek et al. 2013; Sterelny 2014.) Some think this work somehow undermines the idea that there are moral facts, since an evolutionary explanation of moral thinking displaces an account of moral thinking as responding to objective moral facts (see, e.g., Street 2006; Joyce 2006; cf. Harman 1977). Moral judgments are thus false, or probably false, or unjustified, or unjustified if about an objective domain of moral facts. (There are many ways to precisify "undermine.")

However, I think this is too quick. Human moral practices are a complex mosaic, different elements of which have different origins, respond to different selective forces, and depend on different cognitive capacities. There probably is not a unitary explanation of all these elements. Accordingly, the philosophical upshot of the evolutionary explanation for morality may be different for different elements of the moral mosaic. If so, it would be a mistake to see the question of *undermining* versus *vindication* as a stark dichotomy.

Against this background, I will identify one element of the moral mosaic and argue that its evolutionary genealogy supports a version of reductive naturalism. In brief, moral facts will turn out to be facts about human cooperation and the social practices that support cooperation, and moral thinking has evolved in part in response to these facts and to track these facts. However, I will not attempt (nor would I expect to be able) to deliver a full vindication of moral thinking: the tracking of such facts is imperfect, and in any case only one of the functions of moral thinking. Like many folk conceptual systems, moral thinking has only partially been selected as a truth tracking system, but to the extent that its function is to track, it is neither a total failure nor full success.

Moral thinking is linked to the expansion of human cooperation in at least two ways. First, via partner choice: getting the benefits of cooperation often depends on being chosen as an interaction partner (Noë 2001), which in turn often depends on reputation, and often the best way to get a good reputation is to deserve one. Recognizing and internalizing moral norms is typically individually beneficial through its payoff in reputation. Second, via complexity: human social life long ago crossed a complexity threshold, meaning many cooperative challenges could no longer be solved on the go, case by case, via individual interaction (Sterelny 2014). To cope, default patterns of interaction became wired in as social expectations and then as norms. The benefits of coordinating with others, together with the costs of violating others' expectations, gave individuals an incentive to internalize and conform to these norms. Regularities of social interaction and cooperation were not arbitrary but rather reflected (albeit imperfectly) facts about the conditions under which human societies thrive. Given the benefits of cooperation in human social worlds, we have been selected to recognize and respond to these facts. On this adaptationist perspective, moral cognition is a response to selection for capacities that permit stable, long-term cooperation.

From this perspective, a certain realist-friendly notion of moral truth seems appealing. There are objective facts about which conditions make cooperation profitable, and which reduce its profits. There are also objective facts about which conditions stabilize cooperation, and which destabilize it. There is probably no one optimal set of norms: the best norm package for a group will depend on its particular features and circumstances. But, there probably are sets of norms that *satisfice* over a wide range of conditions. So, one might think of the moral truths as members of near-optimal normative packages; sets of norms that if adopted, would help generate high levels of appropriately distributed and hence stable cooperation profits.

Moral facts thus understood would exist independently of any specific agent's recognition of those facts. These are in this respect realist-friendly moral facts. They are also naturalistic and hence epistemically accessible facts, and realist friendly for that reason too. So, this is the reductive naturalistic version of moral realism I will consider below.

The road ahead is as follows. The next section will situate the debunking versus vindicating issue in the broader context of folk versus scientific frameworks for thinking about the world. I will offer a specific case study, folk astronomy, as a good model for moral cognition, insofar as it is an intermediate case that shows how the debunking/vindicating dichotomy can be broken down. The section after that will respond to the argument that moral facts are explanatorily redundant. The final section will argue for the explanatory importance of true moral beliefs, suggesting that true moral beliefs are indeed a "fuel for success" (cf. Godfrey-Smith 1996).

VINDICATION, DEBUNKING, AND MIXED CASES

Humans develop one way of thinking about the world, a folk understanding, via socialization in temporally and geographically variable epistemic communities: we develop folk understandings of the physical, biological, and social world. Humans have also developed another way of thinking via the natural and social sciences: a scientific image of the world and of ourselves.

These views sometimes seem to conflict; for example, the folk view of humans as free-willed rational decision-makers is not obviously compatible with the view that we are apes made of law-bound matter. One major philosophical project is to say what to do when folk and scientific views conflict. One important position here is reductive naturalism.

In science, reduction is the claim that the facts in one domain are less explanatorily fundamental than the facts in another. For example, facts of classical genetics about inheritance patterns are less fundamental than facts of molecular genetics about DNA sequences. The latter explain the former, but not vice versa.

Reductive naturalists extend this idea to folk kinds. Importantly, the reduction relation need not *itself* be part of the folk understanding. The folk can have epistemic access to a system without thereby having epistemic access to its components and organization. Consider the identification of water as H_2O: humans developed a concept of water, and knew many truths about water, without knowing water is a certain configuration of oxygen and hydrogen atoms. So, a reductive naturalist account of a folk kind needs an account of how a reduction relation can hold, without its being known to hold, despite the folk knowing plenty about the reduced domain (and sometimes even the reducing domain).

Norms are part of our folk worldview. We think of actions as right or wrong, good or bad; we think people are admirable, despicable, and so on. How does this normative aspect of the folk image of the world comport with the scientific image? According to certain realists, who extend the reductive naturalist project to normative phenomena, norms turn out to be, or supervene on, natural facts. Contemporary realists take water = H_2O as the paradigm for the relation between natural and normative facts. While this would account for how an identity of supervenience could be true without being apparent to the folk, it is in important respects a misleading model. There is no straightforward compositional relationship between normative kinds and any plausible set of base properties. The reduction base is a highly heterogeneous set of facts about agents, their interests, the social systems of which they are a part, and the deep history of those social systems.

Daniel Dennett's (1991) "intentional stance" offers a better model of the relationship between the normative and the natural. An agent's cognitive and neural organization constrains the belief-desire profiles attributable to that agent but does not uniquely specify an intentional profile. Profiles can be distinct but equally legitimate because, for example, they make different trade-offs between simplicity and accuracy. Also, on this view, specific beliefs and desires do not map directly onto specific elements of an agent's cognitive organization.

The relationship between intentional and cognitive psychology is a better model for the moral naturalist than water = H_2O because it does not imply that there is a unique set of moral truths fixed by the reduction base, nor does it commit one to the view that there is an element by element reduction of the normative to the natural.

This matters because, on the hypothesis I am considering, moral facts reduce to facts about cooperation within human societies, and given the complexity and context-sensitivity of these facts, it is most unlikely that these facts specify a single optimal package of norms, or that there is an element to element mapping between them and the norms in various near-optimal packages. So in thinking about the relation between norms and natural facts, I will work with Dennett's model.

Successful reduction can vindicate a folk conception of the world. But there is a methodological challenge for reductive naturalism: when do we have genuinely successful reduction? When, for example, do we have a reductive account of a folk kind rather than merely an account of an "ersatz" kind, not really the folk kind at all?

The "Canberra Plan" offers a well-developed answer (Jackson 1998; Braddon-Mitchell & Nola 2009). The idea is to take folk opinion and systematize it. Ideally, this captures folk intuitive judgment about specific cases and the folk's general principles. The Canberra Plan recognizes that the folk need not be unanimous, that general principles are rarely explicit, and that not all elements of the folk view are equally important.

There are complications, but the Canberra Plan constructs an implicit definition of a folk kind, and, with that in hand, there is a potential bridge between folk kinds and science. We can ask, from the perspective of our best relevant science, whether anything answers to the implicit definition. If so, we have found the folk kind. If not, a strength of the Canberra Plan is that it allows for error and for many cases intermediate between vindication and error theory. Perhaps something in our best scientific understanding of the world has some but not all features of the folk kind. The Canberra Plan allows for reduction to a best deserver, something that closely enough approximates the folk conception. It also allows for elimination, should nothing answer exactly or even closely enough to the folk concept, and is thus a tool that can be used by the error theorist as well as the moral realist.

Despite recognizing intermediate cases, the Canberra Plan seemingly *over-counts* failure, and in a way is relevant to the prospects for reductive naturalism about moral norms. Take, as a case study, folk astronomy in the ancient world (as systematized in Ptolemaic astronomy). Nearly *all* the general folk astronomical beliefs were mistaken, as were many of the specific identifications (e.g., the moon is not a planet). The Canberra Plan applied to this framework seemingly suggests that we should be error theorists about ancient folk astronomy.

However, agents in the ancient world could use astronomical information adaptively. They could reliably identify specific celestial objects and configurations, and use these identifications to guide navigation, calendar construction, and time keeping. Folk astronomy was not just a set of propositional beliefs; it also involved a complex suite of discriminative capacities which fed into action across several domains. Understanding the folk framework in this way highlights the fact that such frameworks can enable agents to register and respond to environmental features in quite nuanced ways even without an accurate conceptualization of the environmental phenomenon, and indeed, *despite* holding fundamentally mistaken explicit beliefs about the nature of that phenomenon.

The key point here is not just that flawed folk frameworks can be practically useful; they are useful because although they are not simply right, they are not just wrong, either. For contrast, consider a clichéd example, 17th-century witch burning. Even if witch identifications picked out a coherent social subgroup (the friendless, isolated, socially deviant), discriminating in this way did not leverage adaptive behavior: it did not prevent crop failure, storms, or stillbirths. Even if witch discourse generated benefits by strengthening in-group bonds, the details of the folk framework would be irrelevant: burning outsiders for being aliens would work as well. In this case, there are no dependencies between details of the framework, features of the environment, and adaptive behavior. The witch case shows how a folk framework can deserve elimination.

So vindication is possible, as is elimination, but the most important upshot is that mixed cases are possible too, meaning it is a mistake to frame the discussion as a choice between reduction and elimination. Many, perhaps most, folk frameworks will involve a mix of mere causally grounded response to phenomena in the world, partially correct conceptualization of those phenomena, and capacity to support adaptive action.

Folk astronomy is one example of a mixed case. Despite misdescribing the world in genuinely important respects, folk astronomy nevertheless systematically latched onto phenomena in the world in ways that leverage adaptive behavior regularly and non-accidentally. Compare, again, the witch framework, which did neither. Ancient folk astronomy is what I will call a mixed case. (A quick note on terminology: in what follows, I will talk about mixed cases and "partial vindication," but I could equally happily talk about "partial debunking." The language is not intended to favor vindicators over debunkers. The main idea is that the supposed dichotomy between vindication and debunking is itself problematic.) In the next sections, I will argue that the moral case is also a mixed one.

MORALITY AS A MIXED CASE

Folk moral theories will be, at best, a mixed case. First, as noted earlier, human morality is a mosaic comprising various elements with different evolutionary genealogies. Some elements may turn out to be vindicated, others revised, others discarded. There is diversity of moral views, and moral views are often distorted by religion, mythology, and politics. So, no truth-tracking account will vindicate all moral opinions. Second, tracking moral facts is not the only role for moral judgment, which also works to signal between agents, to bond them together, and to shape their behavior. Third, as I will explain in detail below, tracking is at best only partially successful.

In light of this, why think folk morality is even a mixed case—that is, why think there is even partial vindication to be had? To make this case, there are two challenges. One is to show how moral facts are explanatorily potent, contra debunkers who think the truth of moral judgments is redundant when it comes to explaining why those judgments are held. The other is to show how, like in the folk astronomy example, the folk moral conceptual apparatus powers adaptive action flexibly, across a range of contexts, despite embodying significant theoretical misconceptions.

Take the first challenge first. The would-be reductive naturalist realist must draw a reliable causal connection between moral opinions and the corresponding moral facts. (Note: it is not required that every moral view be explained with reference to its truth; realists can allow for individual and collective moral error.) I suggest three ways in which the truth of moral norms can help explain how agents come to endorse those norms: (i) learning guided by prosocial emotions; (ii) vicarious trial and error learning in heterogeneous environments; (iii) cultural group selection.

Begin with (i) prosocial emotions. Jesse Prinz (2007) and Shaun Nichols (2004) argue that while all norms are learned socially, some norms are especially salient—specifically, norms whose learning goes via recognition of others' emotional responses. When our acts affect agents we care about, we notice their emotional responses to our actions and our emotional responses to their responses. So, for example, generosity toward others is reinforced in a feedback loop where others' positive responses induce one's own positive response. While this does not guarantee the acquisition of prosocial norms, it does make the phenomena that fall under those norms salient. And while salience is no guarantee of truth, consider that in the human lineage there is a deep history of both biological and cultural selection favoring cooperation-supporting emotional responses. So, emotionally repugnant behaviors are, all else being equal, likely to be behaviors that would be forbidden by norms in the near-optimal packages for a given social context. Conversely, patterns of behavior that are emotionally appealing are likely to be instances of behaviors that would be endorsed by norms in those packages.

Moving on to (ii), notice that many contemporary societies are normatively heterogeneous. They are composed of subgroups, which can and often do overlap and cross-cut in complex ways, and these groups can have different norms. Such heterogeneous normative contexts offer agents the opportunity to treat each other as a sort of natural experiment. Observing others who live by different normative packages provides some insight into how well or poorly their lives go. One can see whether normatively diverse others live in mutually supportive social networks, or if they are instead exploited by free-loading neighbors, for instance. Of course, an encompassing evolutionary perspective on the origins and stability of moral norms is not part of the folk conception of the moral domain. The idea that norms play a role in promoting fair social interaction may well be part of the folk view, though. In normatively heterogeneous learning environments, folk awareness of that fact may influence which norms agents eventually come themselves to hold. (Moral education does quite often involve noting the effects of norms, and hence norm violations, on cooperative lives.)

Regarding (iii), it is important to note that in the past human social groups were smaller and more homogenous. When it comes to explaining why some of these groups flourished while others did not, it is plausible that one relevant consideration is the extent to which the normative package adopted by a group stabilized and enhanced in-group cooperation. Cultural group selection favors sets of moral norms that promote social peace, regulate conflict, and restrain selfish or destructive behavior. Evolutionary naturalists should see the evolution of norms as an ongoing process, and one which is biased toward finding the near-optimal norm packages (i.e. the true moral norms, on the account being considered here).

An important issue that needs to be raised—if not fully addressed here—is that of in-group versus out-group interactions. On the view proposed, it is at least possible that near-optimal norm packages may include ones that prescribe violence toward out-groups. That is, the moral truths on this view could include what we would pre-theoretically think of as pretty nasty norms. There are two things to say with respect to this issue. The first is that whether the near-optimal norms packages for groups include some packages that are unattractive in this way is of course an empirical matter. And, insofar as one harbors worries along these lines, it is important not to under-estimate the costs of social and outright physical conflict. Perhaps few if any near-optimal norm packages in fact include such xenophobic norms, given those costs and given also the advantages of inter-group cooperation. But, and this is the second thing to say, even if there are near-optimal norm packages that include unappealingly violent xenophobic elements, on the view offered, that would be an interesting—if not entirely welcome—discovery about the nature of the moral truths. The view on offer allows for considerable disconnect between pre-theoretic folk views and the actual facts of the matter. The possibility or even the actuality of near-optimal norm packages including unattractive elements is a feature not a bug, though of course (to mix metaphors) there may be plenty of people unwilling to bite this bullet.

None of the discussion of (i)–(iii) is intended to deny that the mechanisms that encourage moral truth-tracking can be overridden by other processes. And even when these mechanisms are operative in guiding norm acquisition, they are not guaranteed to result in agents coming to adopt true moral norms (meaning, as said in the first section above, norms that are members of one of the near-optimal norm packages). The mechanisms of moral truth tracking are neither omnipresent nor infallible. That said, for any social group in a particular environment, there is some fact of the matter regarding whether their norms are efficient means to stable and profitable cooperation. And, when a group's norms are in fact efficient means to stable and profitable cooperation, it is typically not merely a happy, lucky accident that this is so. There is a tendency for the better norms to be found, however noisy the process, and the fact that these norms *are* better partially explains *why* they are found when they are, and why agents come to hold them.

If the foregoing is on target, then real-world norm packages will most likely be a mixed bag. Some elements will be the legacy of religious or mythological distortions. Some will be levers for exploitation installed by self-interested elites to preserve imbalances of power and wealth. Some will indeed be the result of selective filtering, but for purposes other than tracking facts about cooperation (such as signaling as mentioned above). Despite all this, though, some elements of normative packages will be true and, crucially for current purposes, their truth will have been an important factor in their coming to be endorsed. The truth of these norms is thus not causally idle: it is relevant to their presence, persistence, and learnability. So, to the extent to which the moral norms we hold causally depend on the social and psychological bases of stable cooperation, the truth of those norms is not explanatorily redundant.

MORAL TRUTH AS A FUEL FOR SUCCESS

Recall the lesson of the ancient folk astronomy case study: agents could use the framework to a significant degree as a fuel for successful action, despite the deep conceptual flaws of the framework. Moral beliefs are enmeshed in conceptual frameworks that can be deeply mistaken. The question is now whether, despite that fact, true moral beliefs can also be fuels for success.

There is a case to be made for a partially affirmative answer. Once again, though, the mosaic nature of human morality matters. Given that tracking moral truth is not the only role moral cognition plays, there is scope for conflict between different roles. Consider moral judgment as a coordinating device: once default forms of action establish in a community, agents have incentives to conform to them even if they eliminate or erode cooperation, especially if norm violations are punished (Boyd & Richerson 1992). This contrasts with folk astronomy, which was not particularly relevant to coordination and social interaction (meaning there was little pressure to conform with others' errors). Unlike the folk astronomy case, moral thinking is not a case in which, all else being equal, true belief is automatically rewarded. Tracking moral truth may not be a fuel for success when there is a strong incentive to match one's normative beliefs to the mistaken normative beliefs of one's wider social group. However, even granting this, identifying which norms really do promote profitable and stable cooperation does power adaptive behavior in its own right, in at least two ways.

The first involves partner choice. Sensible partner choice attends not just to the internalization of any old norms, but rather of norms of cooperation, fairness, trustworthiness, commitment, and so on. So, one's prospects of being chosen as a social partner improve the better one grasps the moral facts. As mentioned at the outset, the best way of seeming like a good partner is often to be a good partner, and in that case, tracking the moral facts is indeed a fuel for success.

The second way in which moral truth is a fuel for success is via social engineering. It is true that extant, entrenched norms supply incentives for agents to conform. However, it is important not to think of agents exclusively as passive consumers of local norms: agents also *influence* their local normative environments. This remains true even in the large-scale complex societies that characterize the current social world. Within the vast sweep of modern society, we inhabit smaller-scale social worlds: families, clubs, workplaces, and so on. Individuals in these settings can exert significant positive (and negative) effects via their normative attitudes and actions. Agents who endorse and enforce prosocial, cooperation-sustaining norms can influence their micro-social worlds in ways that make them better for themselves and others. True moral beliefs are tools that can help agents engineer their immediate social environment (again, even if their global environment is largely impervious).

To sum up, then: a version of reductive naturalism about moral norms can be built around one perspective on the evolutionary history of moral thinking. Moral truths are principles of action and interaction which support forms of cooperation that are stable because they are fair enough to give almost everyone an incentive to continue to cooperate. In favorable cases, but only favorable cases, these norms are endorsed because they are true, and, when endorsed, they support successful social interaction. The vindication is at very best partial.

One issue not addressed at all in this piece is the supposed special authority of morality, which debunkers like John Mackie (1977), for example, focus on in developing moral error theory. On the view on offer here, moral facts are not imbued with any special authority. Determining whether such a view could count as even a partial vindication of morality will be an important point of engagement between the proposal laid out here and extant debunking accounts, though admittedly nothing of substance has been said about it here.

To recap the considerations covered here, though, first, moral thinking is not just truth-tracking: it displays community membership and commitment to local mores; norms solve coordination problems in ways that are independent of their truth. Second, to the extent that moral thinking is truth tracking, it is error-prone. Our moral views are roughly analogous to the astronomical lore of the ancient world. Just as ancient astronomy was a response to the celestial world, moral views are a response to the opportunities and challenges of a world in which cooperation is profitable but fraught with potentials for conflict, coordination failure, and misunderstanding. As in the case of premodern astronomy, these responses do not typically identify and solve those challenges ideally. But in a range of cases, the normative practices of individuals and groups are appropriately shaped by these challenges and the available solutions, and they enable individuals and groups to act adaptively in their social environments with some reliability. Moral thinking on this account is not fully vindicated, but nor is it entirely debunked as a mere adaptive fiction.

REFERENCES

Boehm, C. 2012. *Moral Origins: The Evolution of Virtue, Altruism and Shame* (Basic Books).

Boyd, R. & Richerson, P. 1992. "Punishment allows the evolution of cooperation (or anything else) in sizable groups." *Ethology and Sociobiology* 13: 171–195.

Braddon-Mitchell, D. & Nola, R. (eds.). 2009. *Conceptual Analysis and Philosophical Naturalism* (MIT Press).

Chudek, M., Zhao, W., & Henrich, J. 2013. "Culture-gene coevolution, large scale cooperation, and the shaping of human social psychology." In K. Sterelny, R. Joyce, B. Calcott, & B. Fraser (eds.), *Cooperation and Its Evolution* (MIT Press) 425–457.

Dennett, D. 1991. "Real patterns." *Journal of Philosophy* 87: 27–51.

Godfrey-Smith, P. 1996. *Complexity and the Function of Mind in Nature* (Cambridge University Press).

Harman, G. 1977. *The Nature of Morality* (Oxford University Press).

Jackson, F. 1998. *From Metaphysics to Ethics*. (Oxford University Press).

Joyce, R. 2006. *The Evolution of Morality* (MIT Press).

Mackie, J. 1977. *Ethics: Inventing Right and Wrong* (Penguin).

Nichols, S. 2004. *Sentimental Rules: On the Natural Foundations of Moral Judgment* (Oxford University Press).

Noë, R. 2001. "Biological markets: Partner choice as the driving force behind the evolution of cooperation." In R. Noë, J. van Hooff, & P. Hammerstein (eds.), *Economics in Nature: Social Dilemmas, Mate Choice and Biological Markets* (Cambridge University Press) 93–118.

Prinz, J. 2007. *The Emotional Construction of Morals* (Oxford University Press).

Richerson, P. & Boyd, R. 2001. "Institutional evolution in the Holocene: The rise of complex societies." *Proceedings of the British Academy* 110: 197–234.

Richerson, P. & Henrich, J. 2012. "Tribal social instincts and the cultural evolution of institutions to solve collective action problems." *Cliodynamics* 3: 38–80.

Shafer-Landau, R. 2003. *Moral Realism: A Defense* (Oxford University Press).

Sterelny, K. 2014. "A Paleolithic reciprocation crisis: Symbols, signals, and norms." *Biological Theory* 9: 65–77.

Street, S. 2006. "A Darwinian dilemma for realist theories of value." *Philosophical Studies* 127: 109–166.

25

Moral Cheesecake, Evolved Psychology, and the Debunking Impulse

Daniel R. Kelly

INTRODUCTION

In his famous discussion of inductive inferences, David Hume gives one of the most notorious debunking arguments in Western philosophy. His target is causation; despite what we naturally believe or how it intuitively seems to us, Hume is skeptical of the idea that there are necessary connections of cause and effect that can hold between objects and events in the world. Hume also provides an account of the source of mistakes, of where the idea of causation comes from, and why belief in causal connections is so natural and persistent. This account appeals to custom, or mental habit, and the mind's general "propensity to spread itself on external objects" (Hume [1740] 1978: 1.3.14). In the case of cause and effect, he says:

> We remember to have had frequent instances of the existence of one species of objects; and also remember, that the individuals of another species of objects have always attended them, and have existed in a regular order of contiguity and succession with regard to them. Thus we remember to have seen that species of object we call *flame*, and to have felt that species of sensation we call *heat*. We likewise call to mind their constant conjunction in all past instances. Without any farther ceremony, we call the one *cause* and the other *effect*, and infer the existence of the one from that of the other. In all those instances, from which we learn the conjunction of particular causes and effects, both the causes and effects have been perceived by the senses, and are remembered: But in all cases, wherein we reason concerning them, there is only one perceived or remembered, and the other is supplied in conformity to our past experience.
>
> (Hume [1740] 1978: 1.3.6)

Hume's strategy here is instructive even if one rejects it and its conclusion that causation is not a feature of the external world. Perhaps most relevant is his appealing to characteristics of our psychological capacities, and his reasons for doing so. Hume gives an alternative explanation for why we seem to "see" causal connections in the world, why we so naturally interpret the unfolding of events though the lens of cause and effect, even though (he claims) there are no such connections. Instead, Hume locates the source of belief in external causal connections in "a productive faculty" of our minds, a psychological process that, in "gilding and staining all natural objects with the colours, borrowed from internal sentiment, raises in a manner a new creation" (Hume [1751] 1983: 88). Thus, Hume's debunking strategy aspires to give an explanation of its target phenomenon (causation) that (i) illuminates that phenomenon in a new way, and in so doing (ii) shows not just *that* previous ways of construing or accounting for the phenomenon went wrong, but *how* and *why* they did as well.

But did Hume really debunk causation, or did he just put it on a fresh ontological footing by giving a better account of what it really is? Perhaps his appeal to our habits of expectation and the operation of other productive psychological faculties serves rather to clarify and correct some of our intuitive beliefs about the relations between unfolding events, rather than show the whole package to be false. If Hume's account is understood this way, as explaining why causation seems to be out there in the world, but is in fact a "new creation" produced in our own heads and added to experience by our perceptual apparatus, did he debunk causation, or did he supplant an old and mistaken explanation with a newer and better one?

I'll return to questions like these, concerning how to interpret applications of this kind of broad Humean strategy and what counts as "debunking," at the end of this chapter. By way of getting there, in the next section I will canvass some of the recent literature on evolution, psychology, and morality, separating out some of the different ways the debunking impulse as manifested there, and endorsing the recent trend toward a more selective, divide-and-conquer approach. In the following section, I will turn to recent work on the evolution of the human mind, the kinds of mental structure and innate psychological endowment it suggests, and what this implies about our moral capacities and the mechanisms that underpin them. In the penultimate section I will focus on the relationships that can hold between different components of our moral capacities, the environments and contexts in which they operate, and the functions they perform. I argue that assessing those components and their contributions to moral judgment and behavior needs to not only be selective and fine-grained, but also to take into account the fact that human minds are cultural minds (Henrich 2015), and today operate in cognitive niches that are intensely and unprecedently constructed, filled with supernormal stimuli that are intentionally designed to elicit particular responses and shape attitudes, including moral ones, by their designers. I will conclude with some remarks on the Humean lineage of this line of thought, and on how it fits with other recent attempts to square scientific images of the world with their putatively corresponding manifest images.

MORALITY AND THE DEBUNKING IMPULSE

While Hume's treatment of causation is most concise and well known, he was famously skeptical about many other things besides. Indeed, much of his epistemology was focused on how and why we get things wrong, and many of his explanations of our epistemological mishaps appeal in one way or another to the features and functioning of our minds. Here I will be focusing on discussions specifically about morality, and the different ways that the debunking impulse has been expressed and debated in recent work by moral philosophers. Despite their differences, contemporary discussions typically share three related features that set them apart from Hume's. First, the state of science and the scientific picture of the world have developed enormously since Hume's day, and so attempts to locate morality with respect to that picture will be accordingly more sophisticated. Second, a cornerstone of that scientific development and its accompanying naturalistic worldview is evolutionary theory, and discussions of morality have been looking closely at the relationships between the two, trying to discern what evolutionary theory might tell us about human moral nature, and the epistemology and ontology of morality itself. Obviously, Hume, writing over a century before Darwin, was unable to do this. Finally, and related to the two previous points, the models of the mind that feature in many of these contemporary discussions in moral theory depart in fairly radical ways from the broadly empiricist, associationist model Hume worked with (though there are some deep affinities between the two as well; see Fodor 2003). Of particular interest are those models that combine contemporary psychology and contemporary evolutionary theory, on which I will briefly focus in the next section.

A full tour through the recent outgrowth of literature at the intersection of evolutionarily informed behavioral science, empirical moral psychology, and moral philosophy would try most readers' patience, but a few distinctions and landmarks will help orient the discussion to come. In an early *locus classicus*, John Mackie (1977) presents an error theory about morality, defending the view that all moral judgments are false, and gives a rudimentary explanation of our persistence in making the error of believing some of them (usually our own) are true. Other important agenda-setting contributions to these debates, especially those that dwell on the implications of evolutionary theory for questions in metaethics, include those from Allan Gibbard (1990), Sharon Street (2006), and Richard Joyce (2006). The latter two present distinct but complementary versions of arguments that move from premises about the evolution of our moral capacities to conclusions that serve as skeptical challenges to venerable metaethical doctrines; Street's challenge is to moral realist views while Joyce's challenges the idea that any of our moral judgments are properly justified. The discussion, much of it in response to these key texts, fans out in many interesting directions from there. Justin Clarke-Doane (2012) explores an analogy between moral and mathematical beliefs, arguing that beliefs of both types stand or fall together in the face of evolutionary arguments. Brendan Cline (2015) defends a form of moral nativism against a provocative argument made by Edouard Machery and Ron Mallon (2010) that morality is not an adaptation resulting from natural selection. Kelby Mason (2010) focuses less on details of psychology and more on the structure of genealogical debunking arguments of both moral and religious judgments, and lays out an account of whether and when the genealogy of a judgment is "truth-mooting." He explores the possibilities of applying such truth-mooting debunking strategies to both cognitivist as well as non-cognitivist accounts of judgments in those domains (see also Joyce 2013 for discussion of the viability of debunking arguments in terms of genealogy and non-cognitivism).[1]

Alongside this, another interesting and important line of literature is animated largely by the spirit of experimental philosophy, and typically proceeds by investigating folk psychological beliefs about metaethical issues like the objectivity or relativity of moral judgments. Given what is discerned by empirical investigation of folk metaethical beliefs, inferences are typically made about the nature of the cues and psychological mechanisms responsible for such beliefs, or about the implications of those folk beliefs and their sources for positions in metaethical theory itself. (See Nichols 2004; Goodwin & Darley 2008; Wright et al. 2013; and many of the contributions in Knobe & Nichols 2014.) While many of these authors see their empirical results as informing questions about metaethics, sometimes ruling out certain possibilities or serving as evidence against some family of metaethical views, they do not often present their conclusions explicitly in terms of "debunking." For instance, John Doris and Alexandra Plakias's approach (2007) brings a range of empirical psychological and behavioral facts about evaluative diversity to bear on philosophical debates about metaethical realism via the argument from disagreement. They tentatively endorse a patchy realism, arguing that with respect to many but not all moral issues, there is good reason to think that moral disagreement would persist even if all factual disagreement could be resolved, and so claim that with respect to those, but only those, moral issues wherein moral disagreement would persist, the right conclusion to draw is an anti-realist one: there is no fact of the matter. (See Sommers 2012 for a similar approach to issues concerning free will and moral responsibility.) A notable exception is Shaun Nichols (2015), who explores what he explicitly calls "process debunking" approaches, which draw on empirical findings about the psychological processes that form different moral beliefs. Nichols distinguishes between using such an approach to draw implications for metaethical questions, on the one hand, and for normative ethical questions, on the other. He is optimistic about the prospects for the former, but less sanguine about the latter.

Two more distinctions useful in taxonomizing different expressions of the debunking impulse can be drawn from Nichols's discussion. The first is between those that seek to establish metaethical conclusions and those that seek to establish first order normative ethical conclusions. This is more of a continuum than a categorical distinction, but clear and representative instances of each kind can be easily identified. For instance, most of those mentioned in the previous paragraphs—Mackie's error theory, Street's anti-realism, and Joyce's moral skepticism—fall in the first category, and these tend to be the most commonly associated with the idea of moral debunking. However, many authors have recently advanced positions on questions in the second category, about the normative significance of findings in the evolutionary and empirical sciences (e.g., Berker 2009; Kumar & Campbell 2012; Railton 2014; Rini 2013, 2016; Kumar forthcoming).

This leads to the second distinction in Nichols's discussion, this one between global and selective approaches to debunking, or, if not debunking, simply to bringing science into dialogue with moral theory. Global strategies typically seek to treat all of morality at once (or all moral beliefs, or all moral judgments, or all moral norms, or all of our moral capacities, etc.) taking it (or them) to be a kind unified enough to stand or fall together. Most attempts to establish metaethical conclusions are global in this sense (with Nichols 2015 and Doris & Plakias 2007 being notable exceptions), and those attempts to establish normative ethical conclusions noted above proceed with this kind of globalist assumption as well. Selective expressions of the debunking impulse, on the other hand, take more of a divide-and-conquer approach. They find in morality a number of variegated phenomena that differ in ways that are directly relevant to the sorts of empirically and evolutionarily informed considerations that might be brought to bear in attempts to debunk or simply understand them. They see variety in our moral beliefs, expect different components of our moral capacities to be better or worse at producing moral judgments and behaviors, and as such they typically look more closely at the properties of mechanisms and processes that produce different subsets of moral beliefs and behavioral tendencies, or the evolutionary history and selective regimes that shaped different particular components of our moral psychology.

I favor the selective approach, and in previous work (Kelly 2013) have articulated and defended an argument form that underlies many instances of it.[2] Indeed, there has been a recent flowering of philosophical work drawing normative implications for some selective part of morality, or about the influence of some specific component or components of our psychological machinery on behavior and moral judgment. An early instance of this is Peter Singer's argument against a subset of moral intuitions, namely those that drive many objections to utilitarianism (2005). Joshua Greene (2013) advances a similar argument while also expanding the base of cognitive neuroscientific evidence on which the argument draws. (See also Joyce 2016 for discussion.) Other work has singled out for special focus the character and influence of particular emotions. After languishing in relative obscurity for decades, disgust has come in for much recent attention, with both its variously qualified advocates (Kass 1997; Kahan 1998, 1999; Plakias 2013; Fischer 2016) and skeptics (Nussbaum 2004; Kelly 2011; Kelly & Morar 2014). Others have directed their attention to empathy (Prinz 2011; also see Bloom 2013), contempt (Bell 2013), hope (Martin 2014), romantic love (Brogaard 2015), anger (Wiegman 2015), and jealousy (Kristjánsson 2016). Despite differing levels of engagement with the cutting-edge evolutionary and empirical literature, all of these use facts about the respective target of their selective focus to draw out normative implications about if, how, and when that component of our psychology can best inform various aspects of morality and moral theory.

As is evident, when it comes to morality, the debunking impulse can take many forms, and philosophers have recently explored many of them. What they all have in common is that they are based on the idea that facts about the character and evolutionary history of our psychological moral capacities can help us understand those capacities, and perhaps improve their functioning and the judgments and practices they support. Many also suggest a suspicion that in one way or another morality is sometimes not as it seems, and that some portion of our intuitive beliefs about our own moral judgments and their subject matter is mistaken.

EVOLUTION, COGNITIVE ARCHITECTURE, AND THE ADAPTED MIND

Setting aside for a moment the varieties of philosophical debunking, let us consider in more detail the picture of the mind that emerges out of evolutionarily informed cognitive science, and how it construes the mind's component parts, mechanisms, and processes, including those that underpin our moral capacities. Though there are of course differences in detail, emphasis, and terminology, most contemporary approaches to studying the evolution of the human mind share some common assumptions.

Skipping ahead a couple hundred years from Hume to the turn of the millennium brings with it a radical shift not just in our understanding of biology—that is, the rise of evolutionary theory—but also in psychological theorizing. Noam Chomsky and Jerry

Fodor made a general appeal to innate ideas and mental structure intellectually reputable again (see especially Fodor 1983), and while neither of them is particularly enthusiastic about evolution, other theorists took their ideas and developed them in an evolutionary context. The most prominent examples of this are Evolutionary Psychologists (e.g., Barkow et al. 1992; Pinker 1997; and Buss 2005), who hold that the study of human behavior, and more importantly the psychology that produces it, should be informed by adaptationist thinking. Since human minds evolved, they can and should be investigated like the products of natural selection they are. However, the human mind should not be thought of as a single adaptation or one (very) important trait. Rather, it is best conceived of and studied as a collection of loosely affiliated but distinct adaptations.

This is perhaps the most central and common commitment of contemporary evolutionarily informed cognitive science. Rather than seeing the mind as one trait, a single all-purpose, domain-general information processor, proponents share a vision of the mind as being comprised in large part (how large varies from one theorist to another) of a number of distinguishable components or subsystems, distinct traits, many of which have their own particular adaptive history, specific domain, proprietary database and algorithms, and functional profile. So the human mind has, for instance: a suite of mechanisms dedicated to interpreting the mental states of others based on their behavior, mechanisms that evolved to detect predators, to avoid pathogens and parasites, to categorize and make inferences about biological entities, and to predict the movement of objects based on an intuitive understanding of the physics of medium-sized dry goods near the surface of the earth. Each of these evolved to solve a particular problem or related cluster of adaptive problems that were recurrent in our species' history, and each brings to bear a specialized body of innate structure and information to help deal with its associated domain.

Adherents of the Evolutionary Psychology research program have defended a particularly stark version of this kind of picture. Their "massive modularity" hypothesis holds that the core machinery of all human minds is composed of a large number of different and distinguishable modules, psychological subsystems whose functioning is marked by a cluster of properties; they are typically specialized, fast, automatic, domain specific, semi-autonomous, informationally encapsulated, and cognitively impenetrable. (See Machery, Chapter 19 this volume.) They also describe this picture using suggestive metaphors, likening the mind to a Swiss Army Knife, comprising a variety of "elegant machines," each designed by a specific set of selection pressures to efficiently and effectively solve a particular adaptive problem that was recurrent in our species' evolutionary past. Like the distinct tools of a Swiss Army Knife, however, each of the "elegant machines" of the human mind operates fairly autonomously from the rest. Moreover, just as, say, the corkscrew is wonderful for opening a bottle of wine but clumsy if not outright useless for most other tasks, so too is each component of the human mind well-fitted to its associated task, but the performance and effectiveness of a mechanism can fall off abruptly when it operates outside of its evolved domain.

Another area to which researchers have brought this kind of broad picture to bear, and for which they have posited a number of such innate psychological mechanisms is, roughly, that of morality. Though it remains unclear exactly how to delimit the scope of this domain, human moral capacities are often taken to involve mechanisms dedicated to producing altruistic behavior, navigating the social world and facilitating cooperation, distinguishing intentional from unintentional behavior, acquiring and complying with social norms, detecting and punishing norm violators, and recognizing and responding to group membership and status.[3] In the next section, I will focus on two features of these capacities and the mechanisms that underlie them: that they are likely to be especially sensitive to cultural influences and shifting context, and that they can be assessed along a number of orthogonal dimensions that can be easily conflated, but are importantly distinct.

ASSESSING OUR MORAL CAPACITIES IN A HYPER-CONSTRUCTED CULTURAL NICHE: FITNESS, FUNCTIONS, AND SUPERNORMAL STIMULI

The picture of the mind coming out of cognitive science and evolutionary psychology dovetails with and supports the selective, divide-and-conquer strategy identified in the previous section, because it raises the very real possibility that the credentials of different components of our moral cognition need to be assessed separately, and on their own terms. To a first approximation, some components might be equipped with "shoddy software" or bear architectural features that make them ill-suited to contribute to morality in the way they typically do. Others might be well-suited, providing exactly what we would most want in guiding our moral judgments, deliberations, and actions. Indeed, the assessment of these, in this piecemeal way, can turn up many different options regarding how to conceive the relationships between the mechanisms, the contexts in which they operate, and the contributions they are likely to make to morality.

For instance, we might conclude, after considering the operational and evolutionary specifications of some particular component of the moral mind, that it is an "elegant machine." This would suggest that the component evolved specifically to perform a particular function, and that it carries out the job in a concise, efficient, elegant, and largely accurate way. A common alternative to this is that we might see other moral mechanisms as "kludges." Saying a piece of the mind is a kludge is to claim that while it is able to perform whatever function it does, it does so inefficiently, often in a roundabout way; it works, but with all of the streamlined optimality of a Rube Goldberg machine.[4] The connotations of the term "kludge" also suggest that the mechanism itself was not originally designed (or in the case of evolutionary explanations, did not initially evolve) to perform the particular function we find it performing.

Though these two competing concepts are sometimes applied to the mind as a whole, they can easily be transferred to evaluate individual components of it, and are arguably more usefully applied in this more selective way. Both also presuppose a distinction between the function, on the one hand, and the mechanism that performs that function, on the other.[5] This distinction is particularly useful, and can be developed in several ways. It is even more useful when paired with the distinction between the proper function of a trait or mechanism (i.e., the function it was selected to perform, and whose performance explains its existence, why it initially spread through populations and was passed along to subsequent generations) and the actual function of the trait, the function that it currently performs.[6]

Appreciating the difference between proper and actual functions points to a final important distinction. Assessing a mechanism as an elegant machine or a kludge in this way means assessing it in terms of how well it performs a particular function. But functions can change, and with such changes the mechanism itself can be assessed differently. It could very well be the case that a particular mechanism is elegantly designed to perform its proper function, and does so (or did so) efficiently and effectively. However, that same mechanism could acquire other functions, functions that it is able to perform with some degree of effectiveness, but that it does not or cannot perform in anything like an elegant, optimal manner. Thus a mechanism might appear to be an efficient, elegant machine in one context and from the perspective of one standard of evaluation, but appear as a kludgy, clunky Rube Goldberg-esque machine in another. Some Big Mistake or Mismatch hypotheses are extreme versions of this kind of account. Such accounts are often invoked to explain the existence of a trait or tendency that appears to be not just inefficient but outright maladaptive, actively decreasing an organism's fitness. Such putatively maladaptive traits present a puzzle for evolutionary theorists, and Big Mistake and Mismatch hypotheses offer a solution by construing the trait or tendency as producing maladaptive behavior only because it is operating in an environment or context much different than the one in which it initially evolved. While it produced adaptive benefit in that initial environment, the contextual shifts are such that it is mismatched to its current one, so only the behaviors it currently produces are a "mistake" from the point of view of evolution.

These distinctions, between psychological mechanisms and the functions they perform, between the mechanisms and the context in which they perform those functions, between elegant machines and kludges, do not yet say anything about morality, or assessments of the *ethical worth* or *moral significance* of any particular mechanism or the intuitions, deliberations, judgments, and behaviors it contributes to. Rather, they are assessments made from an engineering perspective. Adding an additional, distinct moral standard of evaluation to the mix presents a further complication, but the preceding discussion suggests a network of distinctions into which it can systematically fit. For example, it clarifies the possibility that a single mechanism can be an elegant machine with respect to its proper function, and that its proper function can also be its actual function—that is, it is well designed and efficiently performs the function that it evolved to perform, which it continues to perform in its current environment. Even in light of this, the possibility remains that in performing that function the mechanism is also doing something morally objectionable, making a contribution to moral cognition and behavior that runs afoul of our moral standards of assessment. In such cases, the function the mechanism performs is itself morally undesirable.

Put another way, elements of our moral psychology might, in a particular context, perhaps even in current actual environments, be doing exactly what they were designed to do, and doing it well. However, they could have been selected to generate intuitions, judgments, and actions that our current ethical perspective condemns, that we see in our wisdom are unethical. Our "moral" psychology may have a dark side, and evolution may have equipped us with a set of capacities for navigating our social worlds some of which are evolutionarily effective—fitness enhancing—but morally nasty. Such mechanisms are not "kludges" in the traditional sense at all; they are elegant machines, they are just elegantly designed to, and are actually efficiently doing, something unethical. This is one way to develop and apply the more general point that the categories of engineering efficiency, evolutionary fitness (of whatever sort), and moral worth cross-cut each other. Mechanisms that drive male proprietary attitudes toward women, or those that support in-group and out-group biases, ethnocentrism, cronyism, nepotism, and other types of what we now consider untenable forms of discrimination and prejudice may fit this description: well designed, efficient, fitness enhancing, but immoral.

A slightly different way our moral psychology can go awry is by producing immoral byproducts. Intuitively, a mechanism can attempt to do something that is morally acceptable, to perform an ethically unobjectionable function, but just do it poorly. However, we are now in a position to see that "poorly" has several distinct senses, and the differences can be specified thus: our moral assessment of the task or function is positive (i.e., the aimed at contribution to behavior and judgment is morally acceptable) but the engineering assessment of the mechanism's performance of that function is negative (i.e., the piece of our moral psychology actually performing that function is kludgy and suboptimal from an engineering perspective). Finally, however, the specific inefficiencies and byproducts generated by the kludgy performance of that function by that mechanism are themselves morally undesirable. The ways in which the mechanism fails to optimally perform its task generate psychological and behavioral effects that, for example, impel us to do morally unethical things or induce us to make morally unacceptable judgments.[7]

Finally, appreciating the complexity of the interactions between the evolved features of our minds, on the one hand, and the shifting environments in which they might operate, on the other, can shed light on how components of our moral psychology can also be *actively* led astray. This intriguing possibility has not been much explored, at least not under this description, but seems increasingly relevant to the moral assessment of the deliverances and influences of different components of moral psychology, especially in today's contemporary world. It bears a similarity to mismatch hypotheses and the Evolutionary Psychologists' idea of a Stone Age Mind in a Modern World in focusing on the relationship between the functioning of a psychological mechanism and the environment in which it operates, but its emphasis is slightly different. Consider the general idea of a supernormal stimulus (or superstimuli): an exaggerated version of a stimulus for which there is an existing response tendency, which is embellished in ways that elicit stronger and more exaggerated instances of that response tendency than normal.[8] A familiar example is cheesecake:

> We enjoy strawberry cheesecake, but not because we evolved a taste for it. We evolved circuits that gave us trickles of enjoyment from the sweet taste of ripe fruit, the creamy mouth feel of fats and oils from nuts and meat, and the coolness of fresh water. Cheesecake packs a sensual wallop unlike anything in the natural world because it is a brew of megadoses of agreeable stimuli which we concocted for the express purpose of pressing our pleasure buttons.
> (Pinker 1997: 524 (cited in De Block & Du Laing 2010))

Moving from broadly physiological and sensory "pleasure buttons" to more cognitive and psychological ones, Matthew Hurley et al. (2011) incorporate the idea of superstimuli into their theory of humor, portraying jokes as analogous to a range of other familiar elements of contemporary life:

> Some human artifacts—paintings and sculptures, and pornography, but also music and even aspects of religion—have been devised as supernormal stimuli

that (over) stimulate our instinctual systems, producing more intense reactions than they were designed (by natural selection) to deliver. . . . jokes are prime examples of super-normal stimuli that take advantage of our natural propensity for humor-detection in much the same way that perfumes, makeup, artificial sweeteners, music, and art give us exaggerated experiences with respect to the natural world. Thanks to their refined designs, they tend to have the power to induce in us a far stronger and richer sense of the ludicrous than everyday "found" stimuli, however humorous. (Hurley et al. 2011: 159)

The picture of the mind here should look familiar; it assumes a broadly modularized structure described above, composed of domain-specific mechanisms serving different purposes, producing distinctive responses to a set of proprietary stimuli relevant to the adaptive problems they evolved to deal with in the past. Several more points can help connect this concept back to the main line of discussion about selective debunking.

First, the buttons that supernormal stimuli push need not activate pleasurable responses; they can trigger negative or neutrally valenced responses as well. Second, nothing in the notion of a supernormal stimulus would render it unsuitable for use in thinking about our moral capacities, and the types of cues different mechanisms that underpin those capacities respond to. There is nothing confused or contradictory about the idea of moral cheesecake; though if your daily experience (especially media consumption) is anything like mine you might think that supernormal stimuli that push our moral buttons are much less likely to be designed to trigger positive responses like admiration, elevation, or awe and much more likely to be designed to trigger negative responses like disgust, spite, or outrage.[9] Third, while both cheesecake and jokes are intentionally designed cultural artifacts, supernormal stimuli need not be. Indeed, the concept does not purport to pick out a phenomenon unique to humans, and exaggerated versions of proprietary stimuli might occur naturally, or in the course of normal social interactions (benefitting from a particularly brave or selfless act, or witnessing an especially cruel one). That said, cheesecake and jokes are good illustrations of the concept because most instances of supernormal stimuli pertinent to humans and modern life are indeed designed by people, crafted specifically to activate targeted psychological components and elicit certain responses and behaviors. This leads to a fourth point, which is that the extent to which many of us now live in environments that are massively cultural constructed, littered with intentionally designed supernormal stimuli of this kind, is unprecedented in human history.[10] The cognitive niche in which our psychological machinery, including our moral capacities and their underlying mechanisms, are being asked to operate is media drenched, saturated with advertising and political messaging backed by sophisticated marketing research and funded by entities with virtually bottomless pockets. We—we WEIRDos, anyway[11]—are constantly inundated with moral cheesecake and other supernormal stimuli that, along with other subtle and not so subtle technologies of perception shaping and narrative crafting, are in constant competition to capture and hold our attention, and then to push our psychological buttons in ways that nudge us to form particular beliefs and desires, engage in certain kinds of behaviors, and, often, make specific moral judgments, all of which will serve the ends and agendas of their creators: buy this, click on that, vote for this candidate, support that referendum, be very suspicious of people who look like this.

From the point of view of the discussion above, producing and propagating super-normal stimuli for a particular psychological mechanism is one way of actively changing the environment in which that mechanism operates, and in extreme cases can change or expand the actual function of the mechanism. Failing to take this into account when performing a moral assessment of the deliverances of those mechanisms, or ignoring this possibility when considering the normative significance or ethical implications of what we know about the evolution and functioning of different components of human moral psychology, seems foolish, especially given the informationally engineered contexts in which they are typically operating in the modern world. Philosophers have an understandable tendency to want to formulate and address questions in their most general, universal, and timeless form, but here I think it does them a disservice. Moral cognition "in the wild" today is not just extremely social (cf. Doris & Nichols 2012), but takes place in an increasingly meticulously constructed cognitive niche loaded with meticulously constructed cultural artifacts, pieces of moral cheesecake designed to push hard on our moral buttons, inducing specific actions and attitudes. A selective approach informed by the tools of evolutionary theory and cognitive science will be able to help us understand the different ways in which this happens, what conclusions we might draw from it, and what its moral significance might be.

CONCLUSION: THE DEBUNKING IMPULSE REVISITED

I began this chapter with Hume's argument about causation, and by posing some questions about whether or not that argument actually debunked anything. I then separated out different ways the debunking impulse has manifested in recent debates about morality, particularly those debates that draw on current empirical psychology and evolutionary theory. I advocated a divide-and-conquer approach to drawing implications from science for moral theory, and described a picture of the structure of the human mind and its moral machinery, innate and otherwise, that I hold fits with and supports this selective approach. Particular applications of this selective strategy may result in negative conclusions, in what would be comfortably called the "debunking" of a circumscribed set of claims or views. But other applications might suggest implications that are neutral, positive, or perspective altering in ways not naturally described as "debunking."

Even if Hume's argument is accepted in full, I am not sure whether it would be accurate to say that it debunks causation, or shows causation to be an illusion. I am likewise unsure about what would count as the debunking of different metaethical or normative ethical views, or under what conditions it would be more accurate to say we are coming to a deeper and better appreciation of the subject matter in question. In these domains, as in others, as our scientific understanding of the phenomena increases, the scientific and manifest images of them will become increasingly dissimilar, and grow further apart. The concepts and vocabulary developed to produce a satisfactory explanation for those phenomena will, if they are to be genuinely explanatory, differ from the concepts and vocabulary used to describe those phenomena in the first place. Over time the required explanatory concepts may depart radically from the concepts of the folk theory of the phenomena, and appeal to entities foreign to or at odds with common sense. In many cases it will be useful to allow gains in our scientific understanding of a phenomenon to provide feedback into and inform what we mean by the concepts and words we use to conceive of and talk about it. None of this—none of these observations about the dynamics of classification and the vicissitudes of semantics and terminology—need threaten belief in the existence of the phenomena themselves, though. Explanantia are always pitched at a different level and stated in different terms than their explananda, and no easy ontological or sweeping ethical conclusions can be draw from the fact that the kind of language useful for describing a set of patterns or participating in a set of practices departs, even in radical ways, from the language needed to explain different aspects of those patterns and practices (Fodor 1974, 1997; cf. Stich 1996).

I see no reason to think the phenomena of morality are different from any others in this respect (even if the boundaries demarcating the phenomena that are distinctively moral from those that are not remain difficult to bring into focus (see Sinnott-Armstrong & Wheatley 2012)). As Hume's argument about causation shows, the invocation of the workings of the mind to explain, illuminate, or otherwise inform an area of inquiry is not unique to morality, and invocations of our current, empirically and evolutionarily informed understanding of those workings continue in a range of areas of philosophy to this day (e.g., Maudlin 2010; Goldman 2015; Paul 2016). But whether or not those who draw them wish to frame their conclusions and implications in terms of debunking, moral theory will be best served if philosophers continue to engage deeply with the details of current cognitive science and evolutionary theory, drawing out careful, selectively focused implications for the issues that most concern them.

NOTES

1. The literature in this vein is getting more articulated by the day it seems; for some other highlights see Lillehammer 2003; Enoch 2010; Wielenberg 2010; Shafer 2010; Kahane 2011; Shafer-Landau 2012; Fraser 2014; Buchanan & Powell 2015; FitzPatrick 2015; and Vavova 2015.
2. The key premise of the argument sets up a modus ponens: "If some particular psychological mechanism can be shown to be problematic in a relevant way, and the intuitions or judgments influenced by that psychological mechanism can be identified, then we should disregard, discount or discredit those intuitions and be suspicious of the judgments that they influence, to the extent that we can" (Kelly 2013: 137). This remains neutral on how to individuate mechanisms, and, more importantly, on what "problematic" amounts to; see section "Assessing Our Moral Capacities in a Hyper-Constructed Cultural Niche" of this chapter for more on the latter.
3. In addition to those works cited in the main text, see Doris et al. 2010 and Henrich 2015 for useful overviews of much recent research in this vein.
4. See Marcus (2007) for elaboration of the idea into a general account of the mind, and Stich (2006) for discussion of the distinction between *elegant machine* and *kludge* accounts of human moral psychology.
5. Philosophical debates about functions are vexed; see Neander, Chapter 5 this volume, for discussion.
6. See Sperber 1996 for introduction and elaboration of this distinction.
7. I have argued that the production of concerns about stigma, ethical taint, spiritual pollution, and moral contamination that accompany moral judgments driven by disgust fit this description (Kelly 2011); also see Plakias (2013) for an opposing view of the value of disgust's contamination sensitivity when operating in the social domain.
8. The concept has roots in classic ethology (Tinbergen 1951; Lorenz 1981), but has also appeared in more contemporary discussions concerning humans and human cognition; see De Block & Du Laing 2010 for examples and assessment of different applications of the idea in the social sciences.

9. For recent work on the evolution, psychology, and signaling function of moral outrage see Jordan et al. 2016, and for spite see Forber & Smead 2014.

10. David Foster Wallace (2007) dubbed it the "Total Noise" of contemporary culture (*American* contemporary culture, specifically). See Kelly 2015 for discussion.

11. That is: Western, Educated, Industrialized, Rich, and Democratic. See Henrich et al. 2010.

REFERENCES

Barkow, J., Cosmides, L., & J. Tooby. 1992. *The Adapted Mind: Evolutionary Psychology and the Generation of Culture* (Oxford University Press).

Bell, M. 2013. *Hard Feelings: The Moral Psychology of Contempt* (Oxford University Press).

Berker, S. 2009. "The normative insignificance of neuroscience." *Philosophy and Public Affairs* 37: 293–329.

Bloom, P. 2013. "The baby in the well: The case against empathy." *The New Yorker*, May 20.

Brogaard, B. 2015. *On Romantic Love: Simple Truths about a Complex Emotion* (Oxford University Press).

Buchanan, A. & Powell, R. 2015. "The limits of evolutionary explanations of morality and their implications for moral progress." *Ethics* 126: 37–67.

Buss, D. 2005. *The Handbook of Evolutionary Psychology* (Wiley).

Clarke-Doane, J. 2012. "Morality and mathematics: The evolutionary challenge." *Ethics* 122: 313–340.

Cline, B. 2015. "Nativism and the evolutionary debunking of morality." *Review of Philosophy and Psychology* 6: 231–253.

De Block, A. & Du Laing, B. 2010. "Amusing ourselves to death? Superstimuli and the evolutionary social sciences." *Philosophical Psychology* 23: 821–843.

Doris, J. & Nichols, S. 2012. "Broadminded: Sociality and the cognitive science of morality." In E. Margolis, R. Samuels, & S. Stich (eds.), *The Oxford Handbook of Philosophy and Cognitive Science* (Oxford University Press) 425–453.

Doris, J. & Plakias, A. 2007. "How to argue about disagreement: Evaluative diversity and moral realism." In W. Sinnott-Armstrong (ed.), *Moral Psychology, Vol. 2: The Biology and Psychology of Morality* (Oxford University Press) 303–332.

Doris, J. and The Moral Psychology Research Group (eds.). 2010. *The Moral Psychology Handbook* (Oxford University Press).

Enoch, D. 2010. "The epistemological challenge to metanormative realism: How best to understand it, and how to cope with it." *Philosophical Studies* 148: 413–438.

Fischer, R. 2016. "Disgust as heuristic." *Ethical Theory and Moral Practice* 19: 679–693.

FitzPatrick, W. 2015. "Debunking evolutionary debunking of ethical realism." *Philosophical Studies* 172: 883–904.

Fodor, J. 1974. "Special sciences." *Synthese* 28: 97–115.

Fodor, J. 1983. *The Modularity of Mind* (MIT Press).

Fodor, J. 1997. "Special sciences: Still autonomous after all these years." *Noûs* 31: 149–163.

Fodor, J. 2003. *Hume Variations* (Oxford University Press).

Forber, P. & Smead, R. 2014. "The evolution of fairness through spite." *Proceedings of the Royal Society, Series B: Biological Sciences* 281: 20132439.

Fraser, B. 2014. "Evolutionary debunking arguments and the reliability of moral cognition." *Philosophical Studies* 168: 457–473.

Gibbard, A. 1990. *Wise Choices, Apt Feelings* (Harvard University Press).

Goldman, A. 2015. "Naturalizing metaphysics with the help of cognitive science." In K. Bennett & D. Zimmerman (eds.), *Oxford Studies in Metaphysics, vol. 8* (Oxford University Press) 171–214.

Goodwin, G. & Darley, J. 2008. "The psychology of meta-ethics: Exploring objectivism." *Cognition* 106: 1339–1366.

Greene, J. 2013. *Moral Tribes: Emotion, Reason, and the Gap Between Us and Them* (Penguin).

Henrich, J. 2015. *The Secret of Our Success: How Culture Is Driving Human Evolution, Domesticating Our Species, and Making Us Smarter* (Princeton University Press).

Henrich, J., Heine, S., & Norenzayan, A. 2010. "The weirdest people in the world." *Behavioral and Brain Sciences* 33: 61–135.

Hume, D. [1740] 1978. *A Treatise of Human Nature* (Clarendon).

Hume, D. [1751] 1983. *An Enquiry Concerning the Principles of Morals* (Hackett).

Hurley, M., Dennett, D., & Adams, R. 2011. *Inside Jokes: Using Humor to Reverse Engineer the Mind* (MIT Press).

Jordan, J., Hoffman, M., Bloom, P., & Rand, D. 2016. "Third-party punishment as a costly signal of trustworthiness." *Nature* 530, 473–476.

Joyce, R. 2006. *The Evolution of Morality* (MIT Press).

Joyce, R. 2013. "Irrealism and the genealogy of morals." *Ratio* 26: 351–372.

Joyce, R. 2016. "Reply: Confessions of a modest debunker." In U. Leibowitz & N. Sinclair (eds.), *Explanation in Ethics and Mathematics* (Oxford University Press) 124–145.

Kahan, D. 1998. "The anatomy of disgust in criminal law." *Michigan Law Review* 96: 1621–1657.

Kahan, D. 1999. "The progressive appropriation of disgust." In S. Bandes (ed.), *The Passions of the Law* (New York University Press) 63–80.

Kahane, G. 2011. "Evolutionary debunking arguments." *Noûs* 45: 103–125.

Kass, L. 1997. "The wisdom of repugnance," *The New Republic* 216: 17–26.

Kelly, D. 2011. *Yuck! The Nature and Moral Significance of Disgust* (MIT Press).

Kelly, D. 2013. "Selective debunking arguments, folk psychology and empirical moral psychology." In J. Wright & H. Sarkissian (eds.), *Advances in Experimental Moral Psychology: Affect, Character, and Commitments* (Bloomsbury) 130–147.

Kelly, D. 2015. "David Foster Wallace as American hedgehog." In S. Cahn & M. Eckert (eds.), *Freedom and the Self: Essays in the Philosophy of David Foster Wallace* (Columbia University Press) 109–132.

Kelly, D. & Morar, N. 2014. "Against the yuck factor: On the ideal role for disgust in society." *Utilitas* 26: 153–177.

Knobe, J. & Nichols, S. 2014. *Experimental Philosophy, vol. 2* (Oxford University Press).

Kristjánsson, K. 2016. "Jealousy revisited: Recent philosophical work on a maligned emotion." *Ethical Theory and Moral Practice* 19: 741–754.

Kumar, V. Forthcoming. "The ethical significance of cognitive science." In S. Leslie & S. Cullen (eds.), *Current Controversies in Cognitive Science* (Routledge).

Kumar, V. & Campbell, R. 2012. "On the normative significance of experimental moral psychology." *Philosophical Psychology* 25: 311–330.

Lillehammer, H. 2003. "Debunking morality: Evolutionary naturalism and moral error theory." *Biology and Philosophy* 18: 567–581.

Lorenz, K. 1981. *The Foundations of Ethology* (Springer).

Machery, E. & Mallon, R. 2010. "The evolution of morality." In J. Doris & The Moral Psychology Research Group (eds.), *The Moral Psychology Handbook* (Oxford University Press) 3–46.

Mackie, J. 1977. *Ethics: Inventing Right and Wrong* (Penguin).

Marcus, G. 2007. *Kluge: The Haphazard Construction of the Human Mind* (Mariner).

Martin, A. 2014. *How We Hope: A Moral Psychology* (Princeton University Press).

Mason, K. 2010. "Debunking arguments and the genealogy of religion and morality." *Philosophy Compass* 5: 770–778.

Maudlin, T. 2010. *The Metaphysics Within Physics* (Oxford University Press).

Nichols, S. 2004. *Sentimental Rules: On the Natural Foundations of Moral Judgment* (Oxford University Press).

Nichols, S. 2015. "Processing debunking and ethics." *Ethics* 124: 727–749.

Nussbaum, M. 2004. *Hiding from Humanity: Disgust, Shame, and the Law* (Princeton University Press).

Paul, L. 2016. "Experience, metaphysics, and cognitive science." In J. Sytsma & W. Buckwalter (eds.), *A Companion to Experimental Philosophy* (Blackwell) 417–433.

Pinker, S. 1997. *How the Mind Works* (W.W. Norton).

Plakias, A. 2013. "The good and the gross." *Ethical Theory and Moral Practice* 16: 261–278.

Prinz, J. 2011. "Against empathy." *Southern Journal of Philosophy* 9: 214–233.

Railton, P. 2014. "The affective dog and its rational tale." *Ethics* 124: 813–859.

Rini, R. 2013. "Making psychology normatively significant." *Journal of Ethics* 17: 257–274.

Rini, R. 2016. "Debunking debunking: A regress challenge for psychological threats to moral judgment." *Philosophical Studies* 173: 675–697.

Shafer, K. 2010. "Evolution and normative skepticism." *Australasian Journal of Philosophy* 88: 471–488.

Shafer-Landau, R. 2012. "Evolutionary debunking, moral realism and moral knowledge." *Journal of Ethics and Social Philosophy* 7: 1–37.

Singer, P. 2005. "Ethics and intuitions." *Journal of Ethics* 9: 331–352.

Sinnott-Armstrong, W. & Wheatley, T. 2012. "The disunity of morality and why it matters to philosophy." *Monist* 95: 355–377.

Sommers, T. 2012. *Relative Justice: Cultural Diversity, Free Will, and Moral Responsibility* (Princeton University Press).

Sperber, D. 1996. *Explaining Culture: A Naturalistic Approach* (Blackwell).

Stich S. 1996. *Deconstructing the Mind* (Oxford University Press).

Stich S. 2006. "Is morality an elegant machine or a kludge?" *Journal of Cognition and Culture* 6: 181–189.

Street, S. 2006. "A Darwinian dilemma for realist theories of value." *Philosophical Studies* 127: 109–166.

Tinbergen, N. [1951] 1989. *The Study of Instinct* (Clarendon).

Vavova, K. 2015. "Evolutionary debunking of moral realism." *Philosophy Compass* 10: 104–116.

Wallace, D. F. 2007. "Deciderization 2007—A special report." In D. F. Wallace & R. Atwan (eds.), *The Best American Essays 2007* (Mariner) xii–xxiv.

Wiegman, I. 2015. "The evolution of retribution: Intuitions undermined." *Pacific Philosophical Quarterly.* doi:10.1111/papq.12083.

Wielenberg, E. 2010. "On the evolutionary debunking of morality." *Ethics* 120: 441–464.

Wright, J., Grandjean, P., & McWhite, C. 2013. "The meta-ethical grounding of our moral beliefs: Evidence for meta-ethical pluralism." *Philosophical Psychology* 26: 336–361.

VI

Evolution, Aesthetics, and Art

26

Evolution, Aesthetics, and Art: An Overview

Stephen Davies

PRELIMINARIES

It has become common to explore connections between human evolution and aesthetic and artistic behaviors against the background of a certain framework. We ask if these behaviors are (or were) adaptive. That is, did they give those who adopted them a comparative reproductive advantage over those who did not? Or instead, are they by-products (aka spandrels)? That is, are they adventitious but non-adaptive consequences of adaptations whose biological value lies elsewhere? Or finally, are they mainly cultural, depending on our evolved nature only at a remove and in the most general way? That is, are they technologies that are preserved not mainly by biological inheritance but via deliberate cultural transmission on account of their value to groups and individuals?

I will query the usefulness of this framework later. But even if it is the appropriate one, it has proved intractably difficult to sort aesthetic and art behaviors neatly into these categories.

The biological approach to aesthetic and art behaviors can be variously motivated. The concern might lie with uncovering these behaviors' historical origins, or their original adaptive function if they had one, or to explain their current adaptive function if that carries over, or to consider if they have taken on new adaptive or maladaptive functions in their contemporary setting. Alternatively, if they are regarded as evolutionary spandrels, it should be relevant to identify the adaptations of which they are by-products and to show that they have not become adaptive subsequently. (Advocates of the by-product thesis rarely take the trouble to attempt these demonstrations, however.)

Among the evidence relevant to such matters is data on (the history of) neurological and other biological mechanisms that subserve the relevant behaviors, with special attention to whether these are modularized or task-specific. If some behavior appears to be relatively hard-wired, so that it emerges spontaneously in development, and the relevant circuitry deals only with the behavior in question, that would provide strong evidence that the behaviors were adaptive. Unfortunately, this approach is rarely decisive, however. So, inevitably there is speculation about when the pertinent behavioral capacities were acquired, about the role they played in the natural-cum-cultural history of our ancestors, and in the case of claimed adaptations, about the particular reproduction-enhancing benefits they bestowed.

The model of evolutionary explanation adopted is often plainly supposed to be of the classical kind: individuals are the units of selection and the method of intergenerational transmission is genetic. But when the arguments are presented by humanists rather than by biologists and scientists, as is often the case in this area, they are sometimes careless. Benefits accruing from the behaviors in question are cited as evidence of adaptation with little regard to attendant costs or to the heritability of those behaviors. Sometimes the benefit is identified as learned skills or knowledge, rather than as inherited dispositions that make such acquisitions possible. Where the account makes group benefits central to the story, it might be suggested that group-level selection, with intergenerational cultural transmission, is the evolutionary mechanism. But this does not always go with demonstrating that intergroup competition was evolutionarily more significant with respect to the relevant behavior than intragroup competition, which is the key argument one would hope to find.

Traditionally, the aesthetic has been characterized as the beautiful or sublime, or the experience of these. The meaning of the term "sublime" is perhaps not what it was, but is captured by the notion of awesomeness. Falling within these genera are properties such as elegance, unity, power, splendidness. The beautiful and sublime (or the experiences to which they give rise) are valued positively; their opposites are disvalued.

Some scientists, including Darwin (1879: pt. 1, ch. 3 and pt. 2, ch. 11), make the error of treating all sensorily based pleasurable responses as aesthetic. But claims like "I'm hungry and my food looks so good to eat" or "I'm tired, and my bed looks so welcoming" do not report experiences that usually target the beautiful and the sublime. A possible consequence of this confusion is to extend the notion of aesthetic experience inappropriately to animals, birds, and insects, as Darwin does. The female bowerbird might be intrigued by the male's construction and dancing, but it is not clear that she is moved by their beauty, and it is more likely that her response is lustful rather than aesthetic.

Art has proven difficult to define, though plainly many of the products of sculpture, painting, music, ballet, and literature (including poetry and drama) will qualify. Many Fine Art traditions are found in the Middle East and much of south-east Asia, as well as in the West. But if we confine our attention only to these, a strong connection with evolution is less likely to be plain than if we adopt a broad outlook that includes appropriate domestic, decorative, ritual, and folk practices among the arts, along with high-quality but popular mass entertainments.

Much art aims to be beautiful or sublime—that is, has an aesthetic dimension—but there can be much more to art than this. It can possess important semantic, representational, expressive, and humorous properties, as well as historically conditioned contextual features, including reference, parody, and influence, along with styles and genres. Some philosophers might prefer to distinguish artistic from aesthetic properties. Others (such as Shelley 2003) would expand the notion of the aesthetic to embrace such features.

Even then, the aesthetic is the broader notion, because it can apply to natural or non-artistic events or scenes, such as sunsets, seascapes, sports displays, and rice field terraces, as well as to art.

In considering art, most scientists confine their attention to its sensible aesthetic character and ignore its more abstruse intellectual and symbolic meaning and value. (Among those who are guilty of this are Dissanayake 1988, 1995; Ramachandran & Hirstein 1999.) In doing so, they significantly diminish the achievement that much art displays. Generalizing roughly, scientists often show an impoverished attitude to the comprehension and appreciation of art, one that ignores its cognitive complexity, cultural diversity, and historical embeddedness.

One widespread account, with roots back to Kant ([1790] 1951), regards a lack of functionality as a hallmark of art. Art is for disinterested contemplation for its own sake alone, and not for its usefulness to my ends or those of others. But if art is an evolutionary adaptation, then it somehow improves the biological fitness—that is, potential fecundity—of those who pursue it, which is why art behaviors have been selected over successive generations. How are these views to be reconciled?

The inconsistency is merely apparent. What is in a creature's biological interests is often experienced by it as intrinsically pleasurable and hence the relevant behaviors are self-motivating. Think of food, sleep, and sex. Either the creature in question is not capable of reasoning to what is in its best evolutionary interests or (as in our case) it might not rate evolutionary measures of success above other organism-level goals. The biological agenda can be satisfied without this result being targeted or valued by the creature above the rewards it finds inherent in the relevant behaviors.

That allowed, I think the view that art must be non-functional should be rejected. Art has often explicitly served to polemicize a moral stance, to educate its audience, to elevate the power of ritual, to bond the community, to arouse the group to a shared emotion, and so on. And the generation of useful products is integral to the decorative and domestic arts that were mentioned earlier, while the popular arts are intended as pastimes and entertainments. There need be no incompatibility between appreciating art as art and appreciating it for its functional skill and success. And in the case of Fine Art traditions, rather than describing art as non-functional we might better say that it has a function, namely, to reward close attention that considers it for its own sake.

AESTHETICS

When Did Appreciation of the Beautiful and Sublime Arise?

Beginning about 400 kya, *Homo heidelbergensis*, the progenitor species for the Neanderthals and later for us, lavished special attention on a small minority of the bifacial hand axes they produced. Great care, much more than was necessary for functional efficacy, was taken to make the axes symmetrical. Some axes used rare or special stones, or displayed fossils and other features of the stone. Others were outsize and impractical. Many of the finest examples show no sign of having been used for cutting. Some people (Kohn & Mithen 1999; Berleant 2007) regard these special axes as the first artworks. Even if we do not wish to go so far, these axes surely suggest that some axe making was driven in part by aesthetic motives.

With Neanderthals, our European cousins, there are indications that they sometimes adopted personal decorations (Zilhão & d'Errico 2003; Zilhão 2011), though these might have been selected as insignia or signs of status rather than for their beauty. These apparently took the form of bird feathers (Peresani et al. 2011; Finlayson et al. 2012); also, ochre may have been used to this end. An eagle talon necklace pre-dated the arrival of our species in Europe by 80,000 years (Radovčić et al. 2015). It has also been suggested that Neanderthals may have made art (Choi 2010; Than 2012; Rodríguez 2014). Nevertheless, archaeological evidence of aesthetic behaviors of European Neanderthals and of their Asian counterparts, the Denisovans, is slight at best, though it is conceivable that they took aesthetic delight in nature—in sunsets, impressive vistas, in the sultry warmth of a summer afternoon—as we do.

Artifactual evidence of aesthetic behavior is scarce also for our species following its first emergence about 195 kya. There are tantalizing hints. For instance, treatment of stones with fire to alter their flaking properties at Pinnacle Point, South Africa, along with evidence of cognitively sophisticated behaviors, dates back as far as 164 kya (Stringer 2012; Tattersall 2012). In the Middle Stone Age between 100 and 60 kya, pierced shells (Stringer 2012; Tattersall 2012), engraved ostrich egg-shells (Texier et al. 2010), incised ochre crayons (Henshilwood & d'Errico 2011a), and the ritual use of coloured stone (Coulson et al. 2011) are all suggestive.

A stronger pattern emerges in the Upper Paleolithic (40–12 kya). This saw the flowering of cave art and carved figurines that are widely regarded as art. And at the same time, personal adornments featured as grave goods. Take the spectacular case of the children buried head-to-head about 28 kya at Sunghir in Russia. As well as various mammoth ivory lances and other items, more than 10,000 mammoth ivory beads decorated their clothing and a boy sported a belt adorned with 250 pierced polar fox canines (Trinkaus et al. 2014). Elsewhere in Europe, tools and other artifacts were extensively decorated with incidental depictions or abstract patterns (Cook 2013). Meanwhile, some of the so-called Venuses, carved statues of women, show various styles of clothing and hair design. The 20,000-year-old carving found at Brassempouy survives as a head and neck and the hair is plainly styled (Cook 2013). Indeed, from this time on *Homo sapiens* took its place as the aesthetic decorator par excellence (Dissanayake 1988, 1995).

What is the Evolutionary Function of the Aesthetic Sense?

According to evolutionary psychologists, our aesthetic sense was shaped by biological drivers (Orians 2014). In the case of the environment, those who were attracted to live in waterless deserts or fetid swamps by the beauty they found there did not pass on their genes; those who were drawn instead to the beauty of habitats offering food and shelter, prospect and refuge, bred successfully and passed on their aesthetic preferences for such environments (Kaplan & Kaplan 1989; Orians & Heerwagen 1992). In a more specific version, it is suggested that our hominin ancestors took aesthetic pleasure in the savannah, where they evolved, and that we inherit a vestigial preference for that landscape (Wilson 1984; Dissanayake 1988; Tooby & Cosmides 2001; Dutton 2009; De Smedt & De Cruz 2010). We design parks and gardens to display savannah features, such as long-view lines and scattered clumps of trees. And more generally, we attach a higher real-estate value to sites offering elevated views of parkland and lakes, as well as finding calm and psychological therapy in natural, as against urban, environments.

Even if they are plausible, such views must be qualified (S. Davies 2012). From when we began to live in towns, from about 10 kya, until the 19th century, untamed nature was generally regarded as threatening and hostile. And the most favored "natural" environments are typically humanly constructed, being the product of millennia of grazing, wood-clearing, and the like. Besides, not only have members of our species learned to live in arctic wastes, tropical forests, deserts, and swamps, but also they usually develop aesthetic preferences for the habitats in which they were raised. At least some of our environmental preferences may be bent to look favorably on habitats and landscapes that support a comparatively easy subsistence, but our flexibility and adaptability in face of extreme environmental change might have been more relevant to our survival. Our *Homo sapiens* ancestors faced the consequences of extreme climate instability, both in and out of Africa, for much of our species' existence (Fagan 2010; Stringer 2012; Tomlinson 2015), so fixed aesthetic landscape preferences might have been detrimental.

Another driver of aesthetic preferences identified by evolutionary psychologists is the biological imperative to raise children who will be parents in their turn. Those who were attracted to the infertile or the victims of illness and disease are not our ancestors. Because of the different investments made by fathers and mothers in their children, the sexes are identified as having different aesthetic preferences in potential mates (Trivers 1972). Men value youthfulness and physical markers of health and fecundity (such as symmetry and body shape) in women; women value (the potential for) status and wealth in men (Symons 1979; Buss 1994; Fisher 2004; Chatterjee 2013). Whereas a man might seek relationships with many women, his partner prefers him to invest his time and resources only in their current children.

These views are absurdly crude. It is not clear that sexual attraction always amounts to seeking beauty, as was observed previously, so it is not obvious that we are talking here about aesthetic responses. (I concede, though, that there is a close correlation between judgments of sexual attractiveness and of physical beauty.) And the idea that the best strategy for a man is to maximize the number of his sexual partners, or of a woman that she should get better genes for her children than those of her partner via a casual liaison so long as she can rely on her partner to stick around, ignore how liable to failure these strategies can be. In a world in which most women do not seek random sex (never mind random impregnation), a man might raise more children successfully by devoting himself to the ones he has at home. And in a world in which men do what they can to guarantee the paternity of their children and reject mates who are unfaithful, a woman might do best to provide in her behavior the assurance her partner desires. Serial monogamy is the norm and we are only mildly polygynous (Dixson 2009).

The attempt to be more nuanced is common. It is likely to be mentioned, for instance, that both sexes highly value intelligence, sensitivity, humor, and compassion in those they find attractive. Even this concession seriously underplays the nature and role of beauty in intra-human social relations, however. For a start, evolutionary success is measured in terms of the extent of a person's genetic investment in future generations. Attracting a fecund mate is only the beginning of what will be required for success in this project. The goal is to raise healthy children to adulthood with all the attributes, social as much as physical, that will make them beautiful to others. Parents who lack the relevant qualities—such things as cooperativeness and reciprocity, patience, a sense of justice, care and respect for others, neither too much foolhardiness nor too much timidity, gratitude, even-temperedness, self-esteem and self-care, and the like—are not likely to succeed in cultivating them in their children. So, beauty as a measure of mate attractiveness should take as much account of qualities of character and the skills to impart them as of health and fecundity or status and material resources. Children are comparatively easy to conceive, but raising them to be good members of the community who might, as such, be sought as mates, takes a vastly wider range of aptitudes.

Having extended the notion of human beauty in this way beyond the physical to the social, we can now take it further. Choosing a mate and raising children are important aspects of life, but so are work and broader social interactions. A person might wish to be valued and admired in these domains also. Presenting oneself appropriately, including making the best of one's appearance, is often important in carrying off one's various social roles. To this extent human beauty goes far beyond sexual attractiveness and into the realm of social self-presentation and self-definition (Etcoff 1999; S. Davies 2012). People want to be thought to be attractive, but they do not typically want that to be construed as an invitation to flirting or sexual behavior. Rather, they aim to say something about how they value themselves and their social position, and they hope this esteem will be returned, not as sexual intercourse but as social intercourse of the right kind.

One topic that has not garnered much attention is that of our aesthetic appreciation of non-human animals, both wild and domestic (S. Davies 2012). Among the most sublime or beautiful of experiences can be encounters with animals—the sight of an apex predator closing in on game, of a bird feeding young at the nest, of a troop of monkeys unexpectedly passing overhead. Finding animals aesthetically attractive could be adaptive, for example, where it inclines us to understand the animal's lifeway better, with the result that we might become more skilled in hunting game or avoiding danger. Equally, though, it could be maladaptive. If, in finding animals beautiful or awesome, we are led to over-anthropomorphize them—taking the owl to be wise, the fox to be devious, and the lion to be proud—we falsify their natures, with the result that we comprehend them and their place in the environment less well. In other cases, there may be no evolutionary advantage or disadvantage engendered by the aesthetic frisson they cause. Their behavioral displays present arrays of color that chime with our senses in ways we find beautiful or awesome, say, but this does not otherwise affect our behavior.

The point is this: whatever evolutionary functions might lie behind our aesthetic proclivities, once those aesthetic attitudes and preferences are in place they can be exercised at will. There is no limit to the things or events in which we can seek beauty or sublimity, even if not everything will be equally suited to rewarding that stance. Our search for aesthetic pleasure often involves imaginative modes of engagement that do not narrowly track evolutionary agendas.

ART

What are Art's Origins?

I have already noted that some people think the finest hand axes of 400 kya are works of art. More widely, it is the cave drawing and engravings and the carved figures dating to the Upper Paleolithic (40–12 kya) that are identified unequivocally as art (see Lawson 2012: 10–11). These are associated with Europe, but works of similar antiquity are found in south-east Asia, Africa, and Australia (Henshilwood & d'Errico 2011a; Bednarik 2013; Aubert et al. 2014).

The emergence of art of this kind is frequently associated with the dawn in our species of what is called "behavioral modernity"—art, along with the adoption of symbolic modes of representation, religion, personal adornment, burial with grave goods. (For discussion, see Mellars et al. 2007; Finlayson 2009; Renfrew & Morley 2009; Fagan 2010; Henshilwood & d'Errico 2011b; Stringer 2012; Tattersall 2012.) The current tendency, however, is to push the date of behavioral modernity back (McBrearty & Brooks 2000; Sterelny 2012), prior to the expansion of *H. sapiens* beyond Africa about 60 kya (Wells 2002; Stringer 2012). Along with this, there is a move to identify earlier precedents for art, such as the engraved ostrich shells of Diepkloof rock shelter, South Africa (De Smedt & De Cruz 2011).

One complication is that not all art is artifactual. Song and dance tend not to leave a recognizable archaeological trace. And not all art relies on highly developed tool technologies or on cognitive sophistication. Some art might be more about emotional expression than abstract thought—again, music and dance come to mind. Or again, if art serves as a form of sexual display, it need not require very high skill, but only a level that outperforms that of sexual competitors.

To illustrate the lack of consensus, consider this: Iain Morley (2013) argues that *H. heidelbergensis* had the physiological and neurological resources, along with the behavioral sophistication, necessary for the production of music as much as 500 kya. Gary Tomlinson (2015), by contrast, suggests that the combinatorial and hierarchical thinking necessary for the creation of music with discrete pitches, tonal centers, and metric regularity is apparent only as recently as 30 kya.

One suggestion (Deacon 2010; Chatterjee 2013; Tomlinson 2015) is that art emerged only when the relevant behaviors were freed from practical functions and goals. Only under those circumstances could they become imaginatively creative. An analogy is drawn here with the domesticated Bengalese finch. It is claimed that the bird's song has become elaborated and that this is because the song no longer plays a role in determining which birds mate. The analogy is painful, however. It is surely not the case that the birds have become aware that their song is no longer relevant to attracting a mate since their mates now are selected by their human breeders and, hence, that they now feel liberated to experiment with it! And the assumption behind this model of art—that art is non-functional—had its origins in 18th-century thought. As was explained earlier, most art through most of human history has been viewed explicitly in functional terms. The naïve error of presuming that art must be appreciated for its own sake alone, without regard to any practical functions it might serve, turns up regularly in the scientific literature and (unintentionally?) disenfranchises most non-Western art (Van Damme 1996).

Was Art an Evolutionary Adaptation?

In considering this question we could treat the arts as a group or consider them separately. Ellen Dissanayake (1988, 1995), a pioneer in this field (along with the likes of Grosse 1897, Hirn 1900, and Eibl-Eibesfeldt 1988), identifies the arts in general (along with ritual and play) as forms of "making special" that enhanced the reproductive success of those who engaged in them. The arts may have had their deepest origins in the playful interactions that bond mothers to their babies. In their more public forms, they improve the quality of group life and social interaction. She argues that art behaviors display the characteristic hallmarks of adaptations, being ancient, universal, and a source of intrinsic pleasure. She also thinks that high end Fine Art fails to do for its meager audience what earlier art did when all in the community participated in it. Her account is complex and sophisticated (for recent refinements see Dissanayake 2009, 2013), but may be more compelling as an account of proto-art-behaviors than of art's developed forms.

Another generalist is Geoffrey Miller (2000), who maintains that the arts are all forms of sexual display, primarily made by men to attract women. (See also Dutton 2009.) This was their original evolutionary driver and they continue to retain that same purpose. One concern with this theory is that it lumps the arts with intelligence, humor, a large vocabulary, and all other sexually attractive displays, so what is supposed to make art adaptive is not distinctive to it. And most art takes place in contexts in which mate attraction is not to the fore. Certainly, art (along with much else) can be co-opted for male display, but it is not evident that this is its original, or even current, primary function.

Though variety among the arts—their diverse histories, and their many different functions—does not rule out the possibility that they serve a central evolutionary function, it should give us pause. Not surprisingly, then, theorists often make the case that some specific art form, rather than the arts in general, is adaptive. (Even so, the erratic histories and multifunctionality of individual arts could remain a worry.) Proponents of these views more often come from the art in question than from the biological sciences.

One movement goes under the title of "literary Darwinism." Its proponents argue that fictional storytelling, especially in written form, is adaptive. They disagree about how it is so. Variously it is suggested that it provides status to the storyteller (Boyd 2009), improves mind-reading skills (Zunshine 2006; Vermeule 2010), or "fine tunes" or "calibrates" mind-reading modules (Tooby & Cosmides 2001), or that it otherwise enhances social performance (Scalise Sugiyama 2005; Gottschall 2012). But we can allow the adaptive importance of fictional thinking (which is essential in counterfactual and hypothetical reasoning) and of narrative (which is essential to establishing self- and group-identity and to recording the past) without accepting that fictional literature is the obvious source of the adaptation. Literature could provide valuable information and help hone useful skills without being a biological adaptation to those ends.

A music-specific theory is defended by Ian Cross (2005–2006, 2007, 2012; Cross & Morley 2009), who argues that music's evolutionary function is to assist cognitive development. It is a bearer of non-verbal meaning, and its combination of importance with semantic imprecision allows it to break down barriers between domain-specific mental modules, thereby encouraging the development of general intelligence. As well, he thinks, it models and thereby encourages the development of ethical modes of social behavior. But not everyone would accept the story of cognitive development that here is assumed, and empirical data on the effects of music on intelligence and social skills suggest that it can make positive but only minor improvements in these (S. Davies 2012).

A fairly widespread view among scientists is that art is more likely a non-adaptive by-product of our evolved nature, rather than adaptive in its own right. Darwin implied that music is a non-adaptive spandrel: "As neither the enjoyment nor the capacity of producing musical notes are faculties of the least use to man in reference to his daily habits of life, they must be ranked amongst the most mysterious with which he is endowed"

([1879] 2004: 636). More pointedly, Steven Pinker declared that "music is auditory cheese-cake, an exquisite confection crafted to tickle the sensitive spots of at least six of our mental faculties" (1999: 534), these being language (when the music has lyrics), auditory scene analysis, emotional calls, habitat selection (as expressed in musical tone picturing of the sea, weather, etc.), motor control (when music leads to dancing), and "something else that makes the whole more than the sum of the parts" (1999: 538). It certainly is true that music depends on auditory capacities evolved for ordinary sound processing, but if the musical whole is more than the sum of its derived parts, as Pinker allows, that may be a reason to believe it is not merely an accidental side-effect of non-musical adaptations.

Other, similar suggestions are that music is a by-product of language (Spencer [1857] 1966: vol. 14; Barrow 2005; De Smedt & De Cruz 2010), an offshoot of ancient socio-affective systems (Panksepp 2009), and that it builds on the capacity to understand others as intentional agents with beliefs, desires, and emotions (Livingstone & Thompson 2009).

Now, one might hope that neuroscience could help determine whether music, to stick with that case, is an adaptation or by-product. Some people have argued that there are music-specific neural circuits (e.g., Huron 2003; Peretz & Coltheart 2003; Levitin 2006). Others deny this (such as Patel 2008; Ball 2010; Morley 2013). The matter cannot be decided (McDermott & Hauser 2005). The difficulty lies not only in distinguishing regular sound-processing neural structures from musical ones, but also in the considerable overlap between the use of the brain by music and language, given our uncertainty about which came first (Patel 2008; Bannan 2012; Koelsch 2012; Rebuschat et al. 2012; Morley 2013).

In the case of the other arts, neuroaesthetics has become a burgeoning field. (Recent work includes Kandel 2012; Starr 2013; Chatterjee 2013; Lauring 2014.) For the most part, the primary focus is on the nature of aesthetic experience, rather than on the role of art in the brain's evolution. There is obvious value in work on differences in neural responses between art experts and novices, for instance (see Calvo-Merino et al. 2005). However, the common tendencies in this literature to reduce aesthetic appreciation to hormonal secretions, to equate aesthetic enjoyment with any pleasurable stimulation, and to anthropomorphize the brain, are disappointing. (For other criticisms, see Currie 2003; Stokes 2009; Minissale 2013; D. Davies 2014.) Some interpret the apparent absence of art-specific circuitry as favoring the view that the arts are by-products (De Smedt & De Cruz 2010), while others suggest that such an absence leaves the question open (Merker 2006).

One approach rejects the attempt to tie art closely to evolved behaviors, either as an adaptation or as a by-product. (The position is argued for music in Patel 2008, 2010.) This view regards art as a cultural technology that is transmitted for its value in transforming people's lives. In this it might be compared with the control of fire or with reading and writing. These are important, valuable technologies that deeply affect the lives of those who master them, but they depend only distantly and indirectly on evolved general capacities such as those implicated in intelligence, sociality, emotionality, and learning.

It seems plausible to claim that many artistic behaviors—singing, dancing, play-acting, drawing—emerge robustly in childhood with comparatively little instruction as compared to reading and writing. This implies that they have quite strong biological impellers. We might also question the persuasiveness of the analogy with control of fire (S. Davies 2012). Fire-making is plainly valued as a means to further valuable ends rather than being intrinsically pleasurable and self-motivating as many art-behaviors are. And even if they are not adaptive in their own right, the connection between art-behaviors and intelligence, imagination, emotional expression, and the like seems far more intimate than this view concedes. If biology and culture interact in ways that are mutually altering (as is argued convincingly in Richerson & Boyd 2005), it might not be possible to separate technologies readily from the influence of biology on the capacities they presuppose.

Acknowledging that genes and culture co-evolve, as we just did, permits a new perspective on the earlier debate. It becomes easier to understand why there may be no clear answer to the question framed as one about whether the arts are adaptations or by-products, or whether they are primarily biological or cultural. Culture affects our evolved biology, our biology limits what is culturally possible, and the two are in constant interaction, with feedback in both directions. Above all, we are a niche-constructing species (Odling-Smee et al. 2003; Sterelny 2012). As a result, there is no clear difference between our adapting to our environment and our adapting our environment to us. This undermines the applicability of the notion of adaptation and, with it, the usefulness of distinguishing between adaptations and by-products of adaptations. We make the arts to serve our interests, and those interests are shaped in turn by the arts, which goes on to explain change and development in what we expect of the arts. Because of this complex inter-play, it is more fruitful to ask how this process operates than to assign the arts to the category of adaptation or of by-product (Menary 2014).

REFERENCES

Aubert, M., Brumm, A., Ramli, M., Sutikna, T., Saptomo, E., Hakim, B., Morwood, M., van den Bergh, G., Kinsley, L., & Dosseto, A. 2014. "Pleistocene cave art from Sulawesi, Indonesia." *Nature* 514: 223–227.

Ball, P. 2010. *The Music Instinct* (Bodley Head).

Bannan, N. (ed.). 2012. *Music, Language, and Evolution* (Oxford University Press).

Barrow, J. 2005. *The Artful Universe Expanded* (Oxford University Press).

Bednarik, R. 2013. "Pleistocene palaeoart of Asia." *Arts* 2: 46–76.

Berleant, R. 2007. "Paleolithic flints: Is an aesthetics of stone tools possible?" *Contemporary Aesthetics* 5. www.contempaesthetics.org/newvolume/pages/article.php?articleID=488

Boyd, B. 2009. *On the Origin of Stories* (Harvard University Press).

Buss, D. 1994. *The Evolution of Desire* (Basic Books).

Calvo-Merino, B., Glaser, D., Grèzes, J., Passingham, R., & Haggard, P. 2005. "Action observation and acquired motor skills: An FMRI study with expert dancers." *Cerebral Cortex* 15: 1243–1249.

Chatterjee, A. 2013. *The Aesthetic Brain: How We Evolved to Desire Beauty and Enjoy Art* (Oxford University Press).

Choi, C. 2010. "Heavy brows, high art? Newly unearthed painted shells show Neandertals were *Homo sapiens's* mental equals." *Scientific American* 302: 18–19.

Cook, J. 2013. *Ice Age Art: Arrival of the Modern Mind* (British Museum).

Coulson, S., Staurset, S., & Walker, N. 2011. "Ritualized behavior in the Middle Stone Age: Evidence from Rhino Cave, Tsodilo Hills, Botswana." *PaleoAnthropology* 18–61. doi:10.4207/PA.(2011)ART42.

Cross, I. 2005–2006. "Music and social being." *Musicology Australia* 28: 114–126.

Cross, I. 2007. "Music and cognitive evolution." In R. Dunbar & L. Barrett (eds.), *The Oxford Handbook of Evolutionary Psychology* (Oxford University Press) 649–667.

Cross, I. 2012. "Music as an emergent exaptation." In N. Bannan (ed.), *Music, Language, and Evolution* (Oxford University Press) 263–276.

Cross, I. & Morley, I. 2009. "The evolution of music: Theories, definitions and the nature of the evidence." In S. Malloch & C. Trevarthen (eds.), *Communicative Musicality: Exploring the Basis of Human Companionship* (Oxford University Press) 61–82.

Currie, G. 2003. "Aesthetics and cognitive science." In J. Levinson (ed.), *The Oxford Handbook of Aesthetics* (Oxford University Press) 706–721.

Darwin, C. [1879] 2004. *The Descent of Man and Selection in Relation to Sex* (Penguin).

Davies, D. 2014. "'This is your brain on art': What can philosophy of art learn from neuroscience?" In G. Currie, M. Kieran, A. Meskin, & J. Robson (eds.), *Aesthetics and the Sciences of the Mind* (Oxford University Press) 57–74.

Davies, S. 2012. *The Artful Species: Aesthetics, Art, and Evolution* (Oxford University Press).

Deacon, T. 2010. "A role for relaxed selection in the evolution of the language capacity." *Proceedings of the National Academy of Sciences USA* 107: 9000–9006.

De Smedt, J. & De Cruz, H. 2010. "Toward an integrative approach of cognitive neuroscientific and evolutionary psychological studies of art." *Evolutionary Psychology* 8: 695–719.

De Smedt, J. & De Cruz, H. 2011. "A cognitive approach to the earliest art." *Journal of Aesthetics and Art Criticism* 69: 379–389.

Dissanayake, E. 1988. *What Is Art For?* (University of Washington Press).

Dissanayake, E. 1995. *Homo Aestheticus: Where Art Comes from and Why* (University of Washington Press).

Dissanayake, E. 2009. "The artification hypothesis and its relevance to cognitive science, evolutionary aesthetics, and neuroaesthetics." *Cognitive Semiotics* 5: 148–173.

Dissanayake, E. 2013. "Genesis and development of 'making special': Is the concept relevant to aesthetic philosophy?" *Rivista di Estetica* 54: 83–98.

Dixson, A. 2009. *Sexual Selection and the Origins of Human Mating Systems* (Oxford University Press).

Dutton, D. 2009. *The Art Instinct* (Bloomsbury Press).

Eibl-Eibesfeldt, I. 1988. "The biological foundations of aesthetics." In I. Rentschler, B. Herzberger, & D. Epstein (eds.), *Beauty and the Brain: Biological Aspects of Aesthetics* (Birkhäuser) 29–68.

Etcoff, N. 1999. *Survival of the Prettiest: The Science of Beauty* (Doubleday).

Fagan, B. 2010. *Cro-Magnon: How the Ice Age Gave Birth to the First Modern Humans* (Bloomsbury).

Finlayson, C. 2009. *The Humans Who Went Extinct: Why the Neanderthals Died Out and We Survived* (Oxford University Press).

Finlayson, C., Brown, K., Blasco, R., Rosell, J., Negro, J., Bortolotti, G., Finlayson, G., Marco, A., Pacheco, F., Vidal, J., Carrión, J., Fa, D., & Llanes, J. 2012. "Birds of a feather: Neanderthal exploitation of raptors and corvids." *PLoS ONE* 7: e45927. http://dx.doi.org/10.1371/annotation/5160ffc6-ec2d-49e6-a05b-25b41391c3d1

Fisher, H. 2004. *Why We Love: The Nature and Chemistry of Romantic Love* (Henry Holt).

Gottschall, J. 2012. *The Storytelling Animal: How Stories Make Us Human* (Houghton Mifflin Harcourt).

Grosse, E. 1897. *The Beginnings of Art* (Appleton).

Henshilwood, C. & d'Errico, F. 2011a. "Middle Stone Age engravings and their significance to the debate on the emergence of symbolic material culture." In C. Henshilwood & F. d'Errico (eds.), *Homo Symbolicus: The Dawn of Language, Imagination and Spirituality* (John Benjamins) 75–96.

Henshilwood, C. & d'Errico, F. (eds.). 2011b. *Homo Symbolicus: The Dawn of Language, Imagination and Spirituality* (John Benjamins).

Hirn, Y. 1900. *The Origins of Art: A Psychological and Sociological Inquiry* (Macmillan).

Huron, D. 2003. "Is music an evolutionary adaptation?" In I. Peretz & R. Zatorre (eds.), *The Cognitive Neuroscience of Music* (Oxford University Press) 57–75.

Kandel, E. 2012. *The Age of Insight: The Quest to Understand the Unconscious in Art, Mind, and Brain: From Vienna 1900 to the Present* (Random House).

Kant, I. [1790] 1951. *Critique of Judgment*, J. Bernard (trans.) (Haffner).

Kaplan, R. & Kaplan, S. 1989. *The Experience of Nature: A Psychological Perspective* (Cambridge University Press).

Koelsch, S. 2012. *Brain and Music* (Wiley-Blackwell).

Kohn, M. & Mithen S. 1999. "Handaxes: Products of sexual selection?" *Antiquity* 73: 518–526.

Lauring, J. (ed.). 2014. *Introduction to Neuroaesthetics: The Neuroscientific Approach to Aesthetic Experience, to Creativity* (Museum Tusculanum Press).

Lawson, A. 2012. *Painted Caves: Paleolithic Rock Art in Western Europe* (Oxford University Press).

Levitin, D. 2006. *This is Your Brain on Music: The Science of Human Obsession* (Dutton).

Livingstone, S. & Thompson, W. 2009. "The emergence of music from the theory of mind." *Musicae Scientiae* 13 (suppl.): 83–115.

McBrearty, S. & Brooks, A. 2000. "The revolution that wasn't: A new interpretation of the origin of modern humans." *Journal of Human Evolution* 39: 453–563.

McDermott, J. & Hauser, M. 2005. "The origins of music: Innateness, uniqueness, and evolution." *Music Perception* 23: 29–59.

Mellars, P., Boyle, K., Bar-Yosef, O., & Stringer, C. (eds.). 2007. *Rethinking the Human Revolution* (McDonald Institute for Archaeological Research).

Menary, R. 2014. "The aesthetic niche." *British Journal of Aesthetics* 54: 471–475.

Merker, B. 2006. "The uneven interface between culture and biology in human music." *Music Perception* 24: 95–98.

Miller, G. 2000. *The Mating Mind: How Sexual Choice Shaped the Evolution of Human Nature* (Doubleday).

Minissale, G. 2013. *The Psychology of Contemporary Art* (Cambridge University Press).

Morley, I. 2013. *The Prehistory of Music: Human Evolution, Archaeology, and the Origins of Human Musicality* (Oxford University Press).

Odling-Smee, F., Laland, K. & Feldman, M. 2003. *Niche Construction: The Neglected Process in Evolution* (Princeton University Press).

Orians, G. 2014. *Snakes, Sunrises and Shakespeare: How Evolution Shapes Our Loves and Fears* (Chicago University Press).

Orians, G. & Heerwagen, J. 1992. "Evolved responses to landscapes." In J. Barkow, L. Cosmides, & J. Tooby (eds.), *The Adapted Mind: Evolutionary Psychology and the Generation of Culture* (Oxford University Press) 555–579.

Panksepp, J. 2009. "The emotional antecedents to the evolution of music and language." *Musicae Scientiae* 13 (suppl.): 229–259.

Patel, A. 2008. *Music, Language, and the Brain* (Clarendon).

Patel, A. 2010. "Music, biological evolution, and the brain." In M. Bailar (ed.), *Emerging Disciplines* (Rice University Press) 91–144.

Peresani, M., Fiore, I., Gala, M., Romandini, M., & Tagliacozzo, A. 2011. "Late Neandertals and the intentional removal of feathers as evidenced from bird bone taphonomy at Fumane Cave 44 ky B.P., Italy." *Proceedings of the National Academy of Sciences USA* 108: 3888–3893.

Peretz, I. & Coltheart, M. 2003. "Modularity of music processing." *Nature Neuroscience* 6: 688–691.

Pinker, S. 1999. *How the Mind Works* (Penguin).

Radovčić, D., Sršen, A., Radovčić, J., & Frayer, D. 2015. "Evidence for Neandertal jewelry: Modified white-tailed eagle claws at Krapina." *PLoS ONE* 10. http://dx.doi.org/10.1371/journal.pone.0119802

Ramachandran, V. & Hirstein, W. 1999. "The science of art: A neurological theory of aesthetic experience." *Journal of Consciousness Studies* 6: 15–51.

Rebuschat, P., Rohrmeier, M., Hawkins, J., & Cross, I. (eds.). 2012. *Language and Music as Cognitive Systems* (Oxford University Press).

Renfrew, C. & Morley, I. (eds.). 2009. *Becoming Human: Innovation in Prehistoric Material and Spiritual Culture* (Cambridge University Press).

Richerson, P. & Boyd, R. 2005. *Not by Genes Alone: How Culture Transformed Human Evolution* (University of Chicago Press).

Rodríguez, J. 2014. "A rock engraving made by Neanderthals in Gibraltar." *Proceedings of the National Academy of Sciences USA* 111: 13301–13306.

Scalise Sugiyama, M. 2005. "Reverse engineering narrative." In J. Gottschall & D. Wilson (eds.), *The Literary Animal: Evolution and the Nature of Narrative* (Northwestern University Press) 177–196.

Shelley, J. 2003. "The problem of non-perceptual art." *British Journal of Aesthetics* 43: 363–378.

Spencer, H. 1966. *The Works of Herbert Spencer* (Otto Zeller).

Starr, G. 2013. *Feeling Beauty: The Neuroscience of Aesthetic Experience* (MIT Press).

Sterelny, K. 2012. *The Evolved Apprentice* (MIT Press).

Stokes, D. 2009. "Aesthetics and cognitive science." *Philosophy Compass* 4/5: 715–733.

Stringer, C. 2012. *Lone Survivors: How We Came to Be the Only Humans on Earth* (Times Books).

Symons, D. 1979. *The Evolution of Human Sexuality* (Oxford University Press).

Tattersall, I. 2012. *Masters of the Planet: The Search for our Human Origins* (Palgrave Macmillan).

Texier, J.-P., Porraz, G., Parkington, J., Rigaud, J.-P., Poggenpoel, C., Miller, C., Tribolo, C., Cartwright, C., Coudenneau, A., Klein, R., Steele, T., & Verna, C. 2010. "A Howiesons Poort tradition of engraving ostrich eggshell containers dated to 60,000 years ago at Diepkloof Rock Shelter, South Africa." *Proceedings of the National Academy of Sciences USA* 107: 6180–6185.

Than, K. 2012. "World's oldest cave art found—made by Neanderthals?" *National Geographic* June 14; http://news.nationalgeographic.com/news/2012/06/120614-neanderthal-cave-paintings-spain-science-pike/

Tomlinson, G. 2015. *A Million Years of Music: The Emergence of Human Modernity* (Zone Books).

Tooby, J. & Cosmides, L. 2001. "Does beauty build adapted minds? Toward an evolutionary theory of aesthetics, fiction and the arts." *SubStance* 94/95: 6–27.

Trinkaus, E., Buzhilova, A., Mednikova, M., & Dobrovolskaya, M. 2014. *The People of Sunghir: Burials, Bodies, and Behavior in the Earlier Upper Paleolithic* (Oxford University Press).

Trivers, R. 1972. "Parental investment and sexual selection." In B. Campbell (ed.), *Sexual Selection and the Descent of Man, 1871–1971* (Heinemann) 136–179.

Van Damme, W. 1996. *Beauty in Context: Towards an Anthropological Approach to Aesthetics* (E.J. Brill).

Vermeule, B. 2010. *Why Do We Care about Literary Characters?* (Johns Hopkins University Press).

Wells, S. 2002. *The Journey of Man: A Genetic Odyssey* (Random House).

Wilson, E.O. 1984. *Biophilia* (Harvard University Press).

Zilhão, J. 2011. "The emergence of language, art and symbolic thinking: A Neanderthal test of competing hypotheses." In C. Henshilwood & F. d'Errico (eds.), *Homo Symbolicus: The Dawn of Language, Imagination and Spirituality* (John Benjamins) 111–131.

Zilhão, J. & d'Errico, F. 2003. "A case for Neanderthal culture." *Scientific American* 13: 34–35.

Zunshine, L. 2006. *Why We Read Fiction: Theory of Mind and the Novel* (Ohio State University Press).

27

Music and Human Evolution: Philosophical Aspects

Anton Killin

Music today is ubiquitous, highly valued, multifaceted, and plays many different roles in many different social contexts. Unsurprisingly, then, an increasing number of researchers studying human cognition and evolution have raised questions about the evolutionary role and nature of music. Why did our ancestors spend time, energy, and resources on music, when they could have been performing activities more obviously linked to increases in fitness such as hunting, gathering, or stone tool production? Are music's origins intertwined with the evolution of language, or mother–infant communication, or group vocal "grooming," or sexual selection? Is music a biological adaptation? Is music innate? This chapter reflects on some of this fascinating interdisciplinary research from a philosophical perspective.

In the first section I discuss traces of musical activity that date back to 40 kya (40,000 years ago), connecting them with ideas about human behavioral modernity. I argue that the oldest known flutes do not represent the earliest expressions of our lineage's musicality, and suggest that music may be very ancient indeed. In the second section I reflect on the so-called "music instinct." I query the innate/acquired framework, utilized by many music cognition researchers, for conceptualizing development. In the third section, the heart of this chapter, I outline the debate surrounding the evolutionary status of music. Researchers have conceptualized music as an adaptation, by-product, exaptation, and technology—and debate surrounding music's evolutionary status is heated. I survey leading hypotheses and argue that taking dynamic gene-culture co-evolution and niche construction seriously calls into question the usefulness of those distinctions, making a case for a co-evolutionary perspective of music. (So while the second and third sections adopt a critical stance, my aim here is not to imply that the various empirical research agendas that target the cognition and evolution of music are unreliable or wrong-headed: the intention is to acknowledge methodological and theoretical difficulties and challenges that are priorities for future progress.) In the final section, I briefly outline a socio-cognitive niche construction perspective, drawing on examples from the ethnomusicological literature.

MUSIC IN PREHISTORY

Musical instruments have a deep past. Archaeologists have unearthed an enormous sample of Upper Paleolithic flutes made from bird bone and mammoth ivory—over 100 uncontroversial specimens—dating back to 40 kya.[1] Besides flutes, the prehistoric record contains whistles made from deer phalanges and (alleged) bullroarers and rasps. (For review, see Morley 2013.)

Most of these ancient flutes are made from vulture ulna or radius bones. The proximal ends of these bones have been shaped into concave mouth pieces, the lengths of the bones have been scraped smooth for easier handling, and finger holes have been carefully created. Engraved slits along the body of some flute-bones suggest that the placing of the finger holes was measured, perhaps for practical or pedagogical reasons, perhaps reflecting some scale, mode, or pitch standards. Reconstruction experiments exhibit a wide range of tones and establish them as "fully developed musical instruments" (Conard & Malina 2008: 14).

Mammoth ivory flutes required much more precision work, and the procurement of the raw material would have been a more significant activity than birding. Presumably, obtaining mammoth ivory was a side-effect of hunting or scavenging mammoths—not in itself a primary investment in music—yet making flutes *from* ivory is a serious investment of time and energy:

> Ivory grows in layers (somewhat similar to tree rings), and in order to make it hollow these lamellae (layers) must be separated. To do this, a section of ivory must be sawn to the correct length, it must then be sawn in half along its length, the core lamellae must be removed, and then the two halves of the flute must be refitted and bound together with a bonding substance which must create an airtight seal in order for the pipe to produce a sound . . . This is a technically complex and challenging procedure. (Morley 2013: 50)

It is intriguing that our ancestors spent so much time, energy, and resources on crafting musical instruments.

The sophistication of these flutes' design and construction suggests complex communication and coordination, including teaching and learning, skilled manipulation of raw materials, good episodic and working memory skills, task specialization, and some degree of division of labor: the full gamut of behavioral modernity evident in each flute. Making musical instruments such as these requires an extended learning period, with presumably little to no payoff until the skill is more or less mastered, pointing to advanced mental time travel capabilities, increased forward planning, focus, and impulse control given the future utility of the item under construction (unlike, say, opportunistically knapped stone choppers or flakes). In short, musical instruments epitomize the advanced, modern minds of Upper Paleolithic hominins.[2]

What's more, although these ancient flutes are the earliest musical instruments that we currently have, they could not have been among the first musical technologies. Their sophistication implies a longer musical-technological tradition concealed—perhaps forever—from the material record. The European caves provided excellent preservation houses: if there were much earlier bird-bone flutes, for instance on the expedition into Europe from Africa, we cannot expect them to have lasted unsheltered from the elements. Bird bones typically preserve poorly due to their light structure.[3] And long-standing traditions of music, deeply entrenched in their respective cultures, appear in all known human societies (Nettl 2000). This suggests that music (or at least proto-music) appeared before modern humans left Africa, otherwise music would have had to evolve independently many times over, in populations geographically isolated from each other (Davies 2015).[4] At the very least, the strong possibility that the earliest full-fledged expressions of musicality are (and will remain) invisible to the archaeological record requires theorists to extend the estimate to well before the first traces at 40 kya. Yet the jury is out on how far back it should be extended (see Davies, Chapter 26 this volume).

Indeed, the absence of earlier musical-technological evidence is not evidence of absence. After all, the material source for ancient flutes need not be limited to bone and ivory. The ethnomusicological record details numerous musical traditions that utilize bamboo, wooden, or cane flutes (Titon 1996). Ancient flutes made from these easily worked and easily procured ephemeral materials may well have pre-dated (and co-existed with) those of the material record (Epsi-Sanchis & Bannan 2012). And other musical instruments may have been made from bark, reeds, shells, logs, horns, taut animal skins, stone.[5] Not to mention the human body—singing, humming, sighing, crooning, stomping, slapping, clapping, thwacking—surely the focus of much ancient musical production. Iain Morley's recent review of the paleoanthropological literature (2013) reveals that the biological preconditions for these, including complex vocalization, may have been in place since *Homo erectus/ergaster*, and that by the appearance of *Homo heidelbergensis* around 600 kya, we see cues of a modern vocal tract and auditory system. It is certainly not implausible that archaic sapiens at 200–300 kya, or even their heidelbergensian predecessors, were (proto-)musically active in some respect.

So although the timing of music's origin is unknown, that music has a deep history foregrounds its importance in ancient social life. (Indeed selection may have made this so, building on existing sensory biases.) One upshot is that complete models of hominin evolution and cognition need to say something about music. It cannot be assumed straight off the bat that music is a newcomer to relatively modern human social life and thus superfluous to an explanation of hominin evolution and cognition.

MUSIC AND INNATENESS

Neuroscience has shown that engaging with music, whether through listening or performing, utilizes a broad cartography of the contemporary human brain (see, e.g., Alluri et al. 2012)—unlike reading or doing arithmetic, say, which have more localized neural correlates. Our skulls do not house a unitary music *module*; there is no evolved musical center in the brain. So development of the cognitive capacities that enable music's production and appreciation, and the nature of those capacities, have become hotly contested topics. Several related "music instinct" research agendas have emerged from this debate (Marcus 2012), of which one will be my focus: whether music is innate or acquired.

Dividing an organism's characteristics into those explained by its intrinsic nature and those explained by external influence is a standard move, part of folk-philosophical wisdom. And it is commonplace for especially talented musicians to be described colloquially as "born that way," innately musical—that the talent is for example "in the genes." This kind of talk presupposes an innate/acquired distinction, which has come to form an influential framework for conceptualizing development. One study reports that three-quarters of informally surveyed music educators believe children require innate talent in order to do well (Davis 1994).

Many researchers have famously rejected such an admittedly naïve nativist position (Howe et al. 1998). Musical expertise takes an enormous amount of deliberate practice: it has been estimated that top-level expert violinists clock up an average of over 10,000 hours of individual practice by age 20 (Ericsson et al. 1993). Nonetheless, debate is live concerning the innateness (or otherwise) of various underlying musical capacities; that is, capacities distributed among the general population, such as beat and tone perception.

Consider beat perception (detection of a regular, underlying pulse in music). István Winkler and colleagues (2009) privilege a nativist explanation given the very early stage in ontogeny—two to three days old—that this capacity has been evidenced. Jessica Grahn (2012), though, points out that experience of beat perception might begin prenatally, so the inference to innateness may be premature. Prenatal infants are exposed to a variety of external rhythms, such as the pulse and heartbeat of their mother. So while beat perception presumably relies on the infant's motor networks, it may also rely on her experience of auditory environment.

Consider also tone perception. Very young infants show a preference for consonant over dissonant tones in music, and this has led researchers to suggest that such a preference is innate (see Marcus 2012; Trainor et al. 2002). Interestingly, however, studies have established that neonates show preferences for music that was played to them whilst in the womb (see Sloboda 2005; also Parncutt 2009; Hepper 1991), so prenatal musical exposure influences infant preferences. And Josh McDermott and colleagues suggest that preference for consonance over dissonance is not universal: it appears that the Tsimané people lack this preference (McDermott et al. 2016). As with beat perception, presumably both inborn networks and auditory environment matter in tone perception.

Not only are innate ascriptions of musical capacities too quick, they conceal a deeper issue. There are general reasons to query the usefulness of the innate/acquired framework. First, the concept of innateness is subject to much philosophical scrutiny and debate. Rounding up twenty-six different definitions of "innate," Matteo Mameli and Patrick Bateson (2006) distinguish at least eight distinct concepts of innateness put to use by scientists. This suggests that there are numerous properties (and roles targeted and best played by those properties) that scientists are interested in when investigating innate traits. Mameli (2008) dubs the conflation of these distinct properties the clutter hypothesis of innateness. Paul Griffiths (2002) argues that attempts to operationalize "innateness" might elucidate some property or properties picked out by the broader (i.e., cluttered) concept, which might result in a construct useful for scientific investigation, but will fail to capture innateness *simpliciter*, undermining the general innate/acquired distinction.

Second, the great plasticity of the human brain makes distinguishing innate and acquired neural circuits—even on some "innateness" operationalization—something of a challenge (McDermott & Hauser 2005).

Third (and as the musical examples show), developmental interaction between genotypes and environments is ubiquitous and complex. There is widespread consensus that, apart from perhaps DNA processing's direct molecular products (proteins), a phenotypic trait's development in any individual depends on both genes and environment. This is known as the *interactionist consensus* (Sterelny & Griffiths 1999). So although the literature has moved past false and unhelpful caricatures of "irreconcilable" nature versus nurture perspectives, and although researchers might agree that the concepts are now "a matter of degree," the innate/acquired distinction is misguiding in practice at least for the reason that it focuses attention away from the interaction of developmental resources key to understanding human cognition. In my view, the innate/acquired distinction is an unhelpful framework for conceptualizing development of cognitive traits both generally and applied to music.

Nonetheless, music researchers can avoid the pitfalls of innateness by specifying their research targets explicitly, distinguishing the properties they are interested in from other ("uncluttered") candidates; for example, how developmentally robust a capacity is, whether a capacity can be modified down the line by learning (or other mechanisms for adaptive plasticity), whether it is environmentally canalized relative to some variation range, whether it is highly heritable (or otherwise) relative to some population, whether or not it is statistically universal in some population of normally developed individuals, and so on. There is no reason to prevent the debate over beat perception and tone perception side-stepping the innate/acquired distinction and proceeding along these lines.

We should prefer debate about detailed causal stories of musical development to rough innate/acquired ascriptions. The subsequent development through ontogeny of musical capacities (building on that of beat and pitch perception and so on) presumably relies on a very complex interaction of developmental resources. Consider that identical twins raised in different musical environments are likely to exhibit variation in musical abilities. Twin studies reveal that musical capacities are in part genetically influenced, yet that genetic effects on musical capacities are more pronounced among those in more musical environments (Hambrick & Tucker-Drob 2015). The developmental landscape metaphor may be useful here: causes from a variety of resources (genetic, epigenetic, behavioral, symbolic; Jablonka & Lamb 2005) shape the pathway one takes as one navigates the hills and canals of development from conception to death (Ariew 2007; Waddington [1957] 2014). And as André Ariew suggests, perhaps replacing attributions of innateness with environmental canalization—that is, the degree to which the end-state of a phenotype's development is buffered against environmental effects—will prove to be more useful for scientific modeling and debate than standard innate/acquired theorizing, true of musical capacities as much as any other.

MUSIC AND EVOLUTIONARY THEORY

Debate about music's evolutionary status can be traced back to an exchange between Charles Darwin and Herbert Spencer. Spencer (1857) suggests that music's origins are in the prosodic/emotional elements of language, while Darwin (1871) suggests that music's origins are in sexual selection (see below) and that language emerged from its musical predecessor. (For a lucid historical background, beyond the scope of this chapter, see Cross 2007.) One of the questions dominating the contemporary philosophical literature on the evolution of music is whether or not music is an evolutionary *adaptation* (see Davies, Chapter 26 this volume). That is, does music have a proper function, having evolved due to its fitness benefits in our ancestral environments? Briefly, on Darwin's proposal, music's evolutionary origins might be functional in this sense—an adaptation for courtship/sexual advertising.[6] Spencer's view is different: it conceptualizes music as a by-product of language. There is no shortage of adaptationist hypotheses, and non-adaptationist alternatives, of music's evolution.

I will first consider adaptationist hypotheses. Following Darwin's lead, Geoffrey Miller (2000a), Denis Dutton (2009), and others argue that music evolved primarily through sexual selection, functioning as a courtship signal/fitness indicator. Darwin's own, oft-quoted conjecture is that:

> primeval man, or rather some early progenitor of man, probably used his voice largely, as does one of the gibbon-apes at the present day, in producing true musical cadences, that is in singing; we may conclude from a widely-spread analogy that this power would have been especially exerted during the courtship of the sexes, serving to express various emotions, as love, jealousy, triumph, and serving as a challenge to their rivals. (Darwin 1871: 54)

Darwin's idea is that our ancestors used vocal displays in their successful attempts to "woo" sexual partners—and to distinguish themselves from (and even challenge) their rivals.[7] Miller has been a strong advocate of a neo-Darwinian sexual selection hypothesis not only for music, but for many other aspects of human life (Miller 2000b).

Other adaptationist hypotheses of music's evolution stress aspects of cooperation and sociality rather than competition or sexual display. Some suggest that music reinforced "groupishness," appealing to the notion of multilevel selection (Brown 2000); others suggest that music functioned as a coalitional signaling system (Hagen & Bryant 2003). Robin Dunbar (1993, 1996) argues that "vocal grooming"—affective vocalizations from individuals to members of the group—took hold in our deep past. The idea: once hominin group size increased beyond a threshold, an upgrade from the one-on-one grooming social strategy observable in extant primates was called for (i.e., vocal grooming), which functioned as social glue. Dunbar first applied this hypothesis to proto-language, but arguably it is more plausible as a theory of proto-music (Dunbar 2012; see also Sterelny, Chapter 9 this volume). Ian Cross (2012a, 2012b) suggests that music functioned as a medium for the management of unclear or ambiguous social situations. Ellen Dissanayake (1982) argues that the arts (in which one might lump music) impart "specialness," which enhanced ritual and reinforced social cohesion. In later work (2008, 2009), Dissanayake suggests that music evolved through mother–infant communication, such as lullaby and play song that produces affective changes and arousal in infants and strengthens mother–infant bonds (see also Trehub 2003; Trehub & Trainor 1998). Steven Mithen (2005) postulates a "musilanguage"—a holistic (non-compositional), mimetic, multimodal, and manipulative (affective) communicative system—that he suggests precedes both music and language.

Participants in the debate freely admit that there is some speculation involved here. However, good hypotheses can generate testable predictions, and assessment of the evidence can help adjudicate between rival hypotheses. To illustrate, I briefly focus on and critique the sexual-selection adaptationist hypothesis about music. The following facts seem at odds with the predictions that we would make in order to test that hypothesis: (i) children are capable of demonstrating musical play and even good musical competence well before sexual maturity; (ii) it has not been established that musicians are any more successful at reproducing than other people (despite the sex-god escapades of a few rock stars and music celebrities)—conversely, *non*-musicians are not thereby less successful reproducers; (iii) cooperative, group-based—rather than individually focused—music is ubiquitous in the ethnomusicological record; (iv) the same goes for lullaby and play song between mothers and children, hardly acts of sexual display; (v) experimental findings suggest there is no correlation between female preference for higher levels of complexity in music (as a proxy for male quality) and occurrent fertility (ovulation), which would be consistent with the sexual selection hypothesis (Charlton et al. 2012); and, finally, (vi) findings from a recent twin study, boasting a huge sample of over ten thousand twins, show that "genetic correlations between musical aptitude and the measures of mating success were all nonsignificant . . . findings show that higher musical aptitude or achievement does not lead to increased sexual success (quantitatively) . . . there was no significant association between musical aptitude and number of children" (Mosing et al. 2015: 364).

My point is not that musical skill is ignored in our assessments of—or our attraction to—potential sexual partners, whether long term or one-night stands. It is extremely plausible that once music emerged, the demonstration of fine musical skills and musical creativity played a role in mate selection, and that music continues to be harnessed by some individuals as a means to attract sexual partners. Rather, my point is that an exclusively sexual-selection adaptation hypothesis for the evolution of music is unpersuasive. Even if important, sexual selection cannot be the whole story.

I'll now turn to non-adaptationist hypotheses of music.

First I will rather schematically introduce the additional terms for the uninitiated (Table 27.1). A *by-product* is a non-adaptive consequence of selection for some other adaptation or set of adaptations (Gould & Lewontin 1979). The redness of mammalian blood, for example, is a by-product of blood's biochemistry and function (internal, iron-based oxygen transport); it is not that selection conferred adaptive fitness on red-blooded ancestors versus their (say) green- or blue-blooded competitors. Blood's redness is merely an *indirect* result of selection. Traits that take on some new adaptive role—that is, "features that now enhance fitness but were not built by natural selection for their current role" (Gould & Vrba 1982: 4)—are *exaptations*. For example, bird feathers initially evolved for thermodynamic regulation, and were later

Table 27.1 Explanation of terms

Taxonomy	Description
Adaptation	Product of selection with proper function(s)
By-product (spandrel)	Incidental consequence of selection
Exaptation (adaptive offshoot)	Product of selection with adaptive effect(s)
Technology	Cultural invention; not a product of Darwinian selection/biology

co-opted for use in flight (although many subsequent structural changes in feathers are, adaptations for flight). Finally, a *technology* is a cultural, rather than biological, innovation, though one that can have significant upshots. Technologies are conceived as not caused by biology, at least not in the same sense as blood's redness. Hominin fire behaviors have been conceived as a technology in this sense: a cultural product rather than a biological one (Patel 2010; Davies 2012). Music has been conceptualized as each of these options. To this I now turn.

Many researchers argue that music is a non-adaptive by-product of other long-established cognitive capacities. Steven Pinker has called music an "auditory cheesecake": "a cocktail of recreational drugs that we ingest through the ear to stimulate a mass of pleasure circuits at once" (Pinker 1997: 528); "music is . . . an exquisite confection crafted to tickle the sensitive spots of at least six of our mental faculties" (534).[8] On this view, music merely generates pleasure rewards, the underlying mechanisms of which were originally selected for in other contexts. Similarly, for Dan Sperber, "music . . . is parasitic on a cognitive module the proper domain of which pre-existed music and had nothing to do with it" (1996: 142).

Davies (2012) argues that even if music first emerged as a by-product, it would have taken on a clear adaptive role in due course. Although he does not use the term "exaptation," the idea is captured by his slogan *"form becomes norm"* (Davies 2012: 144). In the context of art behaviors generally speaking, Davies suggests:

> If art behaviors came to us as ancillary evolutionary by-products, they would not remain merely incidental. Their occurrence in the usual manner would become normative because they provide honest, because costly, signals of fitness. As a result, not only the absence of art behaviors but also the degrees to which they are represented can be informationally significant in assessing someone's fitness.
>
> (2012: 145)

Another example of an exaptation may be literacy. The ability to read and write is presumably a by-product of adaptations for language, pattern recognition, and social learning, not some de novo cognitive adaptation—yet once individuals develop proficient literacy skills, being able to read and write might increase their adaptive fitness.[9] Importantly, however, music is ancient enough for its fitness effects to have caused further evolutionary changes through positive feedback. (I return to this thought below.) This would be so *even if* music were as recent as the earliest known flutes at 40 kya. Writing systems, on the other hand, only date back to roughly 5 kya (Houston 2004)—and general literacy competence, broadly distributed through populations, is a strikingly recent phenomenon.

Pinker's by-product view leads him to call music a "technology." Reclaiming this term, Aniruddh Patel (2010) argues that music is a *transformative* technology, although a cultural invention, somewhat akin to the discovery and production of fire. On Patel's view, music is "transformative" because it is not a mere hedonistic, parasitic pleasure device, as per Pinker's by-product hypothesis: it can shape the neural structures of individuals during their lives and have real effects on their development, largely through the effects of neuroendocrine hormone-release and the brain's great plasticity. To be sure, music has been used to some effect in therapy, for instance in aiding stroke rehabilitation (Thaut et al. 1997) and reducing the desire for self-administered pain medication during locally anesthetized surgery (Ayoub et al. 2005); lullabies have a positive influence on the feeding patterns and development of prematurely born neonates (Standley 2003); music learning is correlated with increased general intelligence in children (Schellenberg 2006) as well as mathematical competency and emotion detection (Gardiner 2008; Thompson et al. 2004). Nonetheless, Patel thinks that music has not been targeted by selection so it is neither an adaptation nor an exaptation.

One problem with the debate is that theorists are not always clear on what they mean by "music." Adaptationist hypotheses might target music conceived as a trait or series of traits of individual phenotypes, rather than, say, the acoustic *product* of a collective, social practice, which might be the target of by-product or technology hypotheses. So it could be supposed that several views are compatible. They are about different *explananda* so we just have to disambiguate "music" appropriately. I argue elsewhere that there are multiple, legitimate, non-equivalent concepts of music (Currie & Killin 2016); knowing whether individual phenotypes or social products are being theorized about is important for disambiguating the target of hypotheses. However, *gene-culture co-evolution* under-mines distinguishing between sharp biological and cultural conceptions of music, at least with respect to evolutionary theorizing. Music, on either disambiguation noted above, is a bio-cultural phenomenon; theorists who blackbox one in attempting to explain the other do so at their peril. In my view, this undermines couching music in the adaptation/non-adaptation framework. I will explain first with an analogy to hominin fire behavior.

Material evidence of sustained fire control appears around 790 kya (Goren-Inbar et al. 2004), becoming more continuous (and archaeologically visible) from around 400 kya. Fire contributed to our ancestors' increase in brain size, it enabled the cooking and soft-ening of foods such as meat and tubers (chimpanzees spend hours every day chewing and digesting their uncooked foods), and it extended the light of the day, enabling oppor-tunities for social and other pursuits. Crucially, fire control fed back into hominin anat-omy (Wrangham 2009). Softening food through cooking enabled the reduction of tooth and gut size of our ancestors by lowering the demand on chewing and digestion, which in turn allowed developmental-energy reallocation to brain size increase. Fire culture had biological consequences, generating positive feedback loops. So (contra Patel) it is unhelpful to conceive of fire control as a purely cultural "technology," artificially splitting biological and cultural evolution, because doing so undermines the dynamic, intertwined evolutionary forces here.

Fire is not a one-off. Other examples include the dynamic co-evolution of cattle farm-ing and the increased adult-age tolerance to lactose (Beja-Pereira et al. 2003), and the co-evolution of the hominin hand and tools (Marzke 2013). Crucially, co-evolution does not render it false to say that hands are, say, "exapted for toolmaking and use" versus "adapted for toolmaking and use" or vice versa—the important point is that this is not a productive framework: it does not focus attention on the co-evolutionary dynamics and feedback loops that were key in shaping (say) the hominin hand and its executive control. The same *mutatis mutandis* is true of music.

Recent research suggests that music is very likely to be a result of gene-culture co-evolution. Steven Brown and colleagues (2013) examine the correlations between genes (via mitochondrial DNA haplotype analysis) and music (via musicological analysis of recordings of traditional vocal songs sung in a group performance context) in nine geographically distinct indigenous populations of Taiwan. Their research indicates that musical diversity and genetic diversity are indeed significantly correlated, with musical overlap between that of nearby regions reflecting shared ancestry (genes) and not just inter-group cultural diffusion. (This is evidence of indirect co-evolution of course, not "genes for music.") From the results of their study, Brown and colleagues conclude: "The correlations we observed between musical and genetic diversity support the contention that music and genes may have been coevolving for a significant time period and that music might possess the capacity to track population changes" (Brown et al. 2013: 5). Similarly striking results are found by Patrick Savage and colleagues in examining the correlations between Ainu music and genes (Savage et al. 2015).

Music's co-evolutionary dynamics are important in assessing the adaptation debate's framework. And in my view, human musicality is a co-evolving patchwork of anatomic, cognitive, behavioral, and socio-cultural features. In cases of complex co-evolution, such as fire control and music, the explanatory usefulness of the standard set of distinctions is undermined, and boundaries are blurred.

MUSIC, ETHNOMUSICOLOGY, AND NICHE CONSTRUCTION

In the previous section I emphasized gene-culture co-evolution. In this section I draw on examples from the ethnographic record in order to illustrate some of the myriad ways in which musicality is expressed in hunter-gatherer life, taking a broader niche construction perspective.

Living organisms alter their environments through their activities, shaping the living conditions of themselves, their offspring, and other affected organisms. Beavers, for example, drastically alter their environments, and the local ecosystem at large, by building dams. (For an introduction to niche construction theory, see Odling-Smee et al. 2003.) Through feedback loops, changing living conditions affect the development, actions, and selection of organisms down the evolutionary trail.

Human populations in particular flourish because humans construct the environments that humans experience (Kendal et al. 2011). Human children grow and develop in a world of innovations, norms, and conventions; their *socio-cultural, informational* environments shaped by previous generations:

> [Humans develop through] interaction with family, friends, collaborators, strangers, and domesticated animals, and an environment loaded with human artifacts. . . . All of this social and material infrastructure *scaffolds* the development of human cognitive phenotypes in every culture today and has for many thousands of years.
> (Trestman 2015: 92)

Humans are niche constructers par excellence. Turning to the ethnographic record can help explore the ways in which music plays a role in such processes of cultural niche construction (leading to fitness differences), and its roles in the social group and wider context. Take, for example, Australian Aboriginal society that traditionally maintained a highly symbolic, lyrically driven song tradition tied closely to concepts of geography, land ownership, story, and ritual (*Tjukurrpa*, "The Dreaming"), and which maps their vast, barren, homogenous desert environment, on one hand, and preserves creation mythology and traditional folklore through song, on another. Children that grow up in a social environment such as this, in which much information is expressed through traditional song, develop in a context that is *scaffolded* by these songs and their role in informational and cultural transmission and social learning. In the 1950s, Richard Waterman noted that:

> Throughout his life, the Aboriginal is surrounded by musical events that instruct him about his natural environment and its utilization by man, that teach him his world-view and shape his system of values, and that reinforce his understanding of Aboriginal concepts of status and his own role. More specifically, songs function as emblems of membership in his moiety and lineage, as validation of his system of religious belief, and as symbols of status in the age-grading continuum. They serve on some occasions the purpose of releasing tensions, while other types are used for heightening the emotionalism of a ritual climax. They provide a method of controlling, by supernatural means, sequences of natural events otherwise uncontrollable. Further, some types of songs provide an outlet for individual creativity while many may be used simply to conquer personal dysphoria.
> (Waterman 1956: 41)

Daou Joiris (1996) has carried out fieldwork on the semi-nomadic Baka Pygmy hunter-gatherers of southeastern Cameroon. Joiris describes their big-game hunting rituals, in which musical performance plays a crucial part. Music performed on "the female musical bow (*ngbiti*) with which women charm spirits in order to make them direct animals towards hunters" (1996: 273), in addition to whole-group song-dances and the (solo) yodeling songs of ritual specialists, are intended to entreat the spirits. These ritual musical performances, if performed successfully, are thought to be partially responsible for the hunters' locating and capturing/killing of game, deciphering of animal trackways, and so on. This cultural niche may very well co-evolve with a psychological one. The music and ensuing hype may raise the affect and emotional profile of the hunters, imbuing them with a heightened sense of awareness, confidence, and camaraderie—leading to a cooperative and bountiful hunt.

Traditionally, the Blackfoot and Sioux tribes of the North American plains, nomadic hunter-gatherers, utilized music daily, in ritual activities and puberty rites, social activities (promoting in-group cohesion), and war dances (promoting out-group dissonance). Their traditional music is primarily vocal, comprising "vocables" rather than words—contributing to activities emotionally rather than symbolically (McAllester 1996)—with percussion made from gourds, turtle shells, wood, cocoons, deer hoofs, and the like. Some songs, however, were iconic: a mimetic, "bleating-calf" song was used in a characteristic musical hunting strategy (Kehoe 1999; Morley 2013). The song lured bison towards the top of a cliff face, to be spooked and forced over:

> By skillfully mimicking a lost calf in his actions and by bleating pitifully, the caller attempted to make the bulls that led the herd follow him. Taking a zig-zag course so as not to make them suspicious, he would gradually lure them into the funnel [of an impoundment], quickening his pace all the while. Once the herd was well into the chute, tribes-men concealed behind the stone and brush piles stood up, shouted, and waved their robes to stampede the bison at the rear end of the herd.

These frightened beasts would surge forward, impelling the leaders of the herd into the pound, willing or not. The buffalo caller quickly dodged to one side to save himself from being trampled to death, and made his escape. Sometimes a convenient steep bluff or gully was substituted for the pound, the bison breaking their legs as they fell over the precipice or being trampled by those behind as they crowded into the narrow gully. (Howard 1984: 61)

Notice that here I am looking to the musics of various societies, not because ethnographically recorded foragers are living replicas of our Pleistocene ancestors, but to provide illustrations of music's part in the niche construction of human lifeways. The Blackfoot "bleating calf" musical hunting strategy, for instance, obviously has ecological and cultural consequences. Moreover, looking to the musics of various historic traditions helps us "to examine and illustrate a wider diversity of the musical behaviors that exist" (Morley 2013: 12), allowing us to better understand the ways in which our ancestors could have been musical that are actually instantiated in groups living under some similar conditions. It reveals some similarities in material resources used, predominant use of the voice (rather than musical instruments), and the social/communal/affective nature of much music.

Although many open questions about music and human evolution remain, ethnomusicological research can bolster theories of music's evolution, fuel phylogenetic modeling methods, and allow researchers to test predictions generated from evolutionary hypotheses (for further discussion, see Killin 2016). One task for future research is elucidating how ethnographic research can feed into a niche construction perspective of human musicality, and hominin socio-cognitive evolution generally.

NOTES

1. All from confirmed *sapiens* sites, the oldest from Germany's Swabian Jura ranges. The earlier Divje Babe I "Neanderthal flute" (a femur bone of a juvenile cave bear, dated to 60 kya) turns out not to be a flute at all (Diedrich 2015). There is, as yet, no confirmed evidence of Neanderthal musical technology.
2. For an extended discussion of behavioral modernity in general see, e.g., Sterelny 2011.
3. Instruments made from ivory may have a greater chance at preservation—if regularly made in preexodus Africa we might expect to find some and excavations may yet uncover them—though for now the complex processes required for the construction of ivory flutes (discussed above) perhaps suggests that they were a more recent extension of the prehistoric wind section.
4. The latter is not impossible. There may have been some novel feature of more recent human environments to which music was a response by convergence.
5. Ancient, intentional striking of stalactites and stalagmites has been evidenced (Montelle 2004), suggesting a lithic musical culture.
6. Stephen Davies (Chapter 26 this volume) characterizes Darwin's account of music as a by-product view, appealing to Darwin's reflection that music faculties are "mysterious." I think this characterization is too quick. Darwin's sexual selection conjecture aims to take that mystery out of music, which Darwin thought was perhaps not explained by *natural* selection. (For a closer look at Darwin's view on music, see Bannan 2017.)
7. Sexual selection is also the standard explanation for the evolution of birdsong; see, e.g., Catchpole 1987.
8. For objections to the cheesecake analogy, see, e.g., Carroll 1998; Dutton 2009; Davies 2012.
9. Writing systems allow the literate to "offload" information and ease constraints on semantic memory, problem solving, attention, and focus.

REFERENCES

Alluri, V., Toiviainen, P., Jääskeläinen, I., Glerean, E., Sams, M., & Brattico, E. 2012. "Large-scale brain networks emerge from dynamic processing of musical timbre, key and rhythm." *NeuroImage* 59: 3677–3689.

Ariew, A. 2007. "Innateness." In M. Matthen & C. Stephens (eds.), *Philosophy of Biology* (Elsevier) 567–584.

Ayoub, C., Rizk, L., Yaacoub, C., Gaal, D., & Kain, Z. 2005. "Music and ambient operating room noise in patients undergoing spinal anesthesia." *Anesthesia and Analgesia* 100: 1316–1319.

Bannan, N. 2017. "Darwin, music and evolution: New insights from family correspondence on *The Descent of Man*." *Musicae Scientiae*. 21: 3–25.

Beja-Pereira, A., Luikart, G., England, P., Bradley, D., Jann, O., Bertorelle, G., Chamberlain, A., Nunes, T., Metodiev, S., Ferrand, N., & Erhardt, G. 2003. "Gene-culture coevolution between cattle milk protein genes and human lactase genes." *Nature Genetics* 35: 311–313.

Brown, S. 2000. "Evolutionary models of music: From sexual selection to group selection." In F. Tonneau & N. Thompson (eds.), *Perspectives in Ecology 13: Behavior, Evolution and Culture* (Plenum) 231–281.

Brown, S., Savage, P., Ko, A., Stoneking, M., Ko, Y., Loo, J., & Trejaut, J. 2013. "Correlations in the population structure of music, genes and language." *Proceedings of the Royal Society, Series B: Biological Sciences* 281: 20132072. doi:10.1098/rspb.2013.2072.

Carroll, J. 1998. "Steven Pinker's cheesecake for the mind." *Philosophy and Literature* 22: 478–485.

Catchpole, C. 1987. "Bird song, sexual selection and female choice." *Trends in Ecology and Evolution* 2: 94–97.

Charlton, B., Filippi, P., & Fitch, W. 2012. "Do women prefer more complex music around ovulation?" *PLoS One* 7: e35626. doi:org/10.1371/journal.pone.0035626.

Conard, N. & Malina, M. 2008. "New evidence for the origins of music from the caves of the Swabian Jura." In A. Both, R. Eichmann, E. Hickmann, & L.-C. Koch (eds.), *Orient-Archäologie Band 22: Studien zur Musikarchäologie VI* (Verlag Marie Leidorf GmbH) 13–22.

Cross, I. 2007. "Music and cognitive evolution." In L. Barrett & R. Dunbar (eds.), *Handbook of Evolutionary Psychology* (Oxford University Press) 649–667.

Cross, I. 2012a. "Music and biocultural evolution." In M. Clayton, T. Herbert, & R. Middleton (eds.), *The Cultural Study of Music: A Critical Introduction*, 2nd ed. (Routledge) 15–27.

Cross, I. 2012b. "Music as a social and cognitive process." In P. Rebuschat, M. Rohrmeier, J. Hawkins, & I. Cross (eds.), *Language and Music as Cognitive Systems* (Oxford University Press) 313–328.

Currie, A. & Killin, A. 2016. "Musical pluralism and the science of music." *European Journal for Philosophy of Science* 6: 9–30.

Darwin, C. 1871. *The Descent of Man, and Selection in Relation to Sex* (Appleton & Co.).

Davies, S. 2012. *The Artful Species: Aesthetics, Art, and Evolution* (Oxford University Press).

Davies, S. 2015. "How ancient is art?" *Evental Aesthetics* 4: 22–45.

Davis, M. 1994. "Folk music psychology." *Psychologist* 7: 537.

Diedrich, C. 2015. "'Neanderthal bone flutes': Simply products of Ice Age spotted hyena scavenging activities on cave bear cubs in European cave bear dens." *Royal Society Open Science* 2: 140022. doi:10.1098/rsos.140022.

Dissanayake, E. 1982. "Aesthetic experience and human evolution." *Journal of Aesthetics and Art Criticism* 41: 145–155.

Dissanayake, E. 2008. "If music is the food of love, what about survival and reproductive success?" *Musicae Scientiae* 12 (suppl. vol.): 169–195.

Dissanayake, E. 2009. "Root, leaf, blossom, or bole: Concerning the origin and adaptive function of music." In S. Malloch & C. Trevarthen (eds.), *Communicative Musicality* (Oxford University Press) 17–30.

Dunbar, R. 1993. "Coevolution of neocortical size, group size, and language in humans." *Behavioral and Brain Sciences* 16: 681–735.

Dunbar, R. 1996. *Grooming, Gossip, and the Evolution of Language* (Harvard University Press).

Dunbar, R. 2012. "On the evolutionary function of song and dance." In N. Bannan (ed.), *Music, Language, and Human Evolution* (Oxford University Press) 201–214.

Dutton, D. 2009. *The Art Instinct* (Oxford University Press).

Epsi-Sanchis, P. & Bannan, N. 2012. "Found objects in the musical practices of hunter-gatherers: Implications for the evolution of instrumental music." In N. Bannan (ed.), *Music, Language, and Human Evolution* (Oxford University Press) 173–198.

Ericsson, K., Krampe, R., & Tesch-Römer, C. 1993. "The role of deliberate practice in the acquisition of expert performance." *Psychological Review* 100: 363–406.

Gardiner, M. 2008. "Music training, engagement with sequence, and the development of the natural number concept in young learners." *Behavioral and Brain Sciences* 31: 652–653.

Goren-Inbar, N., Alperson, N., Kislev, M., Simchoni, O., Melamed, Y., Ben-Nun, A., & Werker, E. 2004. "Evidence of hominin control of fire at Gesher Benot Ya'aqov, Israel." *Science* 304: 725–727.

Gould, S. J. & Lewontin, R. 1979. "The spandrels of San Marco and the panglossian paradigm: A critique of the adaptationist programme." *Proceedings of the Royal Society, Series B: Biological Sciences* 205: 581–598.

Gould, S. J. & Vrba, E. 1982. "Exaptation—a missing term in the science of form." *Paleobiology* 8: 4–15.

Grahn, J. 2012. "Neural mechanisms of rhythm perception: Current findings and future perspectives." *Topics in Cognitive Science* 4: 585–606.

Griffiths, P. 2002. "What is innateness?" *The Monist* 85: 70–85.

Hagen, E. & Bryant, G. 2003. "Music and dance as a coalitional signalling system." *Human Nature* 14: 21–51.

Hambrick, D. & Tucker-Drob, E. 2015. "The genetics of music accomplishment: Evidence for gene-environment correlation and interaction." *Psychonomic Bulletin and Review* 22: 112–120.

Hepper, P. 1991. "An examination of fetal learning before and after birth." *Irish Journal of Psychology* 12: 95–107.

Houston, S. (ed.). 2004. *The First Writing* (Cambridge University Press).

Howard, J. 1984. *The Canadian Sioux* (University of Nebraska Press).

Howe, M., Davidson, J., & Sloboda, J. 1998. "Innate talents: Reality or myth?" *Behavioral and Brain Sciences* 21: 399–407.

Jablonka, E. & Lamb, M. 2005. *Evolution in Four Dimensions: Genetic, Epigenetic, Behavioral, and Symbolic Variation in the History of Life* (MIT Press).

Joiris, D. 1996. "A comparative approach to hunting rituals among Baka Pygmies (southeastern Cameroon)." In S. Kent (ed.), *Cultural Diversity among Twentieth-century Foragers: An African Perspective* (Cambridge University Press) 245–275.

Kehoe, A. 1999. "Blackfoot and other hunters on the North American Plains." In R. Lee & R. Daly (eds.), *The Cambridge Encyclopedia of Hunters and Gatherers* (Cambridge University Press) 36–40.

Kendal, J., Tehrani, J., & Odling-Smee, J. 2011. "Human niche construction in interdisciplinary focus." *Philosophical Transactions of the Royal Society, Series B: Biological Sciences* 366: 785–792.

Killin, A. 2016. "Rethinking music's status as adaptation versus technology: A niche construction perspective." *Ethnomusicology Forum* 25: 210–233.

Mameli, M. 2008. "On innateness: The clutter hypothesis and the cluster hypothesis." *Journal of Philosophy* 105: 719–736.

Mameli, M. & Bateson, P. 2006. "Innateness and the sciences." *Biology and Philosophy* 21: 155–188.

Marcus, G. 2012. "Musicality: Instinct or acquired skill?" *Topics in Cognitive Science* 4: 498–512.

Marzke, M. 2013. "Tool making, hand morphology and fossil hominins." *Philosophical Transactions of the Royal Society, Series B: Biological Sciences* 368: 20120414. doi10.1098/rstb.2012.0414.

McAllester, D. 1996. "North America/Native America." In J. Titon (ed.), *Worlds of Music: An Introduction to the Music of the World's People* (Schirmer) 33–82.

McDermott, J. & Hauser, M. 2005. "The origins of music: Innateness, uniqueness, and evolution." *Music Perception* 23: 29–59.

McDermott, J., Schultz, A., Undurraga, E., & Godoy, R. 2016. "Indifference to dissonance in native Amazonians reveals cultural variation in music perception." *Nature* 535: 547–550.

Miller, G. 2000a. "Evolution of human music through sexual selection." In N. Wallin, B. Merker, & S. Brown (eds.), *The Origins of Music* (MIT Press) 329–360.

Miller, G. 2000b. *The Mating Mind: How Sexual Choice Shaped the Evolution of Human Nature* (Doubleday).

Mithen, S. 2005. *The Singing Neanderthals: The Origins of Music, Language, Mind and Body* (Weidenfeld and Nicolson).

Montelle, Y.-P. 2004. "Paleoperformance: Investigating the human use of caves in the Upper Paleolithic." In G. Bergaus (ed.), *New Perspectives on Prehistoric Art* (Praeger) 131–152.

Morley, I. 2013. *The Prehistory of Music: Human Evolution, Archaeology, and the Origins of Musicality* (Oxford University Press).

Mosing, M., Verweij, K., Madison, G., Pederson, N., Zietsch, B., & Ullén, F. 2015. "Did sexual selection shape human music? Testing predictions from the sexual selection hypothesis of music evolution using a large genetically informative sample of over 10,000 twins." *Evolution and Human Behavior* 36: 359–366.

Nettl, B. 2000. "An ethnomusicologist contemplates universals in musical sound and culture." In N. Wallin, B. Merker, & S. Brown (eds.), *The Origins of Music* (MIT Press) 463–472.

Odling-Smee, J., Laland, K., Feldman, M. 2003. *Niche Construction: The Neglected Process in Evolution* (Princeton University Press).

Parncutt, R. 2009. "Prenatal and infant conditioning, the mother schema, and the origins of music and religion." *Musicae Scientiae* 13 (suppl. vol.): 119–150.

Patel, A. 2010. "Music, biological evolution, and the brain." In M. Bailar, C. Field, & C. Henry (eds.), *Emerging Disciplines* (Rice University Press) 91–144.

Pinker, S. 1997. *How the Mind Works* (Allen Lane).

Savage, P., Matsumae, H., Oota, H., Stoneking, M., Currie, T., Tajima, A., Gillian, M., & Brown, S. 2015. "How 'circumpolar' is Ainu music? Musical and genetic perspectives on the history of the Japanese archipelago." *Ethnomusicology Forum* 24: 443–467.

Schellenberg, E. 2006. "Long-term positive associations between music lessons and IQ." *Journal of Educational Psychology* 98: 457–468.

Sloboda, J. 2005. *Exploring the Musical Mind* (Oxford University Press).

Spencer, H. 1857. "The origin and function of music." *Fraser's Magazine* 56: 396–408.

Sperber, D. 1996. *Explaining Culture: A Naturalistic Approach* (Blackwell).

Standley, J. 2003. "The effect of music-reinforced non-nutritive sucking on feeding rate of premature infants." *Journal of Pediatric Nursing* 18: 169–173.

Sterelny, K. 2011. "From hominins to humans: How *sapiens* became behaviourally modern." *Philosophical Transactions of the Royal Society, Series B: Biological Sciences* 366: 809–822.

Sterelny, K. & Griffiths, P. 1999. *Sex and Death: An Introduction to Philosophy of Biology* (University of Chicago Press).

Thaut, M., McIntosh, G., & Rice, R. 1997. "Rhythmic facilitation of gait training in hemiparetic stroke rehabilitation." *Journal of Neurological Sciences* 151: 207–212.

Thompson, W., Schellenberg, E., & Husain, G. 2004. "Decoding speech prosody: Do music lessons help?" *Emotion* 4: 46–64.

Titon, J. (ed.). 1996. *Worlds of Music: An Introduction to the Music of the World's People* (Schirmer).

Trainor, L., Tsang, C., & Cheung, V. 2002. "Preference for sensory consonance in 2- and 4-month old infants." *Music Perception* 20: 187–194.

Trehub, S. 2003. "The developmental origins of musicality." *Nature Neuroscience* 6: 669–673.

Trehub, S. & Trainor, L. 1998. "Singing to infants: Lullabies and play songs." *Advances in Infancy Research* 12: 43–77.

Trestman, M. 2015. "Clever Hans, Alex the parrot, and Kanzi: What can exceptional animal learning teach us about human cognitive evolution?" *Biological Theory* 10: 86–99.

Waddington, C. [1957] 2014. *The Strategy of the Genes* (Routledge).

Waterman, R. 1956. "Music in Australian aboriginal culture—some sociological and psychological implications." In E. Gaston (ed.), *Music Therapy 1955* (Proceedings of the National Association for Music Therapy) 40–49.

Winkler, I., Háden, G., Ladinig, O., Sziller, I., & Honing, H. 2009. "Newborn infants detect the beat in music." *Proceedings of the National Academy of Sciences USA* 106: 1479–1489.

Wrangham, R. 2009. *Catching Fire: How Cooking Made us Human* (Profile).

28

Emotional Responses to Fiction: An Evolutionary Perspective

Helen De Cruz and Johan De Smedt

INTRODUCTION: THE PARADOX OF FICTION

Across cultures, humans create fictional worlds. Storytelling is a cross-cultural phenomenon, taking various forms, such as narrative dances that act out passages of the Rāmāyaṇa and Mahābhārata, Latin American telenovelas, recitations by West African *griots* (troubadours) accompanied by a *kora* (twenty-one-string lute), and intricate Russian novels. Narratives elicit emotions. Like with other artworks, some of these emotions are evaluative, directed at the artwork as artwork. We may find a story beautiful, intriguing, exhilarating, or merely bland, predictable, or boring. Other emotions are directed at elements within the narrative, such as empathy with its characters. Chekhov's play *Uncle Vanya* elicits empathy for Vanya and Sonya. When king Stannis Baratheon sacrificed his only daughter Shireen to ask for divine help in battle, viewers of the HBO show *Game of Thrones* watched in horror how the girl pleaded in vain to be spared.

There is an enduring discussion about the *paradox of fiction*. One version of this paradox concerns the possibility of emotions elicited by fiction. We know that Vanya and Sonya are not real persons, yet we feel empathy for them; we know that Shireen is not a real girl, yet we grieve for her. As Jerrold Levinson (1990: 79) summarizes the paradox: "Since fictional characters do not exist, and we know this, it seems we cannot, despite appearances, literally have towards them bona fide emotions—ones such as pity, love, or fear—since these presuppose belief in the existence of the appropriate objects." More formally, the possibility paradox of fiction holds that there is an inconsistency between the following three statements:

Possibility Paradox
PF1   We have genuine emotional responses to fiction.
PF2   We do not believe that the characters and situations in fiction exist.
PF3   We are genuinely moved only by things we believe exist.

Another version of the paradox of fiction, discussed by Colin Radford (1975), focuses on the rationality of emotional responses to fiction.

Rationality Paradox[1]
PF1*   Our emotional responses to fiction are sometimes rational.
PF2*   We do not believe that the characters and situations in fiction exist.
PF3*   Emotions are rational only when we believe their objects exist.

Radford's original conclusion was that these emotions are irrational, even if they feel natural and intelligible to us. In the extensive literature on the paradox of fiction, the majority of authors do not think that emotional responses to fiction are irrational, and also believe that emotional responses occur frequently (see Davies 2009 for review), so one or more of the central assumptions of the possibility paradox must be wrong. Kendall Walton (1990) denies PF1: emotions elicited by artworks are merely pretend emotions.[2] Fictions are props that people use in games of make-believe, similar to how children use props in pretend games, such as playing house. When watching the movie *Alien*, we do not genuinely fear the monstrous alien roaming the spaceship, nor are we genuinely concerned for the crew's safety. While this proposal resolves the paradox (since there are no bona fide emotions involved), it creates its own difficulties. As Noël Carroll (1991) has remarked, for children's games the prop itself does not matter a great deal when creating the make-believe. A patch of leaves can be a castle, a stick can be a cannon. But for fiction, the artwork does make a great difference in whether or not we are moved. Many horror movies do not elicit fear but ridicule. By contrast, if the emotions are genuine, we can easily explain why some horror movies work (because they genuinely horrify) and others do not (because they fail to move us in this way). We cannot at will turn off emotions experienced in response to fiction, something that is not explained by Walton's account.

The second type of response denies PF2: people temporarily suspend their disbelief in the fictional situation they are encountering (Hurka 2001). For the time being, they really believe that, say, Gollum desires the ring. This position seems implausible. Under controlled experimental conditions, fiction tends to elicit higher degrees of emotional transportation (feeling absorbed by the story) than non-fictional (newspaper) reports—and only transportation in fiction, but not in non-fiction, results in higher feelings of empathy (Bal & Veltkamp 2013). Melanie Green and Timothy Brock (2000) presented identical stories as either factual or purely fictitious. Respondents reported no difference in transportation between these conditions, yet one would expect that suspension of disbelief is easier with stories labeled as factual. Moreover, suspension of disbelief cannot explain why some delight in genres like tragedy or horror that aim to elicit negative emotions. Under experimental conditions, participants who watch a sad film enjoy themselves more when they are more transported by the story (Ahn et al. 2012). This correlation between enjoyment and transportation does not hold up when we learn about real-world sad events, where the opposite is the case.

Given that the first and second statements are plausibly true, most responses to the possibility paradox of fiction involve a denial of PF3, that is, they argue that existence beliefs are not necessary for emotional engagement with artworks. Peter Lamarque (1981), for instance, proposes that we can really be moved by artworks: we do not enter pretend worlds, as Walton argues, but rather, we let the artworks enter our world.

By considering fictional characters like Desdemona, or the nameless woman who is grieving for her dead child in Picasso's *Guernica*, we form thoughts; although such thoughts cannot be objects of pity (and other emotions), they "can be pitiful and can fill us with pity" (Lamarque 1981: 294). Lamarque's proposal solves the possibility paradox but not the rationality paradox: even if it can be shown *why* we have these responses, we have thereby not shown they are rational. Richard Joyce (2000) argues that seeking out movies, books, and other media that elicit emotions can be practically rational: when someone watches *Doctor Zhivago*, they plausibly know they will be sad, but if they watch the movie wanting to have this experience, believing it will serve their ends, they are rational for watching it.

Supposing the paradox of fiction can be resolved in a way that is philosophically satisfying, it remains "an interesting psychological question why certain kinds of affect persist after one learns of and forms the belief that objects of an emotional response are fictional" (Tullman & Buckwalter 2014: 794). Emotional engagement with fiction is indeed a peculiar feature of human cognition, which, given its cross-cultural ubiquity, requires an explanation in evolutionary terms. In this chapter, we will use the tools of evolutionary psychology, broadly construed, to explain emotional responses to fiction. In the next section, we look at some attempts to explain emotional responses in an evolutionary context. The following section proposes that fiction is a form of cognitive engineering. We propose two forms of successful cognitive engineering through fiction which both rely on emotions: the aims of 19th-century social reformers to change public opinion by making readers empathize with characters that were not normally part of their social circles (penultimate section), and the desires of readers and watchers of fiction to achieve transportation, a sense of being absorbed by a story, and in this way, to experience greater well-being and happiness (final section).

EXPLAINING EMOTIONAL RESPONSES TO FICTION IN AN EVOLUTIONARY CONTEXT

Many authors (e.g., Aristotle, Hume, James) have proposed comprehensive theories about what emotions are, and how they could be categorized (e.g., basic emotions, social emotions). There is no consensus about what emotions are—for instance, whether they are mainly consciously experienced physiological changes or whether they have also cognitive content—nor is there consensus about what affective states count as emotions. In spite of this lack of definitional clarity, emotions have become an important field of study. Most evolutionary psychologists in the broad sense understand emotions as relatively short-duration states that involve physiological factors, such as muscle tension and cardiovascular changes, facial expressions, as well as attention and higher cognition. Emotions are crucial for how people approach social relationships, and for long-term health and well-being.

Darwin (1872) was the first to propose an evolutionary, functional explanation for emotions: they help to prepare an animal for appropriate actions and to effectively communicate inner states, such as distress or anger, to others. Some evolutionary thinkers (e.g., Nesse 1990) argue that emotions are cognitive and physiological states that are shaped through natural selection. They contribute to fitness by helping an organism respond appropriately to threats and take advantage of opportunities. For example, anger directs blood away from internal organs toward arms and hands, and increases blood pressure, which is useful for direct confrontation in combat. Others (e.g., Keltner & Haidt 1999) have stressed the social effects of emotions: emotions, such as anger and embarrassment, influence other people; perceiving emotions plays a key role in social interactions, such as courtship and reconciliation. Displays of emotions, such as pride and embarrassment, help to negotiate group status and social roles. For example, embarrassment signals that someone is aware they have made a social gaffe that they are unlikely to repeat in the future, which may prompt others in the group to forgive them. Robert Frank (1988) has hypothesized that moral emotions serve as "commitment devices," compelling people to cooperate in social situations where there is a temptation to defect, and signaling to others that one can be trusted in doing so.

Emotions elicited by fiction, and by art more broadly, do not fit neatly in these evolutionary scenarios. The situations depicted do not really occur, so how could emotional responses to fiction be adaptive? This problem can be termed *the evolutionary paradox of fiction*, which consists of a tension between the following three claims:

EPF1   We have genuine emotional responses to fiction.
EPF2   Emotions are functional adaptations that help us respond to threats and opportunities in our environment which impact our fitness.
EPF3   Social and other situations depicted in fiction do not impact our fitness.

There have been several attempts to resolve this paradox. One is to deny EPF1—that is, to argue that there is a qualitative difference between emotions elicited by artworks and those evoked by analogous real situations. Thalia Goldstein (2009) noticed a large difference between imagined and real (remembered) experiences. Adults who compared their experience of a gloomy film to an unhappy personal event experienced similar levels of sadness, but felt considerably less anxious watching the movie compared to recollecting personal experiences. This does not make the emotional responses to fiction any less puzzling though, since levels of sadness were similar, which seems prima facie not an adaptive response to a fictional situation (why be sad if nothing sad really happened?).

Other authors qualify EPF2 by examining what happens when people emotionally respond to fiction. For example, Garry Young (2010) draws on theoretical work by Paul Ekman and Paul Griffiths to argue that some emotional responses are involuntary and unconscious, and as a result do not always agree with our consciously held beliefs. Such emotions are elicited when the appropriate stimulus is present, regardless of our higher-order beliefs. In this way, emotions are indeed functional adaptations that help us respond to our environment, but as they are triggered automatically and, in the case of fiction, inappropriately, they do not always impact fitness. Emotions are adaptive responses to type events (e.g., jealousy makes us vigilant for potential partner infidelity), but things can misfire in particular token events (e.g., Othello's wrongly believing his wife to be unfaithful). Perhaps emotional responses to fiction fall in this category of evolved emotions "misfiring" in particular token situations.

The most common way to resolve the evolutionary paradox of fiction is to deny EPF3. Michelle Scalise Sugiyama (2001) and John Tooby and Leda Cosmides (2001) have proposed some form of the following thesis: narratives provide a simulation of real-world experiences that allow listeners to engage in vicarious learning. As Raymond Mar and Keith Oatley (2008: 176) put it: "Works of imaginative literature—stories—are one means by which we make sense of our history and our current life and by which we make predictions and decisions regarding our future world." (See also Steen & Owens 2001 who connect fiction to mammalian adaptations for play and imagination; see further Bateson, Chapter 30 this volume.) Emotional responses help to make the learning experience more compelling. In this way, the emotions elicited through fiction are adaptive. Scalise Sugiyama (2001, 2011) compiled cross-cultural evidence showing that folktales in hunter-gatherer societies often contain foraging information; there may have been multiple selective pressures for the transfer of information between older and younger members of a group through means of narrative—children and adolescents can vicariously learn about subsistence strategies through stories (e.g., what to do if there is a drought).

Tooby and Cosmides (2001) also argue that fiction provides a risk-free environment where we can engage in vicarious learning. Confronting a lion in the real is a frightening experience; hearing a story about a predator (*Little Red Riding Hood*) or watching a predator in a movie (*Jaws*) is scary too, but we do not have the associated flight reactions. In this way, such stories are helping us assess the dangers of predators and offer appropriate responses. Tooby and Cosmides point to several features of storytelling, including its emotional involvement, as evidence that fiction is adaptive. The emotions elicited by fictional accounts augment our involvement with stories and increase our capacity to learn from them. Readers of Jane Austen's *Pride and Prejudice* feel relief when Elizabeth rejects Mr. Collins, who stands to inherit a fine estate but is a rather silly man (resources but no mate quality), and rejoice when she finally accepts the rich and high-principled Mr. Darcy, sentiments that support judicious partner choice (resources as well as mate quality). Stories may also help to instill correct social behaviors and moral values, as in the Hindu *Panchatantra* (compiled around the 3rd century BCE), animal fables that are meant to instill *nīti* (wise conduct). For example, in the story *Aparīkṣitakārakaṃ* ("Rash action"), a Brahmin leaves his infant son in the care of his mongoose friend. Upon his return, the child has vanished, and the mongoose has blood on its snout. The Brahmin immediately assumes the mongoose has savaged his son, and kills the animal. However, he later finds that it defended his infant, who is alive and well, from a snake attack. The vicarious remorse the reader or listener of this tale feels instills a state of mind that guards against drawing hasty conclusions and acting rashly upon them. Variations of this story are widespread; for example, the Welsh legend of Gelert the dog.

While there is significant corroboration for the claim that fiction supports vicarious learning, this proposed function seems quite narrow when we consider the breadth and scope of fiction. Next to learning, people also engage in fiction for other reasons. For example, most readers of young adult fiction (fiction ostensibly aimed at readers aged 12 to 17, including such titles as *The Hunger Games* and *Twilight*) are adults. Presumably, adults enjoy these novels, which typically have less complicated plotlines, lots of dialogue, and engaging and straightforward storylines and characters, as a form of escapism from the drudgeries of work, commuting, and childcare. There is undoubtedly some vicarious learning at work in reading young adult literature. For example, one consistent theme in the *Harry Potter* heptalogy is helping misfits. Loris Vezzali et al. (2015) found that children and adults who read these books exhibit less prejudice toward outgroup members (e.g., immigrants, gay people), an effect mediated by the extent to which they identify with its eponymous character. But engaging in young adult fiction, and the emotions it elicits, is more than vicarious learning—as we will argue, emotions in fiction can be marshaled for several functional goals, one of which is pure enjoyment.

EMOTIONAL RESPONSES TO FICTION AS A COGNITIVE TECHNOLOGY

To understand the role of emotions in fiction, it is important to look at the motivations of both authors and consumers (readers, listeners, watchers, etc.). Why do they produce or consume fiction; what, if anything, do they hope to accomplish with it? Is the ability of works of fiction to move their audience a design feature? If it is, we can conceptualize the ability of narratives to elicit emotions as a form of cognitive engineering or cognitive technology.

Technologies are ways in which organisms alter their environment. Pragmatic technologies meet pragmatic, practical ends, that is, transformations of the physical surroundings. Cognitive technologies are not specifically aimed at altering our physical surroundings (though they may sometimes result in this), but at transforming our cognitive environment, including its epistemic and affective properties. For example, number words form a cognitive technology. As our evolved capacities for number discrimination are limited to three or four, number words allow us to do something we would not be able to do otherwise; namely, denoting precisely the cardinality of larger collections (Frank et al. 2008). Or take the use of calendrical notation systems as a way to predict cyclically occurring events. Our evolved systems for keeping track of time are quite fine-grained in the short term, but it is hard to keep track of relatively rare events that occur cyclically over the long term, such as the timing of herd migrations, the spawning of fish, or the flowering of trees (De Smedt & De Cruz 2011). Calendars help to lift these cognitive limitations by keeping track of such cyclical events, which may explain their prevalence in human cultures since 30 kya, long predating the invention of writing.

The examples of number words and calendars indicate two important ways in which cognitive engineering is achieved: language and material culture. Language can provide a "handle" for attention, allowing us to focus on properties of the environment that would otherwise elude us (Jackendoff 1996). Material culture can help overcome cognitive constraints caused by limitations in memory (by providing external memory) and in conceptual stability (helping to stabilize concepts that would otherwise fluctuate, e.g., depictions of supernatural beings). Fiction not only uses the cognitive engineering potential of language, but also often of material culture. For example, in order to enhance the pageantry and immersiveness[3] of a play, actors often use costumes, masks, makeup, and decors.

Regarding fiction as a cognitive technology is consistent with a wide range of motivations for why people consume and produce stories, ranging from deriving unadulterated enjoyment to effecting societal change. We will discuss how fiction can be seen as a cognitive technology by focusing on two motivations for engaging in fiction. The first case study looks at fiction from the perspective of an author, showing how writers such as Charles Dickens and Elizabeth Gaskell used their novels to sway public opinion in the direction of social reforms. The second case study looks at why readers and watchers enjoy fiction that transports them into fictional narrative worlds.

EXPANDING EMPATHY AND PROSOCIAL BEHAVIOR
THROUGH THE SOCIAL NOVEL

Social novels (also known as "social protest novels") highlight problematic social circumstances, such as extreme poverty, slavery, or animal cruelty. By making their public aware of dire circumstances, they aim to marshal public opinion for social change. The genre particularly flourished during the 19th century. Dickens's novels *Bleak House* and *The Life and Adventures of Nicholas Nickleby*, which highlighted abject living conditions and lack of medical care for the poor, helped support numerous social reforms that led to better living conditions in the UK. Harriet Beecher Stowe's *Uncle Tom's Cabin* played a pivotal role in the abolitionist movement in the United States in the 1850s. Anna Sewell's *Black Beauty* has as narrator and protagonist a horse that goes from a relatively carefree life as a farm animal to the strenuous life of being a taxicab horse. Black Beauty is unabashedly anthropomorphic, yet the book is also rich in details about horse behavior and handling. As a direct result of the novel, bearing reins (which forced a horse's head in a constant high position that was regarded as aesthetically pleasing, but that was awkward and painful for the animal) were forbidden in Victorian England.

How do novels effect social change? Literature can be used as a cognitive technology to decrease the limitations of empathizing. Empathy is an important catalyst for prosocial behavior, but it is limited in scope. Humans find it easier to empathize with single identifiable individuals than with large numbers of nameless victims, and they are subject to similarity bias, being less able or willing to empathize with those who are different from themselves. In one study (Xu et al. 2009), Chinese and Caucasian college students witnessed people receiving a painful stimulation (a needle prick) or a neutral control stimulus (a touch with a Q-tip). Participants had to judge how painful the stimulus was while their brain was scanned using fMRI. The anterior cingulate cortex (ACC), which is involved in perceiving one's own pain as well as that of others, was more activated when participants saw the needle prick compared to the touch with the Q-tip. However, the ACC showed larger responses when seeing pain inflicted on people from the same racial group—this effect was similar in Chinese and Caucasian participants. In another study (Stürmer et al. 2006), German male Muslim and non-Muslim college students received a plea for help (to find short-term accommodation in town) from either a Muslim ("Mohammed") or a non-Muslim ("Markus") fellow student. In both groups, empathy predicted helping intentions only when the helpee was an ingroup member, but not when he was an outgroup member. Such findings can explain why people who are in a position to help fail to aid those who are part of marginalized outgroups.

Social novels reduce similarity bias by decreasing the perceived difference between ingroup and outgroup members through several narrative techniques. One particularly effective, yet simple technique is to cast an outgroup member as the protagonist. Drawing on empirical research on discourse processing, the literature scientist Mary-Catherine Harrison (2011) finds that people tend to empathize more with the protagonist of a story than with secondary characters. Novelists frequently use the technique of foregrounding, "a kind of privileged focus that establishes the status of a protagonist within a text" (Harrison 2011: 266), for instance, by devoting more text to describing the thoughts and actions of that character, or by granting more introspective access to her inner states.

As a result of this focus, readers can overcome the perceived otherness of outgroup members, and realize that they have a lot in common with characters who do not belong to their social circle. Authors do this by de-emphasizing differences, such as race, gender, or class, and by focusing on shared emotions, such as hopes and fears. For example, in Gaskell's *Mary Barton*, middle-class readers are invited to empathize with working-class characters through the highlighting of common concerns, such as dealing with unrequited love, losing a child, and choosing to marry for material convenience or for love.

TRANSPORTATION AND ENJOYMENT

When looking at fiction as a cognitive technology, we need not only consider the producers of fiction, but also its consumers. Reading or watching a story is not a passive affair, but an active, reconstructive process. As fantasy and science fiction author Lois McMaster Bujold puts it:

> The book, if you like, is not the story but merely the blueprint of the story, like the architect's drawing of a house. The reader, then, is the contractor, the guy who does the actual sweat-work of building the dwelling. From the materials in his or her head, the ideas, the images, the previous knowledge, each one actively reconstructs the story experience. (Bujold 1996: 176)

We will now briefly look at two motivations that readers have for engaging in fiction: experiencing different times, places, characters, and events, and escaping from their everyday lives. Both are achieved by transportation. "Transportation" is a metaphor coined by the cognitive scientist Richard Gerrig (1993) to describe the subjective sense of being absorbed and immersed in a story. Transportation into a narrative results in reduced attention for one's surroundings and a diminished focus on oneself, and greater attention for the narrative. It involves a complex amalgam of emotions, mental imagery, and attributions of mental states (Green & Brock 2000). There is convergent evidence (see Green et al. 2004 for a review) that transportation contributes to enjoyment, which is an important motivation for readers and viewers to engage in fiction. Negative reviews of books on Goodreads frequently bemoan a lack of transportation ("I just couldn't get into the story"), whereas positive reviews hail it ("The book gripped me from the beginning; I couldn't put it down"). If transportation is indeed a desirable state, and if fiction can help to accomplish it, two questions arise. How does fiction result in transportation, and why does transportation contribute to happiness and well-being?

Narratives create a sense of transportation by keying in on our evolved ability for self-projection. Neuroimaging studies have identified a common neural network (including medial prefrontal, medial temporal, and medial and lateral parietal regions) involved in retrieving personal memories, predicting personal future events, attributing mental states (theory of mind), and navigation (Buckner & Carroll 2007). These cognitive faculties are usually studied separately, but a meta-analysis (Spreng et al. 2009) revealed that there is an extensive functional overlap between them. Intriguingly, the same network is also active when participants are in a conscious resting state, the so-called *default mode network* (DMN) (Buckner et al. 2008). Randy Buckner and Daniel Carroll (2007) propose that all the tasks carried out by the DMN require some form of self-projection. The self-projection theory provides a unified account for why an integrated functional network can perform such seemingly diverse tasks as remembering, predicting, navigating, and attributing mental states. When we remember an event in our personal past, we place ourselves in that situation and reimagine visual, tactile, olfactory, and other features of the event. For example, when Marcel Proust (1913) ate a madeleine cake dipped in tea, it brought a host of childhood memories back to mind. When we think about our personal future, we project ourselves in a future state. When we imagine a hypothetical situation, such as visiting a museum, we project ourselves spatially, emotionally, temporally. A study with patients who were unable to recall personal memories found that they were also unable to project themselves in hypothetical situations, such as lying on a tropical beach (Hassabis et al. 2007).

Several neuroimaging studies indicate that narrative comprehension, such as reading Aesop's fables (Xu et al. 2005) or nursery rhymes or vignettes made up by researchers (AbdulSabur et al. 2014), engage the DMN. Diana Tamir et al. (2016) looked specifically at the components of fiction that contribute to activity in the DMN. They found two types of features that increased activity in brain areas that are part of the DMN: vivid descriptions of scenery led to increased activation compared to generic texts, especially in the medial temporal lobe subnetwork, and narratives that described social interactions resulted in greater activity in the dorsomedial prefrontal cortex. Additionally, several studies indicate a deep and sustained involvement of theory of mind in narrative processing; for example, performance of theory of mind tasks is improved in frequent readers. Mar et al. (2006) found a positive correlation between being a long-time reader of fiction, especially someone who is often transported into stories, and social acumen and empathy. It is difficult to tease apart cause and effect in this study—maybe people with a keen sense of social interactions are more drawn to literature. Under more controlled conditions, Jessica Black and Jennifer Barnes (2015) found that after reading fiction participants perform better on theory of mind tasks, but not on intuitive physics tasks, compared to participants who read a non-fiction piece of similar length.[4]

Taken together, these results support our following tentative hypothesis: the experience of being transported into a story is a result of an increased engagement of the DMN, which is otherwise mainly engaged in everyday self-projection activities, which happen spontaneously when one is not overtly focused on the external world. As originally coined by Gerrig, "transportation" was a spatial metaphor, pointing to fiction's ability to mentally place readers in a different location. But if the activity of the DMN results in the phenomenological sense of transportation, it not only involves a spatial component but also emotional and other elements. While better theory of mind comprehension is a salutary effect of reading fiction, it is not the main motivation for consumers of fiction. Instead, they enjoy transportation, and read or watch in order to be transported, even for a short while, in another world, inhabited with fictional characters—as one respondent to a Pew Forum survey[5] on reading put it: "being able to experience so many times, places, and events."

A sustained engagement of the DMN contributes to feelings of well-being and happiness as it counters rumination and other forms of self-reflection, which are also subserved by the DMN. Several studies (see Mor & Winquist 2002 for a meta-analysis) show an overall negative effect of self-directed thought on well-being: reflecting on one's past, future, or things one could have done differently (counterfactual thinking) on the whole results in lower happiness and increased anxiety, the only exception being when one thinks of oneself in a very positive light, following a positive life event (e.g., a promotion at work). On the whole, self-directed thought includes a lot of negative elements, which even deliberate attempts to think positively about oneself cannot completely avoid (Nolen-Hoeksema et al. 2008). Transportation can enhance well-being by directing the functional activities of the DMN to the fictional world and its characters, away from one's own situation and life. When we are absorbed in thinking counterfactually and theorize about fictional characters (What if Emma Bovary hadn't married her boring country doctor?), we cannot at the same time ponder our own situation and life (What if *I* didn't have this boring job?). While we are consciously aware of the fact we are reading a novel or watching a movie, the DMN is not aware of this, and is instead playing out what we are reading or watching. In this way, transportation can make consumers of fiction happier, as their attention is drawn away from negative self-directed ruminations. This explanation of the functional role of transportation also clarifies why emotions elicited through stories, including negative ones, may contribute to a positive evaluation of artworks. Since Aristotle, philosophers have wondered why tragedies are enjoyable to watch, in spite of their sad situations and the feelings of empathy one has for the unfortunate characters. Fiction that elicits deeper emotions achieves higher degrees of transportation, regardless of whether such emotions are negative or positive. This may explain the observation of Dohyun Ahn et al. (2012) that watchers of sad movies feel more enjoyment when they are more transported in a narrative.

CONCLUSION

The paradox of fiction queries why people are emotionally moved by fictional situations and characters, while they know these are imaginary. Some versions of the paradox (notably Radford 1975) ask whether being emotionally moved by fiction can ever be rational. In this chapter, we have examined the paradox of fiction from an evolutionary point of view. Emotions elicited by fiction do not neatly fit evolutionary explanations of emotional responses. It would seem that fictional situations do not directly impact our fitness, so how can emotional responses to fiction be adaptive? In this chapter we have argued against this supposition (i.e., we reject the claim that emotional responses to fiction do not impact fitness), but argue that such responses are deliberately sought or engineered by producers and consumers of fiction. Fiction can be regarded as a cognitive technology that engenders emotions that are pursued by the readers or watchers. Some authors use narrative techniques to elicit empathy in order to transform the attitudes of readers to outgroup members. Some readers use fiction to achieve transportation, which helps them to be mentally situated in a different realm, or to escape their everyday existence and associated ruminations.

NOTES

1. We thank Richard Joyce for this formulation of the paradox.
2. Walton's response also diffuses the paradox as it pertains to rationality, since by denying that we have genuine emotional responses to fiction, he in effect denies PF1*.

3. "Immersiveness" is a term from game design, meaning more than immersion; it refers to an interactive, multisensory experience of the narrative.

4. This latter study is a replication and extension of Kidd & Castano (2013), which reported improved performance in theory of mind tasks when reading fiction. However, it also proposed that so-called high literature is superior in improving theory of mind performance compared to popular fiction. This part of the study could not be replicated.

5. www.pewinternet.org/2012/04/05/why-people-like-to-read/

REFERENCES

AbdulSabur, N., Xu, Y., Liu, S., Chow, H., Baxter M., Carson J., & Braun, A. 2014. "Neural correlates and network connectivity underlying narrative production and comprehension: A combined fMRI and PET study." *Cortex* 57: 107–127.

Ahn, D., Jin, S., & Ritterfeld, U. 2012. "Sad movies don't always make me cry." *Journal of Media Psychology* 24: 9–18.

Bal, P. & Veltkamp, M. 2013. "How does fiction reading influence empathy? An experimental investigation on the role of emotional transportation." *PLoS ONE* 8: e55341.

Black, J. & Barnes, J. 2015. "The effects of reading material on social and non-social cognition." *Poetics* 52: 32–43.

Buckner, R. & Carroll, D. 2007. "Self-projection and the brain." *Trends in Cognitive Sciences* 11: 49–57.

Buckner, R., Andrews-Hanna, J., & Schacter, D. 2008. "The brain's default network." *Annals of the New York Academy of Sciences* 1124: 1–38.

Bujold, L. 1996. "The unsung collaborator." In S. Lewis (ed.), *Dreamweaver's Dilemma* (NESFA Press) 175–179.

Carroll, N. 1991. "On Kendall Walton's *Mimesis as Make-believe.*" *Philosophy and Phenomenological Research* 51: 383–387.

Darwin, C. 1872. *The Expression of the Emotions in Man and Animals* (John Murray).

Davies, S. 2009. "Responding emotionally to fictions." *Journal of Aesthetics and Art Criticism* 67: 269–284.

De Smedt, J. & De Cruz, H. 2011. "The role of material culture in human time representation: Calendrical systems as extensions of mental time travel." *Adaptive Behavior* 19: 63–76.

Frank, M., Everett, D., Fedorenko, E., & Gibson, E. 2008. "Number as a cognitive technology: Evidence from Pirahã language and cognition." *Cognition* 108: 819–824.

Frank, R. 1988. *Passions within Reason: The Strategic Role of the Emotions* (W.W. Norton).

Gerrig, R. 1993. *Experiencing Narrative Worlds: On the Psychological Activities of Reading* (Yale University Press).

Goldstein, T. 2009. "The pleasure of unadulterated sadness: Experiencing sorrow in fiction, nonfiction, and 'in person.'" *Psychology of Aesthetics, Creativity, and the Arts* 3: 232–237.

Green, M. & Brock, T. 2000. "The role of transportation in the persuasiveness of public narratives." *Journal of Personality and Social Psychology* 79: 701–721.

Green, M., Brock, T., & Kaufman, G. 2004. "Understanding media enjoyment: The role of transportation into narrative worlds." *Communication Theory* 14: 311–327.

Harrison, M.-C. 2011. "How narrative relationships overcome empathic bias: Elizabeth Gaskell's empathy across social difference." *Poetics Today* 32: 255–288.

Hassabis, D., Kumaran, D., Vann, S., & Maguire, E. 2007. "Patients with hippocampal amnesia cannot imagine new experiences." *Proceedings of the National Academy of Sciences USA* 104: 1726–1731.

Hurka, T. 2001. *Virtue, Vice, and Value* (Oxford University Press).

Jackendoff, R. 1996. "How language helps us think." *Pragmatics and Cognition* 4: 1–34.

Joyce, R. 2000. "Rational fear of monsters." *British Journal of Aesthetics* 40: 209–224.

Keltner, D. & Haidt, J. 1999. "Social functions of emotions at four levels of analysis." *Cognition and Emotion* 13: 505–521.

Kidd, D. & Castano, E. 2013. "Reading literary fiction improves theory of mind." *Science* 342: 377–380.

Lamarque, P. 1981. "How can we fear and pity fictions?" *British Journal of Aesthetics* 21: 291–304.

Levinson, J. 1990. "The place of real emotion in response to fiction." *Journal of Aesthetics and Art Criticism* 48: 79–80.

Mar, R. & Oatley, K. 2008. "The function of fiction is the abstraction and simulation of social experience." *Perspectives on Psychological Science* 3: 173–192.

Mar, R., Oatley, K., Hirsh, J., dela Paz, J., & Peterson, J. 2006. "Bookworms versus nerds: Exposure to fiction versus non-fiction, divergent associations with social ability, and the simulation of fictional social worlds." *Journal of Research in Personality* 40: 694–712.

Mor, N. & Winquist, J. 2002. "Self-focused attention and negative affect: A meta-analysis." *Psychological Bulletin* 128: 638–662.

Nesse, R. 1990. "Evolutionary explanations of emotions." *Human Nature* 1: 261–289.

Nolen-Hoeksema, S., Wisco, B., & Lyubomirsky, S. 2008. "Rethinking rumination." *Perspectives on Psychological Science* 3: 400–424.

Proust, M. 1913. *Du Côté de Chez Swann* (Bernard Grasset).

Radford, C. 1975. "How can we be moved by the fate of Anna Karenina?" *Proceedings of the Aristotelian Society* 49: 67–80.

Scalise Sugiyama, M. 2001. "Food, foragers, and folklore: The role of narrative in human subsistence." *Evolution and Human Behavior* 22: 221–240.

Scalise Sugiyama, M. 2011. "The forager oral tradition and the evolution of prolonged juvenility." *Frontiers in Psychology* 2: 133.

Spreng, R., Mar, R., & Kim, A. 2009. "The common neural basis of autobiographical memory, prospection, navigation, theory of mind, and the default mode: A quantitative meta-analysis." *Journal of Cognitive Neuroscience* 21: 489–510.

Steen, F. & Owens, S. 2001. "Evolution's pedagogy: An adaptationist model of pretense and entertainment." *Journal of Cognition and Culture* 1: 289–321.

Stürmer, S., Snyder, M., Kropp, A., & Siem, B. 2006. "Empathy-motivated helping: The moderating role of group membership." *Personality and Social Psychology Bulletin* 32: 943–956.

Tamir, D., Bricker, A., Dodell-Feder, D., & Mitchell, J. 2016. "Reading fiction and reading minds: The role of simulation in the default network." *Social Cognitive and Affective Neuroscience* 11: 215–224.

Tooby, J. & Cosmides, L. 2001. "Does beauty build adapted minds? Toward an evolutionary theory of aesthetics, fiction, and the arts." *SubStance* 30: 6–27.

Tullmann, K. & Buckwalter, W. 2014. "Does the paradox of fiction exist?" *Erkenntnis* 79: 779–796.

Vezzali, L., Stathi, S., Giovannini, D., Capozza, D., & Trifiletti, E. 2015. "The greatest magic of Harry Potter: Reducing prejudice." *Journal of Applied Social Psychology* 45: 105–121.

Walton, K. 1990. *Mimesis as Make-believe. On the Foundations of the Representational Arts* (Harvard University Press).

Xu, J., Kemeny, S., Park, G., Frattali, C., & Braun, A. 2005. "Language in context: Emergent features of word, sentence, and narrative comprehension." *NeuroImage* 25: 1002–1015.

Xu, X., Zuo, X., Wang, X., & Han, S. 2009. "Do you feel my pain? Racial group membership modulates empathic neural responses." *Journal of Neuroscience* 29: 8525–8529.

Young, G. 2010. "Virtually real emotions and the paradox of fiction: Implications for the use of virtual environments in psychological research." *Philosophical Psychology* 23: 1–21.

29

Evolution and Literature: Theory and Example

Brian Boyd

Can and should evolution inform the study of literature? Evolutionary theory, like philosophy, prompts us to step outside what we take for granted and look at it afresh. It can raise new questions and deepen old answers.

No one would doubt that literary studies should investigate the responses that writers like Samuel Butler, H.G. Wells, George Bernard Shaw, A.S. Byatt, and Margaret Atwood have made to evolutionary themes. But exploring evolutionary perspectives in literature uninfluenced by the theory of evolution has seemed troubling to many.

Obviously, literary creation and comprehension do not require an awareness of evolution: Homer and Kālidāsa composed classic stories, and Aristotle and Abhinavagupta classic criticism, millennia before Darwin. But although we *can* engage richly in literature without knowing about evolution, this does not mean we cannot engage more richly still if we bring evolution to bear.

Many object, nevertheless: What can evolutionary theory offer literary inquiry when it deals with species, with human universals, not with what is unique to particular periods, traditions, contexts, and texts? (See Deresiewicz 2009.) Is not literature a matter of culture rather than biology? Doesn't evolutionary literary criticism ignore language and style? (See Kramnick 2011; for a reply, see Boyd 2012c.) Doesn't evolutionary theory presuppose ruthless competition?

These questions reflect prejudice rather than knowledge. Evolution often deals with "universals" (species-wide or higher-level phenomena), including the human propensity for fiction, but it also studies individual variation, without which evolution could not take place. Behavioral ecology analyzes the effects of local conditions on bodies, behaviors, and societies. Behavioral genetics tests individual variation. Evolutionary personality psychology and evolutionary social psychology investigate, respectively, individuals and groups. "Evo-devo" studies the complex relations between genes and their environment's modification of the way they are expressed. Evolutionary cultural theory explores culture in other animals, its preconditions and operations in humans, and the emergence of cumulative culture as one of the handful of Major Transitions in evolutionary biology (Richerson & Boyd 2005; Sterelny 2012; Henrich 2015). In Darwin's own work and in the last forty years of evolutionary biology and evolutionary economics, evolutionary theory has had a central preoccupation with explaining and even encouraging the evolution of cooperation (Axelrod 1990; Singer 1999; Fehr & Henrich 2003; Wilson 2011, 2015).

Evolutionary considerations can make readers more sensitive not only to local contexts but also to individual and intra-individual factors, to the artistic problems writers pose themselves in particular works at particular moments of their career, and to their stylistic means of earning and holding the attention of audiences (Boyd 2009, 2010, 2012b). But evolutionary factors can also explain why we can be deeply moved by literature of other times or places, in ways that would seem to be impossible in view of the insistence, within post-structuralism and since, only on historical and local difference: the emphasis on *différance* in Derrida, on each distinct *episteme* in Foucault, on the supposed emergence of unprecedented new attitudes to, say, childhood, or madness, or laughter, or this or that emotion, in the 18th century or whatever period a New Historicist or a historian of the emotions has focused on (see Boyd 2010b for a critique).

PHILOSOPHY, THEORY, AND EMPIRICAL RESEARCH

Philosophy recognizes that we always act on theories, whether rudimentary and implicit or elaborate and explicit, and that we should test them if we can rather than leave them unacknowledged and untested. But "Theory," a smorgasbord of continental philosophies from Derrida to Žižek that was dominant in literary studies for the last decades of the 20th century, has always been, despite its name, "decidedly anti-theoretical" (Mulligan et al. 2006: 63), and psychoanalysis, its most common fellow-traveler, has always both lacked empirical support and been hostile to empirical testing (Grünbaum 1984, 1993). For many in literary studies, Theory merely wore itself out as its literary applications became more and more repetitive.[1] In reaction, some turned to empirical work in archives; others drew on empirical work in related fields, especially the human sciences. But such work often suffers from unacknowledged theoretical assumptions, as in the archival research that often subscribes more or less uncritically to the assumption that only historical difference matters, an assumption that evolutionary theory can cogently critique (Boyd et al. 2010: 2).

Or literary theory drawing on empirical results involves applying them without testing against broader theory. Philosopher of art Greg Currie, for instance, has argued from findings in social psychology that our sense of character in literature must be a false construct, because social psychology experiments show that our choices can be conditioned by circumstances in ways we do not recognize: circumstance rather than "character" shapes what we do. Currie admits, however, to being "very prone to draw conclusions about people's Characters" (Currie 2010: 215)—and indeed how could we (or Prince Hal) not discriminate between and respond differently to the personalities of a Hotspur or a Falstaff? Had Currie kept in mind evolutionary theory, and the range of its branching manifestations, he could not have argued himself from a single set of results into dismissing his intuitive responses. Other results from evolutionary personality psychology, behavioral genetics, ethology, and even animal breeding show both the presence of character differences and their importance for individuals and those interacting with them. (Even woodlice have recently been discovered to have personality differences: Tuf et al. 2015.) Many social animals assess character in their conspecifics, and relate to them according to their assessments, which coincide with objective measures taken by ethologists. The evidence of personality psychology is that we form trait impressions rapidly, they tend to remain durable, and they tend to be accurate. Evolutionary considerations suggest that if we were not "very prone to draw conclusions about people's Characters" we would very rapidly find ourselves socially bewildered and at risk of exploitation.

Unlike Theory, psychoanalysis, or isolated empirical results, evolutionary studies offer a comprehensive, fertile, intensely self-critical, and empirically testable set of theories for explaining life, and therefore sociality, which exists only within living things, and culture, which exists only in social animals (J. Carroll 2004, 2011; Boyd 2009; Wilson 2016). It has much to offer literary study.

EXTRA TOOLS IN THE EVOLUTIONARY KIT

Evolution by natural selection has three criteria: the variation of phenotypes (bodies and behaviors), the differential fitness of the variant forms, and the heritability of the variations (Lewontin 1970). Apart from the theory of evolution by natural selection and its many extensions within biology and beyond (into anthropology, economics, history, psychology, and sociology), I suggest three facets of evolutionary theory of particular power and relevance for literature: continuity as well as change; problem-solving; and costs and benefits, especially in terms of attention.

The first facet, continuity, suggests that what we do now is likely to have been shaped importantly by our deep past. Narrative comprehension, for instance, draws on our capacity to understand events through our major senses, sight and hearing, and thereby extends capacities common to other animals in understanding events in *their* worlds. There will also be discontinuities, of course, but discontinuities threaded through with continuities. We humans are probably unique in using language to deepen our understanding of events, and to *share* events, observed or merely imagined. But that capacity for language has itself been developing for many thousands of generations, and communication of varying degrees of richness exists in organisms from microbes and plants to ants, parrots, and dolphins. Deep continuity makes it as unlikely that our repertoire of emotions has radically changed in short periods of historical time as that we cannot understand and empathize with the anger of Achilles, the sadness of Genji, or the mixed mirth of Rosalind.

Continuity also makes an evolutionary perspective likely to connect literature strongly with other capacities, dispositions, and behaviors: with the capacity to understand events in other species, as mentioned; with dispositions to play, laughter, and social and emotional attunement; with other arts, in different modes, like music, dance, and the visual arts; with other forms of expression, like orature and everyday speech, gossip, and jokes; with other forms of narrative, like film and television, comics and video games; and with other developmental stages, like the pretend play of children. Connecting literature with other aspects of human lives does not *require* an evolutionary perspective (such connections have been brilliantly analyzed within the philosophy of art by Noël Carroll, for instance),[2] but the principle becomes all the more central when we take evolution into account.

To select one example of continuity applied to literary theory: many literary theorists suppose all narrative must have a narrator somehow distinct from the author of the story, not only in the case of a dramatized narrator like Pip or David Copperfield, but even in the case of, say, *Hard Times* or *Our Mutual Friend*, which would seem to be in Dickens's own voice. But when a friend relates gossip or tells a joke we do not posit a narrator between the friend and ourselves. The facet of continuity suggests the relevance of this ordinary narrative to literary narrative, and that positing a narrator anywhere where an author has not sought to create one would appear quite superfluous. It therefore adds to the other telling arguments (Stecker 1987; Walsh 1997; Kania 2005; Patron 2013; Currie 2010) against literary studies' pervasive assumption—its near-dogma—that we cannot have a story without a narrator distinct from the author.[3]

A second facet of evolutionary theory that allows tight purchase on both universals and particulars is that of problems and solutions. Problems arise only with life, and all life is problem-solving (Popper 1999; Boyd 2008). Biological features—not only in bodies, but also in brains and behaviors—become evolutionary adaptations because they solve problems better than the other available alternatives that have arisen within that lineage. Among the problems that arise for humans is how to make best use of the time left over from coping with survival. When we started to become major predators, we might have solved the problem of what to do with our surplus time in the way lions, tigers, and bears have: by sleeping and conserving energy. Instead, our most successful ancestors fared better by using their spare time to exercise the minds and the sociality that offered them their competitive edge. Outside what we need to do for immediate survival, therefore, we continue to develop our sociality, intelligence, and imagination through our attunement to and interest in each other, through our curiosity and creativity, through socializing, play, learning, and art.

Animals can successfully interact with others by understanding their actions through intuiting their goals and intentions: roughly, across species; more finely tuned, within social species; very finely tuned indeed, within ultrasocial and linguistically endowed humans. We infer the goals and intentions of others, including their verbal intentions, by intuiting the problems they are trying to solve.

At the end of Shakespeare's *All's Well That Ends Well*, Helen has solved a whole succession of problems standing in the way of her deep but unrequited love for Bertram. She has returned before the King and Bertram, having satisfied Bertram's seemingly impossible conditions for fulfilling the marriage to her that the King has forced on him. Yet she now declares herself "but the shadow of a wife . . . The name and not the thing." "Both, both, O, pardon!" responds Bertram. Are his words his attempt to make up for his past disdain for Helen, and his belated recognition of the force of her character and of her desire for him? Or are they merely his attempt to wriggle out of the public humiliation involved in the backstory of how his waywardness and dishonor led to his being duped into unwittingly allowing Helen to fulfill the conditions he had set down? His bare words are not enough to answer these questions. And are Shakespeare's intentions in setting up this situation, and in having Bertram voice those words, to resolve the play happily despite Bertram's former misdemeanors, or to undermine the usual promise of the happy ending? In order to solve such interpretive problems, we ourselves need to understand both the problems of the characters and the problems of the author, as we infer them from all their relevant actions and verbal choices. In that sense, problems of literary interpretation are deeply continuous with the problems we face in understanding one another as social beings.

A third facet of evolutionary theory, following closely on that of problems and solutions, and also allowing purchase on both large-scale evolutionary trends and on particular moments, is that of costs and benefits (Boyd 2009). All biological processes, bodily and behavioral, exact costs in energy, resources, and time. If they do not on average offer benefits that outweigh the costs, nature will select against them: if the time, energy, and resources spent in constructing mounds, hives, or dams did not earn sufficient rewards for termites, bees, or beavers, respectively, the disposition to build them would have been bred out. Determining costs against benefits in different environments or economies proves a demanding task for behavioral ecologists, but nature has evolved efficient ways for organisms to track information and react to it in ways that raise benefits and reduce costs: in plants, through more or less automatic but context-sensitive responses to changing heat, light, and nutrients; in simpler animals, through more or less fixed but context-sensitive motivations; and in higher animals also through highly context-sensitive emotions.

A major cost in higher animals is attention (Boyd 2009). Neural space is always expensive and remains limited even in big-brained humans: one sensory input must compete against a myriad other such inputs and many other neural processes. Working memory is still more tightly limited, even in humans, to around only four chunks of information, according to the best current estimates (Bor 2012). Attention inhibits irrelevant information and amplifies cognitive resources dedicated to task-relevant information in the global workspace of consciousness. But attending to one thing distracts from other things our minds may need or wish to focus on. Competition for attention within brains is fierce.

So is competition for attention within groups. If we can earn positive attention, we gain a chance to influence others within our group. We therefore compete for attention and influence, and we resist attempts by others to usurp our attention in ways we see offer us little benefit.

Art depends on its capacity to earn attention, to persuade potential audiences it is in their interest to attend. A particular work of art has to compete against other works of its kind, against other modes of art, against other possible uses of down time, like conversation or play. Like all optional activities it usually follows only after necessities have been attended to: in early human conditions, after time for hunting, gathering, cooking, tool-making, and in time not interrupted by the need for vigilance toward predators, enemies, or rivals. During the day, Ju/'hoansi bushmen report on pragmatic activities and gossip to regulate social relations, but at night around the campfire tell stories for entertainment (Wiessner 2014). Similar patterns tend to prevail from fireside or bedtime stories to modern theaters, novels, and film, radio, and television drama.

Philosopher Stephen Davies notes, as a criticism of the role of attention in art: "In many contexts, reciting Shakespearean sonnets, painting pictures, or singing the latest hit song would attract attention only because of their social inappropriateness" (Davies 2012: 127). Indeed, but who has argued that the attention art can earn overrides all other considerations? Evolution attunes organisms to respond in context-sensitive ways: humans do not paint while shopping or perform plays while hunting or gathering any more than trees bud in autumn. We know when we can safely and fruitfully attend to literature when the anticipated benefits of doing so are likely to exceed the costs.

Davies also observes that "there are many ways of attracting attention to oneself. . . . One way is by approaching strangers, introducing oneself, and offering a hand as a preliminary to a conversational exchange" (Davies 2012: 127). Apart from the fact that we resist having our attention commandeered by strangers with nothing to offer us except a drain on our time—we have a shrewd emotional appraisal of costs and benefits—a Shakespeare, a Mozart, or a Hokusai could hardly shake hands with the hundreds of millions whose attention he had earned, held, and rewarded.

Art can earn attention in uniquely efficient and contagious ways. Storytellers therefore compete to hold the floor or to fill bookshelves. They need to assess audiences' readiness to attend, and to adjust their stories to audience composition, circumstances, and expectations. They will seek to reduce their own cognitive effort in relating or inventing stories by adopting prior models (in *All's Well*, the Boccaccio story of Giuletta of Narbonne, the genre of tragicomedy, the device of the double plot) while injecting attention-worthy novelty (in *All's Well*, a thousand adjustments of structure, character, and emphasis, including making an ending which is shallowly resolved in Boccaccio instead deeply and fascinatingly problematic). Audiences will seek to reduce search costs for new reading by choosing from their favored genres or favored authors, or by following reviews or recommendations, or classic status, bestseller lists, or Twitter feeds. Once engaged in reading they will seek to reduce comprehension costs by understanding works in terms of known models, while rejecting stale or redundant information, costs with insufficient benefit.

An evolutionary approach to literary art that focuses on the costs and benefits for artists of earning the attention of audiences, within the current conditions of their art, and on the cost and benefits for audiences in choosing and attending to work that appeals, can offer a more fine-tuned account of literary style and effect than even a traditional close reading, and one that integrates art, artists, works, and details within a larger framework.

QUESTIONS AND EXAMPLES

Philosophy, like literary theory, including evolutionary literary theory, usually asks general questions. Questions that come naturally to philosophers and theorists of literature include: How do we distinguish literature from other uses of language or story? Is it a human universal? Why do we engage in it so much, and esteem it so much? How do we explain the processes of composition and comprehension, of intention, meaning, interpretation, response, and evaluation? Does literature yield truth or only other kinds of value? How do we explain literary continuities, traditions, and innovations? And in considering the role evolutionary perspectives may have to offer: Can literature as an activity be explained better, can questions of these kinds be answered better, by incorporating evolutionary considerations? Can specific literary works be more fruitfully discussed by incorporating or by excluding evolution?

Most literary criticism, including evolutionary literary criticism, tends to prefer particular examples to general questions: the work as the equivalent of the sample, the test case, and the experiment. To satisfy this disciplinary drive, and to answer those who think evolutionary considerations preclude language and particulars, let me choose one compact literary text, a 2005 poem by British poet Carol Ann Duffy, as a focus for asking the kinds of questions the philosophy of literature and literary theory ask, and as a way of suggesting how some of the emphases of evolutionary literary criticism and theory might differ subtly from those of non-evolutionary modes.

Cuba

No getting up from the bed in this grand hotel
and getting dressed, like a work of art
rubbing itself out. No lifting the red rose
from the room service tray when you leave,
as though you might walk to the lip of a grave
and toss it down. No glass of champagne, left
to go flat in the glow of a bedside lamp,
the frantic bubbles swimming for the light. No white towel,
strewn, like a shroud, on the bathroom floor.
No brief steam on the mirror there for a finger
to smudge in a heart, an arrow, a name. No soft soap
rubbed between four hands. No flannel. No future plans.
No black cab, sad hearse, on the rank. No queue there.
No getting away from this. No goodnight kiss. No Cuba.

(Duffy 2005: 19)

Now to our questions. First:

How do we distinguish literature from other uses of language and story?

I suggest that keeping evolutionary continuity in mind can highlight the closeness of literature, as of other arts, to the widespread behavior of play (Boyd 2009). Play allows serious, pragmatic modes of action to be decoupled from their usual consequences and engaged in intense, exploratory ways (see Bateson, Chapter 30 this volume). Play can be hard, fierce, testing. So can literature. Literature incorporates language and story in play mode, not tied to necessity, but in free exploration of possibility. The more engaged in the play and the more intense the exploration, the more centrally it is literature.

Human play no doubt began to take on narrative and poetic dimensions long before there were discrete stories and poems. Play becomes ritualized: like a dog's play-bow or a story's "Once upon a time," poetry's fenced-off lines signal a special range of play. The poet focuses the attention of audiences a line at a time, for roughly the duration that we can hold sounds in working memory, and usually repays that tightly focused attention with the play of heightened pattern, including rhythm, rhyme, imagery, and syntactic parallelism (Boyd 2012b: 13–23; for the uniquely intense human appetite for pattern, see Bor 2012). And just as ordinary chase- or fight-play allows skills developed there to be deployed in other, serious modes, so literature allows skills of language and story to be used in other modes, for reflection, instruction, or persuasion.

Play may be decoupled from serious behavior, but it need not be relaxed or light-hearted: it can be focused, absorbing, with everything at stake so long as we choose to remain involved. In "Cuba," Duffy plays grimly with pattern and possibility in language and story: thirteen "sentences" in fourteen lines, all starting "No," all lacking a main verb. She tells a story—or she invites us to infer it—without affirmation or action, with only negation and deactivation, the affirmations of love and anticipation all negated in rejection and renunciation.

Is literature a human universal?

Evolution of course takes a particular interest in species-typical traits, and evolutionary anthropology finds indeed that literature is a human universal: all human groups have fiction and poetry (Brown 1991). After all, play is universal across not only humans, but also apparently all mammals and at least many other vertebrates and even the most intelligent invertebrates (Burghardt 2005; Graham & Burghardt 2010). But only humans have language as a main tool for thought and narrative as a main mode of understanding, especially of the volatile social world.

This does not mean that any particular work or mode will be universal. But music and visual arts can often appeal around the world, and stories from *The Arabian Nights* to *Star Wars* easily cross language barriers. To appreciate "Cuba" one needs not only English or excellent translation, but background knowledge of grand hotels and other features of modern life. The more we know of modern verse conventions, too, the easier we can respond. On the other hand, lyrics have been with us for thousands of years, in many cultures, and the concentration of verse has often made poems riddling, across millennia and languages. And disappointment in love has been and remains an endless subject for literature and song.

Why do we engage in literature so much, and in the ways we do, and value it so much?

One answer evolutionary literary and art critics have focused on intently, perhaps too intently, is: because literature is adaptive, because it is part of our evolved species design. There have been many proposals to explain why art in general, or the art of literature, or the art of fiction, is adaptive—what benefits it offers, and how it recoups its cost—but these have only partially converged, and have rarely convinced others.[4] But evolutionary adaptiveness need not be the only answer, and as biologist Richard Prum has noted, the mere co-evolution of signals and preferences can suffice to establish traditions of art in humans and other biological lines (Prum 2013). These signals may at times offer advantages, as I argue they often do in human art,[5] but they need not always do so for art practices to be maintained. The advantages I see are like those of play in social animals. As play became compulsive across many species, because it trains them in secure moments in key physical skills that they may need in urgent non-play modes, so art's play with patterns, by harnessing our key senses, sight and sound, and our key modes of understanding, through language and narrative, helps extend our imaginations[6] and our possibilities, individually and collectively, and deepens our social cohesion in multiple ways. We will consider how this may be reflected in "Cuba" when we come to the question of the truth of literature.

How do we explain as richly as we can the processes of composition and comprehension, of intention, meaning, interpretation, response and evaluation?

We intuitively understand one another's actions in terms of intentions (we react to what we see as an accidental bump very differently from a deliberate shove), and we can understand intentions better in terms of the problems we infer, from context and action, that an individual is trying to solve. The continuity guideline suggests we will carry that over into interpreting works of art.

We can understand actions still better if we factor in costs and benefits as well as problems and solutions. Authors, like others, reduce search costs through social learning, in their case through the example of other authors. Duffy lowers invention costs by drawing on the sonnet form in "Cuba," and in *Rapture*, the volume from which it comes, on the sonnet sequence, especially Shakespeare's *Sonnets*, which move from idealized love to disenchantment. Mere repetition of earlier modes, however, risks earning diminished interest and attention;[7] Duffy lowers costs and increases benefits (unlike many of the Elizabethan sonnet-sequence poets) by drawing on her own life and its novel particulars, like, here, the foregrounded trip to Cuba.

Audiences, too, reduce costs through social learning. We acquire skills in reading poetry through others' guidance (in line with Kim Sterelny's model of what makes us unique: the human as "evolved apprentice" (Sterelny 2012)). We reduce search time by following reputations (which guided me years ago to Duffy).

We should expect a partial convergence of individual interpretations, through what we share as human beings (including, in this case: our inclination to understand the actions of others, such as their utterances, in terms of their intentions; our knowledge of poetry; our experience of romantic disappointment and of canceled hopes), but also partial divergence, through individual variation in dispositions, experience, and the problem situations in which we find ourselves as we read (for distraction, elevation, consolation, study, or as teachers, reviewers, or critics, for instance).

With the help of our extreme sociality as well as our linguistic skills, we decode utterances by way of rapid inferential elaboration, within a criterion of relevance: we assume others' words directed our way will offer some relevance to us; our seeking the relevance presupposes benefits and reduces search costs (Sperber & Wilson 1986; Wilson & Sperber 2012). In speech, relevance will usually be supplied by a context more or less known by speaker and listener; in a poem, since the poet cannot know the circumstances of readers, the relevance will tend to be supplied internally.

"Cuba" poses us a problem: what do all these negated actions, all these non-statements, amount to? We enjoy the pleasure of solving the riddle, of being able to infer so much more than what is explicitly said, a narrative of love's hopes abruptly ended. The poem has not a single emotion term, except for "sad" in the penultimate line—even then referring not to the poet but to "black cab, sad hearse"—yet we infer the poet's emotion, her devastating disappointment, as we infer her situation.

Poems invite us to reread. If the invitation is worded invitingly enough, we may read again and perhaps even again and again, so long as our sense of the likely benefits of rereading outweighs the additional costs in time and effort we invest. A poet's core problem will therefore often be to load her poem sufficiently for readers to wish to return to discover more. On a rereading of "Cuba" we may notice, for instance, the way the negations and erasures loop from first to last: "like a work of art / rubbing itself out" not only offers a flattering image of getting dressed, in the poet's now abandoned anticipation of the erotic charge of the shared trip, but prefigures the poem itself rubbing itself out, the "Cuba" of its title being dismissed in its final words, "No Cuba."

But readers' problems differ. We read to maximize the benefit for ourselves, appropriating a work according to the problems we set ourselves or have imposed on us (in my case, for instance, on these latest rereadings, partly to tease out elements of literary theory). Different search tasks will yield different search patterns and tend to produce somewhat different results.

An evolutionary approach to questions of literary response suggests what philosophers of art call a modified intentionalism (the intentions of the artist matter, but not exclusively or unproblematically: see N. Carroll 1992), and a partial convergence and a partial divergence of responses. "Convergent responses . . . are natural, and so are divergent responses, and so are ongoing critical discussions . . . and ongoing disagreements, from which we can all learn" (Boyd 2012a: 75).

Does literature yield truth or other kinds of value?

Fiction has been likened to a flight simulator, putting us safely through the ups and downs of social experience, even at their most turbulent (Oatley 2011). Is this poem, even if it seems closer to lived personal experience than to fiction, a kind of dramatization of the poet's thoughts, and in that sense a simulation, as we enter and entertain her thoughts and feelings in time with her utterance? Not quite: the poem does not simulate in us the poet's experience (after all, that includes the challenges, surprises and satisfactions of composing the poem), but it does spur us to deduce and reconstruct her wrecked and empty state; to feel its force by activating our own memories and anticipations of hotel rooms, of shared beds and bathrooms and intimacies, and by calling on our own experience of disappointment in love; to empathize with her in her shocked new grief; to admire the discipline she has imposed on the experience; and to recall the way we too have eventually made sense of life's jarring surprises and found a way to recover a sense of order.

Literature is not a direct record or even a direct simulation of experience, but an imaginative, active construction of experiences, first by an author, then by readers. As readers we have a sense of sharing experience with characters (in this case, with the poet as heartbroken lover), with authors (in this case, with the poet as craftsperson), and with other readers who will discover and feel much as we have. Since print became cheaply available and literacy widespread, reading has tended to be solitary, and understanding a poem like this requires individual imaginative effort, yet it evokes thoughts and feelings that are deeply social, even here in the sudden sinkhole of new loneliness. It draws on what we share—memories and hopes of romantic travel—in order to share still more, to attune ourselves in feeling.

Literature serves multiple functions. It extends our imaginations, by harnessing our primary mode of understanding experience—narrative—and our primary mode of articulating experience—language.[8] It challenges us to adopt rapid perspectival shifts, such as, in "Cuba," from negation to evocation to re-negation; from memory to anticipation to exclusion; from literal to analogical; from poet as suffering character to poet as shaping author. It serves as a means for social attunement, for sharing what is common in our experience, and therefore as a means for social cohesion.

To do this, literature does not depend on literal truth. We feel that Duffy's "Cuba" arises from personal experience, but for all that most readers know, Cuba could be a mere stand-in for another destination, or thwarted travel plans a mere painted backdrop for the disappointments of love Duffy was experiencing at the time, or even her present disappointments in love a mere fiction or a receding memory. She offers not a particular record of her experience or a universal truth of experience but a particular series of moments that cue us as readers to reactivate our own experience (of hotels, memories, anticipations, love, repudiation) to recreate what we imagine to be her experience and empathize with her suffering, to test what a sudden rebuff to love can be like. At the same time, in order to shape the poem with as much attention-earning power as she can muster, line by line, Duffy also makes it inevitable that attentive poetry-loving readers will enjoy, with a different part of our minds, the craft, the expert play, that has enabled her to concentrate and sharpen the prompts that allow us both to taste the immediate rawness of her experience and to savor her control. We recreate and empathize with her pain, and we recreate and participate in her artistry. We acknowledge the roughness of life, and the power of human minds, appealing to one another and mastering it themselves, to smooth it out again, individually and together.

Literature is not just simulation, or not simulation just of experience, but simulation also of the ways we cope with experience, recreate it, go beyond it, share it. We self-assemble the simulator, with our own experience as partial input, and part of the pleasure of doing so is admiring how authors design their work for our self-assembly and our individual inputs. These will inevitably be diverse and to some extent idiosyncratic, yet we are likely to share or have shared not only particular kinds of experience, like the plunge in "Cuba" from romantic rapture to despair, but also the capacity to imagine, intuit, empathize, with characters, authors, and other readers: a strong affirmation of our interconnectedness.

I have identified continuity as a factor that evolution can highlight for literary studies. But evolution is a theory of change.

Can evolution help to explain the continuities and changes, the traditions and innovations, in literature?

It can, in many ways. There are multiple evolutionary models of cultural change: the meme, as an analogue of the gene, a cultural product that replicates itself more or less independently of those who host it (Dawkins 1989; Blackmore 1999); or perhaps better, an epidemiological model, in which different individuals are differently susceptible to cultural viruses (Sperber 1996); or an arbitrary coevolution of signals and preferences (Prum 2013), which may explain some of the mere fashion in the arts, and the lack of the rapid ratcheted progress that occurs in the sciences over similar periods; or a model that emphasizes our genetic predisposition to copy from some people rather than others (Henrich 2015). Or still better, a model that dwells on the problems and solutions of individuals, in authors and audiences, and on the costs and benefits they assess, including the reduction of costs possible through imitating successful others.

As I have observed elsewhere, "Unlike other species, we can imitate closely and therefore follow established forms. Crucially, we *need* to imitate in order to innovate. Building on what came before underlies all creativity, in biology and culture. Starting again from scratch wastes too much accumulated effort: far better to recombine existing design successes. Even adopting a high mutation rate, changing many features at once, would rapidly dismantle successful design" (Boyd 2009: 122).

Shakespeare, perhaps the most singularly creative of all artists, borrowed most of his plots, and Elizabethan stage conventions, and the sonnet form, and the structure of sonnet sequences, but he adapted as he appropriated and innovated as he imitated.[9]

Duffy in turn has drawn on the example of Shakespeare's sonnets in "Cuba" and of his sonnet sequence in *Rapture*. She even draws—and this is typical—on one of her own previous solutions to the problem of writing sonnets so long after Shakespeare. She does not play with the patterns of lines, sounds, images, and structures as intensely as Shakespeare, because that did not quite work even for him, in many of his less memorable sonnets, and because the taste for tight rhyme has weakened in serious modern poetry in English. But she still adopts features of the Shakespearean sonnet, as in "Cuba." Shakespeare's sonnets have alternating rhymes throughout the three quatrains (*abab cdcd efef*) but switch to adjacent rhymes in the couplet (*gg*). "Cuba" has no rhymes at all until it switches to an adjacent near-rhyme in the final two lines: *No queue here, No Cuba*. That combination of a basically rhymeless sonnet with a final rhyming couplet itself reworks a solution Duffy had found earlier, in her otherwise rhymeless sonnet "Anne Hathaway" in *The World's Wife* (1999), which ends "I hold him in the casket of my widow's head / as he held me upon that next best bed."

In his sonnets, Shakespeare's shift of rhymes in the couplet marks a shift in gear or direction. Duffy too shifts, from inside the hotel to outside. Shakespeare's couplets, unusually, gather up elements from earlier in the sonnet,[10] as Duffy's does here: *No* five times in two lines; the first line of the sonnet, "No getting up from the bed in this grand hotel" returns in the last, "No getting away from this," with its wider metaphorical sense adding to the literal stasis, sad enough in itself; the repetition of the first line's "this," with its seeming promise, by now thoroughly negated, rhymes internally with "No goodnight kiss"; "No Cuba" picks up on both the title and the first word, reversing their order, the promise of the title finally negated completely in the poem's last words.

CONCLUSION

An evolutionary approach to literature does not diminish our interest in particulars. It even sharpens the interest, by focusing tightly on authors' problem situations, on their solutions to earning and holding attention, on their intuitive awareness of the costs and benefits for readers in engaging with their work, and their own costs and benefits in imitating, innovating and inventing. But it also extends the scope of literary theory, by seeing literature as continuous with other human and non-human activities, and yet as deeply dependent on our uniquely human ultrasociality, our uniquely vivid and avid imaginations, and our uniquely intense appetite for pattern. Evolutionary perspectives allow us to understand literature within the widest context, and one that continually tests and revises its own theories and data. And we need not sacrifice an iota of detail and difference.

ACKNOWLEDGMENTS

I would like to thank Emily Parke for providing feedback on an earlier version of this chapter.

NOTES

1. For a critique of the claim that although Theory is worn out, there is no other option, see Boyd 2005; for a critique of the notion that we are post-Theory, see Boyd 2006.
2. N. Carroll (1992) offers a particularly striking example, although what I have called the "continuity principle" pervades his work.
3. See Boyd 2017 for a comprehensive critique, fortified by but not dependent on evolutionary considerations, of the Necessary Narrator thesis.
4. See Dissanayake 1988, 2000; J. Carroll 2004, 2011; Boyd 2005, 2009; Dutton 2009. For critiques, see Davies 2012 and especially Verpooten 2015.
5. Others have begun to investigate empirically the benefits of literature: see Djikic et al. 2013; Bal & Veltkamp 2013; Kidd & Castano 2013.
6. J. Carroll (2004, 2011) foregrounds imagination as central to the evolution of literature as an adaptation.
7. See Martindale 1990, 2009; Boyd 2009.
8. See Changizi 2011 for a theory of how language and music have harnessed natural features; Dehaene 2009 for a similar claim for writing.
9. For a discussion of Shakespeare's innovations within sonnet sequence conventions, in terms of the precise problems he posed himself at a particular moment of his career, see Boyd 2012b.
10. As Helen Vendler pointed out (Vendler 1997). Poets read Vendler with care, and I suspect Duffy noted Vendler's insight after she wrote "Anne Hathaway" (by 1999) and before she wrote "Cuba" (by 2005).

REFERENCES

Axelrod, R. 1990. *The Evolution of Cooperation* (Penguin).

Bal, P. & Veltkamp, M. 2013. "How does fiction reading influence empathy? An experimental investigation on the role of emotional transportation." *PLoS ONE* 8(1): e55341. doi:10.1371/journal.pone.0055341.

Blackmore, S. 1999. *The Meme Machine* (Oxford University Press).

Bor, D. 2012. *The Ravenous Brain: How the New Science of Consciousness Explains our Insatiable Search for Meaning* (Basic Books).

Boyd, B. 2005. "Evolutionary theories of art." In J. Gottschall & D. Wilson (eds.), *The Literary Animal: Evolution and the Nature of Narrative* (Northwestern University Press) 149–178.

Boyd, B. 2006. "Theory is dead—like a zombie." *Philosophy and Literature* 30: 289–298.

Boyd, B. 2008. "Art and evolution: Spiegelman in *The Narrative Corpse.*" *Philosophy and Literature* 32: 31–57.

Boyd, B. 2009. *On the Origin of Stories: Evolution, Cognition, and Fiction* (Harvard University Press).

Boyd, B. 2010a. "On the origins of comics: New York double-take." *Evolutionary Review* 1: 97–111.

Boyd, B. 2010b. "Review of Indira Ghose, *Shakespeare and Laughter: A Cultural History.*" *Review of English Studies* 61: 301–303.

Boyd, B. 2012a. "Evolution and literary response." In C. Gansel & D. Vanderbeke (eds.), *Telling Stories: Literature and Evolution* (De Gruyter) 64–76.

Boyd, B. 2012b. *Why Lyrics Last: Evolution, Cognition, and Shakespeare's Sonnets* (Harvard University Press).

Boyd, B. 2012c. "On contra-evidential criticism." *Diacritics* 40: 97–100.

Boyd, B. 2017. "Does Austen need narrators? Does anyone?" *New Literary History* 48: 285–308.

Boyd, B., Carroll, J., & Gottschall, J. 2010. "Introduction." In B. Boyd, J. Carroll, & J. Gottschall (eds.), *Evolution, Literature, and Film: A Reader* (Columbia University Press) 1–17.

Brown, D. 1991. *Human Universals* (Temple University Press.)

Burghardt, G. 2005. *The Genesis of Animal Play: Testing the Limits* (MIT Press).

Carroll, J. 2004. *Literary Darwinism: Evolution, Human Nature and Literature* (Routledge).

Carroll, J. 2011. *Reading Human Nature: Literary Darwinism in Theory and Practice* (SUNY Press).

Carroll, N. 1992. "Art, intention, and conversation." In G. Iseminger (ed.), *Intention and Interpretation* (Temple University Press) 97–131.

Changizi, M. 2011. *Harnessed: How Language and Music Mimicked Nature and Transformed Ape to Man* (Benbella).

Currie, G. 2010. *Narratives and Narrators: A Philosophy of Stories* (Cambridge University Press).

Davies, S. 2012. *The Artful Species: Aesthetics, Art, and Evolution* (Oxford University Press).

Dawkins, R. 1989. *The Selfish Gene*, 2nd ed. (Oxford University Press).

Dehaene, S. 2009. *Reading in the Brain: The Science and Evolution of a Human Invention* (Viking).

Deresiewicz, W. 2009. "Adaptation: On literary Darwinism." *The Nation*, June 8: 26–31.

Dissanayake, E. 1988. *What is Art For?* (University of Washington Press).

Dissanayake, E. 2000. *Art and Intimacy: How the Arts Began* (University of Washington Press).

Djikic, M., Oatley, K., & Moldoveanu, M. 2013. "Opening the closed mind: The effect of exposure to literature on the need for closure." *Creativity Research Journal* 25: 149–154.

Duffy, C. 1999. *The World's Wife* (Picador).

Duffy, C. 2005. *Rapture* (Picador).

Dutton, D. 2009. *The Art Instinct: Beauty, Pleasure, and Human Evolution* (Bloomsbury).

Fehr, E. & Henrich, J. 2003. "Is strong reciprocity a maladaptation? On the evolutionary foundations of human altruism." In P. Hammerstein (ed.), *Genetic and Cultural Evolution of Cooperation* (MIT Press) 55–82.

Graham, K. & Burghardt, G. 2010. "Current perspectives on the biology of play: Signs of progress." *Quarterly Review of Biology* 85: 393–418.

Grünbaum, A. 1984. *The Foundations of Psychoanalysis: A Philosophical Critique* (University of California Press).

Grünbaum, A. 1993. *Validation in the Classical Theory of Psychoanalysis: A Study in the Philosophy of Psychoanalysis* (International Universities Press).

Henrich, J. 2015. *The Secret of Our Success: How Culture Is Driving Human Evolution, Domesticating Our Species, and Making Us Smarter* (Princeton University Press).

Kania, A. 2005. "Against the ubiquity of fictional narrators." *Journal of Aesthetics and Art Criticism* 63: 47–54.

Kidd, D. & Castano, E. 2013. "Reading literary fiction improves theory of mind." *Science* 342: 377–380.

Kramnick, J. 2011. "Against literary Darwinism." *Critical Inquiry* 37: 315–347.

Lewontin, R. 1970. "The units of selection." *Annual Review of Ecology and Systematics* 1: 1–8.

Martindale, C. 1990. *The Clockwork Muse: The Predictability of Artistic Change* (Basic Books).

Martindale, C. 2009. "The evolution and end of art as Hegelian tragedy." *Empirical Studies of the Arts* 27: 141–145.

Mulligan, K., Symons P., & Smith B. 2006. "What's wrong with contemporary philosophy?" *Topoi* 25: 63–67.

Oatley, K. 2011. *Such Stuff as Dreams: The Psychology of Fiction* (Wiley-Blackwell).

Patron, S. 2013. "Discussion: 'Narrator.'" *Living Handbook of Narratology*. www.lhn.uni-hamburg.de, 2011, rev. 2013

Popper, K. 1999. *All Life is Problem-Solving* (Routledge).

Prum, R. 2013. "Coevolutionary aesthetics in human and biotic worlds." *Biology and Philosophy* 28: 811–832.

Richerson, P. & Boyd, R. 2005. *Not by Genes Alone: How Culture Transformed Human Evolution* (University of Chicago Press).

Singer, P. 1999. *A Darwinian Left: Politics, Evolution, and Cooperation* (Weidenfeld and Nicolson).

Sperber, D. 1996. *Explaining Culture: A Naturalistic Approach* (Blackwell).

Sperber, D. & Wilson, D. 1986. *Relevance: Communication and Cognition* (Blackwell).

Stecker, R. 1987. "Apparent, implied, and postulated authors." *Philosophy and Literature* 11: 258–271.

Sterelny, K. 2012. *The Evolved Apprentice: How Evolution Made Humans Unique* (MIT Press).

Tuf, I., Drábková, L., & Šipoš, J. 2015. "Personality affects defensive behaviour of *Porcellio scaber* (Isopoda, Oniscidea)." *ZooKeys* 515: 159–171.

Vendler, H. 1997. *The Art of Shakespeare's Sonnets* (Harvard University Press).

Verpooten, J. 2015. *Art and Signaling in a Cultural Species*. Dissertation. KU Leuven and University of Antwerp.

Walsh, R. 1997. "Who is the narrator?" *Poetics Today* 18: 495–514.

Wiessner, P. 2014. "Embers of society: Firelight talk among the Ju/'hoansi Bushmen." *Proceedings of the National Academy of Sciences* 111: 14027–14035.

Wilson, D.S. 2011. *The Neighborhood Project: Using Evolution to Improve My City, One Block at a Time* (Little, Brown).

Wilson, D.S. 2015. *Does Altruism Exist?: Culture, Genes, and the Welfare of Others* (Yale University Press).

Wilson, D.S. 2016. "The One Culture: Four new books indicate that the barrier between science and the humanities is at last breaking down." https://evolution-institute.org/focus-article/the-one-culture/

Wilson, D.S. & Sperber, D. 2012. *Meaning and Relevance* (Cambridge University Press).

30

Play and Evolution

Patrick Bateson

Those who regard play as frivolous and without function will doubtless be incredulous to be told that playful play had a role in the evolution of cognition, particularly in mammals and birds. Most animals are active and, in principle, their behavior can impact on the evolution of their descendants. Charles Darwin (1871) suggested such a role with his ideas about the evolutionary process of sexual selection. His co-discoverer of the principle of natural selection, Alfred Russel Wallace, never liked Darwin's proposal for sexual selection and, for a large part of the 20th century, little empirical work was done on how individuals' choices could affect evolution. Sexual selection was not thought to be an important component of evolutionary processes. Nowadays more than a thousand papers on the process are published each year. The role of adaptability, which had its origins in a paper by Douglas Spalding (1873), was once again dismissed as relatively unimportant until recently, when interest in plasticity during development took firm root (West-Eberhard 2003; Bateson 2015). Though its origins were much later, Richard Lewontin's (1983) ideas about niche construction have similarly excited a growing amount of attention (Laland et al. 2014). One aspect of the way in which animals' activities might affect evolution is through their play. This chapter considers how this process might have come about.

I first describe how play is defined and then consider how the behavior of individuals can affect the evolution of their descendants. Sometimes this can have a ratchet-like effect in evolution. I then consider the various functions attributed to play, since they can lend suggestions about how play evolved—even though current use is not necessarily a good guide to how the behavior evolved. Finally, I consider the idea that playful play facilitates creativity and drives the evolution of cognitive ability.

WHAT IS PLAY?

I use "play" in a more restricted sense than when it is used colloquially. I shall not refer to competitive games or theatrical performances. Like most psychologists and biologists

414

who study the subject, I regard play behavior as spontaneous and rewarding to the individual; it is intrinsically motivated and its performance serves as a goal in itself. The player is to some extent protected from the normal consequences of serious behavior. The behavior appears to have no immediate practical goal or benefit. Social forms of the behavior may be preceded or accompanied by specific signals or facial expressions indicating that the behavior is not to be taken as a threat. Play is the antithesis of "work" or "serious" behavior. The behavior consists of actions or, in the case of humans, thoughts, expressed in novel combinations. Social forms of the behavior may be accompanied by role reversals, in which a normally dominant individual may become temporarily subordinate while playing, and vice versa. Individual actions or thoughts are performed repeatedly; they may also be incomplete or exaggerated relative to non-playful behavior in adults; play looks different. Play is sensitive to prevailing conditions and occurs only when the player is free from illness or stress. Play is an indicator of well-being. Playful play is accompanied by a particular positive mood state in which the individual is more inclined to behave in a spontaneous and flexible way. These definitions are much the same as those introduced by Gordon Burghardt (2005).

EVOLUTIONARY RATCHET

Accepting the definitions of play, most examples in the animal kingdom are to be found in birds and mammals. Burghardt (2005) suggests that the extended parental care of these animals provided the lift-off for the evolution of play. It may have introduced instability into evolutionary processes that led on to increased complexity in playful behavior and the appearance of a variety of biological functions. His ideas provided much of the stimulus for this chapter.

The idea of evolutionary instabilities has a long history. R.A. Fisher (1930) suggested that mate choice may set up a runaway process so that the act of choice can affect the evolution of a characteristic in the chosen that then affects what is chosen. In another context, Alison Jolly (1966) suggested that the evolution of cognitive complexity was driven by the need to catch up with the most cognitively advanced individuals. Independently, Nick Humphrey (1986) developed the thesis that the explosion of intellect in the hominin line had a ratchet-like quality. The result is that individuals evolve to understand and predict what other members of their social group are about to do. As Humphrey put it, they become better psychologists, thereby becoming better able to compete with others that did not have this ability.

For the young bird and mammal, while it has time when it does not have to find food for itself and is well protected, those individuals that are able to engage in adult-like behavior may have an advantage over the others. In effect they go through a period of training that perfects the behavior that they will need when adult. The disparity between them and the others, who do not have this training experience, provides the next step. As a consequence of the disparity, the non-playful individuals' offspring that behave like the playful individuals will be more likely to survive and, by degrees, the whole population plays more. Some theorists argue that that is the end of the evolutionary process (Paenke et al. 2009). However, if some individuals are able to profit in ways other than merely improving their motor skills during play, the upward movement toward greater complexity will continue. They might, for example, become more aware of environmental contingencies than others and gain advantage by doing so. Once again this drives the evolution of the same abilities in the rest of the population. Where does such a process stop? Eventually the costs of evolving new forms of behavior or the sheer difficulties of doing so become limiting. Nevertheless, the hypothetical process could generate some individuals that are able to solve new problems creatively.

The unstable evolutionary processes could be generated by spontaneous alterations in the genome, but the likelihood of this happening diminishes with the number of components in a behavioral sequence necessary to produce an overall change. The resulting stasis can be overcome since the plasticity of the individuals may allow the evolutionary process to occur piecemeal. This hypothetical process was first suggested by Douglas Spalding (1873). His proposed mechanism comprised a sequence of learned behavior patterns followed by the differential survival of those individuals that expressed the behavior more efficiently *without* learning. To give one hypothetical example, Galapagos woodpecker finches use cactus spines or small sticks to probe into holes for insect larvae. This behavior could have been learned initially in the species' history through trial and error. But subsequently, in the course of evolution, the behavior of picking up and probing with small sticks could have been expressed spontaneously, because those individuals that did so expended less time and effort than those that continued to learn the behavior again in each generation. The birds that came to express the behavior spontaneously were therefore more likely to survive and reproduce than those relying on individual learning. Spalding's idea was advanced again independently by James Baldwin (1896), Conwy Lloyd Morgan (1896), and Henry Osborn (1896), all publishing in the same year, and subsequently became known as the "Baldwin effect." Seemingly their ideas were proposed independently of Spalding and, indeed, of each other, although they may have unconsciously assimilated what Spalding had written twenty-three years before. Baldwin became solely associated with the conjecture because of what he wrote in a widely read book (Baldwin 1902). To avoid confusion about terminology and precedence, I have suggested (in Bateson 2006) that the proposed process be described by a descriptive term: "the adaptability driver."

Lloyd Morgan's (1896) account of the adaptability driver was particularly clear. He suggested that if a group of organisms respond adaptively to a change in environmental conditions, the modification will recur generation after generation under the changed conditions, but the modification will not be genetically inherited. However, any genetic variation in the ease of expression of the modified characteristic is liable to favor those individuals that express it most readily. Consequently, an inherited predisposition to express the modification will tend to evolve. The longer the evolutionary process continues, the more marked will be such a predisposition. The process starts through learning or some other form of plastic modification within individuals, but this paves the way for a longer-term change in the genes. The Galapagos woodpecker finch that pokes small sticks into holes for insect larvae appears to have a strong predisposition to pick up sticks, but learns from others what to do with those sticks (Tebbich et al. 2010). Eventually the spontaneously expressed behavior of picking up sticks may be accompanied by the spontaneously expressed behavior of poking those sticks into holes.

In principle, then, behavior patterns that were initially acquired through the animal's plasticity could be expressed spontaneously, without employing such plasticity, in subsequent generations. The hypothesis of the adaptability driver has been repeatedly modeled, both analytically (e.g., Paenke et al. 2009) and by simulation (e.g., Red'ko et al. 2005). The outcome of this theoretical activity has been variable, sometimes supporting the adaptability driver hypothesis and sometimes not. Paenke et al. (2009) proposed a general framework that explained both effects. Spontaneously expressing a behavior pattern that had been learned in previous generations could be costly if it means that the animal loses all of its ability to learn. Some evidence from fruit flies suggests that this might well be the case, at least in simpler organisms (Kawecki 2010). The benefit of expressing a behavior pattern spontaneously was found to be outweighed by the cost of losing the capacity to learn about other things. The argument is much less cogent when applied to big-brained animals like birds and mammals with multiple parallel pathways involved in learning. In these animals, the loss of capacity to learn in one way has no effect on the capacity to learn in other ways (Bateson 2004). Inasmuch as it has been taken seriously, the hypothetical process has usually been regarded as providing a mechanism for the slow accretion of spontaneously expressed phenotypic elements in the course of evolution.

Emphasis was initially placed on how particular behavior patterns initially acquired by learning could be expressed spontaneously without learning in the course of subsequent evolution. Recent developments have shifted the focus to other issues, such as the way in which plasticity can accelerate the rate at which challenges set by the environment can be met, and the advantages of plasticity in a changing environment. The effect of plasticity on evolution may have become increasingly powerful as animals, in particular, became more complex. The general idea is that a system that enhances the fitness of an individual depends on a number of elements, such as the capacity to use information contained in the energy impinging on a sense organ, specific biochemical reactions, and particular effectors that respond adaptively to the stimulation (Anokhin 1974). Elements may be recombined in different ways to perform different functions. This evolutionary process can lead to the establishment of increasingly elaborate organization and patterns of behavior. When such complexity entails a greater ability to discriminate between different features of the environment or a greater ability to manipulate the environment, the organism will benefit and will be more likely to survive and reproduce in the face of multiple challenges during its lifetime. A new adaptation would emerge in evolution when the accumulated phenotypic effects of genomic reorganization were added to the existing phenotype. Although these phenotypic effects are specific to the new function, existing parts of the phenotype would also be recruited for this function. As a result, phenotypic elements established earlier in evolution would be incorporated in more adaptive systems than later evolved elements. Plasticity would promote much more rapid genetic evolution of complex sets of adaptive systems than can be accomplished by mutation alone. This occurs as previously plastic elements are replaced by inherited elements and the model organism is able to fill by its plasticity missing elements in subsequent systems. The exposure to novel environments would be likely to lead to the subsequent evolution by means of classical Darwinian processes of morphological, physiological, and biochemical adaptations to those niches.

An important empirical demonstration of adaptability driving evolutionary change may be that of the house finch (*Carpodacus mexicanus*). In the middle of the 20th century, the finch was introduced to eastern regions of the USA far from where it was originally found on the west coast. It was able to adapt to the new and extremely different climate and spread up into Canada. The finch also extended its western range north into Montana, where it has been extensively studied. After a period involving a great deal of plasticity, the house finch populations spontaneously expressed the physiological characteristics that best fitted them to their new habitats without the need for developmental plasticity (Badyaev 2009). Initially the adaptive onset of incubation and the sex bias in the order of ovulation were affected by ambient temperature in the more northerly climes, but as evolution in the population occurred, these behavioral and physiological effects were no longer dependent on the external cues for their expression. After using their adaptability to respond to the new environmental conditions, the house finch populations spontaneously expressed the physiological characteristics that best fitted them to their new habitats.

Play leads to a form of plasticity, since by playing an individual is able to acquire skills and understanding of its physical and social environment. Those aspects of play that are creative in solving a problem, or breaking out of local optima, are beneficial to the individual. Such improvements in what could be perceived as cognitive ability would not occur readily by genetic recombination or random mutation, since the probability of all the necessary changes occurring simultaneously would have been small. For instance, a squirrel might have discovered while playing that swinging on a branch enabled it to reach nuts that were previously inaccessible, but this beneficial change in its behavior remains a learned modification. Similarly, dolphins playfully blowing bubbles might have learned that a curtain of bubbles can trap fish. The next step could occur in one of two ways. The discovery made through play by one individual could then spread by social learning. Alternatively, the discovery could be made separately by many individuals through play, all of them benefiting in the same way. Then, as is postulated in the adaptability driver hypothesis, those individuals that were able to express spontaneously the beneficial trait (swinging on branches, blowing bubbles, or whatever) would be able to compete more successfully.

FUNCTIONS OF PLAY

One approach to understanding the evolution of play is to consider what benefits derive from the behavior in the present. Nikolaas Tinbergen (1963) was careful to point out that an answer to the question of current utility does not necessarily provide an answer to why it evolved (see also Bateson & Laland 2013). Nevertheless, a functional explanation can suggest what might have been involved in the course of evolution. In the history of thinking about the function of play a very large number of hypotheses have been offered to explain how the various aspects of play increase the individual's chances of survival and reproducing itself.

When young animals playfully practice the stereotyped movements they will use in earnest later in life, they are often thought to improve the coordination and effectiveness of these behavior patterns. The short dashes and jumps of a young gazelle when it is playing bring benefits that may be almost immediate, as it faces the threat of predation from cheetah or other carnivores intent on a quick meal, and needs considerable skill when escaping (Gomendio 1988). Even though the benefits may be immediate in such cases, they may also persist into adult life. Most theories of the functions of play have continued to focus on its role in enabling the developing individual to acquire and practice complex physical skills and, by so doing, to fine tune neuromuscular systems. Others, observing how much young animals play with each other, have emphasized that the individual also develops social skills and cements its social relationships; play may also serve to improve its capacity to compete and cooperate with other members of its own species. Play can make an individual more resistant to stress and enlarge its repertoire. Play may enhance an individual's resourcefulness and flexibility and make it able to adjust to new conditions. Play may enhance its ability to cooperate with others and to co-exist with older members of its own species. Play may increase its knowledge of its home range. Play, or at least some components of it, allows young animals to simulate, in a relatively safe context, potentially dangerous situations that will arise in their adult life. On this view, play exerts its most important developmental effects on risky adult behavior such as fighting, mating in the face of serious competition, catching dangerous prey, and avoiding becoming someone else's prey. Indeed, the behavior patterns of fighting and prey-catching are especially obvious in the play of cats and other predators, whereas safe activities such as grooming, defecating, and urinating have no playful counterparts. None of the suggestions about the function of play are mutually exclusive.

When differences between the sexes arise in play, as they often do, these are reflected in differences between the sexes in the activities of adults (Meaney & Stewart 1985). For instance, young female chimpanzees seem to behave maternally toward sticks, doing so more than males and ceasing to do so when they have real offspring to care for (Kahlenberg & Wrangham 2010). Stick-carrying consists of holding or cradling detached sticks pieces of bark, small logs, or woody vine, with their hand or mouth, underarm, or, most commonly, tucked between the abdomen and thigh. Individuals carry sticks for periods of one minute to more than four hours during which they rest, walk, climb, sleep, and feed as usual. The occurrence of stick-carrying peaks among juveniles and is higher in females than males. This sex difference cannot be explained by a general propensity for females to play with objects more than males, because several types of object such as weapons are played with more by males. Males in many species, including humans, do more rough and tumble play than females, and engage in more agonistic behavior when adult (Auger & Olesen 2009).

Play has features that are likely to make it especially suitable for finding the best way forward in a world of conflicting demands. In acquiring cognitive skills, individuals are in danger of finding sub-optimal solutions to the many problems that confront them. In deliberately moving away from what might look like the final resting point, each individual may get somewhere that is better. Play may, therefore, fulfill an important probing role that enables the individual to escape from false end-points or "local optima." An analogy is a mountain surrounded by lesser peaks. A climber might get to the top of a lesser peak only to discover that she had to descend before scaling a higher one. When on a metaphorical lower peak, active ways of getting off it can be highly beneficial. In practice what this could mean is that the activities involved in play discover possibilities that are better than those obtained without play.

TESTING THE HYPOTHESES ABOUT FUNCTION

The utility to an individual of having a characteristic that enhances its chances of surviving and reproducing is testable in principle but much less often so in practice. Evidence

for the utility of play has not been readily forthcoming (Martin & Caro 1985). Nevertheless, play has real biological costs. Animals expend more energy and expose themselves
to greater risks of injury and predation when they are playing than when they are resting.
Play makes them more conspicuous and less vigilant. For example, young Southern fur
seals are much more likely to be killed by sea lions when they are playing in the sea than
at other times when they are in the sea and not playing (Harcourt 1991). The costs of play
must presumably be outweighed by its benefits, otherwise animals that played would
be at a disadvantage compared with those that did not. Among the suggested costs for
human children are poor socialization, poor physical fitness, poor mental health, and
poor creativity. If play is beneficial, then it follows that depriving the young animal of
opportunities for play should have harmful effects on the outcome of its development
other things being equal.

A different approach to understanding the functions of play relies on correlations
between the behavior of young animals and their subsequent lives. A study of play and
survival in the offspring of eleven families of individually identified, free-ranging brown
bears in Alaska (Fagen & Fagen 2004) revealed that cubs which played more during their
first summer survived better from their first summer to the end of their second summer.
This could have been for a variety of reasons. Potential confounding factors such as cub
condition, prenatal and first-year salmon availability (an important resource for bears),
and maternal characteristics were examined statistically (Fagen & Fagen 2009). The
researchers controlled for these factors and confirmed that the more the bears had played
when they were cubs, the more likely they were to survive to their first year. The association between amount of play and survival persisted into subsequent years of the bears'
lives when they reached independence. The amount of play accounted for 35 percent of the
variance in the percentage that survived. Just how play benefited the bear cubs could not
be determined but the pre-adult mortality might result from events occurring during the
stressful environmental conditions of winter hibernation and early spring. Resistance to
exposure and infectious disease might be involved. Possibly play produces a more resilient
individual both behaviorally and immunologically. If so, it would be capable of withstand-
ing stress in ways that physical condition alone would not predict. In other populations
or species, these same factors could still be important, but in different ways—in mediating development and performance of behavior patterns involving predator avoidance and
defense, for example. Predator avoidance and defense necessarily involve cognition and
emotion, whether the argument is made in physiological terms or in behavioral terms.

In a study of feral horses, maternal condition influenced play behavior only in males,
with sons of mothers in good condition playing more (Cameron et al. 2008). When a
son and a daughter of the same mother were compared, the daughter played more when
its mother was in poor condition and sons played more when their mother was in good
condition. Mothers of foals that played more lost more condition and weaned their foals
earlier, indicating that play behavior was affected by maternal investment. An important
finding of the study was that those individuals that played more survived better and had
better body condition as yearlings despite being weaned earlier.

In both the studies of free-living bears and horses, it remains possible that an unmea-
sured third variable explains the results. The individuals that played less may have been
less healthy from the outset. Nevertheless, the results are consistent with the hypothesis
that playing when young reaps benefits later in life.

Studies of function do not necessarily inform about evolution. As Daniel Sol (2015) notes, play and its consequences might have evolved through the cooption of a variety of other behavior patterns. Burghardt (2005, 2015) proposes that the surplus energy available to young birds and mammals in particular might create optimal conditions for the evolution of the initial appearance of the playful behavior patterns. These could then enhance neural processing, physical fitness, behavioral coordination, and behavioral flexibility. The next step in evolution could, among other things, have led to the ability to generate novel behavior that would have provided the basis for creativity and innovation. These suggestions provide explanations for the heterogeneity of play. In my book with Paul Martin, we noted how different forms of play involve different behavior patterns and may have different developmental time courses (Bateson & Martin 2013).

TESTING THE IDEA

Martin and I have suggested that playful play can provide a necessary condition for subsequent creativity. We argue that a broad distinction may be drawn between creativity and innovation (Bateson & Martin 2013). In human behavior, creativity refers to coming up with a new idea, whereas innovation refers to changing the way things are done. The big question is whether creativity could accelerate the evolution of cognitive ability.

The difficulty in testing such an idea is that evolution has already happened. It cannot be replayed. However, the animal kingdom consists of many species so it is possible to use a comparative approach to discover whether a correlation is detectable in the incidence of play and the discovery of novel solutions. Mammals are on the whole more playful than individuals in other taxonomic groups, and they are also more creative. Among the birds, the parrots and crows are the most playful and also astonishingly creative (see Bateson & Martin 2013). On the whole, generalists are more playful and creative than those species with specialized feeding habits.

Most of the people working on innovation in animals have not considered creativity as being distinct from innovation. Nor have they related their studies to the extensive work on creativity in humans. Equally, most of the people working on creativity in humans have ignored what might be relevant studies of animals. The gap has been closed in an excellent book edited by Allison and James Kaufman (Kaufman & Kaufman 2015). Each chapter on non-human creativity and innovation is followed by a commentary from a specialist working on human creativity. Moreover, many of the contributors explore the links between playfulness and creativity. Irene Pepperberg (2015) describes in the Grey parrot play with sounds which leads to the creative use of new sounds. Robert Mitchell (2015) describes the way in which games between dogs and humans can be creative. In an extensive and excellent review of the corvids and parrots, Alice Auersperg (2015) considers how their play boosts their creativity. She also considers how normal behavior constrains the different ways in which these birds interact with their environments. In another splendid review Eric Patterson and Janet Mann (2013) describe the remarkable playfulness of dolphins and whales, and also note how these animals develop novel behavior. Burghardt (2015) notes that many of us who have worked on play have overlooked the work of Michael Huffman and his colleagues (e.g., Huffman 1984 and reviewed by Leca et al. 2012), who have worked for many years on the stone-play of young Japanese macaques. Many uses of stone toys were observed, including gathering, carrying, scattering them together, rolling, rubbing, and cuddling them. Here again playfulness seems to have been involved in the production of so much novel behavior.

The methods used by researchers comparing play and creativity in different species can also be used for comparisons of individuals within a species. Daniel Nettle and I explored this possibility in humans (Bateson & Nettle 2014). In humans, five main dimensions have been used to describe the variation in personalities (Nettle 2007). Many of these are not usually regarded as attributes of cognitive ability, but the dimension "Openness to Experience" is one that could have evolved as the result of play. The dimension ranges from Creativity to Analytical Ability. We asked people in an online survey whether they viewed themselves as playful and creative (Bateson & Nettle 2014). The respondents were presented with a series of statements and asked to state whether each one was very characteristic of themselves (and, if so, to score 1) or very uncharacteristic of themselves (and, if so, to score 7). "Acting playfully" and "Coming up with new ideas" were the statements in which we were particularly interested. These statements were embedded in a number of other statements designed to assess different dimensions of personality. A total of 1536 people responded to the survey. The individuals who reckoned that they were playful also reckoned that they were creative.

The link between "acting playfully" and "coming up with new ideas" emerged very strongly in our survey. To validate this finding, the respondents were asked to offer ideas for the uses of two items: a jam jar and a paperclip. In the literature on creativity, those individuals who produce few answers are referred to as "convergers" and those who produce many suggestions are known as "divergers." The typical sole response from a converger when asked for uses for a paper clip was "Clip paper together." The remarkable response from one diverger in our survey (presumably a woman) was: "Clip papers, unfold to clean fingernails, clip bra, general clothes fixing in an emergency, put on a magnet for a science experiment for children, make a mobile with lots of them, make a sculpture with one or more of them, earrings, pick a lock."

In fact, the sharp dichotomy between convergers and divergers was not borne out in reality. Just under 10 percent offered only one use for the paperclip. Double that percentage suggested three uses, and the percentage dropped off thereafter. Those suggesting ten uses comprised 12 percent of the population and some of them would undoubtedly have gone on to offer more if they had not been restricted to ten uses. So the population, far from being bimodal in the suggested number of uses, resembled a Poisson distribution. Most of the respondents provided a relatively small number of uses for the objects and only a few offered many uses. However, the respondents who regarded themselves as playful and producers of new ideas were much more likely to give lots of uses for a jam jar and a paper clip.

CONCLUSION

What is it about fooling around while having fun that enhances creativity? By rearranging actions or thoughts, play generates novel ways of dealing with the environment, most of which lead nowhere but some of which may turn out to be useful—often at a much later date. Play is also about breaking away from established patterns and combining actions or thoughts in new ways. Play is an effective mechanism, therefore, for encouraging creativity since creativity also involves breaking away from established patterns of thought and behavior. Creative people perceive new relations between thoughts, or things, or forms of expression that would normally seem utterly different. They are able to combine them into new forms, connecting the seemingly unconnected.

Whether or not uncreative people can be made more creative has been much debated. Mihalyl Csikszentmihalyi (1996) argues that if they are allowed more space and time, they can do so. He offers a do-it-yourself guide on how to do it. At any one time, though, the number of creative people in a human population is likely to be low, and the same may also be true for the number of innovators. The frequency of each capacity might depend on the frequency of the other, such that if a population consists of too many creative people and too few innovators, the evolution of more innovators would be favored. And vice versa. The process is comparable to the evolution of sex ratios except that the majority of the population might be neither creative nor innovative. Does the population benefit from possessing people who come up with new ideas and those who implement productively the ideas of others? Did populations that had some creatives and some innovators benefit over others which lacked one or the other type? The question that relates back to the main theme of this chapter is whether creativity in a population was driven by the playfulness of its members.

The role that I have attributed to play in evolution would not have been popular in an era when the organism was thought to be essentially passive in evolution. However, the Extended Evolutionary Synthesis welcomes thinking about the active role of behavior in the evolutionary process (see Laland et al. 2015). Playful play, with its way of creatively finding links between previously unrelated events and thereby affecting the evolution of descendants, is very much part of the new wave of thinking.

ACKNOWLEDGMENTS

I am grateful for comments on an earlier version of this chapter from Gordon Burghardt with whom I share a number of the same ideas.

REFERENCES

Anokhin, P. 1974. *Biology and Neurophysiology of the Conditioned Reflex and its Role in Adaptive Behavior* (Pergamon Press).

Auersperg, A. 2015. "Exploration technique and technical innovations in corvids and parrots." In A. Kaufman & J. Kaufman (eds.), *Animal Creativity and Innovation* (Elsevier) 45–68.

Auger, A. & Olesen, K. 2009. "Brain sex differences and the organisation of juvenile social play behaviour." *Journal of Neuroendocrinology* 21: 519–525.

Badyaev, A. 2009. "Evolutionary significance of phenotypic accommodation in novel envirinment: An empirical test of the Baldwin effect." *Philosophical Transactions of the Royal Society of London, Series B: Biological Sciences* 364: 1125–1141.

Baldwin, J. 1896. "A new factor in evolution." *American Naturalist* 30: 441–451, 536–553.

Baldwin, J. 1902. *Development and Evolution* (Macmillan).

Bateson, P. 2004. "The active role of behaviour in evolution." *Biology and Philosophy* 19: 283–298.

Bateson, P. 2006. "The adaptability driver: Links between behaviour and evolution." *Biological Theory* 1: 342–345.

Bateson, P. 2015. "Why are individuals so different from each other?" *Heredity* 115: 285–292.

Bateson, P. & Laland, K. 2013. "Tinbergen's four questions: An appreciation and an update." *Trends in Ecology & Evolution* 28: 712–718.

Bateson, P. & Martin, P. 2013. *Play, Playfulness, Creativity and Innovation* (Cambridge University Press).

Bateson, P. & Nettle, D. 2014. "Playfulness, ideas and creativity: A survey." *Creativity Research Journal* 26: 219–222.

Burghardt, G. 2005. *The Genesis of Animal Play: Testing the Limits* (MIT Press).

Burghardt, G. 2015. "Creativity, play, and the pace of evolution." In A. Kaufman & J. Kaufman (eds.), *Animal Creativity and Innovation* (Elsevier) 129–159.

Cameron, E., Linklater, W., Stafford, K., & Minot, E. 2008. "Maternal investment results in better foal condition through increased play behaviour in horses." *Animal Behaviour* 76: 1511–1518.

Csikszentmihalyi, M. 1996. *Creativity: Flow and the Psychology of Discovery and Invention* (HarperCollins).

Darwin, C. 1871. *The Descent of Man, and Selection in Relation to Sex* (Murray).

Fagen, R. & Fagen, J. 2004. "Juvenile survival and benefits of play behaviour in brown bears, *Ursus arctos*." *Evolutionary Ecology Research* 6: 89–102.

Fagen, R. & Fagen, J. 2009. "Play behaviour and multi-year juvenile survival in free-ranging brown bears, *Ursus arctos*." *Evolutionary Ecology Research* 11: 1–15.

Fisher, R. 1930. *The Genetical Theory of Natural Selection* (Clarendon Press).

Gomendio, M. 1988. "The development of different types of play in gazelles: Implications for the nature and functions of play." *Animal Behaviour* 36: 825–836.

Harcourt, R. 1991. "Survivorship costs of play in the South American fur seal." *Animal Behaviour* 42: 509–511.

Huffman, M. 1984. "Stone-play of *Macaca fuscata* in Arashiyama B: Transmission of a non-adaptive behavior." *Journal of Human Evolution* 13: 725–735.

Humphrey, N. 1986. *The Inner Eye* (Faber & Faber).

Jolly, A. 1966. "Lemur social behavior and primate intelligence." *Science* 153: 501–506.

Kahlenberg, S. & Wrangham, R. 2010. "Sex differences in chimpanzees' use of sticks as play objects resemble those of children." *Current Biology* 20: R1067–R1068.

Kaufman, A. & Kaufman, J. 2015. *Animal Creativity and Innovation* (Elsevier).

Kawecki, T. 2010. "Evolutionary ecology of learning: Insights from fruit flies." *Population Ecology* 52: 15–25.

Laland, K., Odling-Smee, J., & Turner, S. 2014. "The role of internal and external constructive processes in evolution." *Journal of Physiology* 592: 2413–2422.

Laland, K., Uller, T., Feldman, M., Sterelny, K., Müller, G., Moczek, A., Jablonka, E., & Odling-Smee, J. 2015. "The extended evolutionary synthesis: Its structure, assumptions and predictions." *Proceedings of the Royal Society, Series B: Biological Sciences* 282: 20151019.

Leca, J.-B., Gunst, N., & Huffman, M. 2012. "Thirty years of stone handling tradition in Arashiyama-Kyoto macaques: Implications for cumulative culture and tool use in non-human primates." In J. Leca, N. Gunst, & M. Huffman (eds.), *The Monkeys of Stormy Mountain: 60 years of Primatological Research on the Japanese Macaques of Arahiyama* (Cambridge University Press) 223–257.

Lewontin, R. 1983. "Gene, organism and environment." In D. Bendall (ed.), *Evolution from Molecules to Men* (Cambridge University Press) 273–285.

Lloyd Morgan, C. 1896. "On modification and variation." *Science* 4: 733–740.

Martin, P., & Caro, T. 1985. "On the functions of play and its role in behavioral development." *Advances in the Study of Behavior* 15: 59–103.

Meaney, M. & Stewart, J. 1985. "Sex differences in social play: The socialization of sex roles." *Advances in the Study of Behavior* 15: 1–58.

Mitchell, R. 2015. "Creativity in the interaction: The case of dog-human play." In A. Kaufman & J. Kaufman (eds.), *Animal Creativity and Innovation* (Elsevier) 31–42.

Nettle, D. 2007. *Personality: What Makes You the Way You Are* (Oxford University Press).

Osborn, H. 1896. "Ontogenic and phylogenic variation." *Science* 4: 786–789.

Paenke, I., Kawecki, T., & Sendhoff, B. 2009. "The influence of learning on evolution: A mathematical framework." *Artificial Life* 15: 227–245.

Patterson, E. & Mann, J. 2013. "Cetacean innovation." In A. Kaufman & J. Kaufman (eds.), *Animal Creativity and Innovation* (Elsevier) 73–125.

Pepperberg, I. 2015. "Creativity and innovation in the Grey Parrot (*Psittacus erithacus*)". In A. Kaufman & J. Kaufman (eds.), *Animal Creativity and Innovation* (Elsevier) 3–25.

Red'ko, V., Mosalov, O., & Prokhorov, D. 2005. "A model of evolution and learning." *Neural Networks* 18: 738–745.

Sol, D. 2015. "The evolution of innovativeness: Exaptation or specialized adaptation?" In A. Kaufman & J. Kaufman (eds.), *Animal Creativity and Innovation* (Elsevier) 163–182.

Spalding, D. 1873. "Instinct with original observations on young animals." *Macmillan's Magazine* 27: 282–293.

Tebbich, S., Sterelny, K., & Teschke, I. 2010. "The tale of the finch: Adaptive radiation and behavioural flexibility." *Philosophical Transactions of the Royal Society, Series B: Biological Sciences* 365: 1099–1109.

Tinbergen, N. 1963. "On aims and methods of ethology." *Zeitschrift für Tierpsychologie* 20: 410–433.

West-Eberhard, M. 2003. *Developmental Plasticity and Evolution* (Oxford University Press).

Index